Coenzyme Q in Aging

Guillermo López Lluch
Editor

Coenzyme Q in Aging

Springer

Editor
Guillermo López Lluch
Physiology, Anatomy and Cell Biology Department
Andalusian Center of Developmental Biology (CSIC-UPO), CIBERER
Carlos III Health Institute
Pablo de Olavide University
Sevilla, Spain

ISBN 978-3-030-45641-2 ISBN 978-3-030-45642-9 (eBook)
https://doi.org/10.1007/978-3-030-45642-9

This Springer imprint is published by the registered company Springer Nature Switzerland AG.
The registered company address is: Gewerbestrasse 11, 6330 Cham, Switzerland

To my wife Araceli and my daughters Helena and Lucía.
To the memory of my father and to my mother.
Thanks for your effort to permit my education and formation.

Dedications

This book is dedicated to the memory of Fred L. Crane (December 3, 1925 – August 11, 2016). He discovered one of the most important molecules in organisms, Coenzyme Q_{10}. After that, he worked hard in many other subjects to highlight the importance of this molecule till the end of his fruitful scientific life.

In 1953, Fred Crane obtained his PhD in Botany from the University of Michigan, Ann Harbor. After that, he discovered coenzyme Q at the Institute of Enzyme Research, the University of Wisconsin, in 1957 (Crane et al, 1957 Isolation of a quinone from beef heart mitochondria. BBA 25: 220-221). His discovery started a long and productive scientific career with more than 400 papers. But the main characteristics of Fred Crane were his friendly character, his ability to influence young scientists and to offer his enormous capacity to teach science, and his disposition to visit research institutes around the world sharing his expertise with other researchers.

This great capacity helped him receive many awards including the American Chemical Society Eli Lilly Award in Biochemistry in 1961, the Fulbright Award in Australia in 1971, the NIH Career Award between 1964 and 1994, the Silver Medal in Biochemistry from the University of Bologna in 1989, and the Folkers Foundation Award for Research on Coenzyme Q in 1996. All these awards and other merits such as the Doctor honoris causa of Medicine in the Karolinska Institute in 1989 are a faithful reflection of the research capacity of Dr. Fred L. Crane. A great scientist and an enormous person.

Fred L. Crane at Purdue University, 1984 (Picture from P. Navas).

Karl Folkers (left), Plácido Navas (centre), and Fred L. Crane (right) at Ancona, 1997 (from P. Navas).

Preface

This book shows the importance of Coenzyme Q_{10} (CoQ_{10}) in aging progression and age-related diseases. CoQ_{10} is an essential factor involved in two main aspects of cell physiology: bioenergetics and antioxidant protection. Primary deficiency of CoQ_{10} is associated with severe and lethal diseases but secondary deficiency can be associated with the progression of mitochondrial dysfunction linked to the lessening of biological activities during aging.

In its four parts, this book tries to show all the essential age-related aspects in which CoQ_{10} has been associated. The first part offers a wide overview of the main functions of CoQ_{10} in physiology from the complex synthesis mechanism of this molecule to the regulation of the expression of the proteins involved in this synthesis. We also highlights the two main functions of CoQ_{10} in cells, the essential role as electron transport chain member in mitochondria and the protection of cell membranes against oxidation as one of the main endogenous-synthesized antioxidants.

The second part includes a revision of the different research in which CoQ has been involved in aging studies. Invertebrate models for longevity studies indicated that levels of CoQ could be involved in the regulation of the longevity of the organisms, *C. elegans* and *D. melanogaster* permitted advances in the knowledge of CoQ-related activities and aging progression. Mammal models permitted to introduce some controversy since knocking down of clk1/COQ7, a member of the CoQ synthome has been associated with higher longevity. This model and the role of CoQ in senescence-accelerate mice can highlight the importance of CoQ in mammals' longevity.

In humans, CoQ levels have been associated with many age-related diseases. This is the subject of the third part of this book. In this part, we associate reduction of CoQ levels with mitochondrial dysfunction, metabolic syndrome, neurodegenerative disorders, immunosenescence, and fertility and reproduction. It seems clear that the progression of these age-associated diseases is aggravated when CoQ_{10} levels decrease in individuals, indicating the importance of CoQ_{10} buffering the different physiological dysfunction associated with these diseases.

The importance of CoQ in aging and longevity is finally linked to different pro-longevity interventions such as calorie restriction, modulation of CoQ_{10} levels by induction of synthesis, or by supplementation. This is the objective of the three final chapters of this book. These chapters open the possibility to use CoQ_{10} levels as target to improve functionality of cells during aging, delaying age-related diseases progression and, probably, aging as a whole.

Sevilla, Spain Guillermo López Lluch

Contents

Part I Basics on Coenzyme Q

1 **The Current Coenzyme Q Science and Knowledge** 3
Plácido Navas

2 **Molecular Structure, Biosynthesis, and Distribution of Coenzyme Q** 11
L. Vázquez-Fonseca, I. González-Mariscal, and C. Santos-Ocaña

3 **Coenzyme Q Function in Mitochondria** 51
Maria Luisa Genova and Giorgio Lenaz

4 **Extramitochondrial Coenzyme Q_{10} in Aging** 91
Guillermo López-Lluch

5 **Regulation of Synthesis of Coenzyme Q_{10}** 113
María Victoria Cascajo-Almenara and Guillermo López-Lluch

Part II Coenzyme Q in Aging Studies

6 **Coenzyme Q and Aging in *C. elegans*** 131
Claudio Asencio

7 **Coenzyme Q and Aging in the Fruit Fly *Drosophila melanogaster*** 141
Daniel J. M. Fernández-Ayala and Alberto Sanz

8 **The Aging Process and Coenzyme Q: Clk-1 Mouse Models** 157
Mayumi Takahashi, Kazuhide Takahashi, and Takuji Shirasawa

9 **Reduced Coenzyme Q_{10} Decelerates Senescence and Age-Related Hearing Loss in Senescence-Accelerated Mice by Activating Mitochondrial Functions** 169
Jinko Sawashita, Xu Zhe, and Keiichi Higuchi

Part III Age-Related Diseases and Coenzyme Q

10 Coenzyme Q, mtDNA and Mitochondrial Dysfunction During Aging ... 191
José L. Quiles, Alfonso Varela-López, María D. Navarro-Hortal, and Maurizio Battino

11 Coenzyme Q_{10} and Metabolic Syndrome ... 227
Juan Diego Hernández-Camacho

12 Coenzyme Q and Age-Related Neurodegenerative Disorders: Parkinson and Alzheimer Diseases ... 241
Francisco J. Alcaín, Javier Domínguez, Mario Durán-Prado, and Julia Vaamonde

13 Immunosenescence and CoQ_{10} ... 269
Juan Diego Hernández-Camacho, Catherine Meza-Torres, and Guillermo López-Lluch

14 Coenzyme Q_{10} in Fertility and Reproduction ... 283
Diana Vaamonde, Anthony C. Hackney, Carolina Algar-Santacruz, María José Garcia-Moreno, and Juan Manuel García-Manso

Part IV Prolongevity Effectors and Coenzyme Q

15 Caloric Restriction, Longevity and Coenzyme Q ... 311
María Isabel Burón, Cristina Parrado, José Antonio González-Reyes, Lucía Fernández del Río, Elena Gutiérrez-Casado, Miguel Calvo-Rubio, Sandra Rodríguez-López, and José Manuel Villalba

16 Age Dependent Changes of Coenzyme Q Levels and its Induction in Experimental Systems ... 329
Magnus Bentinger, Gustav Dallner, Kerstin Brismar, Ewa Swiezewska, and Michael Tekle

17 Effects of Coenzyme Q_{10} Supplementation on Elderly People ... 347
Elena M. Yubero-Serrano, Francisco M. Gutierrez-Mariscal, and Jose Lopez-Miranda

Index ... 367

Part I
Basics on Coenzyme Q

Chapter 1
The Current Coenzyme Q Science and Knowledge

Plácido Navas

Abstract Coenzyme Q (CoQ) is a redox lipid essential for aerobic respiration and antioxidant protection. It is synthesized in each cell by a multiprotein complex inside mitochondria and incorporated in all cellular membranes. It is however an amazing molecule whose homeostasis not only depends on the proper function of biosynthesis complex but also on age, diet and the wholeness of mitochondria functions.

Keywords Coenzyme Q · Ubiquinone · Mitochondria · Mitochondrial diseases

1.1 CoQ Function

Since the discovery of CoQ (ubiquinone) as the essential redox quinone component of the respiratory chain in beef heart (Crane et al. 1957), knowledge of properties and functions of CoQ, and its role in aging and disease has highly evolved. CoQ is present in all cells in which acts as an essential electron carrier in the respiratory chain, transporting electrons from either complex I or complex II to complex III, but also reducing complex III from other redox enzymes (Alcazar-Fabra et al. 2016). The efficiency of electron chain is due to the integration of respiratory complexes including CoQ into supercomplexes that dynamically organize the electron flux to optimize the use of available substrates (Lapuente-Brun et al. 2013). This efficiency depends strictly on the ratio of reduced to oxidized forms of CoQ (Guaras et al. 2016). It is still unknown the mechanisms that provide the appropriate amount levels of CoQ to the supercomplex structure and if total or partial contents of this carrier are integrated into the supercomplex or a free pool exists.

P. Navas (✉)
Centro Andaluz de Biología del Desarrollo, Universidad Pablo de Olavide-CSIC-JA, Seville, Spain

CIBERER, Instituto de Salud Carlos III, Sevilla, Spain
e-mail: pnavas@upo.es

G. López Lluch (ed.), *Coenzyme Q in Aging*,
https://doi.org/10.1007/978-3-030-45642-9_1

1.2 CoQ Synthesis

CoQ is synthesized by a multiprotein complex (synthome) inside mitochondria (Stefely and Pagliarini 2017) with a high level of regulation (Gonzalez-Mariscal et al. 2018). Synthome is a complex pathway which use the initial substrate polyprenyl-4-hydroxybenzoate (pp-4-HB), which is produced by the attachment of polyprenyl-PP, produced in the mevalonate pathway, to the precursor 4-hydroxybenzoate (4-HB), produced from tyrosine, by the p-hydroxybenzoate polyprenyl transferase encoded by *COQ2* gene (Desbats et al. 2016). This synthome modifies the benzoquinone ring until CoQ is produced although different reactions are still unknown as decarboxylation and hydroxylation of pp-4-HB (Acosta Lopez et al. 2019).

The CoQ biosynthesis pathway is regulated at the transcriptional level through, e.g. NF-κB activated by mitochondrial stress (Brea-Calvo et al. 2009), at translational level by COQ7 mRNA binding to proteins (Cascajo et al. 2016), and at the post-translational levels through phosphorylation cycles of COQ encoded proteins involving PPTC7 phosphatase (Gonzalez-Mariscal et al. 2018; Niemi et al. 2019). There are accumulated evidence that synthome would be mainly associated to the endoplasmic reticulum (ER)-mitochondria contacts domains (Subramanian et al. 2019). In this sense, mutations of genes that encode proteins located in ER-mitochondria contact sites, which are related to lipids transport into mitochondria can cause defects in CoQ biosynthesis. Mfn2 and Adck2 colocalise in the mitochondrial-associated membranes and have been demonstrated to contribute to the maintenance of the mitochondrial CoQ (Mourier et al. 2015; Vazquez-Fonseca et al. 2019).

1.3 CoQ and Longevity

Since the discovery of clk-1/COQ7 function in regulating C. elegans and mouse longevity through CoQ biosynthesis regulation (Wang and Hekimi 2016), there are recent evidence that CoQ would extend longevity through endogenous stress signaling and protein synthesis downregulation (Molenaars et al. 2018; Scialo et al. 2016). It has been proposed that CoQ levels in tissues and organs are depleted during aging and depend on dietary interventions (Parrado-Fernandez et al. 2011). For example, clasical studies showed that calorie restriction increased CoQ content in plasma membrane providing antioxidant protection and modulating cytosolic NAD^+ levels in different tissues and cells (De Cabo et al. 2004; Hyun et al. 2006; Lopez-Lluch et al. 2005), recently supported because the overexpression of extramitochondrial CoQ-dependent enzymes, $CytB_5R_3$ and NQO1, increases health- and life-span in mice (Diaz-Ruiz et al. 2018; Martin-Montalvo et al. 2016). This plasma membrane antioxidant system, in which $CytB_5R_3$ and NQO1 reduce CoQ, was demonstrated to protect cells from phospholipid peroxidation and ceramide-dependent apoptosis

(Villalba and Navas 2000). It has been recently described additional components of the plasma membrane antioxidant system, which acts through ferroptosis suppressor protein1 (FSP1) reduction of CoQ (Bersuker et al. 2019; Doll et al. 2019) preventing phospholipid peroxidation and ferroptosis.

1.4 Essential Role of CoQ in Physiology

The main functions of CoQ as electron carrier in respiratory chain and as antioxidant have been studied as regulator of mitochondrial dysfunction in aging and directly involved in glucose homeostasis by regulating insulin sensitivity (Fazakerley et al. 2018). Also, supplementation of CoQ has been also successfully used to improve senescence-accelerated mice aerobic metabolism (Tian et al. 2014), and female mice fertility by improving mitochondrial efficiency (Ben-Meir et al. 2015). Interestingly, clinical trials of CoQ supplementation showed an increase in IGF-1 and postprandial IGFBP-1 levels in elderly (Alehagen et al. 2017), and plasma levels of CoQ positively correlated with the insulin level, homeostatic model assessment-insulin resistance, and quantitative insulin sensitivity check index in diabetic patients (Fallah et al. 2018; Gholnari et al. 2018; Yen et al. 2018). Also, CoQ supplementation showed the improvement of chronic inflammation by decreasing inflammatory markers as TNF-α and IL-6 (Farsi et al. 2019; Zhai et al. 2017). In fact, cardiovascular diseases are potentiated by age and is one important major cause of death in western countries, which show improvement under calorie restriction as indicated in the review by de Cabo and Mattson (de Cabo and Mattson 2019).

Mitochondrial diseases are genetic heterogeneous and tissue-specific disorders that show different severities and time-onset forms of age-associated diseases as sarcopenia and neurodegeneration (Frazier et al. 2019). A group of these mitochondrial diseases are included in the CoQ deficiency syndrome caused by a significant low level of CoQ, which is developed by mutations in *COQ* genes that encode for the synthome proteins, named primary CoQ deficiency (Salviati et al. 2017). However, mostly of the diseases included in this syndrome are considered secondary deficiencies because the low level of CoQ is induced as an adaptation to either electron chain defects or defective mitochondrial functions (Yubero et al. 2016). These secondary defects can cause human deterioration of muscle with weakness and exercise intolerance that finally can cause rhabdomyolysis (Jou et al. 2019), severe necropsy of neurons associated to mutations in the mitochondrial PARL protease with complex III defect (Spinazzi et al. 2019), or multiorgan affectation caused with autosomal recessive *TANGO2* mutations (Mingirulli et al. 2019). However, although it has been demonstrated that mouse models with defects in nuclear encoded factors that regulate mitochondrial DNA functions developed secondary CoQ deficiency (Kuhl et al. 2017), there is not described any mechanism to demonstrate these adaptations.

The current knowledge of CoQ functions and biosynthesis have highly evolved since its discovery in 1957 (Crane et al. 1957) demonstrating its essential function in respiration and its important nuclear function as antioxidant. CoQ homeostasis is involved in essential bioenergetics pathway, plasma membrane mediated cellular redox homeostasis and age-dependent disorders but still unknown the complete biosynthesis pathway and how CoQ levels adaptate to different pathological conditions to guarantee cell survival.

References

Acosta Lopez MJ, Trevisson E, Canton M, Vazquez-Fonseca L, Morbidoni V, Baschiera E, Frasson C, Pelosi L, Rascalou B, Desbats MA, Alcazar-Fabra M, Rios JJ, Sanchez-Garcia A, Basso G, Navas P, Pierrel F, Brea-Calvo G, Salviati L (2019) Vanillic acid restores coenzyme Q biosynthesis and ATP production in human cells lacking COQ6. Oxidative Med Cell Longev 2019:3904905–3904911. https://doi.org/10.1155/2019/3904905

Alcazar-Fabra M, Navas P, Brea-Calvo G (2016) Coenzyme Q biosynthesis and its role in the respiratory chain structure. Biochim Biophys Acta 1857(8):1073–1078. https://doi.org/10.1016/j.bbabio.2016.03.010

Alehagen U, Johansson P, Aaseth J, Alexander J, Brismar K (2017) Increase in insulin-like growth factor 1 (IGF-1) and insulin-like growth factor binding protein 1 after supplementation with selenium and coenzyme Q10. A prospective randomized double-blind placebo-controlled trial among elderly Swedish citizens. PLoS One 12(6):e0178614. https://doi.org/10.1371/journal.pone.0178614

Ben-Meir A, Burstein E, Borrego-Alvarez A, Chong J, Wong E, Yavorska T, Naranian T, Chi M, Wang Y, Bentov Y, Alexis J, Meriano J, Sung HK, Gasser DL, Moley KH, Hekimi S, Casper RF, Jurisicova A (2015) Coenzyme Q10 restores oocyte mitochondrial function and fertility during reproductive aging. Aging Cell 14(5):887–895. https://doi.org/10.1111/acel.12368

Bersuker K, Hendricks JM, Li Z, Magtanong L, Ford B, Tang PH, Roberts MA, Tong B, Maimone TJ, Zoncu R, Bassik MC, Nomura DK, Dixon SJ, Olzmann JA (2019) The CoQ oxidoreductase FSP1 acts parallel to GPX4 to inhibit ferroptosis. Nature 575(7784):688–692. https://doi.org/10.1038/s41586-019-1705-2

Brea-Calvo G, Siendones E, Sanchez-Alcazar JA, de Cabo R, Navas P (2009) Cell survival from chemotherapy depends on NF-kappaB transcriptional up-regulation of coenzyme Q biosynthesis. PLoS One 4(4):e5301. https://doi.org/10.1371/journal.pone.0005301

Cascajo MV, Abdelmohsen K, Noh JH, Fernandez-Ayala DJ, Willers IM, Brea G, Lopez-Lluch G, Valenzuela-Villatoro M, Cuezva JM, Gorospe M, Siendones E, Navas P (2016) RNA-binding proteins regulate cell respiration and coenzyme Q biosynthesis by post-transcriptional regulation of COQ7. RNA Biol 13(7):622–634. https://doi.org/10.1080/15476286.2015.1119366

Crane FL, Hatefi Y, Lester RL, Widmer C (1957) Isolation of a quinone from beef heart mitochondria. Biochim Biophys Acta 25(1):220–221. https://doi.org/10.1016/0006-3002(57)90457-2

de Cabo R, Mattson MP (2019) Effects of intermittent fasting on health, aging, and disease. N Engl J Med 381(26):2541–2551. https://doi.org/10.1056/NEJMra1905136

De Cabo R, Cabello R, Rios M, Lopez-Lluch G, Ingram DK, Lane MA, Navas P (2004) Calorie restriction attenuates age-related alterations in the plasma membrane antioxidant system in rat liver. Exp Gerontol 39(3):297–304. https://doi.org/10.1016/j.exger.2003.12.003

Desbats MA, Morbidoni V, Silic-Benussi M, Doimo M, Ciminale V, Cassina M, Sacconi S, Hirano M, Basso G, Pierrel F, Navas P, Salviati L, Trevisson E (2016) The COQ2 genotype predicts the severity of coenzyme Q10 deficiency. Hum Mol Genet 25(19):4256–4265. https://doi.org/10.1093/hmg/ddw257

Diaz-Ruiz A, Lanasa M, Garcia J, Mora H, Fan F, Martin-Montalvo A, Di Francesco A, Calvo-Rubio M, Salvador-Pascual A, Aon MA, Fishbein KW, Pearson KJ, Villalba JM, Navas P, Bernier M, de Cabo R (2018) Overexpression of CYB5R3 and NQO1, two NAD(+)

-producing enzymes, mimics aspects of caloric restriction. Aging Cell 17(4):e12767. https://doi.org/10.1111/acel.12767

Doll S, Freitas FP, Shah R, Aldrovandi M, da Silva MC, Ingold I, Grocin AG, Xavier da Silva TN, Panzilius E, Scheel CH, Mourao A, Buday K, Sato M, Wanninger J, Vignane T, Mohana V, Rehberg M, Flatley A, Schepers A, Kurz A, White D, Sauer M, Sattler M, Tate EW, Schmitz W, Schulze A, O'Donnell V, Proneth B, Popowicz GM, Pratt DA, Angeli JPF, Conrad M (2019) FSP1 is a glutathione-independent ferroptosis suppressor. Nature 575(7784):693–698. https://doi.org/10.1038/s41586-019-1707-0

Fallah M, Askari G, Soleimani A, Feizi A, Asemi Z (2018) Clinical trial of the effects of coenzyme Q10 supplementation on glycemic control and markers of lipid profiles in diabetic hemodialysis patients. Int Urol Nephrol 50(11):2073–2079. https://doi.org/10.1007/s11255-018-1973-z

Farsi F, Heshmati J, Keshtkar A, Irandoost P, Alamdari NM, Akbari A, Janani L, Morshedzadeh N, Vafa M (2019) Can coenzyme Q10 supplementation effectively reduce human tumor necrosis factor-α and interleukin-6 levels in chronic inflammatory diseases? A systematic review and meta-analysis of randomized controlled trials. Pharmacol Res 148:104290. https://doi.org/10.1016/j.phrs.2019.104290

Fazakerley DJ, Chaudhuri R, Yang P, Maghzal GJ, Thomas KC, Krycer JR, Humphrey SJ, Parker BL, Fisher-Wellman KH, Meoli CC, Hoffman NJ, Diskin C, Burchfield JG, Cowley MJ, Kaplan W, Modrusan Z, Kolumam G, Yang JY, Chen DL, Samocha-Bonet D, Greenfield JR, Hoehn KL, Stocker R, James DE (2018) Mitochondrial CoQ deficiency is a common driver of mitochondrial oxidants and insulin resistance. elife 7. https://doi.org/10.7554/eLife.32111

Frazier AE, Thorburn DR, Compton AG (2019) Mitochondrial energy generation disorders: genes, mechanisms, and clues to pathology. J Biol Chem 294(14):5386–5395. https://doi.org/10.1074/jbc.R117.809194

Gholnari T, Aghadavod E, Soleimani A, Hamidi GA, Sharifi N, Asemi Z (2018) The effects of coenzyme Q10 supplementation on glucose metabolism, lipid profiles, inflammation, and oxidative stress in patients with diabetic nephropathy: a randomized, double-blind, placebo-controlled trial. J Am Coll Nutr 37(3):188–193. https://doi.org/10.1080/07315724.2017.1386140

Gonzalez-Mariscal I, Martin-Montalvo A, Vazquez-Fonseca L, Pomares-Viciana T, Sanchez-Cuesta A, Fernandez-Ayala DJ, Navas P, Santos-Ocana C (2018) The mitochondrial phosphatase PPTC7 orchestrates mitochondrial metabolism regulating coenzyme Q10 biosynthesis. Biochim Biophys Acta Bioenerg 1859(11):1235–1248. https://doi.org/10.1016/j.bbabio.2018.09.369

Guaras A, Perales-Clemente E, Calvo E, Acin-Perez R, Loureiro-Lopez M, Pujol C, Martinez-Carrascoso I, Nunez E, Garcia-Marques F, Rodriguez-Hernandez MA, Cortes A, Diaz F, Perez-Martos A, Moraes CT, Fernandez-Silva P, Trifunovic A, Navas P, Vazquez J, Enriquez JA (2016) The CoQH2/CoQ ratio serves as a sensor of respiratory chain efficiency. Cell Rep 15(1):197–209. https://doi.org/10.1016/j.celrep.2016.03.009

Hyun DH, Emerson SS, Jo DG, Mattson MP, de Cabo R (2006) Calorie restriction up-regulates the plasma membrane redox system in brain cells and suppresses oxidative stress during aging. Proc Natl Acad Sci U S A 103(52):19908–19912. https://doi.org/10.1073/pnas.0608008103

Jou C, Ortigoza-Escobar JD, O'Callaghan MM, Nascimento A, Darling A, Pias-Peleteiro L, Perez-Duenas B, Pineda M, Codina A, Arjona C, Armstrong J, Palau F, Ribes A, Gort L, Tort F, Navas P, Ruiz-Pesini E, Emperador S, Lopez-Gallardo E, Bayona-Bafaluy P, Montero R, Jimenez-Mallebrera C, Garcia-Cazorla A, Montoya J, Yubero D, Artuch R (2019) Muscle involvement in a large cohort of pediatric patients with genetic diagnosis of mitochondrial disease. J Clin Med 8(1). https://doi.org/10.3390/jcm8010068

Kuhl I, Miranda M, Atanassov I, Kuznetsova I, Hinze Y, Mourier A, Filipovska A, Larsson NG (2017) Transcriptomic and proteomic landscape of mitochondrial dysfunction reveals secondary coenzyme Q deficiency in mammals. elife 6. https://doi.org/10.7554/eLife.30952

Lapuente-Brun E, Moreno-Loshuertos R, Acin-Perez R, Latorre-Pellicer A, Colas C, Balsa E, Perales-Clemente E, Quiros PM, Calvo E, Rodriguez-Hernandez MA, Navas P, Cruz R, Carracedo A, Lopez-Otin C, Perez-Martos A, Fernandez-Silva P, Fernandez-Vizarra E,

Enriquez JA (2013) Supercomplex assembly determines electron flux in the mitochondrial electron transport chain. Science 340(6140):1567–1570. https://doi.org/10.1126/science.1230381

Lopez-Lluch G, Rios M, Lane MA, Navas P, de Cabo R (2005) Mouse liver plasma membrane redox system activity is altered by aging and modulated by calorie restriction. Age (Dordr) 27(2):153–160. https://doi.org/10.1007/s11357-005-2726-3

Martin-Montalvo A, Sun Y, Diaz-Ruiz A, Ali A, Gutierrez V, Palacios HH, Curtis J, Siendones E, Ariza J, Abulwerdi GA, Sun X, Wang AX, Pearson KJ, Fishbein KW, Spencer RG, Wang M, Han X, Scheibye-Knudsen M, Baur JA, Shertzer HG, Navas P, Villalba JM, Zou S, Bernier M, de Cabo R (2016) Cytochrome b5 reductase and the control of lipid metabolism and healthspan. NPJ Aging Mech Dis 2:16006. https://doi.org/10.1038/npjamd.2016.6

Mingirulli N, Pyle A, Hathazi D, Alston CL, Kohlschmidt N, O'Grady G, Waddell L, Evesson F, Cooper SBT, Turner C, Duff J, Topf A, Yubero D, Jou C, Nascimento A, Ortez C, Garcia-Cazorla A, Gross C, O'Callaghan M, Santra S, Preece MA, Champion M, Korenev S, Chronopoulou E, Anirban M, Pierre G, McArthur D, Thompson K, Navas P, Ribes A, Tort F, Schluter A, Pujol A, Montero R, Sarquella G, Lochmuller H, Jimenez-Mallebrera C, Taylor RW, Artuch R, Kirschner J, Grunert SC, Roos A, Horvath R (2019) Clinical presentation and proteomic signature of patients with TANGO2 mutations. J Inherit Metab Dis 43:297–308. https://doi.org/10.1002/jimd.12156

Molenaars M, Janssens GE, Santermans T, Lezzerini M, Jelier R, MacInnes AW, Houtkooper RH (2018) Mitochondrial ubiquinone-mediated longevity is marked by reduced cytoplasmic mRNA translation. Life Sci Alliance 1(5). https://doi.org/10.26508/lsa.201800082

Mourier A, Motori E, Brandt T, Lagouge M, Atanassov I, Galinier A, Rappl G, Brodesser S, Hultenby K, Dieterich C, Larsson NG (2015) Mitofusin 2 is required to maintain mitochondrial coenzyme Q levels. J Cell Biol 208(4):429–442. https://doi.org/10.1083/jcb.201411100

Niemi NM, Wilson GM, Overmyer KA, Vogtle FN, Myketin L, Lohman DC, Schueler KL, Attie AD, Meisinger C, Coon JJ, Pagliarini DJ (2019) Pptc7 is an essential phosphatase for promoting mammalian mitochondrial metabolism and biogenesis. Nat Commun 10(1):3197. https://doi.org/10.1038/s41467-019-11047-6

Parrado-Fernandez C, Lopez-Lluch G, Rodriguez-Bies E, Santa-Cruz S, Navas P, Ramsey JJ, Villalba JM (2011) Calorie restriction modifies ubiquinone and COQ transcript levels in mouse tissues. Free Radic Biol Med 50(12):1728–1736. https://doi.org/10.1016/j.freeradbiomed.2011.03.024

Salviati L, Trevisson E, Doimo M, Navas P (2017) Primary coenzyme Q10 deficiency. In: Adam MP, Ardinger HH, Pagon RA et al (eds) GeneReviews((R)), Seattle

Scialo F, Sriram A, Fernandez-Ayala D, Gubina N, Lohmus M, Nelson G, Logan A, Cooper HM, Navas P, Enriquez JA, Murphy MP, Sanz A (2016) Mitochondrial ROS produced via reverse electron transport extend animal lifespan. Cell Metab 23(4):725–734. https://doi.org/10.1016/j.cmet.2016.03.009

Spinazzi M, Radaelli E, Horre K, Arranz AM, Gounko NV, Agostinis P, Maia TM, Impens F, Morais VA, Lopez-Lluch G, Serneels L, Navas P, De Strooper B (2019) PARL deficiency in mouse causes complex III defects, coenzyme Q depletion, and Leigh-like syndrome. Proc Natl Acad Sci U S A 116(1):277–286. https://doi.org/10.1073/pnas.1811938116

Stefely JA, Pagliarini DJ (2017) Biochemistry of mitochondrial coenzyme Q biosynthesis. Trends Biochem Sci 42(10):824–843. https://doi.org/10.1016/j.tibs.2017.06.008

Subramanian K, Jochem A, Le Vasseur M, Lewis S, Paulson BR, Reddy TR, Russell JD, Coon JJ, Pagliarini DJ, Nunnari J (2019) Coenzyme Q biosynthetic proteins assemble in a substrate-dependent manner into domains at ER-mitochondria contacts. J Cell Biol 218(4):1353–1369. https://doi.org/10.1083/jcb.201808044

Tian G, Sawashita J, Kubo H, Nishio SY, Hashimoto S, Suzuki N, Yoshimura H, Tsuruoka M, Wang Y, Liu Y, Luo H, Xu Z, Mori M, Kitano M, Hosoe K, Takeda T, Usami S, Higuchi K (2014) Ubiquinol-10 supplementation activates mitochondria functions to decelerate senescence in senescence-accelerated mice. Antioxid Redox Signal 20(16):2606–2620. https://doi.org/10.1089/ars.2013.5406

Vazquez-Fonseca L, Schaefer J, Navas-Enamorado I, Santos-Ocana C, Hernandez-Camacho JD, Guerra I, Cascajo MV, Sanchez-Cuesta A, Horvath Z, Siendones E, Jou C, Casado M, Gutierrez P, Brea-Calvo G, Lopez-Lluch G, Fernandez-Ayala DJM, Cortes-Rodriguez AB, Rodriguez-Aguilera JC, Matte C, Ribes A, Prieto-Soler SY, Dominguez-Del-Toro E, Francesco AD, Aon MA, Bernier M, Salviati L, Artuch R, Cabo R, Jackson S, Navas P (2019) ADCK2 Haploinsufficiency reduces mitochondrial lipid oxidation and causes myopathy associated with CoQ deficiency. J Clin Med 8(9). https://doi.org/10.3390/jcm8091374

Villalba JM, Navas P (2000) Plasma membrane redox system in the control of stress-induced apoptosis. Antioxid Redox Signal 2(2):213–230. https://doi.org/10.1089/ars.2000.2.2-213

Wang Y, Hekimi S (2016) Understanding ubiquinone. Trends Cell Biol 26(5):367–378. https://doi.org/10.1016/j.tcb.2015.12.007

Yen CH, Chu YJ, Lee BJ, Lin YC, Lin PT (2018) Effect of liquid ubiquinol supplementation on glucose, lipids and antioxidant capacity in type 2 diabetes patients: a double-blind, randomised, placebo-controlled trial. Br J Nutr 120(1):57–63. https://doi.org/10.1017/S0007114518001241

Yubero D, Montero R, Martin MA, Montoya J, Ribes A, Grazina M, Trevisson E, Rodriguez-Aguilera JC, Hargreaves IP, Salviati L, Navas P, Artuch R, Qdsg C, Jou C, Jimenez-Mallebrera C, Nascimento A, Perez-Duenas B, Ortez C, Ramos F, Colomer J, O'Callaghan M, Pineda M, Garcia-Cazorla A, Espinos C, Ruiz A, Macaya A, Marce-Grau A, Garcia-Villoria J, Arias A, Emperador S, Ruiz-Pesini E, Lopez-Gallardo E, Neergheen V, Simoes M, Diogo L, Blazquez A, Gonzalez-Quintana A, Delmiro A, Dominguez-Gonzalez C, Arenas J, Garcia-Silva MT, Martin E, Quijada P, Hernandez-Lain A, Moran M, Rivas Infante E, Avila Polo R, Paradas Lope C, Bautista Lorite J, Martinez Fernandez EM, Cortes AB, Sanchez-Cuesta A, Cascajo MV, Alcazar M, Brea-Calvo G (2016) Secondary coenzyme Q10 deficiencies in oxidative phosphorylation (OXPHOS) and non-OXPHOS disorders. Mitochondrion 30:51–58. https://doi.org/10.1016/j.mito.2016.06.007

Zhai J, Bo Y, Lu Y, Liu C, Zhang L (2017) Effects of coenzyme Q10 on markers of inflammation: a systematic review and meta-analysis. PLoS One 12(1):e0170172. https://doi.org/10.1371/journal.pone.0170172

Chapter 2
Molecular Structure, Biosynthesis, and Distribution of Coenzyme Q

L. Vázquez-Fonseca, I. González-Mariscal, and C. Santos-Ocaña

Abstract Coenzyme Q is a very old molecule in evolutionary terms that has accumulated numerous functions in the cellular metabolism beyond its primordial function, the electron transport. In all organisms, coenzyme Q maintains a highly conserved structure allowing a localization inside cell membranes in a hydrophobic environment thanks to having an isoprenoid tail, and at the same time allows the polar ring benzene to interact with acceptors and electron donors. Coenzyme Q deficiency constitutes a group of mitochondrial diseases. Affected patients suffer mainly a decrease in energy production that induces dysfunctions in most organs and body systems. Current therapeutic alternatives are based on increasing coenzyme Q levels either through induction of endogenous mechanisms or exogenous supplementation. This chapter includes both aspects, the mechanisms associated with the coenzyme Q supplementation and the regulatory mechanisms of coenzyme Q biosynthesis. In terms of synthesis, the structure of coenzyme Q is complicated since it requires the participation of two well-differentiated pathways that must be carefully regulated. The synthesis is carried out through the participation of a multienzyme complex located in the inner mitochondrial membrane and controlled by different levels of regulation that at this time are not well-known.

Keywords Coenzyme Q · Mitochondria · Mitochondrial diseases · Lipids

L. Vázquez-Fonseca
CABD-Universidad Pablo de Olavide, CIBERER Instituto de Salud Carlos III, Sevilla, Spain

Clinical Genetics Unit, Department of Woman and Child Health, University of Padova, Padova, Italy

IRP Città della Speranza, Padova, Italy

I. González-Mariscal
CABD-Universidad Pablo de Olavide, CIBERER Instituto de Salud Carlos III, Sevilla, Spain

National Institute on Aging, NIH Laboratory of Clinical Investigation, Baltimore, MD, USA

C. Santos-Ocaña (✉)
CABD-Universidad Pablo de Olavide, CIBERER Instituto de Salud Carlos III, Sevilla, Spain
e-mail: csanoca@upo.es

G. López Lluch (ed.), *Coenzyme Q in Aging*,
https://doi.org/10.1007/978-3-030-45642-9_2

2.1 Introduction

Coenzyme Q or ubiquinone (CoQ) is an old molecule in terms of the evolution. CoQ can be found in most living organisms from mammalian to prokaryotic cells (Morton 1958). This presence is an indicator of it extremely important function as an electron carrier in mitochondria but also another quinone molecule, the plastoquinone, plays a similar function in plant chloroplasts. This function makes CoQ as a crucial molecule to provide energy to the cell, being the central element of the proton motive cycle converting the energy accumulated in redox molecules to ATP through the transient generation of an electrochemical gradient. However, in addition to this older function along the evolution, CoQ has been responsible for other additional and well-known functions (Bentinger et al. 2010) such as antioxidant, pyrimidine synthesis, brown adipose tissue heat production, β-oxidation, mPTP opening and for other new functions such the involvement in life span extension (López-Lluch et al. 2010), regulation of bioenergetics metabolism (Gonzalez-Mariscal et al. 2017) and mitophagy (Cotan et al. 2011; Gonzalez-Mariscal et al. 2017; Rodríguez-Hernández et al. 2009) that promise the rejuvenation of the role of this molecule on living organisms.

All previously commented functions of CoQ have a common origin in its special chemical structure; a combination of an aromatic polar ring condensed with an isoprenoid and therefore hydrophobic tail. That combination makes CoQ an amphipathic molecule able to be embedded in membranes by the isoprenoid tail and ready to interact exchanging electrons with other redox molecules of the electron transport chain thanks to the aromatic ring. That is the first aim of this chapter; a description of the chemical structure of CoQ analyzing in parallel how this structure explains the diversity of CoQ functions.

This complex structure is the combination of two separated general biochemical pathways; the synthesis of p-hydroxybenzoate (or p-aminobenzoate, pABA in *S. cerevisiae*) from tyrosine or phenylalanine and the synthesis of the poly-isoprenoid tail from the mevalonate pathway. Both pathways converge as a Y shaped pathway in the specific CoQ biosynthesis that is localized in mitochondria (with some discussion about that idea) thanks to a multi-enzymatic complex. That is the aim of the second part of this chapter, shows the available information about CoQ biosynthesis.

Some of the CoQ functions are produced in mitochondria, where is synthesized but other functions are produced in other cell membranes indicating that the new synthesized CoQ must be distributed among membranes. This topic together the effect of CoQ supplementation on cells and organs is the third aim of this chapter.

2.2 The Chemical Structure Serving the Biological Function of CoQ

Coenzyme Q (CoQ) is probably the most important member of the chemical class known as polyprenyl benzoquinones (Fig. 2.1). These compounds contain a polyisoprene chain attached to a quinone ring at position 6. Although chemically is possible naming coenzyme Q_{10} (CoQ_{10}) using several ways, probably the most descriptive is 2,3-dimethoxy-5-methyl-6-decaprenyl-benzoquinone. Both components, the isoprenoid chain, and the polar quinone ring are quite different attending its chemical properties to understand the CoQ functions. Comparing all kind of CoQ molecules found in living organisms, all share the same quinone ring but differs in the length of the isoprenoid chain, from six units (30 carbons) in *S. cerevisiae* to ten units in human (50 carbons) (Fig. 2.1). Since the isoprenoid chain is responsible for the CoQ localization in membranes is necessary to explain how the isoprenoid chain interacts with phospholipids and the effect of the chain length in this interaction.

2.2.1 *The Isoprenoid Chain*

The isoprenoid chain of CoQ is responsible for the location of the molecule in hydrophobic environments such as the inner space of the lipid bilayer. However, CoQ is an amphipathic molecule that interacts with the polar component of phospholipids. It is therefore fundamental to determine the topology of the molecule, on which much of its function depends. The inclusion of CoQ_{10} molecules in phospholipid monolayers is difficult and such molecules are usually expelled from the interface with air (Maggio et al. 1977). However, this effect is dependent on the length of the isoprenoid chain since the solubility of CoQ_3 in the phospholipid monolayers is much higher (Asai and Watanabe 1999). These data are reinforced in the case of lipid bilayers, in which using several analytical techniques CoQ_{10} is localized in the

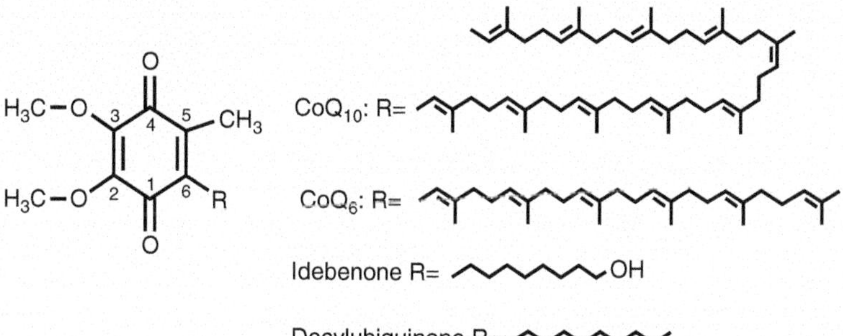

Fig. 2.1 Structure of coenzyme Q_{10} and other analogs. The quinone ring is common for all kinds of ubiquinones found in different organisms and also for some coenzyme Q analogs such as idebenone and decylubiquinone

internal hydrophobic plane generated by the lipidic layers whereas CoQ_3 is mostly mixed with phospholipids (Alonso et al. 1981; Hoyo et al. 2013; Katsikas and Quinn 1982a, b; Kaurola et al. 2016). This location of CoQ_{10} can be altered as a function of concentration; at concentrations lower than 1% (usual in cell membranes), aggregates can be formed (Katsikas and Quinn 1982a, b) mainly between lipid bilayers (Hoyo et al. 2013). This effect also occurs when comparing CoQ_{10} and CoQ_2, both show a different location and only CoQ_{10} is aggregated. This effect explains a possible biological function allowing the formation of a CoQ_{10} pool (Roche et al. 2006). Although the stabilizing and destabilizing effects of CoQ_{10} on membranes have been discussed, recent studies indicate that CoQ can perform a stabilizing function especially on membranes that do not contain cholesterol (Agmo Hernández et al. 2015), albeit by different mechanisms.

It is impossible to separate the location of CoQ from its most important biological functions as an electron carrier and antioxidant. The oxidized form of CoQ (ubiquinone) is mostly located at the hydrophobic interface of the artificial lipid bilayer while the reduced form (ubiquinol) is mixed with phospholipids near their polar groups (Aranda and Gómez-Fernández 1985) (Fig. 2.2 right). This differential localization not only occurs in artificial membranes but also occurs in cell

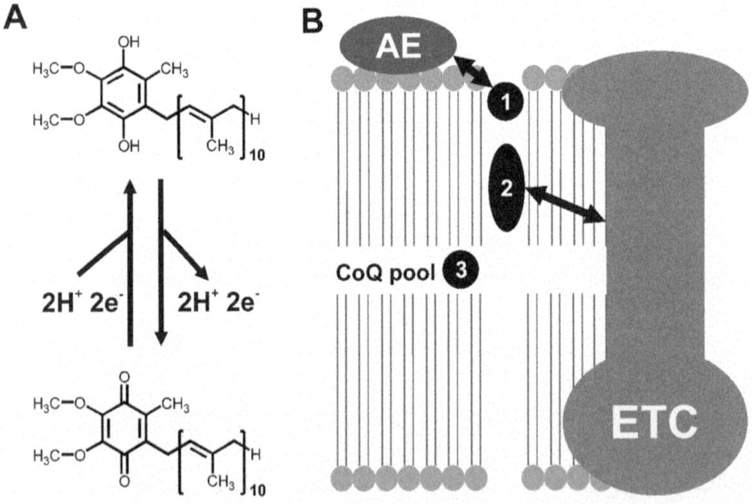

Fig. 2.2 Coenzyme Q_{10} location in membranes (**a**) Redox states of coenzyme Q_{10}. Quinone groups (bottom) can be reduced to quinol groups (top) by 2 electrons and 2 protons, being the two-electron reduction. It is possible a double one-electron reduction of coenzyme Q that produces an intermediate named semiquinone radical, a molecule that can generated superoxide in presence of oxygen. (**b**) Coenzyme Q depending of the hydrophobic tail length and its concentration can be located at three different locations. At the position 1 CoQ is located close to the water-lipid interphase allowing the interactions with metabolites (antioxidant recycling) or soluble proteins (to reduce CoQ). In this position is found the reduced state of CoQ. At the position 2, CoQ is embebed between the phospholipid tails ready to accept/release electrons from/to mitochondrial respiratory complexes. Also, at the position 2 can be found the oxidized state of CoQ. The position 3 corresponds to a pool required to maintain regulates the CoQ concentration on membranes

membranes (Fato et al. 1986; Samori et al. 1992), including mitochondria. Under these conditions, several studies have shown that the lateral diffusion of CoQ_{10} is not a limiting factor in the interaction with enzymes (Lenaz et al. 1999; Ondarroa and Quinn 1986). Since the lateral diffusion in long and short-chain CoQ molecules is the same, it has been shown that long-chain molecules undergo a folding that adjusts their length to short chains (Di Bernardo et al. 1998). From a functional point of view, it has been determined that the polar group of CoQ_{10} is located near the polar phase without interacting with the glycerol whereas the isoprenoid tail is included in the hydrophobic phase (Grzybek et al. 2005; Lenaz et al. 1992). This location is necessary to explain the antioxidant function. The access of the polar head of the quinone state of short-chain molecules (CoQ_2 or CoQ_3) during the reducing treatment with BH_4 is greater than for long-chain forms (CoQ_{10}) (Ulrich et al. 1985), indicating greater proximity to the aqueous phase. In contrast, short-chain molecules in the reduced state inhibit more effectively the lipid peroxidation compared to long-chain ones in microsomes of hepatocytes (Kagan et al. 1990) but some studies do not show the same results with CoQ_3 and CoQ_7 in liposomes (Fiorentini et al. 1993). A summary of coenzyme Q location in membranes is shown in Fig. 2.2b.

For the electron transport function, there is an inverse relationship compared to the antioxidant function. Oxidized forms of CoQ_2 and CoQ_3 inhibit the NADH oxidation but not of succinate in bovine mitochondria compared to CoQ_{10}. Also, the oxidation of ubiquinol increases with the isoprene length (Lenaz et al. 1978). There is evidence against the effectiveness of short-chain quinones to transport electrons in membranes; short-chain quinones (CoQ_1) have a lower affinity for the Q_B sites of photosynthetic bacterial reaction centers (Diner et al. 1984) and in the NADH-coenzyme Q oxidoreductase of the bovine heart the NADH binding produces a conformation change that allows the interaction of long-chain but not short-chain quinones (Hano et al. 2003). In purified mitochondria of *S. cerevisiae* deficient in the synthesis of CoQ_6, the reduction of short-chain exogenous quinones (CoQ_0 and CoQ_1) with succinate was decreased to less than 10% compared to the control, CoQ_2 or decylubiquinone (Zhu and Beattie 1988). However, the reduction with NADH showed a complete recovery of the activity. Comparing with the bovine mitochondria assays (Lenaz et al. 1978) the results of yeasts are contradictory. However, there are two important differences, the bovine mitochondria possesses endogenous CoQ_{10} and therefore the assay is performed in an intact mitochondrion while the CoQ_6-deficient yeast show a mitochondrion with defects in the expression of respiratory complexes (Santos-Ocaña et al. 2002) because some of them require endogenous CoQ_6 for a right assembly as is the case for the bc_1 complex (Bartoschek et al. 2001). This cause would explain the null effect of these molecules on succinate-quinone reductase (SQR). The recovery of NADH-quinone reductase (NQR) activity can be explained by the particular nature of NADH-dehydrogenases of *S. cerevisiae*, they do not form a complex, but are monomeric enzymes (Joseph-Horne et al. 2001) located at the inner mitochondrial membrane on the cytosolic face (NDE1 and NDE2) or on the matrix face (NDI1), and therefore may have easier access to short-chain quinones that are more amphipathic than long-chain quinones.

Another option to study the role of the hydrophobic tail in CoQ_{10} is the use of CoQ analogs that have been used as a potential therapeutic agent such as idebenone. Basically, it is a decylubiquinone with a hydroxyl group at the end of the aliphatic chain of 10 carbons (Okamoto et al. 1988; Sugiyama et al. 1985) (Fig. 2.1) which gives it a higher polarity. Idebenone is an efficient electron acceptor from succinate (SQR) and an electron donor to complex III (UCR) but does not accept electrons from the NADH (NQR), producing the inhibition of the proton pump of the complex I (Esposti et al. 1996). As compared to other ubiquinones condensed with isoprenoid chains, their effect is like CoQ_2, which also has 10 carbons, thus the presence of the hydroxyl group confers a similar degree of polarity. In quinones substituted with aliphatic tails, an increase of the electron transport efficiency was correlated with an increase in the length of the aliphatic tail. However, with the same number of carbons decylubiquinone shows more activity than idebenone. Further studies have shown that the more hydrophobic ubiquinone aliphatic derivatives support a higher activity than idebenone and that idebenone is functional with SQR but inhibit NQR activity (Brière et al. 2004).

The ubiquinone alkylated analogs have been well studied as their synthesis is much less complex than the isoprenoids. To act as acceptors, alkylated quinones require at least 6 carbons for maximal SQR or NQR activity while requiring 10 carbons to act efficiently as donors in UCR (Yu et al. 1985; Yu and Yu 1993). The addition of 1 or 2 double bonds does not substantially affect the function of quinones but a conjugated double bond system significantly decreases activity.

The most accepted model for localization of CoQ_{10} in membranes involves the interaction of the isoprenoid tail with the fatty acid moiety of phospholipids ivn the middle plane of the bilayer, while the polar head approaches to the polar moiety of phospholipids (Lenaz et al. 1992). This arrangement allows the interaction with enzymes embedded in the membrane as the respiratory complexes and in other cases with enzymes located at the aqueous phase. In this sense, studies carried out using molecular dynamics simulations and free energy calculations show that the electron translocation of the oxidized form of CoQ is ten times faster than the translocation of the reduced form (Kaurola et al. 2016). This implies that the location of coenzyme Q would ensure maximum occupancy of binding on entry sites in complexes of the respiratory chain to maintain its optimal activity. The existence of several pools of CoQ have been described in several membranes and organisms (Enriquez and Lenaz 2014; Heron et al. 1978; Jørgensen et al. 1985; Michaelis and Moore 1985) that has recently been associated with the organization of respiratory complexes in supercomplexes or respirasome (Acin-Perez et al. 2008; Enríquez 2016). The clustering of CoQ in pools has structural parallelism in the CoQ localization models, where equilibrium aggregations of molecules occur with a homogeneous distribution in membranes (Roche et al. 2006).

According to Fig. 2.2, the isoprenoid tail is the hydrophobic moiety of CoQ that supports its location in the inner middle plane of the phospholipid bilayer. This location is required for the expected interaction with larger respiratory complexes in mitochondria such as Complex I or bc_1 complex that show CoQ binding sites located in contact with the hydrophobic component of lipidic membranes. Complex I shows a CoQ binding site and a CoQ reduction site in contact with the membrane but electron from NADH must be channeled from a domain in contact with the aqueous phase of the matrix (Lenaz et al. 2006; Wirth et al. 2016). A similar situation is common to the *bc*$_1$ complex, CoQ binding sites are in contact with the membrane and electrons are channeled to the cytochrome *c* to a non-intermembrane protein domain (Xia et al. 1997). However, at the same time the isoprenoid tail is flexible to support the access of aromatic ring to the interphase lipid-water of membranes when CoQ works like antioxidant (Beyer 1994; Gómez-Díaz et al. 1997; Kagan et al. 1998a, b) or when is able to gain electrons from peripherally located reducing agents (Jiménez-Hidalgo et al. 2009; Santos-Ocaña et al. 1998; Villalba et al. 1995).

It is possible that the variations found at the isoprenoid tail length correspond to the specific features of enzymes involved in redox reactions of CoQ. In this way, *Saccharomyces cerevisiae* does not show a typical Complex I but synthesize a shorter CoQ molecule with only six isoprene units (CoQ_6). Considering that the polar ring is similar in all organisms probably the optimal adjust of CoQ to membranes can be caused by the isoprenoid tail nature.

2.2.2 *The Quinone Ring*

The quinone ring is directly responsible for the electron transfer characteristic of CoQ. Like the full structure of the CoQ molecule, the quinone ring structure is a balance between its location in a hydrophobic environment and a redox interaction with a polar environment. Both quinone groups are polar, allow the formation of hydrogen bonds, solvation and the interaction with dipoles, but the rest of the substituents (methoxy and methyl groups) give to the ring a hydrophobic character.

The reduction of both quinone groups occurs through two mechanisms, the two-electron mechanism, and the one-electron mechanism. The main difference is that the one-electron reduction of CoQ produces the semiquinone radical as an intermediary (Matsuda et al. 2000; Nakamura and Hayashi 1994), which may be a potential source of oxidative stress. In fact, the reduction of acetylated cytochrome *c* by semiquinone is used to determine the oxidative stress generated in cell membranes (Azzi et al. 1975). The mechanism of one-electron is characteristic of the NADH-cytochrome b_5 reductase family in *S. cerevisiae* (Jiménez-Hidalgo et al. 2009; Santos-Ocaña et al. 1998), plants (Serrano et al. 1995) or pig liver (Villalba et al. 1995, 1997). The two-electrons mechanism does not imply the formation of a semiquinone intermediate, it is the mechanism of type DT-diaphorase typical of an enzyme involved in detoxification (Anusevičius et al. 2002; Nakamura and Hayashi 1994).

Beyond the redox function of the quinone ring groups, the function of the remaining substituents, the C2 and C3 methoxy groups and the methyl group at the C5 position, have been analyzed. After studying the effect of modifications of these groups on quinones with alkyl or geranyl substituents on the succinate-cytochrome *c* reductase CoQ-dependent activity (SQCR), it was found that in general, the methoxy groups were more important than the methyl group (Gu et al. 1990). Switching C5-CH_3 to C5-H increases activity while switching to methoxy decreases activity. This implies that hydrophobicity is crucial at the C5 position, the elongation of the alkyl group beyond the methyl group results in a decrease in activity (He et al. 1994). In C2 and C3 positions, the change of methoxy group to hydrogen drastically decreases the activity, although C3 is more specific than C2. The loss of the methoxy group at the C3 position has been studied extensively *in vivo* because it has physiological connotations. During synthesis of CoQ_6 in *S. cerevisiae,* it has been shown that the non-methoxylated form of ubiquinone, demethoxy-Q_6 or DMQ_6 accumulates naturally (Padilla et al. 2004). This data together with the massive accumulation in missense mutants of the *COQ7* gene and subsequent studies of the CoQ_6 biosynthetic complex assembly have shown that DMQ_6 is an intermediate affected by processes of CoQ_6 synthesis regulation (Busso et al. 2015; Gonzalez-Mariscal et al. 2017; Martin-Montalvo et al. 2011; Padilla et al. 2009). However, in yeasts, it has been shown that DMQ_6 is unable to support electron transport in mitochondria or antioxidant defense when it is produced alone (Padilla et al. 2004). In the case of the nematode *C. elegans*, the same mutations that in yeast produced the accumulation of DMQ_6 (e2519) (Padilla et al. 2004) produces the accumulation of DMQ_9 (Ewbank et al. 1997; Felkai et al. 1999). Since nematodes carrying the *e2519* allele showed extended longevity compared to wild-type controls, it was concluded that DMQ_9 could have an anti-aging effect (Felkai et al. 1999). In this sense, in nematodes, this allele produces a lower respiratory chain activity and a lower production of ROS, which was used to explain its beneficial function as a caloric restriction mimetic (Branicky et al. 2000; Miyadera et al. 2002). However, subsequent analysis of quinone content in mitochondrial nematode carrying the e2519 allele samples analyzed by HPLC-ECD showed that there was an accumulation of CoQ_9 lower than 10% of the control. This small amount of CoQ_9 is responsible for the maintenance of respiratory activity in the mutant nematodes, decreased compared to wild-type and therefore not generating endogenous oxidative stress (Arroyo et al. 2006).

The selectivity of additional ring substituents is more relevant in UQCR activity compared to SQR activity (He et al. 1994). In SQR activity, ubiquinone interacts by producing a hydrogen bond between the oxygen of the C1 quinone group and tryptophan and tyrosine at the active site (Maklashina and Cecchini 2010). The arrival of an electron from the cluster (3Fe-4S) causes a pendular movement of the ubiquinone that brings it closer to the catalytic position stabilizing it by the formation three hydrogen bonds; (a) between the oxygen of the C4 quinone group and serine, (b) between the C3 methoxy and arginine and a water molecule and (c) between the C2 methoxy and aspartate.

The conservation of the structure and composition of both the ring and hydrophobic tail respond to functional needs in the access of CoQ to the aqueous phase to accept electrons and the lipid phase to donate them. In addition, the conservation of the CoQ polar ring responds specifically to the design of its target enzymatic complexes, much more selective at the time of oxidation than for reduction.

2.3 Biosynthesis of Coenzyme Q

As was indicated previously, CoQ biosynthesis is composed by two non-specific pathways (the synthesis of activated poly-isoprenoid and the aromatic polar ring) that converge in one specific pathway in mitochondria (CoQ_{10} in Fig. 2.3). This synthesis requires a large number of proteins that must be imported from the cytosol to mitochondria. The regulation of CoQ not only requires a precise synthesis of proteins but also need the formation of a biosynthetic complex (Q-synthome) inside the mitochondria that must be regulated by post-translational regulation. Most of the information about CoQ synthesis was obtained from the *Saccharomyces cerevisiae* model and in this section, the nomenclature of this model will be used (Table 2.1). In some cases, will be included in references to human cells or other models.

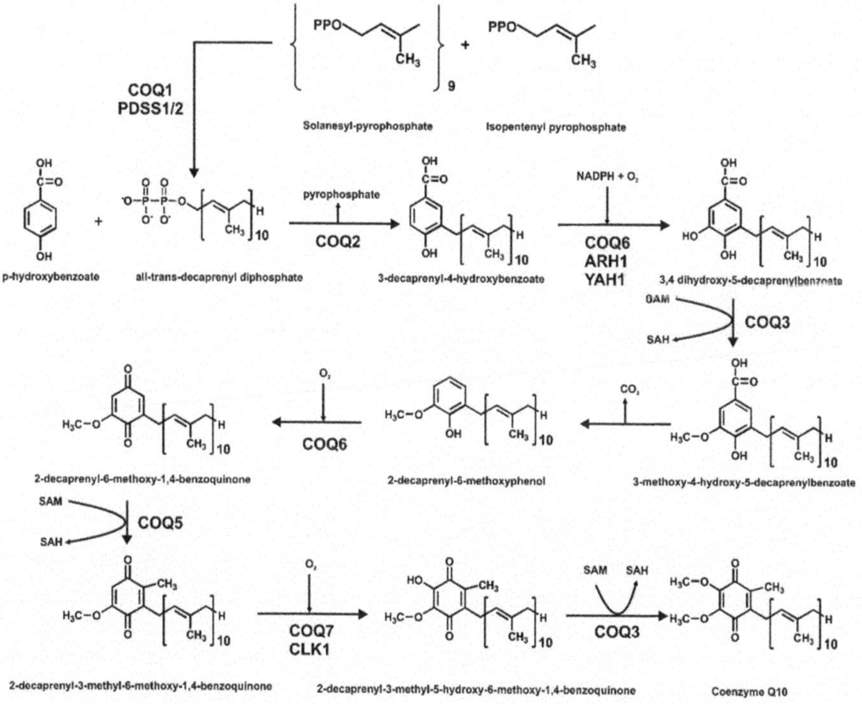

Fig. 2.3 Biochemical reactions involved on the coenzyme Q biosynthetic pathway. In the figure was indicated the reaction required for the coenzyme Q_{10} biosynthesis. The name of enzymes catalyzing these reactions appears in bold and also includes the name of cofactors or substrate more relevants. At the bottom are indicated the name of the intermediates. In some cases the name of human gene/enzyme are similar in human and yeast but in others cases are included both names

Table 2.1 Genes and proteins involved in coenzyme Q biosynthesis in yeast and human

Yeast name	CDS (bp)	Chr	aa	Location	Human name	mRNA (bp)	Chr	aa	Location
COQ1/ YBR003W	1422	II	473	Matrix	**PDSS1**	1657	10p12.2	415	Matrix
					PDSS2	3568	6q21	399	Cytosol/ matrix
COQ2/ YNR041C	1119	XI	372	MIM	**COQ2**	1641	4q21.23	421	MIM
COQ3/ YOL096C	1939	XV	312	Matrix	**COQ3**	1265	6q16.2	369	Matrix
COQ4/ YDR204W	1008	IV	335	MIM?/ Matrix	**COQ4**	1597	9q34.11	265	Matrix
COQ5/ YML110C	924	XIII	307	Matrix	**COQ5**	1530	12q24.31	327	Matrix
COQ6/ YGR255C	1440	VII	479	Matrix	**COQ6**	1615	14q24.3	468	Matrix
ARH1/ YDR376W	1482	IV	493	Matrix					
YAH1/ YPL252C	512	XVI	172	Matrix					
COQ7/ YOR125C	702	XV	233	Matrix	**COQ7**	1618	16p12.3	443	Matrix
PTC7/ YHR076W	1115	VIII	343	Matrix	**PPTC7**	2667	12q24.11	217	Matrix
COQ8/ YGL119W	1506	VII	501	Matrix	**ADCK3**	2616	1q42.13	179	Matrix
COQ9/ YLR201C	783	XII	260	Matrix	**COQ9**	3748	16q21	304	Matrix
COQ11/ YLR290C	834	XII	277	Matrix					

2.3.1 Genes and Proteins Involved

The isoprenoid precursor shares some reactions with the mevalonate pathway (Grünler et al. 1994; Olson and Rudney 1983). The activated polyprenyl-pyrophosphate is the product of the condensation of three units of farnesyl-pyrophosphate that also are produced by a previous condensation of three isopentenyl-pyrophosphate molecules. Until farnesyl-PP synthesis, this pathway shares some reactions with cholesterol synthesis. A trans-prenyl transferase is responsible for the last condensation. Organisms that synthesize CoQ molecules with 6 o 9 isoprene units such as *S. cerevisiae* or *C. elegans* show only one trans-prenyl transferase protein (*COQ1* in yeast) but human cells showing 10 isoprene units require a couple of protein with trans-prenyl transferase activity (PDSS1/2) (Saiki et al. 2003). This enzyme has been proposed as responsible for the isoprenoid tail length of CoQ (Okada et al. 1996). From the earliest years of research on $CoQ_{10,}$ it was demonstrated that the origin of the aromatic ring in animal cells are the amino acids phenylalanine and tyrosine (Olson and Rudney 1983). However, an alternative pathway has been described in bacteria, the synthesis from chorismate using the shikimate pathway (Zhou et al. 2013). In yeast, both pathways are present being the shikimate pathway the most important involving the participation of two genes/proteins (*ARO1* and *ARO2*). However, in *ARO1/2* mutants p-hydroxybenzoate (p-HB) is synthesized and CoQ_6 synthesis is not affected indicating that the phenylalanine-tyrosine pathway is functional in yeast (Goewert 1980). There are three possible pathways for the phenylalanine-tyrosine pathway, although remains some doubts about the reactions involved in the first step, the conversion of tyrosine to 4-hydroxyphenyl lactate. Recently, in the yeast model it has been demonstrated that genes *ARO8/9* are responsible for the tyrosine deamination and also that the gene *HFD1* is responsible for the last reaction, the 4-hydroxybenzaldehyde conversion to p-HB (Payet et al. 2016; Pierrel 2017; Stefely et al. 2016a, b). A possible orthologous of *HFD1* is the human gene ALDH3A1 that is an aldehyde dehydrogenase (Pappa et al. 2003; Payet et al. 2016; Stefely et al. 2016a, b).

The condensation of the activated isoprenoid (polyprenyl diphosphate) and the polar ring of CoQ (p-hydroxybenzoic acid) are catalyzed by the product of *COQ2*, a polyprenyl transferase (Ashby et al. 1992; Forsgren et al. 2004). The COQ2 protein (Coq2) is an enzyme with 5 transmembrane domains located at the mitochondrial inner membrane that produces the first molecule of CoQ biosynthesis that resembles the final product of the pathway, the 3-polyprenyl 4-hydroxybenzoate. This molecule is hydroxylated (C5 hydroxylation) by the product of gene *COQ6*, encoding for one of three monooxygenases involved in CoQ biosynthesis. Coq6 is a peripheral membrane protein on the matrix side of the inner mitochondrial membrane. Initially, Coq6 was been proposed as responsible for C1 hydroxylation step being the mitochondrial ferredoxin Yah1 and ferredoxin reductase Arh1 responsible for the C5 hydroxylation as electron donors (Pierrel et al. 2010). Yah1 is a ferredoxin (Alves et al. 2004) that together with Arh1 participates in the synthesis of sulfoferric complexes. Arh1 was described as a homolog of the mitochondrial human adrenodoxin reductase (Manzella et al. 1998). Yeast mutants strains of both genes are deficient in CoQ_6 but Coq6 must be present given that accumulates 4-hydroxy 6-hexaprenyl phenol, a molecule already hydroxylated in C1 position (Pierrel et al. 2010). However, in *COQ6* null mutants (*coq6Δ*) or missense *COQ6* mutants of FAD-binding domain accumulated 4-hydroxy 6-hexaprenyl phenol indicating that C5 hydroxylation but not C1 requires the participation of Yah1 and Arh1 as electron donors for Coq6 (Ozeir et al. 2011). At this time, the gene/enzyme responsible for C1 hydroxylation remains unknown.

The results of the Coq6 activity, 3-polyprenyl 4,5-dihydroxybenzoate is methylated by the product of *COQ3* gene in C5. Coq3 encodes for an O-methyl transferase that catalyzes (at least in yeast) two reactions in the biosynthetic pathway; the indicated previously and the methylation of demethyl coenzyme Q. This reaction requires the presence of S-adenosyl methionine (SAM) as a cofactor (Jonassen and Clarke 2000). The human COQ3 gene can complement the null mutant in yeast (Jiang et al. 2002b).

The next step is the conversion of 3-polyprenyl 4,5-dihydroxybenzoate in 2-polyprenyl 6-methoxyphenol after the C1 decarboxylation. In *E. coli* has been found that this step was catalyzed by the product of UbiX and UbiD genes, two isofunctional genes (Zhang and Javor 2003). A deletion strain for the UbiX gene (LL1), which show a low synthesis of CoQ_8 can be rescued by complementation with the *S. cerevisiae* ortholog gene *PAD1* (Gulmezian et al. 2007). UbiD show also an ortholog in *S. cerevisiae*, the *FDC1* gene but any of them are responsible for the decarboxylation step in yeast because even null mutants show wild-type levels of CoQ_6 (Mukai et al. 2010). At this time was not found the gene/protein responsible for this step. Recently it has been suggested the gene *COQ11* as responsible for the decarboxylation step (Allan et al. 2015) based on some sequence homology but it was discarded because null *COQ11* mutants (*coq11Δ*) maintains the CoQ_6 synthesis.

The next reaction is well characterized, the *COQ5* gene encodes a C-methyl transferase that catalyzes the methylation of the 2-polyprenyl 6-methoxy 1,4 benzoquinone at the C3 carbon (Barkovich et al. 1997; Dibrov et al. 1997; Nguyen et al. 2014), to produced 6-demethoxy coenzyme Q or DMQ. Also, SAM is required as a methyl group donor. Coq5 protein is peripherally associated with the inner mitochondrial membrane on the matrix side (Baba et al. 2004). Recently it has been described that the protease Oct1, required for the Coq5 processing (Veling et al. 2017), is required to synthesize CoQ in yeast.

The third hydroxylation reaction is catalyzed by the Coq7 protein, a monooxygenase. This enzyme is responsible for the C6 hydroxylation step, converting of 6-demethoxy coenzyme Q into 6-demethyl coenzyme Q (Clarke 1996; Jonassen et al. 1996). Coq7 belongs to a family of di-iron hydroxylases based on the existence of a conserved motif for iron ligands found also in bacteria such as *Pseudomonas aeruginosa* and *Thiobacillus ferrooxidans* (Stenmark et al. 2001). This point, the Coq7 product, 6-demethyl coenzyme Q, is methylated again in C6 by Coq3 to produce CoQ.

However, other proteins with structural and regulatory functions are required to make possible the CoQ synthesis. That is the case of proteins encoded by the *COQ4*, *COQ8*, *COQ9*, and *COQ11* genes. The *COQ4* gene encodes for a protein that supports the assembly of the CoQ biosynthetic complex in yeast or Q-synthome. It has been related Coq4 with the complex assembly because in bacteria no homologous proteins to Coq4 have yet been detected (Marbois et al. 2005). The polypeptide encoded by *COQ4* contains 335 amino acids with a calculated molecular mass of 38.6 kDa. Coq4 was localized to the matrix side in contact with the mitochondrial inner membrane. The human ortholog of yeast *COQ4* gene restored both growth in glycerol and the CoQ_6 status in the yeast strain after its expression in *COQ4* null yeast (Casarin et al. 2008).

The *COQ8* gene has received several names but the actual function of this protein is yet unknown. *ABC1/COQ8* is a component of a family of unusual kinases (Leonard et al. 1998) which also includes proteins from *Providencia stuarti* (AarF) and *E. coli* (yGR) that have been reported to be involved in CoQ biosynthesis (Macinga et al. 1998). This family shows several motifs typical of protein kinases but Coq8 may be catalytically inactive because the critical Asp residue was not detected in the triplet (DFG) which is required for catalytic activity (Leonard et al. 1998). Recently it has been demonstrated that Coq8 does not show a canonical protein kinase activity although can be autophosphorylated in cis mode (Stefely et al. 2015). However, this kinase function was not required to CoQ biosynthesis being a new ATPase function of Coq8 responsible for the stabilizing effect of Coq8 on the Q-synthome (Stefely et al. 2016a, b) The human ortholog is CABC1/ADCK3 but there are two paralogous genes in human for COQ8, ADCK3 and ADCK4, now termed COQ8A and COQ8B (Vazquez Fonseca et al. 2017). Patients with mutations found in COQ8A show CoQ_{10} deficiency and cerebellar ataxia (Chretien et al. 2008; Gerards et al. 2010). However, patients with mutations found in COQ8B show only renal affectation with the steroid-resistant nephrotic syndrome (Ashraf et al. 2013). In both cases, the patient cells show a CoQ_{10} deficiency.

The Coq9 protein is a component of the biosynthetic complex with several functions related to Q-synthome stability. Coq9 is a peripheral membrane protein on the matrix side of the mitochondrial inner membrane (Hsieh et al. 2007), which co-migrates with Coq3 and Coq4 at a molecular mass of approximately 1 MDa. Also, it was detected by direct interaction by immunoprecipitation of the HA-tagged Coq9 polypeptide with Coq4, Coq5, Coq6 and Coq7 demonstrating that Coq9 is a component of the Q-synthome. Specifically, it has been proposed that Coq9 is a lipid-binding protein that helps Coq7 to interact with the Q-synthome (Lohman et al. 2014).

COQ11 is the last gene that has been included in the list of genes required for CoQ biosynthesis (Allan et al. 2015). After a proteomic analysis, Coq11 has been demonstrated to be a Q-synthome component. The function remains obscure but according to its similarity with some decarboxylases, it has been proposed such as the not-discovered decarboxylase of Q-synthome.

2.3.2 *Biosynthesis Regulation*

Considering that the most important function of CoQ is as electron carrier of the respiratory chain it is required to study a possible association between regulation of synthesis and respiratory metabolism. *Saccharomyces cerevisiae* is a good model because is a facultative aerobic fermenter that grows in glucose even in presence of oxygen but can grow as a respiratory organism in the absence of fermentable carbon sources (Gancedo 1998). This makes possible to study the transition from fermentation to respiration because this molecule is needed preferentially during respiration. Initial studies demonstrated this relationship, higher levels of glucose in the culture media clearly decrease the amount of CoQ_6 in yeasts that was inversely correlated with the amount of 3,4-hydroxy hexaprenyl benzoate (Sippel et al. 1983), an early precursor of CoQ biosynthesis (Fig. 2.3) (Goewert et al. 1981). The catabolic repression produced by glucose in *Saccharomyces cerevisiae* is mediated by the repression of the Snf1p kinase (Gancedo 1998), a complex belonging to the family of AMP-kinases (Santangelo 2006). In the active state (low glucose) Snf1p induces the expression of downstream genes that are required for the use of other carbon sources such as ethanol, lactate or glycerol. Low levels of glucose correlate with high levels of cAMP and CoQ_6 (Sippel et al. 1983) supporting the idea that phosphorylation may be a regulatory element to consider in CoQ_6 biosynthesis.

In a typical yeast culture, the glucose exhaustion is associated with the raising of ethanol altering the metabolic state of yeast. Ethanol can be consumed but it requires a modification of metabolism named post-diauxic-shift or PDS (Pedruzzi et al. 2000). This growth stage is dependent on respiratory metabolism that needs a higher amount of CoQ_6. Under these conditions, DMQ_6 is accumulated until PDS initiation and then DMQ_6 decreases and CoQ_6 becomes the predominant quinone in yeasts (Padilla et al. 2009). This change of quinone concentration is correlated with the expression of several COQ genes such as *COQ3*, *COQ4*, *COQ5*, *COQ7,* and *COQ8*. A second intermediate, the 3-hydroxy-hexaprenyl benzoate (HHB) is also accumulated at a high proportion in YPD, being an 80% against total quinones during the exponential phase of growth (Poon et al. 1995), previous to PDS. HHB becomes a minor component at stationary phase. CoQ_6 biosynthesis shows at least two different intermediates, the first is HHB accumulated during fermentation and a second molecule (DMQ_6) close to the PDS. DMQ_6 can be converted to CoQ_6 when it is required.

A simple method to study gene expression related to CoQ_6 biosynthesis is the use of glycerol-based media (YPG). Glycerol induces the expression of *COQ5* (Hagerman et al. 2002). Glycerol growth increases the accumulation of CoQ_6 in wild-type yeast (Padilla et al. 2009) with the simultaneous reduction of DMQ_6 and also induces the expression of several COQ genes such as *COQ5*, *COQ7,* and *COQ8*. The gene *COQ4* is up-regulated by the growth in YPG as same as other mitochondrial proteins such as subunits of the complex IV (Belogrudov et al. 2001). However, less is known about the transcription factors involved in this regulation. *COQ5* was regulated by Hap2, a subunit of the HAP complex (heme activated-glucose repressed CCAAT-binding complex) and down-regulated by Mig1. Hap2 that is a global regulator of respiratory gene expression and Mig1 is a transcription factor that collaborates in glucose repression inactivated by phosphorylation by Snf1p (Hagerman and Willis 2002). *COQ2* is up-regulated by Hap1, a transcription factor that produces a response based on the levels of heme and oxygen (Harbison et al. 2004; Hickman and Winston 2007). YEASTRACT (Abdulrehman et al. 2011; Monteiro et al. 2008) is a platform to find potential promoter sequences recognized by transcription factors involved in the mitochondrial respiratory metabolism such as Hap3, Hap4, Hap5, and Adr1. Most of COQ genes show putative sequences of these transcription factors.

The second function of CoQ is antioxidant protection (Bentinger et al. 2007; Frei et al. 1990). In yeast, CoQ_6 protects plasma membrane for phospholipid peroxidation in *COQ3* null mutants (*coq3Δ*) produced by linolenic acid (Do et al. 1996). All null mutant or not-CoQ_6 producer strains of yeast are affected by linolenic acid (Schultz and Clarke 1999). When wild-type cells are incubated with linolenic acid several genes such as *COQ3* and *COQ7* are up-regulated (Padilla et al. 2009) and boost the DMQ_6 conversion to CoQ_6. Since this conversion requires the hydroxylase activity of Coq7 and the O-methyl transferase activity of Coq3, this effect supports that translational or post-translational mechanism of regulation is required. The effect of linolenic in gene regulation is specific because other oxidant compounds such as hydrogen peroxide did not affect COQ genes expression (Padilla et al. 2009). Again, using the platform YEASTRACT (Abdulrehman et al. 2011; Monteiro et al. 2008) it is possible to find potential promoter sequences recognized by transcription factors involved in the induction of antioxidant defenses such as Msn2, Msn4, Yap1, and Hsf1. All COQ genes are up-regulated (documented or potential regulation) by a general situation of stress but only *COQ1* shows a specific response to H_2O_2 and only *COQ4* and *COQ6* for hyperthermia.

The oxidative stress also promotes CoQ biosynthesis up-regulation in mammalian cells. Serum deprivation produces mild oxidative stress increasing mitochondrial ROS production and Ca^{2+} release (Kuznetsov et al. 2008). This method is comparable to methods based on oxidative stress generation to induce cell death (Kuznetsov et al. 2011), being apoptosis the most important consequence of serum deprivation. Exogenous CoQ_{10} shows a protective role against oxidative stress. CEM-C7H2 cells that undergo apoptosis after serum deprivation conditions are protected in presence of exogenous CoQ_{10} (Fernández-Ayala et al. 2000). The protection is caused by the inhibition of ceramide release and caspase-3 activation through the inhibition of plasma membrane-bound neutral sphingomyelinase (Navas et al. 2002). In parallel, serum deprivation generates mild oxidative stress that not only induces apoptosis but also increases CoQ_{10} levels in the plasma membrane (Barroso et al. 1997).

CoQ_{10} biosynthesis up-regulation can be obtained by oxidative stress generated by the chemotherapeutic drug camptothecin (CPT). This topoisomerase inhibitor induces DNA damage and oxidative stress (Gorman et al. 1997) followed by apoptosis. It is difficult to associate oxidative stress with the DNA alteration caused by the CPT, but the effect of CPT is inversely related to reduced glutathione content (Troyano et al. 2001). The amount of total CoQ_{10} was increased by CPT treatment in several cancer cell lines (Brea-Calvo et al. 2006). The CoQ_{10} increase is dose-dependent and is inhibited by the presence of anti-oxidants. The mechanism of CoQ_{10} biosynthesis up-regulation by CPT requires the activation of the COQ7 gene by the transcription factor NF-kappaB (Brea-Calvo et al. 2009). In the yeast model, the *COQ7* gene is a key component of the CoQ biosynthetic complex with a relevant function in the regulation of the CoQ biosynthetic complex activity. This gene COQ7 in human cells suffers a second layer of regulation at the post-transcriptional level. The UTR region of mRNA COQ7 is recognized by two RNA-binding proteins (RBPs) HuR and hnRNP C1/C2 (Schepens et al. 2007) inducing a modest mRNA COQ7 stabilization (Cascajo et al. 2015). However, it is most important the effect of HuR down-regulation caused by silencing or serum deprivation that decreases not only the stability of mRNA COQ7 but also shows physiological consequences decreasing COQ7 protein expression and the CoQ_{10} biosynthesis.

2.3.3 *The Q-Synthome*

Biosynthesis of CoQ requires the participation of at least 12 genes in yeast directly involved in the synthesis of the quinone ring. Although bacteria (*E. coli*) adopt a linear pathway (Gibson and Young 1978) several lines of evidence strongly support the existence of a biosynthetic complex of CoQ_6, the Q-synthome.

2.3.3.1 The Nature of Precursors in Null and Point Mutants

A common observation about null COQ mutants (*COQ3-COQ9*) is the accumulation of HHB (4-hydroxy-3-hexaprenyl benzoate) (Gin et al. 2003; Johnson et al. 2005; Poon et al. 1997). However, some point mutants in COQ genes accumulate the expected intermediate or diagnostic precursor. This effect suggests that most Coq proteins show a dual role, catalytic and structural, namely some mutations can remove the enzymatic function but the presence of protein in mitochondria supports the complex assembly. This behavior has been demonstrated in the *COQ7* gene; several point mutants such as *qm30* or *e2519* (Padilla et al. 2004) or *E194K* (Tran et al. 2006) accumulate DMQ_6 instead HHB. A similar case was the *COQ5* gene (Baba et al. 2004).

2.3.3.2 The Steady-State Levels of Coq Proteins in Mitochondria

The steady-state levels of Coq proteins can detect if a possible component of a multi-enzymatic complex is absent in mitochondria by a null mutation in another gene. It constitutes an indicator of the complex existence. Coq3 protein was the first analyzed with this approach basically by the existence of specific antibodies and the possibility to measure the O-methyl transferase activity (Hsu et al. 2000). Other studies were performed with other Coq proteins in the laboratory of Dr. C. F. Clarke (Baba et al. 2004; Gin and Clarke 2005; Hsieh et al. 2007; Marbois et al. 2009; Xie et al. 2012). With these data, it was possible to establish the existence of proteins always present in mitochondria independently of the null mutant analyzed, such as Coq1, Coq2, Coq5 and Coq8 and the rest of proteins that partially or totally disappear in null mutants. If we consider that if a subunit of a complex cannot be assembled it will be removed and degraded from mitochondria (Arlt 1998; Rugarli and Langer 2012), the proteins unaffected in null mutants (Coq1, Coq2, Coq5, and Coq8) must be independent to the complex formation.

2.3.3.3 Detection of Components of Coenzyme Q_6 Biosynthesis Complex

The presence of a Coq protein in mitochondria itself does not demonstrate the participation in a biosynthetic complex and does not allow to deduce the complex composition. To afford the analysis of the composition of the coenzyme Q_6 biosynthesis complex was used size exclusion chromatography (SEC) and BN-PAGE. Initially, it was detected a 700 kDa complex containing Coq3p, Coq4p, Coq6p, and Coq9p but no Coq1p or Coq5p using SEC (Hsieh et al. 2007; Marbois et al. 2005) and BN-PAGE analysis (Marbois et al. 2009; Tauche et al. 2008). Again, using SEC a second complex with a high molecular weight (1300 kDa) containing Coq3, Coq4, and Coq7 (Marbois et al. 2009; Tran et al. 2006). The CoQ_6 complex assembly is produced in two steps; the first is the formation of a pre-complex of 700 kDa containing all Coq proteins with enzymatic activity except Coq7. This pre-complex accumulates the precursor of Coq7, DMQ_6. The second step is the full complex assembly after the addition of Coq7 producing CoQ6.

2.3.3.4 The Overexpression Studies

The analysis of Q-synthome composition may be analyzed by overexpression of specific components to suppresses the structural defects found in COQ mutants. This approach is applicable mainly to regulatory Coq proteins with an unknown function. That is the case of *COQ8* gene. In *COQ7* null mutants (*coq7Δ*) was no assembled the Q-synthome and these strains does not accumulate the expected precursor, DMQ_6. *COQ8* overexpression restores in *coq7Δ*strains DMQ_6 accumulation (Padilla et al. 2009). That effect is relevant because the Q-synthome stabilization (even was impossible to recover the CoQ_6 synthesis) can facilitate the discovery of bypassing molecules (Pierrel 2017). The addition of vanillic (4-hydroxy-3-methoxybenzoate) acid together with *COQ8* overexpression restores the CoQ_6 synthesis in a *COQ6* null mutant strain (*coq6Δ*) (Ozeir et al. 2011). The vanillic acid bypassed the lack of Coq6 but *COQ8* overexpression made possible the complex stabilization required to modify vanillic acid to CoQ_6. Using this method, it was possible to analyze the accumulation of diagnostic intermediates and the steady-state levels of Coq proteins after *COQ8* overexpression (Xie et al. 2012).

As a summary, a protein complex is required in mitochondria of yeast and human to synthesize CoQ, the Q-synthome (Allan et al. 2015; González-Mariscal et al. 2014; Stefely and Pagliarini 2017). This Q-synthome is composed for a subset of Coq proteins (Coq3, Coq4, Coq5, Coq6, Coq7, and Coq9) that excludes some proteins involved in the synthesis of hydroxy-hexaprenyl benzoic acid (Coq1 and Coq2) and regulatory proteins such Coq8 and Coq11. The initial assembly of Q-synthome can be nucleated around Coq4 (Gin and Clarke 2005; Marbois et al. 2005, 2009) but we do not have direct evidence of this function. The structure of cyanobacteria protein Alr8543 has been solved and contains the Coq4 fold (Rea et al. 2010). With this structure as a template, it was possible to obtain a homology model for yeast (Rea et al. 2010). One interesting observation is that together with

Alr8543 protein co-crystallized one molecule of geranylgeranyl monophosphate in a binding pocket. This protein also shows a putative zinc finger chelating a magnesium ion. The mapping of several nonsense mutations found in yeast around the zinc finger predicted the function of Coq4 as an anchor protein for the poly-isoprenoid tail of HHB. Around this starting point, the rest of catalytical proteins must be arranged resulting in the full Q-synthome. By the Coq4 action, will be nucleated a pre-complex of 700 kDa that converts the initial precursor in DMQ_6 that is accumulated even in wild-type strains (Padilla et al. 2004, 2009). DMQ_6 accumulation must be the effect of a regulatory step, no other intermediate is accumulated in yeast. The conversion of DMQ_6 to CoQ_6 requires the assembly of the full complex or Q-synthome (of 1300 kDa) after the binding of Coq7. At this point can be introduced Coq9 in the complex (Hsieh et al. 2007; Johnson et al. 2005). Recently, it has been demonstrated that Coq9 can recognize isoprenoids lipids and also help Coq7 to be redirected to the pre-complex (Lohman et al. 2014). However, this picture is not complete without a mechanism explaining the regulation of this last step of the Q-synthome assembly, a phosphorylation mechanism.

2.3.4 Phosphorylation Events on Coenzyme Q Biosynthesis

Several studies have demonstrated the existence of COQ phosphoproteins; Coq3, Coq5 and Coq7 (Martin-Montalvo et al. 2011; Tauche et al. 2008; Xie et al. 2011) and also proposed a protein kinase function for Coq8 because in *COQ8* null mutants strains (*coq8Δ*) the phosphorylation of Coq3, Coq5, and Coq7 cannot be detected and was recovered after the expression of ADCK3, the human homolog of yeast *COQ8* gene (Xie et al. 2011). ADCK3 rescued phosphorylation but also recover the assembly of the Q-synthome and the CoQ_6 synthesis (Xie et al. 2012). These data supported the hypothesis that Coq8 is a protein kinase required for the Q-synthome assembly. However, recent studies demonstrated that human Coq8 ortholog, ADCK3 is an unusual kinase that although show cis-autophosphorylation, this activity is not relevant for CoQ_{10} synthesis (Stefely et al. 2015). It was required to CoQ_{10} synthesis but being responsible for Q-synthome assembly. That is a controversial point that can be solved by hard speculation, Coq8/ADCK3 can be required to allow the access of an unknown kinase to Q-synthome. This idea is in agreement with the recruiting effect of Coq8 on COQ proteins (Padilla et al. 2009; Xie et al. 2011, 2012), but requires a new gene/protein with kinase activity. Because at this time it was not detected, the hypothesis is that kinase activity corresponds probably to a general mitochondrial kinase responsible for mitochondrial metabolic regulation and therefore impossible to associate specifically to CoQ_6 biosynthesis.

However, we have more information about the regulation by phosphorylation of Coq7 in yeast. It has been demonstrated that Coq7 contains three phospho-amino acids (S20, S28, and T32) detected by a NetPhos analysis (Ingrell et al. 2007). These phosphosites can be phosphorylated *in vitro* and the yeast metabolic state modify the *in vivo* phosphorylation state (Martin-Montalvo et al. 2011). When modifications in Coq7 to remove phosphosites (changes to alanine) were introduced, its expression in a *COQ7* null mutant strain increases CoQ_6 up to 250%. The modification to a permanently charged state (to aspartic and glutamic acids) decrease the CoQ_6 level to a 50%.

While the putative kinase remains obscure, the phosphatase has been found and with extra functions not only in the regulation of CoQ_6 biosynthesis. In *S. cerevisiae* the number of phosphatases, 40, is far to the kinase abundance (Sharmin et al. 2014). Seven of these 40 phosphatases (Sakumoto et al. 1999) show mitochondrial localization (Claros and Vincens 1996). In most cases, the function is known except for Ptc7, which belongs to the PPM family. (Jiang et al. 2002a) and participates in the osmotic stress response (Runner and Brewster 2003; Sharmin et al. 2014). Ptc7 show a double localization originated by the presence of an intron (Juneau et al. 2009). The non-spliced form of Ptc7 is located in the nuclear envelope and it is expressed on fermentative conditions of growth while the spliced form is located in the mitochondrial matrix after growth with non-fermentative carbons sources. This location supports the proposed function as a regulator of CoQ_6 biosynthesis. Ptc7p dephosphorylates Coq7p *in vivo* and in *vitro* (Martin-Montalvo et al. 2013). Ptc7 dephosphorylated Coq7 phosphorylated *in vitro* by PKA and the lack of Ptc7 in null mutant strains results in the accumulation of phosphorylated Coq7. According to previous studies, Pct7 is expressed in non-fermentable carbon sources and in oxidative stress conditions. This supports completely the function of Ptc7 as Coq7-phosphatase and points out the Ptc7 function as a regulator of CoQ_6 biosynthesis. However, other studies suggest a function as a general regulator of mitochondrial metabolism. Recently, we have demonstrated that Ptc7 is required to support a normal chronological lifespan extension (CLS) (Gonzalez-Mariscal et al. 2017). However, the shorter lifespan measured in *ptc7Δ* could not be recovered by the addition of exogenous CoQ_6 supporting that a lower CoQ_6 content is not responsible for the defect on CLS. The shorter CLS can be produced by a lack mitophagy activation detected in *ptc7Δ* since mitophagy is required to an adequate CLS extension (Abeliovich 2011; Sampaio-Marques et al. 2014). Other functions have been added to Ptc7 such as regulator of enzymes involved in the TCA cycle. Ptc7 regulates the activity of Cit1 but also affect other mitochondrial enzymes, specifically or not (Guo et al. 2017). Recently, the double location of Ptc7 has been unveiled, it is regulated by the chromatin-remodeling SWI/SNF complex. This complex down-regulates the expression of ribosomal protein gene (RPG), rich in introns, to release spliceosome complexes to other intron-containing genes poorly spliced such as Ptc7 (Awad et al. 2017). The SWI/SNF complex works as a nutrient sensor, the spliced Ptc7 is located in mitochondria and activate mitochondrial metabolism (CoQ_6 synthesis among others) but the non-spliced form is located at the nuclear envelope and produces active repression of CoQ_6 synthesis and mitochondrial metabolism. In human, the *PTC7* orthologous gene is PPTC7. Recently, it has been demonstrated that PPTC7 is a mitochondrial protein-phosphatase that not only participate in CoQ_{10} synthesis regulation but also may be a key regulatory element of mitochondrial metabolism (González-Mariscal et al. 2018). This mitochondrial regulatory function has been corroborated after the characterization of a knock-out mouse model of PPTC7 (Niemi et al. 2019), showing typical markers of mitochondrial dysfunction such as hypoketotic hypoglycemia, elevated acylcarnitines, and lactic acidosis, accompanied by a significant smaller mitochondrial size and perinatal death.

2.4 CoQ Distribution, Transport, and Supplementation

According to its second name (ubiquinone), CoQ is a molecule distributed among all organs, tissues, and endomembranes. That fact arises in several questions around the CoQ localization, what is the synthesis localization? How can be transported among cells and membranes? How can escape from mitochondria? And finally, in the opposite direction, what is the nature of the CoQ import in cells?

The above-mentioned questions are crucial not only to understand the function of CoQ but also are important to decide general approaches to increase CoQ level in patients with CoQ deficiency caused by mitochondrial diseases or in people where the CoQ decline with the age. Both approaches are the direct CoQ supplementation or the increase of endogenous synthesis.

2.4.1 In Organs and Tissues

2.4.1.1 Distribution

The normal distribution of CoQ among the different organs and tissues was a topic analyzed after the discovery of CoQ. The first example of this study (Gale et al. 1961) analyzed the CoQ_{10} content in human samples from three donors, being liver and heart the organs with the highest amount of CoQ_{10} while muscle contained a moderate amount. In later studies performed in rats (Beyer et al. 1985), again liver and heart contained the highest amount of CoQ_9 but with aging CoQ_9 are increased in young rats until 10 months where declines significantly in skeletal muscle tissues and heart. A comparative study of lipid levels during aging in humans and rats (Kalén et al. 1989) showed in both organisms that CoQ content increases until reaching a maximum in adulthood (30 days for rats and 20 years for human), is stabilized and then there is a significant decline in aged organisms in all organs analyzed.

In general, it is accepted that CoQ biosynthesis localization in mitochondria. Some studies analyzed CoQ levels in mitochondrial samples purified from different organs of rats (Battino et al. 1990) demonstrating that mitochondria show a similar profile of CoQ_9 and CoQ_{10} that in whole cells, being heart the organ containing the highest amount of CoQ. In this case, mitochondrial from muscle show increased levels of CoQ_9 and CoQ_{10} compared to other studies. The same study performed in mitochondrial pig samples was comparable even in absolute numbers. According to the function of CoQ in cell membranes, others studies were focused on the redox state of CoQ in rats and human (Åberg et al. 1992). The distribution in rats organs was similar to other studies, being higher in heart, kidney and liver and one third in other organs. The amount of CoQ_9 was about 80% of total CoQ. In human, the distribution was like rats but CoQ_9 was only a 2–5% of total CoQ. In general, and in both organisms CoQ can be found mainly reduced (70%). In rabbits, CoQ_{10} was distributed in a similar way compared to other mammals (Matsura et al. 1991) with the highest accumulation in heart and kidney.

In mammals, CoQ is widely distributed in all organs and tissues, but its concentration is not homogeneous and is related to the function of the organ and better with its specific energy requirements.

2.4.1.2 Transport and Supplementation with Exogenous CoQ

Coenzyme Q is differentially distributed in organs and tissues, and this fact can be explained by two different mechanisms, (a) by local synthesis, each organ and tissue synthesize CoQ and (b) by transport from one biosynthetic organ and from the diet. The number of studies about the CoQ synthesis in whole organs is low. Using radiolabeled precursors it was possible to determine the rate of synthesis in liver and brain slices from rat (Andersson et al. 1990). In both cases, there is a net synthesis but in the liver was 7.2 times higher. In the same study was reported that the amount of CoQ in the liver is double that found in the brain. This study points out a mixed mechanism. However, two interesting studies about the pharmacologic regulation of CoQ synthesis support the local synthesis mechanism. Lovastatin is a molecule that inhibits cholesterol synthesis but also CoQ synthesis because both molecules share some biochemical steps of the mevalonate pathway (Goldstein and Brown 1990). Lovastatin decreases significantly CoQ_{10} and CoQ_9 levels in heart, blood and liver in rats (Willis et al. 1990). In a different sense, peroxisomal inducers such diethylhexylphthalate or DHEP increase the CoQ level in several organs and tissue (Turunen et al. 1999a, b). Taken together, these studies agree with a local ability of organs to synthesize CoQ: However, it is impossible to discard the existence of a CoQ transport mechanism between organs.

Traditionally, CoQ transport has been analyzed by exogenous CoQ_{10} supplementation but the murine model of study synthesizes CoQ_9 and lesser amounts of CoQ_{10}. That is an advantage because exogenous CoQ_{10} can be differentiated easily from endogenous CoQ_9. Initial studies in rat demonstrate that exogenous CoQ_{10} administrated orally was accumulated mainly in blood and liver (Scalori et al. 1988). A similar result was obtained by the oral and intraperitoneal administration after a study of 10 weeks in rats (Rodríguez-Hernández et al. 2009). The administration by gastric intubation duplicated the level of CoQ in blood plasma and liver after 4 days but exogenous CoQ_{10} was not detected in other organs such as the heart or kidney (Zhang et al. 1995). The previous study was a short-term study but a long-term one in rats and mice obtained similar results, a significant increase in blood and liver but not in other organs (Lönnrot et al. 1998). Additionally, this CoQ_{10} long-term administration did not improve mice lifespan. In other studies in mice, CoQ increase was detected in isolated mitochondria such as liver, kidney, and heart but not in whole organs (Lass et al. 1999). The derivatization of exogenous CoQ_{10} to succinyl-CoQ_{10} or acetyl-CoQ_{10} improves the solubility of these molecules compared to CoQ_{10} and also increased significantly the uptake by blood plasma and liver but again no evidence of CoQ_{10} uptake was found in other organs (Turunen et al. 1999a, b). In another supplementation study in rats, CoQ_{10} plasma concentration was increased dramatically while the increase in organs such as kidney, liver, and brain was slightly but significant and no uptake was detected in the muscle (Kwong et al. 2002). Supplementation is a useful method to increase levels of CoQ_{10} in plasma and liver but fail as mechanisms to increase CoQ_{10} in other organs. One explanation of this effect is that exogenous CoQ_{10} can inhibit the endogenous synthesis of CoQ_{10} that also is synthesized by murine cells, being impossible to differentiate exogenous and endogenous CoQ_{10}. Although several studies of CoQ_{10} supplementation discarded the inhibitory effect on CoQ_9 synthesis (Turunen et al. 1999a, b; Zhang et al. 1995) a study of supplementation with radiolabeled CoQ_{10} (Bentinger et al. 2003) demonstrated that higher amounts of C^{14}-CoQ_{10} were found in blood and liver and lower concentrations in adrenals glands, ovaries, thymus, and heart. No uptake was detected in kidney, muscle, and brain. That supports the idea that liver seems to be the endpoint of CoQ_{10} uptake in mammals. The liver may work as a buffer system to accumulate the excess of diet CoQ_{10}. In fact, in some studies, the exogenous CoQ was accumulated in lysosomes (Zhang et al. 1995) and transport vesicles of hepatocytes (Bentinger et al. 2003). However, supplementation of CoQ_{10} is a therapeutic strategy to the treatment of a CoQ_{10} deficiency in humans with neurological (Quinzii et al. 2005; Salviati et al. 2005) and muscular disorders (Di Giovanni et al. 2001; Horvath et al. 2006). This fact demonstrated that exogenous CoQ_{10} orally administered can be redirected to cerebellum or muscle. The main difference between studies of CoQ_{10} uptake in a murine model and human patients is the existence of a deficiency. It is possible that in organs with plenty of CoQ_9 or CoQ_{10} was impossible to relocate more CoQ in membranes being blood plasma and liver, the storage system to accumulate the excess of CoQ_{10}. Mice heterozygous mutant of the COQ7 gene shows mitochondrial defects but no CoQ deficiency was detected (Lapointe et al. 2012). Only was found a low level of CoQ at inner membranes. CoQ_{10} supplementation improves mitochondrial function and inner membrane CoQ_{10} recovery, which demonstrates that CoQ_{10} uptake was functional. Some authors suggest that the lack of CoQ_{10} uptake in murine models is produced by a negative interaction between the native CoQ_9 and the exogenous CoQ_{10} (Miles 2007). In a study using radiolabeled CoQ_{10} in the Guinea pig model CoQ_{10} uptake was detected in all tissues and organs analyzed (Yuzuriha et al. 1983) after intravenous supplementation. That results are completely different compared to murine model but in this case, Guinea pigs show a more balanced CoQ production with a 53% of CoQ_{10} and a 47% of CoQ_9 (Lass et al. 1997).

A final question is the role of the liver in the distribution of CoQ_{10} synthesized in the same liver, accumulated from the diet or after supplementation. CoQ_{10} of diet or supplemented can be absorbed in the intestine as other lipids by a well-known way for CoQ_{10} (Miles 2007). This CoQ_{10} was accumulated finally in chylomicrons as part of plasma lipoproteins. CoQ_{10} can be detected in plasma 1–2 h after the ingestion with a peak of about 6 h (Miles et al. 2002). There are not abundant studies about CoQ distribution in the different type of lipoproteins. Endogenous CoQ can be located in lipoproteins both in rat and human plasma, being the proportion of CoQ_{10} in human HDL of a 63% and of CoQ_9 in the rat of a 18% (Elmberger et al. 1989) although this data is not uniform because in other studies the proportion of CoQ_{10} in HDL from human plasma is of 26% (Tomasetti et al. 1999). However, HDL carries free cholesterol from peripheral cells to the liver (Zhou et al. 2015) while the lipoprotein responsible to transport lipids from the liver to peripheral organs are LDL and VLDL. The different fractions of LDL increased significantly (350%) the amount of CoQ_{10} after the oral supplementation (Alleva 1995), which demonstrate that CoQ_{10} becomes ready to be transported to the rest of organs. The reason that explains the ineffectiveness of CoQ_{10} uptake as therapy in some patients with CoQ_{10} deficiency may be probably caused by the specific features of the pathology better than by a general problem of CoQ_{10} transport.

2.4.2 *In Cell Membranes*

The most important function of CoQ is the electron transport between mitochondrial complexes and in fact, it is one of the endomembranes in which it is accumulated largely. However, CoQ appears distributed among cell membranes with a variable concentration. In the yeast *Saccharomyces cerevisiae*, several studies showed that CoQ_6 can be detected in all membranes analyzed (Padilla et al. 2004; Santos-Ocaña et al. 1998, 2002). The highest concentration can be detected in mitochondria, mitochondria-associated membranes (MAM) and vacuole. In mammalian models such as rat liver (Kalén et al. 1987; Takahashi et al. 1993) was found a similar profile with an important difference, Golgi complex showed the highest CoQ_9 concentration (2.62 μg/ mg protein) followed by lysosome (1.86 μg/ mg protein) and mitochondria (1.4 μg/mg protein). In a second study performed in rats (Takahashi et al. 1993) mitochondria are the most enriched membrane in CoQ_9 far away from lysosome and Golgi complex. The same analysis was done in kidney and in this case mitochondria is again the most enriched membrane followed by the nuclear envelope. In mice, again mitochondria showed the highest concentration and lysosome was the second location (Bentinger et al. 2012). In rabbits, mitochondria showed the highest accumulation of CoQ_{10} followed by nuclei (Matsura et al. 1991).

Mitochondria, in general, seems to be the richest membrane in CoQ. This fact can be explained by the CoQ functions in mitochondria; electron transport, β-oxidation (Frerman 1988), pyrimidine synthesis (Lass et al. 1997), UCP activation (Echtay et al. 2000), and mPTP opening (Fontaine et al. 1998) but more functions have been demonstrated in other membranes such as plasma membrane and lysosome. In plasma membrane CoQ is required as antioxidant as per se

(Fernández-Ayala et al. 2000; Villalba et al. 1997) or by recycling other antioxidant molecules such as ascorbate (Arroyo et al. 2004; Gómez-Díaz et al. 1997; Rodriguez-Aguilera et al. 1993; Santos-Ocaña et al. 1995) and in several type of cells α-tocopherol (Kagan et al. 1998a, b). The high sensitivity of the oxidative stress generated by linolenic acid is a common feature of yeast mutants in CoQ_6 biosynthesis (Do et al. 1996; Poon et al. 1997). In plasma membrane from different sources, some enzymes can maintain CoQ in its reduced state, the protective state, using NAD(P)H as the electron donor (Santos-Ocaña et al. 1998; Villalba et al. 1995, 1997). Examples are the NADPH-CoQ reductase or the NADH-cytochrome b_5 reductase, that can block apoptosis by the inactivation of neutral-sphingomyelinase (Navas et al. 2002). Since yeast has been demonstrated that a plasma membrane NADH-cytochrome b_5 reductase can regulate respiratory metabolism reducing CoQ and oxidizing NADH and a the same time produces chronological lifespan extension (Jiménez-Hidalgo et al. 2009). In the lysosome, CoQ is a component of a second membrane redox system (Gille and Nohl 2000). This system contains a *b*-type cytochrome reductase that transfers electrons from cytosolic NADH to CoQ (Nohl and Gille 2002) with the function of maintaining a low pH inside pumping protons from the cytosol.

This functional location of CoQ in several membranes put in evidence two new questions; the subcellular location of CoQ biosynthesis and in just in case of a single synthesis location, the mechanisms required to transport CoQ from the site of synthesis to cellular membranes.

2.4.3 Biosynthesis Localization

As was indicated in the previous section, CoQ biosynthesis is produced by several proteins assembled in a large enzymatic complex located in the mitochondria. It is necessary to indicate that CoQ synthesis is the final product of a large pathway that share branches with the mevalonate pathway (synthesis of active poly-isoprenoid) and with the metabolisms of aromatic amino acids (synthesis of aromatic ring). Here we are going to analyze only the CoQ specific pathway that starts with the condensation of the poly-isoprenoid tail with the precursor of the aromatic ring, the 4-hydroxy benzoate or p-HB. Now, several pieces of evidence support that mitochondria are the location of CoQ biosynthesis in eukaryotic cells:

(a) In all models, mitochondria are the most CoQ-enriched membrane in the cell.
(b) Biochemical studies: Most of the data about CoQ synthesis come from the yeast model. All proteins in yeast involved in CoQ synthesis were found in mitochondria by classic biochemical methods. The exact submitochondrial localization was previously published. Most human CoQ proteins are orthologous of yeast CoQ-proteins that can complement the defects of respiratory growth showed by yeast mutants. Also, these human CoQ-proteins has been located in mitochondria.

(c) Topological prediction studies: Most of the proteins show a typical mitochondrial leader sequence and also the protease target in yeast (Jonassen and Clarke 2001).
(d) In yeast, the multi-enzymatic complex responsible for CoQ synthesis (Q-synthome) is located in mitochondria (Allan et al. 2015; Marbois et al. 2009).
(e) In yeast, the Q-synthome complex is an example of the interaction of several mitochondrial proteins (interactome) (Floyd et al. 2016).
(f) In human cells, the CoQ_{10} synthesis labeling performed with [^{14}C]-pHB demonstrated that the first signal of synthesis is located in mitochondria (Fernández-Ayala et al. 2005).
(g) In rat liver slices was synthesized the first ubiquinone-like intermediate of the pathway at the mitochondrial inner membrane (Momose and Rudney 1972).

In the past, the CoQ synthesis location was subject to controversy. Using radiolabeled precursors of the isoprenoid tail and the aromatic ring it was possible to detect the CoQ_9 synthesis in rat liver (Kalén et al. 1987). In this study was analyzed the location of the new synthesized CoQ_9 by a time-chase experiment. Although radiolabeled CoQ_9 was located initially in mitochondria and endoplasmic reticulum (ER), in the short term a higher rate was detected in the ER while long-term was accumulated in mitochondria. This idea of ER synthesis was supported by studies that detected the highest nonaprenyl-4-hydroxybenzoate transferase activity in the endoplasmic reticulum and Golgi complex (Kalén et al. 1990; Teclebrhansp et al. 1993), activity for which protein COQ2 is responsible for human (Forsgren et al. 2004; López-Martín et al. 2007) and yeast (Ashby et al. 1992). To support this location on endoplasmic reticulum was argued that was detected the full molecule (no intermediates) and that was not defined a way of transport between mitochondria and ER (Kalén et al. 1987). Also, it was argued a dual location mitochondria-ER to support both CoQ requirements. However, two objections can be raised. The first is that have been described a way of lipid transport between mitochondria and ER, the membrane contacts (Rusiñol et al. 1994) associated to a new membrane fraction named mitochondrial-associated membranes (MAM) (Achleitner et al. 1999; Gaigg et al. 1995). This MAM fraction is a membrane with a mixed membrane composition that contains a high concentration of CoQ_6 in yeast (Padilla-López et al. 2009), and that participate in phospholipid transport between both organelles (Rowland and Voeltz 2012; Scharwey et al. 2013). It is possible that the origin of initial accumulation of radiolabeled CoQ found in ER was the transient accumulation in MAM in transit to the endomembrane system. The ER fractions purified in this study probably contained the MAM fraction that was not yet defined at the time of the study. A second objection is the methodology approach of the study, the detection of new radiolabeled CoQ_9 that is produced in a mitochondrion plenty of CoQ_9. It is possible that the new CoQ_9 was readily exported to the endomembrane system better that accumulated on mitochondria that probably contains already a balanced amount of CoQ_9.

The evidence of a way to transport CoQ from mitochondria to the endomembrane system open a new question, how CoQ can reach the different membranes? A study with a radiolabeled precursor of CoQ ([C^{14}]-pHB) demonstrated that endomembranes are sequentially enriched in CoQ following the typical direction of vesicle movement of from ER to the plasma membrane (Fernández-Ayala et al. 2005). The participation of the vesicle membrane traffic system was demonstrated by the inhibition obtained after the treatment with brefeldin E. In the yeast model has been demonstrated also the participation of the vesicle membrane traffic system. A general feature of yeast mutants of CoQ_6 synthesis is the rescue with exogenous CoQ_6 (Jonassen and Clarke 2001) but not all genetic backgrounds can be rescued (Santos-Ocaña et al. 2002). These non-rescued backgrounds (EG103 and FY250) show defects on the vesicle traffic pathway. In rescuable backgrounds such as BY4741, null mutant strains in genes involved in endo/exocytosis (*VPS45*, *TLG2*, *PEP12*, and *ERG2*) the rescue with exogenous CoQ_6 is not possible (Padilla-López et al. 2009). Together, these data support the requirement of the endomembrane vesicle traffic system to the export/import of CoQ from/to endomembranes.

References

Abdulrehman D, Monteiro PT, Teixeira MC, Mira NP, Lourenço AB, dos Santos SC, Cabrito TR, Francisco AP, Madeira SC, Aires RS, Oliveira AL, Sá-Correia I, Freitas AT (2011) YEASTRACT: providing a programmatic access to curated transcriptional regulatory associations in Saccharomyces cerevisiae through a web services interface. Nucleic Acids Res 39:D136–D140. https://doi.org/10.1093/nar/gkq964

Abeliovich H (2011) Stationary-phase mitophagy in respiring *Saccharomyces cerevisiae*. Antioxid Redox Signal 14:2033–2011. https://doi.org/10.1089/ars.2010.3807

Åberg F, Appelkvist ELL, Dallner G, Ernster L (1992) Distribution and redox state of ubiquinones in rat and human tissues. Arch Biochem Biophys 295:230–234. https://doi.org/10.1016/0003-9861(92)90511-T

Achleitner G, Gaigg B, Krasser A, Kainersdorfer E, Kohlwein SD, Perktold A, Zellnig G, Daum G (1999) Association between the endoplasmic reticulum and mitochondria of yeast facilitates interorganelle transport of phospholipids through membrane contact. Eur J Biochem 264:545–553. https://doi.org/10.1046/j.1432-1327.1999.00658.x

Acin-Perez R, Fernandez-Silva P, Peleato ML, Perez-Martos A, Enriquez JA (2008) Respiratory active mitochondrial supercomplexes. Mol Cell 32:529–539. https://doi.org/10.1016/j.molcel.2008.10.021

Agmo Hernández V, Eriksson EK, Edwards K (2015) Ubiquinone-10 alters mechanical properties and increases stability of phospholipid membranes. Biochim Biophys Acta-Biomembr 1848:2233–2243. https://doi.org/10.1016/j.bbamem.2015.05.002

Allan CM, Awad AM, Johnson JS, Shirasaki DI, Wang C, Blaby-Haas CE, Merchant SS, Loo JA, Clarke CF (2015) Identification of Coq11, a new coenzyme Q biosynthetic protein in the CoQ-synthome in *Saccharomyces cerevisiae*. J Biol Chem 290(12):7517–7534. https://doi.org/10.1074/jbc.M114.633131

Alleva R (1995) The roles of coenzyme Q10 and vitamin E on the peroxidation of human low density lipoprotein subfractions. Proc Natl Acad Sci 92:9388–9391. https://doi.org/10.1073/pnas.92.20.9388

Alonso A, Gomez-Fernandez JC, Aranda FJ, Belda FJF, Goñi FM (1981) On the interaction of ubiquinones with phospholipid bilayers. FEBS Lett 132:19–22. https://doi.org/10.1016/0014-5793(81)80418-8

Alves R, Herrero E, Sorribas A (2004) Predictive reconstruction of the mitochondrial iron-sulfur cluster assembly metabolism: I. the role of the protein pair ferredoxin-ferredoxin reductase (Yah1-Arh1). Proteins 56:354–366. https://doi.org/10.1002/prot.20110

Andersson M, Elmberger PO, Edlund C, Kristensson K, Dallner G (1990) Rates of cholesterol, ubiquinone, dolichol and dolichyl-P biosynthesis in rat brain slices. FEBS Lett 269:15–18. https://doi.org/10.1016/0014-5793(90)81107-Y

Anusevičius Ž, Šarlauskas J, Čènas N (2002) Two-electron reduction of quinones by rat liver NAD(P)H:quinone oxidoreductase: quantitative structure–activity relationships. Arch Biochem Biophys 404:254–262. https://doi.org/10.1016/S0003-9861(02)00273-4

Aranda FJ, Gómez-Fernández JC (1985) The interaction of ubiquinone-10 and ubiquinol-10 with phospholipid bilayers. A study using differential scanning calorimetry and turbidity measurements. Biochim Biophys Acta Biomembr 820:19–26. https://doi.org/10.1016/0005-2736(85)90210-X

Arlt H (1998) The formation of respiratory chain complexes in mitochondria is under the proteolytic control of the m-AAA protease. EMBO J 17:4837–4847. https://doi.org/10.1093/emboj/17.16.4837

Arroyo A, Rodríguez-Aguilera JC, Santos-Ocaña C, Villalba JM, Navas P (2004) Stabilization of extracellular ascorbate mediated by coenzyme Q transmembrane electron transport. In: Methods in enzymology, pp 207–217. https://doi.org/10.1016/S0076-6879(04)78017-9

Arroyo A, Santos-Ocaña C, Ruiz-Ferrer M, Padilla S, Gavilán Á, Rodríguez-Aguilera JC, Navas P (2006) Coenzyme Q is irreplaceable by demethoxy-coenzyme Q in plasma membrane of Caenorhabditis elegans. FEBS Lett 580:1740–1746. https://doi.org/10.1016/j.febslet.2006.02.025

Asai Y, Watanabe S (1999) The interaction of ubiquinone-3 with phospholipid membranes. FEBS Lett 446:169–172. https://doi.org/10.1016/S0014-5793(99)00203-3

Ashby MN, Kutsunai SY, Ackerman S, Tzagoloff A, Edwards P (1992) COQ2 is a candidate for the structural gene encoding para-hydroxybenzoate: polyprenyltransferase. J Biol Chem 267:4128–4136

Ashraf S, Gee HY, Woerner S, Xie LX, Vega-Warner V, Lovric S, Fang H, Song X, Cattran DC, Avila-Casado C, Paterson AD, Nitschké P, Bole-Feysot C, Cochat P, Esteve-Rudd J, Haberberger B, Allen SJ, Zhou W, Airik R, Otto EA, Barua M, Al-Hamed MH, Kari JA, Evans J, Bierzynska A, Saleem MA, Böckenhauer D, Kleta R, El Desoky S, Hacihamdioglu DO, Gok F, Washburn J, Wiggins RC, Choi M, Lifton RP, Levy S, Han Z, Salviati L, Prokisch H, Williams DS, Pollak M, Clarke CF, Pei Y, Antignac C, Hildebrandt F, El Desoky S, Hacihamdioglu DO, Gok F, Washburn J, Wiggins RC, Choi M, Lifton RP, Levy S, Han Z, Salviati L, Prokisch H, Williams DS, Pollak M, Clarke CF, Pei Y, Antignac C, Hildebrandt F (2013) ADCK4 mutations promote steroid-resistant nephrotic syndrome through CoQ_{10} biosynthesis disruption. J Clin Invest 123:5179–5189. https://doi.org/10.1172/JCI69000

Awad AM, Venkataramanan S, Nag A, Galivanche AR, Bradley MC, Neves L, Douglass S, Clarke CF, Johnson TL (2017) Chromatin-remodeling SWI/SNF complex regulates coenzyme Q 6 synthesis and a metabolic shift to respiration in yeast. J Biol Chem 292:14851–14866. https://doi.org/10.1074/jbc.M117.798397

Azzi A, Montecucco C, Richter C (1975) The use of acetylated ferricytochrome C for the detection of superoxide radicals produced in biological membranes. Biochem Biophys Res Commun 65:597–603. https://doi.org/10.1016/S0006-291X(75)80188-4

Baba SW, Belogrudov GI, Lee JC, Lee PT, Strahan J, Shepherd JN, Clarke CF (2004) Yeast Coq5 C-methyltransferase is required for stability of other polypeptides involved in coenzyme Q biosynthesis. J Biol Chem 279:10052–10059. https://doi.org/10.1074/jbc.M313712200

Barkovich RJ, Shtanko A, Shepherd JA, Lee PT, Myles DC, Tzagoloff A, Clarke CF (1997) Characterization of the COQ5 gene from *Saccharomyces cerevisiae*. EVIDENCE for A C-methyltransferase in ubiquinone biosynthesis. J Biol Chem 272:9182–9188. https://doi.org/10.1074/jbc.272.14.9182

Barroso MP, Gómez-Díaz C, Villalba JM, Burón MI, López-Lluch G, Navas P (1997) Plasma membrane ubiquinone controls ceramide production and prevents cell death induced by serum withdrawal. J Bioenerg Biomembr 29:259–67. https://doi.org/10.1023/a:1022462111175

Bartoschek S, Johansson M, Geierstanger BH, Okun JG, Lancaster CR, Humpfer E, Yu L, Yu C, Griesinger C, Brandt U (2001) Three molecules of ubiquinone bind specifically to mitochondrial cytochrome bc1 complex. J Biol Chem 276:35231–35234. https://doi.org/10.1074/jbc.C100365200

Battino M, Ferri E, Gorini A, Villa RF, Huertas JFR, Fiorella P, Genova ML, Lenaz G, Marchetti M (1990) Natural distribution and occurrence of coenzyme Q homologues. Membr Biochem 9:179–190. https://doi.org/10.3109/09687689009025839

Belogrudov GI, Lee PT, Jonassen T, Hsu AY, Gin P, Clarke CF (2001) Yeast COQ4 encodes a mitochondrial protein required for coenzyme Q synthesis. Arch Biochem Biophys 392:48–58. https://doi.org/10.1006/abbi.2001.2448

Bentinger M, Dallner G, Chojnacki T, Swiezewska E (2003) Distribution and breakdown of labeled coenzyme Q10 in rat. Free Radic Biol Med 34:563–575. https://doi.org/10.1016/S0891-5849(02)01357-6

Bentinger M, Brismar K, Dallner G (2007) The antioxidant role of coenzyme Q. Mitochondrion 7(Suppl):S41–S50. https://doi.org/10.1016/j.mito.2007.02.006

Bentinger M, Tekle M, Dallner G (2010) Coenzyme Q – biosynthesis and functions. Biochem Biophys Res Commun 396:74–79. https://doi.org/10.1016/j.bbrc.2010.02.147

Bentinger M, Tekle M, Dallner G, Brismar K, Gustafsson J-Å, Steffensen KR, Catrina S-B (2012) Influence of liver-X-receptor on tissue cholesterol, coenzyme Q and dolichol content. Mol Membr Biol 29:299–308. https://doi.org/10.3109/09687688.2012.694484

Beyer RE (1994) The role of ascorbate in antioxidant protection of biomembranes: interaction with vitamin E and coenzyme Q. J Bioenerg Biomembr 26:349–358. https://doi.org/10.1007/BF00762775

Beyer RE, Burnett BA, Cartwright KJ, Edington DW, Falzon MJ, Kreitman KR, Kuhn TW, Ramp BJ, Rhee SYS, Rosenwasser MJ, Stein M, An LCI (1985) Tissue coenzyme Q (ubiquinone) and protein concentrations over the life span of the laboratory rat. Mech Ageing Dev 32:267–281. https://doi.org/10.1016/0047-6374(85)90085-5

Branicky R, Bénard C, Hekimi S (2000) clk-1, mitochondria, and physiological rates. BioEssays 22:48–56. https://doi.org/10.1002/(SICI)1521-1878(200001)22:1<48::AID-BIES9>3.0.CO;2-F

Brea-Calvo G, Rodríguez-Hernández A, Fernández-Ayala DJM, Navas P, Sánchez-Alcázar JA (2006) Chemotherapy induces an increase in coenzyme Q10 levels in cancer cell lines. Free Radic Biol Med 40:1293–1302. https://doi.org/10.1016/j.freeradbiomed.2005.11.014

Brea-Calvo G, Siendones E, Sanchez-Alcazar JA, de Cabo R, Navas P (2009) Cell survival from chemotherapy depends on NF-kappaB transcriptional up-regulation of coenzyme Q biosynthesis. PLoS One 4:e5301. https://doi.org/10.1371/journal.pone.0005301

Brière J-J, Schlemmer D, Chretien D, Rustin P (2004) Quinone analogues regulate mitochondrial substrate competitive oxidation. Biochem Biophys Res Commun 316:1138–1142. https://doi.org/10.1016/j.bbrc.2004.03.002

Busso C, Ferreira-Júnior JR, Paulela JA, Bleicher L, Demasi M, Barros MH (2015) Coq7p relevant residues for protein activity and stability. Biochimie 119:92–102. https://doi.org/10.1016/j.biochi.2015.10.016

Casarin A, Jimenez-Ortega JC, Trevisson E, Pertegato V, Doimo M, Ferrero-Gomez ML, Abbadi S, Artuch R, Quinzii C, Hirano M, Basso G, Ocaña CS, Navas P, Salviati L (2008) Functional characterization of human COQ4, a gene required for coenzyme Q10 biosynthesis. Biochem Biophys Res Commun 372:35–39. https://doi.org/10.1016/j.bbrc.2008.04.172

Cascajo MV, Abdelmohsen K, Noh JH, Fernandez-Ayala DJ, Willers IM, Brea G, Lopez-Lluch G, Valenzuela-Villatoro M, Cuezva JM, Gorospe M, Siendones E, Navas P (2015) RNA-binding proteins regulate cell respiration and coenzyme Q biosynthesis by post-transcriptional regulation of COQ7. RNA Biol 13:622–634. https://doi.org/10.1080/15476286.2015.1119366

Chretien D, Schlemmer D, Mollet J, De Baulny O, Lombes A, Boddaert N, Desguerre I, De Lonlay P, Delahodde A, Serre V, de Lonlay P, de Baulny HO, Munnich A, Rotig A (2008) CABC1 gene mutations cause ubiquinone deficiency with cerebellar ataxia and seizures. Am J Hum Genet 82:623–630. https://doi.org/10.1016/j.ajhg.2007.12.022

Clarke CF (1996) The COQ7 gene encodes a protein in *Saccharomyces cerevisiae* necessary for ubiquinone biosynthesis. J Biol Chem 271:2995–3004. https://doi.org/10.1074/jbc.271.6.2995

Claros MG, Vincens P (1996) Computational method to predict mitochondrially imported proteins and their targeting sequences. Eur J Biochem 241:779–786. https://doi.org/10.1111/j.1432-1033.1996.00779.x

Cotan D, Cordero MD, Garrido-Maraver J, Oropesa-Avila M, Rodriguez-Hernandez A, Gomez Izquierdo L, De la Mata M, De Miguel M, Bautista Lorite J, Rivas Infante E, Jackson S, Navas P, Sanchez-Alcazar JA (2011) Secondary coenzyme Q10 deficiency triggers mitochondria degradation by mitophagy in MELAS fibroblasts. FASEB J 25:2669–2687. doi:fj.10-165340 [pii]. https://doi.org/10.1096/fj.10-165340

Di Bernardo S, Fato R, Casadio R, Fariselli P, Lenaz G (1998) A high diffusion coefficient for coenzyme Q10 might be related to a folded structure. FEBS Lett 426:77–80. https://doi.org/10.1016/S0014-5793(98)00313-5

Di Giovanni S, Mirabella M, Spinazzola A, Crociani P, Silvestri G, Broccolini A, Tonali P, Di Mauro S, Servidei S (2001) Coenzyme Q10 reverses pathological phenotype and reduces apoptosis in familial CoQ_{10} deficiency. Neurology 57:515–518. https://doi.org/10.1212/WNL.57.3.515

Dibrov E, Robinson KM, Lemire BD (1997) The COQ5 gene encodes a yeast mitochondrial protein necessary for ubiquinone biosynthesis and the assembly of the respiratory chain. J Biol Chem 272:9175–9181. https://doi.org/10.1074/jbc.272.14.9175

Diner BA, Schenck CC, De Vitry C (1984) Effect of inhibitors, redox state and isoprenoid chain length on the affinity of ubiquinone for the secondary acceptor binding site in the reaction centers of photosynthetic bacteria. Biochim Biophys Acta-Bioenerg 766:9–20. https://doi.org/10.1016/0005-2728(84)90211-1

Do TQ, Schultz JR, Clarke CF (1996) Enhanced sensitivity of ubiquinone-deficient mutants of *Saccharomyces cerevisiae* to products of autoxidized polyunsaturated fatty acids. Proc Natl Acad Sci U S A 93:7534–7539. https://doi.org/10.1073/pnas.93.15.7534

Echtay KS, Winkler E, Klingenberg M (2000) Coenzyme Q is an obligatory cofactor for uncoupling protein function. Nature 408:609–613. https://doi.org/10.1038/35046114

Elmberger PG, Kalen A, Brunk UT, Dallner G (1989) Discharge of newly-synthesized dolichol and ubiquinone with lipoproteins to rat liver perfusate and to the bile. Lipids 24:919–930. https://doi.org/10.1007/BF02544535

Enríquez JA (2016) Supramolecular organization of respiratory complexes. Annu Rev Physiol 78:533–561. https://doi.org/10.1146/annurev-physiol-021115-105031

Enriquez JA, Lenaz G (2014) Coenzyme Q and the respiratory chain: coenzyme Q pool and mitochondrial supercomplexes. Mol Syndromol 5:119–140. https://doi.org/10.1159/000363364

Esposti MD, Ngo A, Ghelli A, Benelli B, Carelli V, McLennan H, Linnane AW (1996) The interaction of Q analogs, particularly hydroxydecyl benzoquinone (Idebenone), with the respiratory complexes of heart mitochondria. Arch Biochem Biophys 330:395–400. https://doi.org/10.1006/abbi.1996.0267

Ewbank JJ, Barnes TM, Lakowski B, Lussier M, Bussey H, Hekimi S (1997) Structural and functional conservation of the Caenorhabditis elegans timing gene clk-1. Science 275:980–983. https://doi.org/10.1126/science.275.5302.980

Fato R, Battino M, Degli Esposti M, Parenti Castelli G, Lenaz G (1986) Determination of partition and lateral diffusion coefficients of ubiquinones by fluorescence quenching of n-(9-anthroyloxy)stearic acids in phospholipid vesicles and mitochondrial membranes. Biochemistry 25:3378–3390. https://doi.org/10.1021/BI00359A043

Felkai S, Ewbank JJ, Lemieux J, Labbé JC, Brown GG, Hekimi S (1999) CLK-1 controls respiration, behavior and aging in the nematode Caenorhabditis elegans. EMBO J 18:1783–1792. https://doi.org/10.1093/emboj/18.7.1783

Fernández-Ayala DJ, Martín SF, Barroso MP, Gómez-Díaz C, Villalba JM, Rodríguez-Aguilera JC, López-Lluch G, Navas P (2000) Coenzyme Q protects cells against serum withdrawal-induced apoptosis by inhibition of ceramide release and caspase-3 activation. Antioxid Redox Signal 2:263–275. https://doi.org/10.1089/ars.2000.2.2-263

Fernández-Ayala DJM, Brea-Calvo G, López-Lluch G, Navas P (2005) Coenzyme Q distribution in HL-60 human cells depends on the endomembrane system. Biochim Biophys Acta 1713:129–137. https://doi.org/10.1016/j.bbamem.2005.05.010

Fiorentini D, Cabrini L, Landi L (1993) Ubiquinol-3 and ubiquinol-7 exhibit similar antioxidant activity in model membranes. Free Radic Res Commun 18:201–209. https://doi.org/10.3109/10715769309145869

Floyd BJ, Wilkerson EM, Veling MT, Minogue CE, Xia C, Beebe ET, Wrobel RL, Cho H, Kremer LS, Alston CL, Gromek KA, Dolan BK, Ulbrich A, Stefely JA, Bohl SL, Werner KM, Jochem A, Westphall MS, Rensvold JW, Taylor RW, Prokisch H, Kim JJP, Coon JJ, Pagliarini DJ (2016) Mitochondrial protein interaction mapping identifies regulators of respiratory chain function. Mol Cell 63:621–632. https://doi.org/10.1016/j.molcel.2016.06.033

Fontaine E, Ichas F, Bernardi P (1998) A ubiquinone-binding site regulates the mitochondrial permeability transition pore. J Biol Chem 273:25734–25740. https://doi.org/10.1074/jbc.273.40.25734

Forsgren M, Attersand A, Lake S, Grünler J, Swiezewska E, Dallner G, Climent I (2004) Isolation and functional expression of human COQ2, a gene encoding a polyprenyl transferase involved in the synthesis of CoQ. Biochem J 382:519–526. https://doi.org/10.1042/BJ20040261

Frei B, Kim MC, Ames BN (1990) Ubiquinol-10 is an effective lipid-soluble antioxidant at physiological concentrations. Proc Natl Acad Sci U S A 87:4879–4883. https://doi.org/10.1073/pnas.87.12.4879

Frerman FE (1988) Acyl-CoA dehydrogenases, electron transfer flavoprotein and electron transfer flavoprotein dehydrogenase. Biochem Soc Trans 16:416–418. https://doi.org/10.1042/bst0160416

Gaigg B, Simbeni R, Hrastnik C, Paltauf F, Daum G (1995) Characterization of a microsomal subfraction associated with mitochondria of the yeast, *Saccharomyces cerevisiae*. Involvement in synthesis and import of phospholipids into mitochondria. Biochim Biophys Acta-Biomembr 1234:214–220. https://doi.org/10.1016/0005-2736(94)00287-Y

Gale PH, Koniuszy FR, Page AC, Folkers K, Siegel H (1961) Coenzyme Q. XXIV. On the significance of coenzyme Q 10 in human tissues. Arch Biochem Biophys 93:211–213. https://doi.org/10.1016/0003-9861(61)90251-X

Gancedo JM (1998) Yeast carbon catabolite repression. Microbiol Mol Biol Rev 62:334–361

Gerards M, van den Bosch B, Calis C, Schoonderwoerd K, van Engelen K, Tijssen M, de Coo R, van der Kooi A, Smeets H (2010) Nonsense mutations in CABC1/ADCK3 cause progressive cerebellar ataxia and atrophy. Mitochondrion 10:510–515. https://doi.org/10.1016/j.mito.2010.05.008

Gibson F, Young IG (1978) Isolation and characterization of intermediates in ubiquinone biosynthesis. Methods Enzymol 53:600–609

Gille L, Nohl H (2000) The existence of a lysosomal redox chain and the role of ubiquinone. Arch Biochem Biophys 375:347–354. https://doi.org/10.1006/abbi.1999.1649

Gin P, Clarke CF (2005) Genetic evidence for a multi-subunit complex in coenzyme Q biosynthesis in yeast and the role of the Coq1 hexaprenyl diphosphate synthase. J Biol Chem 280:2676–2681. https://doi.org/10.1074/jbc.M411527200

Gin P, Hsu AY, Rothman SC, Jonassen T, Lee PT, Tzagoloff A, Clarke CF (2003) The *Saccharomyces cerevisiae* COQ6 gene encodes a mitochondrial flavin-dependent monooxygenase required for coenzyme Q biosynthesis. J Biol Chem 278:25308–25316. https://doi.org/10.1074/jbc.M303234200

Goewert RR (1980) Studies on the biosynthesis of ubiquinone. Saint Louis University

Goewert RR, Sippel CJ, Olson RE (1981) Identification of 3,4-dihydroxy-5-hexaprenylbenzoic acid as an intermediate in the biosynthesis of ubiquinone-6 by *Saccharomyces cerevisiae*. Biochemistry 20:4217–4223

Goldstein JL, Brown MS (1990) Regulation of the mevalonate pathway. Nature 343:425–430. https://doi.org/10.1038/343425a0

Gómez-Díaz C, Rodríguez-Aguilera JC, Barroso MP, Villalba JM, Navarro F, Crane FL, Navas P (1997) Antioxidant ascorbate is stabilized by NADH-coenzyme Q10 reductase in the plasma membrane. J Bioenerg Biomembr 29:251–257

González-Mariscal I, García-Testón E, Padilla S, Martín-Montalvo A, Pomares-Viciana T, Vazquez-Fonseca L, Gandolfo-Domínguez P, Santos-Ocaña C (2014) Regulation of coenzyme Q biosynthesis in yeast: a new complex in the block. IUBMB Life 66:63–70. https://doi.org/10.1002/iub.1243

Gonzalez-Mariscal I, Martin-Montalvo A, Ojeda-Gonzalez C, Rodriguez-Eguren A, Gutierrez-Rios P, Navas P, Santos-Ocana C (2017) Balanced CoQ6 biosynthesis is required for lifespan and mitophagy in yeast. Microb Cell 4:38–51. https://doi.org/10.15698/mic2017.02.556

González-Mariscal I, Martin-Montalvo A, Vazquez-Fonseca L, Pomares-Viciana T, Sánchez-Cuesta A, Fernández-Ayala DJ, Navas P, Santos-Ocana C (2018) The mitochondrial phosphatase PPTC7 orchestrates mitochondrial metabolism regulating coenzyme Q_{10} biosynthesis. Biochim Biophys Acta-Bioenerg 1859:1235–1248. https://doi.org/10.1016/j.bbabio.2018.09.369

Gorman A, McGowan A, Cotter TG (1997) Role of peroxide and superoxide anion during tumour cell apoptosis. FEBS Lett 404:27–33. https://doi.org/10.1016/S0014-5793(97)00069-0

Grünler J, Ericsson J, Dallner G (1994) Branch-point reactions in the biosynthesis of cholesterol, dolichol, ubiquinone and prenylated proteins. Biochim Biophys Acta Lipids Lipid Metab 1212:259–277. https://doi.org/10.1016/0005-2760(94)90200-3

Grzybek M, Stebelska K, Wyrozumska P, Grieb P, Langner M, Jaszewski A, Jezierski A, Sikorski AF (2005) ESR and monolayer study of the localization of coenzyme Q10 in artificial membranes. Gen Physiol Biophys 24:449–460

Gu L-Q, Yu L, Yu C-A (1990) Effect of substituents of the benzoquinone ring on electron-transfer activities of ubiquinone derivatives. Biochim Biophys Acta Bioenerg 1015:482–492. https://doi.org/10.1016/0005-2728(90)90082-F

Gulmezian M, Hyman KR, Marbois BN, Clarke CF, Javor GT (2007) The role of UbiX in Escherichia coli coenzyme Q biosynthesis. Arch Biochem Biophys 467:144–153. https://doi.org/10.1016/j.abb.2007.08.009

Guo X, Niemi NM, Hutchins PD, Condon SGF, Jochem A, Ulbrich A, Higbee AJ, Russell JD, Senes A, Coon JJ, Pagliarini DJ (2017) Ptc7p dephosphorylates select mitochondrial proteins to enhance metabolic function. Cell Rep 18:307–313. https://doi.org/10.1016/j.celrep.2016.12.049

Hagerman RA, Willis RA (2002) The yeast gene COQ5 is differentially regulated by Mig1p, Rtg3p and Hap2p. Biochim Biophys Acta 1578:51–58. https://doi.org/10.1016/S0167-4781(02)00496-7

Hagerman RA, Trotter PJ, Willis RA (2002) The regulation of COQ5 gene expression by energy source. Free Radic Res 36:485–490. https://doi.org/10.1080/10715760290021360

Hano N, Nakashima Y, Shinzawa-Itoh K, Yoshikawa S (2003) Effect of the side chain structure of coenzyme Q on the steady state kinetics of bovine heart NADH: coenzyme Q oxidoreductase. J Bioenerg Biomembr 35:257–265

Harbison CT, Gordon DBB, Lee TI, Rinaldi NJ, Macisaac KD, Danford TW, Hannett NM, Tagne J-B, Reynolds DB, Yoo J, Jennings EG, Zeitlinger J, Pokholok DK, Kellis M, Rolfe PA, Takusagawa KT, Lander ES, Gifford DK, Fraenkel E, Young RA (2004) Transcriptional regulatory code of a eukaryotic genome. Nature 431:99–104. https://doi.org/10.1038/nature02800

He DY, Yu L, Yu CA (1994) Protein ubiquinone interaction. Synthesis and biological properties of 5-alkyl ubiquinone derivatives. J Biol Chem 269:27885–27888

Heron C, Ragan CI, Trumpower BL (1978) The interaction between mitochondrial NADH-ubiquinone oxidoreductase and ubiquinol-cytochrome c oxidoreductase. Restoration of ubiquinone-pool behaviour. Biochem J 174:791–800. https://doi.org/10.1042/bj1740791

Hickman MJ, Winston F (2007) Heme levels switch the function of Hap1 of *Saccharomyces cerevisiae* between transcriptional activator and transcriptional repressor. Mol Cell Biol 27:7414–7424. https://doi.org/10.1128/MCB.00887-07

Horvath R, Schneiderat P, Schoser BGH, Gempel K, Neuen-Jacob E, Plöger H, Müller-Höcker J, Pongratz DE, Naini A, DiMauro S, Lochmüller H (2006) Coenzyme Q10 deficiency and isolated myopathy. Neurology 66:253–255. https://doi.org/10.1212/01.wnl.0000194241.35115.7c

Hoyo J, Guaus E, Oncins G, Torrent-Burgués J, Sanz F (2013) Incorporation of ubiquinone in supported lipid bilayers on ITO. J Phys Chem B 117:7498–7506. https://doi.org/10.1021/jp4004517

Hsieh EJ, Gin P, Gulmezian M, Tran UC, Saiki R, Marbois BN, Clarke CF (2007) *Saccharomyces cerevisiae* Coq9 polypeptide is a subunit of the mitochondrial coenzyme Q biosynthetic complex. Arch Biochem Biophys 463:19–26. https://doi.org/10.1016/j.abb.2007.02.016

Hsu AY, Do TQ, Lee PT, Clarke CF (2000) Genetic evidence for a multi-subunit complex in the O-methyltransferase steps of coenzyme Q biosynthesis. Biochem Biophys Acta 1484:287–297. https://doi.org/10.1016/S1388-1981(00)00019-6

Ingrell CR, Miller ML, Jensen ON, Blom N (2007) NetPhosYeast: prediction of protein phosphorylation sites in yeast. Bioinformatics 23:895–897

Jiang L, Whiteway M, Ramos CW, Rodriguez-Medina JR, Shen S-HH (2002a) The YHR076w gene encodes a type 2C protein phosphatase and represents the seventh PP2C gene in budding yeast. FEBS Lett 527:323–325

Jiang L, Whiteway M, Ramos CW, Rodriguez-Medina JR, Shen S-H (2002b) The YHR076w gene encodes a type 2C protein phosphatase and represents the seventh PP2C gene in budding yeast. FEBS Lett 527:323–325. https://doi.org/10.1016/S0014-5793(02)03247-7

Jiménez-Hidalgo M, Santos-Ocaña C, Padilla S, Villalba JM, López-Lluch G, Martín-Montalvo A, Minor RK, Sinclair DA, de Cabo R, Navas P (2009) NQR1 controls lifespan by regulating the promotion of respiratory metabolism in yeast. Aging Cell 8:140–151. https://doi.org/10.1111/j.1474-9726.2009.00461.x

Johnson A, Gin P, Marbois BN, Hsieh EJ, Wu M, Barros MH, Clarke CF, Tzagoloff A (2005) COQ9, a new gene required for the biosynthesis of coenzyme Q in *Saccharomyces cerevisiae*. J Biol Chem 280:31397–31404. https://doi.org/10.1074/jbc.M503277200

Jonassen T, Clarke CF (2000) Isolation and functional expression of human COQ3, a gene encoding a methyltransferase required for ubiquinone biosynthesis. J Biol Chem 275:12381–12387. https://doi.org/10.1074/jbc.275.17.12381

Jonassen T, Clarke CF (2001) Genetics analysis of coenzyme Q biosynthesis. In: Kagan VE, Quinn PJ (eds) Coenzyme Q: molecular mechanisms in health and disease. CRC Press LLC, pp 185–208

Jonassen T, Marbois BN, Kim L, Chin A, Xia Y-RR, Lusis AJ, Clarke CF (1996) Isolation and sequencing of the rat Coq7 gene and the mapping of mouse Coq7 to chromosome 7. Arch Biochem Biophys 330:285–289. https://doi.org/10.1006/abbi.1996.0255

Jørgensen BM, Rasmussen HN, Rasmussen UF (1985) Ubiquinone reduction pattern in pigeon heart mitochondria. Identification of three distinct ubiquinone pools. Biochem J 229:621–629. https://doi.org/10.1042/bj2290621

Joseph-Horne T, Hollomon DW, Wood PM (2001) Fungal respiration: a fusion of standard and alternative components. Biochim Biophys Acta Bioenerg 1504:179–195. https://doi.org/10.1016/S0005-2728(00)00251-6

Juneau K, Nislow C, Davis RW (2009) Alternative splicing of PTC7 in *Saccharomyces cerevisiae* determines protein localization. Genetics 183:185–194. https://doi.org/10.1534/genetics.109.105155

Kagan VE, Serbinova EA, Koynova GM, Kitanova SA, Tyurin VA, Stoytchev TS, Quinn PJ, Packer L (1990) Antioxidant action of ubiquinol homologues with different isoprenoid chain length in biomembranes. Free Radic Biol Med 9:117–126. https://doi.org/10.1016/0891-5849(90)90114-X

Kagan VE, Arroyo A, Tyurin VA, Tyurina YY, Villalba JM, Navas P (1998a) Plasma membrane NADH-coenzyme Q 0 reductase generates semiquinone radicals and recycles vitamin E homologue in a superoxide-dependent reaction. FEBS Lett 428:43–46. https://doi.org/10.1016/S0014-5793(98)00482-7

Kagan VE, Tyurina YY, Witt E (1998b) Role of coenzyme Q and superoxide in vitamin E cycling. Subcell Biochem 30:491–507

Kalén A, Norling B, Appelkvist EL, Dallner G (1987) Ubiquinone biosynthesis by the microsomal fraction from rat liver. BBA Gen Subj 926:70–78. https://doi.org/10.1016/0304-4165(87)90183-8

Kalén A, Appelkvist EL, Dallner G (1989) Age-related changes in the lipid compositions of rat and human tissues. Lipids 24:579–584

Kalén A, Appelkvist EL, Chojnacki T, Dallner G (1990) Nonaprenyl-4-hydroxybenzoate transferase, an enzyme involved in ubiquinone biosynthesis, in the endoplasmic reticulum-Golgi system of rat liver. J Biol Chem 265:1158–1164

Katsikas H, Quinn PJ (1982a) The polyisoprenoid chain length influences the interaction of ubiquinones with phospholipid bilayers. Biochim Biophys Acta Biomembr 689:363–369. https://doi.org/10.1016/0005-2736(82)90270-X

Katsikas H, Quinn PJ (1982b) The distribution of ubiquinone-10 in phospholipid bilayers. A study using differential scanning calorimetry. Eur J Biochem 124:165–169. https://doi.org/10.1111/j.1432-1033.1982.tb05920.x

Kaurola P, Sharma V, Vonk A, Vattulainen I, Róg T (2016) Distribution and dynamics of quinones in the lipid bilayer mimicking the inner membrane of mitochondria. Biochim Biophys Acta Biomembr 1858:2116–2122. https://doi.org/10.1016/j.bbamem.2016.06.016

Kuznetsov AV, Smigelskaite J, Doblander C, Janakiraman M, Hermann M, Wurm M, Scheidl SF, Sucher R, Deutschmann A, Troppmair J (2008) Survival signaling by C-RAF: mitochondrial reactive oxygen species and Ca2+ are critical targets. Mol Cell Biol 28:2304–2313. https://doi.org/10.1128/MCB.00683-07

Kuznetsov AV, Kehrer I, Kozlov AV, Haller M, Redl H, Hermann M, Grimm M, Troppmair J (2011) Mitochondrial ROS production under cellular stress: comparison of different detection methods. Anal Bioanal Chem 400:2383–90. https://doi.org/10.1007/s00216-011-4764-2

Kwong LK, Kamzalov S, Rebrin I, Bayne A-C V, Jana CK, Morris P, Forster MJ, Sohal RS (2002) Effects of coenzyme Q10 administration on its tissue concentrations, mitochondrial oxidant generation, and oxidative stress in the rat. Free Radic Biol Med 33:627–638. https://doi.org/10.1016/S0891-5849(02)00916-4

Lapointe J, Wang Y, Bigras E, Hekimi S (2012) The submitochondrial distribution of ubiquinone affects respiration in long-lived Mclk1+/− mice. J Cell Biol 199:215–224. https://doi.org/10.1083/jcb.201203090

Lass A, Agarwal S, Sohal RS (1997) Mitochondrial ubiquinone homologues, superoxide radical generation, and longevity in different mammalian species. J Biol Chem 272:19199–19204. https://doi.org/10.1074/jbc.272.31.19199

Lass A, Forster MJ, Sohal RS (1999) Effects of coenzyme Q10 and alpha-tocopherol administration on their tissue levels in the mouse: elevation of mitochondrial alpha-tocopherol by coenzyme Q10. Free Radic Biol Med 26:1375–1382

Lenaz G, Landi L, Cabrini L, Pasquali P, Sechi AM, Ozawa T (1978) On the sidedness of the ubiquinone redox cycle. Kinetic studies in mitochondrial membranes. Biochem Biophys Res Commun 85:1047–1053. https://doi.org/10.1016/0006-291X(78)90648-4

Lenaz G, Samori B, Fato R, Battino M, Castelli GP, Domini I (1992) Localization and preferred orientations of ubiquinone homologs in model bilayers. Biochem Cell Biol 70:504–514. https://doi.org/10.1139/092-078

Lenaz G, Fato R, Di Bernardo S, Jarreta D, Costa A, Genova ML, Castelli GP (1999) Localization and mobility of coenzyme Q in lipid bilayers and membranes. Biofactors 9:87–93. https://doi.org/10.1002/biof.5520090202

Lenaz G, Fato R, Genova ML, Bergamini C, Bianchi C, Biondi A (2006) Mitochondrial complex I: structural and functional aspects. Biochim Biophys Acta Bioenerg 1757:1406–1420. https://doi.org/10.1016/j.bbabio.2006.05.007

Leonard CJ, Aravind L, Koonin EV (1998) Evolution of the eukaryotic protein kinase superfamily novel families of putative protein kinases in bacteria and archaea: evolution of the ' ' eukaryotic protein kinase superfamily. Genome Res 8:1038–1047. https://doi.org/10.1101/8.10.1038

Lohman DC, Forouhar F, Beebe ET, Stefely MS, Minogue CE, Ulbrich A, Stefely JA, Sukumar S, Luna-Sánchez M, Jochem A, Lew S, Seetharaman J, Xiao R, Wang H, Westphall MS, Wrobel RL, Everett JK, Mitchell JC, López LC, Coon JJ, Tong L, Pagliarini DJ (2014) Mitochondrial COQ9 is a lipid-binding protein that associates with COQ7 to enable coenzyme Q biosynthesis. Proc Natl Acad Sci U S A 111(44):E4697–E4705. https://doi.org/10.1073/pnas.1413128111

Lönnrot K, Holm P, Lagerstedt A, Huhtala H, Alho H (1998) The effects of lifelong ubiquinone Q10 supplementation on the Q9 and Q10 tissue concentrations and life span of male rats and mice. Biochem Mol Biol Int 44:727–737

López-Lluch G, Rodríguez-Aguilera JC, Santos-Ocaña C, Navas P (2010) Is coenzyme Q a key factor in aging? Mech Ageing Dev 131:225–235. https://doi.org/10.1016/j.mad.2010.02.003

López-Martín JM, Salviati L, Trevisson E, Montini G, DiMauro S, Quinzii C, Hirano M, Rodriguez-Hernandez A, Cordero MD, Sánchez-Alcázar JA, Santos-Ocaña C, Navas P (2007) Missense mutation of the COQ2 gene causes defects of bioenergetics and de novo pyrimidine synthesis. Hum Mol Genet 16:1091–1097. https://doi.org/10.1093/hmg/ddm058

Macinga DR, Cook GM, Poole RK, Rather PN (1998) Identification and characterization of aarF, a locus required for production of ubiquinone in Providencia stuartii and Escherichia coli and for expression of 2'-N-acetyltransferase in P. stuartii. J Bacteriol 180:128–135. https://doi.org/10.1128/AEM.68.12.6371

Maggio B, Diplock AT, Lucy JA (1977) Interactions of tocopherols and ubiquinones with monolayers of phospholipids. Biochem J 161:111–121. https://doi.org/10.1042/bj1610111

Maklashina E, Cecchini G (2010) The quinone-binding and catalytic site of complex II. Biochim Biophys Acta Bioenerg 1797:1877–1882. https://doi.org/10.1016/j.bbabio.2010.02.015

Manzella L, Barros MH, Nobrega FG (1998) ARH1 of *Saccharomyces cerevisiae*: a new essential gene that codes for a protein homologous to the human adrenodoxin reductase. Yeast 14:839–846. https://doi.org/10.1002/(SICI)1097-0061(19980630)14:9<839::AID-YEA283>3.0.CO;2-A

Marbois B, Gin P, Faull KF, Poon WW, Lee PT, Strahan J, Shepherd JN, Clarke CF (2005) Coq3 and Coq4 define a polypeptide complex in yeast mitochondria for the biosynthesis of coenzyme Q. J Biol Chem 280:20231–20238. doi:M501315200 [pii]. https://doi.org/10.1074/jbc.M501315200

Marbois B, Gin P, Gulmezian M, Clarke CF (2009) The yeast Coq4 polypeptide organizes a mitochondrial protein complex essential for coenzyme Q biosynthesis. Biochim Biophys Acta Mol Cell Biol Lipids 1791:69–75. https://doi.org/10.1016/j.bbalip.2008.10.006

Martin-Montalvo A, Gonzalez-Mariscal I, Padilla S, Ballesteros M, Brautigan DL, Navas P, Santos-Ocaña C (2011) Respiratory-induced coenzyme Q biosynthesis is regulated by a phosphorylation cycle of Cat5p/Coq7p. Biochem J 440:107–114. https://doi.org/10.1042/BJ20101422

Martin-Montalvo A, Gonzalez-Mariscal I, Pomares-Viciana T, Padilla-Lopez S, Ballesteros M, Vazquez-Fonseca L, Gandolfo P, Brautigan DL, Navas P, Santos-Ocana C (2013) The phosphatase Ptc7 induces coenzyme Q biosynthesis by activating the hydroxylase Coq7 in yeast. J Biol Chem 288:28126–28137. https://doi.org/10.1074/jbc.M113.474494

Matsuda H, Kimura S, Iyanagi T (2000) One-electron reduction of quinones by the neuronal nitric-oxide synthase reductase domain. Biochim Biophys Acta Bioenerg 1459:106–116. https://doi.org/10.1016/S0005-2728(00)00117-1

Matsura T, Yamada K, Kawasaki T (1991) Changes in the content and intracellular distribution of coenzyme Q homologs in rabbit liver during growth. Biochim Biophys Acta Lipids Lipid Metab 1083:277–282. https://doi.org/10.1016/0005-2760(91)90083-T

Michaelis L, Moore MJ (1985) Location of ubiquinone-10 (CoQ-10) in phospholipid vesicles. Biochim Biophys Acta Biomembr 821:121–129. https://doi.org/10.1016/0005-2736(85)90162-2

Miles MV (2007) The uptake and distribution of coenzyme Q(10). Mitochondrion 7:S72–S77. https://doi.org/10.1016/j.mito.2007.02.012

Miles MV, Horn P, Miles L, Tang P, Steele P, DeGrauw T (2002) Bioequivalence of coenzyme Q10 from over-the-counter supplements. Nutr Res 22:919–929. https://doi.org/10.1016/S0271-5317(02)00402-5

Miyadera H, Kano K, Miyoshi H, Ishii N, Hekimi S, Kita K (2002) Quinones in long-lived clk-1 mutants of Caenorhabditis elegans. FEBS Lett 512:33–37. https://doi.org/10.1016/S0014-5793(02)02282-2

Momose K, Rudney H (1972) 3-polyprenyl-4-hydroxybenzoate synthesis in the inner membrane of mitochondria from p-Hydroxybenzoate and Isopentenylpyrophosphate. A demonstration of isoprenoid synthesis in rat liver mitochondria. J Biol Chem 247:3930–3940

Monteiro PT, Mendes ND, Teixeira MC, d'Orey S, Tenreiro S, Mira NP, Pais H, Francisco AP, Carvalho AM, Lourenço AB, Sá-Correia I, Oliveira AL, Freitas AT (2008) YEASTRACT-DISCOVERER: new tools to improve the analysis of transcriptional regulatory associations in *Saccharomyces cerevisiae*. Nucleic Acids Res 36:D132–D136. https://doi.org/10.1093/nar/gkm976

Morton RA (1958) Ubiquinone. Nature 182:1764–1767

Mukai N, Masaki K, Fujii T, Kawamukai M, Iefuji H (2010) PAD1 and FDC1 are essential for the decarboxylation of phenylacrylic acids in *Saccharomyces cerevisiae*. J Biosci Bioeng 109:564–569. https://doi.org/10.1016/j.jbiosc.2009.11.011

Nakamura M, Hayashi T (1994) One- and two-electron reduction of quinones by rat liver subcellular fractions. J Biochem 115:1141–1147

Navas P, Fernandez-Ayala DM, Martin SF, Lopez-Lluch G, De Cabo R, Rodriguez-Aguilera JC, Villalba JM (2002) Ceramide-dependent caspase 3 activation is prevented by coenzyme Q from plasma membrane in serum-deprived cells. Free Radic Res 36:369–374

Nguyen TPT, Casarin A, Desbats MA, Doimo M, Trevisson E, Santos-Ocaña C, Navas P, Clarke CF, Salviati L (2014) Molecular characterization of the human COQ5 C-methyltransferase in coenzyme Q10 biosynthesis. Biochim Biophys Acta Mol Cell Biol Lipids 1841:1628–1638. https://doi.org/10.1016/j.bbalip.2014.08.007

Niemi NM, Wilson GM, Overmyer KA, Vögtle F-N, Myketin L, Lohman DC, Schueler KL, Attie AD, Meisinger C, Coon JJ, Pagliarini DJ (2019) Pptc7 is an essential phosphatase for promoting mammalian mitochondrial metabolism and biogenesis. Nat Commun 10:1–14. https://doi.org/10.1038/s41467-019-11047-6

Nohl H, Gille L (2002) The bifunctional activity of ubiquinone in lysosomal membranes. Biogerontology 3:125–131

Okada K, Suzuki K, Kamiya Y, Zhu X, Fujisaki S, Nishimura Y, Nishino T, Nakagawad T, Kawamukai M, Matsuda H (1996) Polyprenyl diphosphate synthase essentially defines the length of the side chain of ubiquinone. Biochim Biophys Acta Lipids Lipid Metab 1302:217–223. https://doi.org/10.1016/0005-2760(96)00064-1

Okamoto K, Watanabe M, Morimoto H, Imada I (1988) Synthesis, metabolism, and in vitro biological activities of 6-(10-hydroxydecyl)-2,3-dimethoxy-5-methyl-1,4-benzoquinone (CV-2619)-related compounds. Chem Pharm Bull (Tokyo) 36:178–189

Olson RE, Rudney H (1983) Biosynthesis of ubiquinone. Vitam Horm 40:1–43

Ondarroa M, Quinn PJ (1986) A difference infrared-spectroscopic study of the interaction of ubiquinone-10 with phospholipid bilayers. Biochem J 240:325–331

Ozeir M, Mühlenhoff U, Webert H, Lill R, Fontecave M, Pierrel F (2011) Coenzyme q biosynthesis: coq6 is required for the c5-hydroxylation reaction and substrate analogs rescue coq6 deficiency. Chem Biol 18:1134–1142. https://doi.org/10.1016/j.chembiol.2011.07.008

Padilla S, Jonassen T, Jiménez-Hidalgo MA, Fernández-Ayala DJM, López-Lluch G, Marbois B, Navas P, Clarke CF, Santos-Ocaña C (2004) Demethoxy-Q, an intermediate of coenzyme Q biosynthesis, fails to support respiration in *Saccharomyces cerevisiae* and lacks antioxidant activity. J Biol Chem 279:25995–26004. https://doi.org/10.1074/jbc.M400001200

Padilla S, Tran UC, Jiménez-Hidalgo M, López-Martín JM, Martín-Montalvo A, Clarke CF, Navas P, Santos-Ocaña C (2009) Hydroxylation of demethoxy-Q6 constitutes a control point in yeast coenzyme Q6 biosynthesis. Cell Mol Life Sci 66:173–186. https://doi.org/10.1007/s00018-008-8547-7

Padilla-López S, Jiménez-Hidalgo M, Martín-Montalvo A, Clarke CF, Navas P, Santos-Ocaña C (2009) Genetic evidence for the requirement of the endocytic pathway in the uptake of coenzyme Q6 in *Saccharomyces cerevisiae*. Biochim Biophys Acta Biomembr 1788:1238–1248. https://doi.org/10.1016/j.bbamem.2009.03.018

Pappa A, Estey T, Manzer R, Brown D, Vasiliou V (2003) Human aldehyde dehydrogenase 3A1 (ALDH3A1): biochemical characterization and immunohistochemical localization in the cornea. Biochem J 376:615–623. https://doi.org/10.1042/BJ20030810

Payet L, Leroux M, Willison JC, Kihara A, Pelosi L, Pierrel F (2016) Mechanistic details of early steps in coenzyme Q biosynthesis pathway in yeast. Cell Chem Biol 23:1241–1250. https://doi.org/10.1016/j.chembiol.2016.08.008

Pedruzzi I, Bürckert N, Egger P, De Virgilio C (2000) *Saccharomyces cerevisiae* Ras/cAMP pathway controls post-diauxic shift element-dependent transcription through the zinc finger protein Gis1. EMBO J 19:2569–2579. https://doi.org/10.1093/emboj/19.11.2569

Pierrel F (2017) Impact of chemical analogs of 4-Hydroxybenzoic acid on coenzyme Q biosynthesis: from inhibition to bypass of coenzyme Q deficiency. Front Physiol 8:436. https://doi.org/10.3389/fphys.2017.00436

Pierrel F, Hamelin O, Douki T, Kieffer-Jaquinod S, Mühlenhoff U, Ozeir M, Lill R, Fontecave M (2010) Involvement of mitochondrial ferredoxin and para-aminobenzoic acid in yeast coenzyme Q biosynthesis. Chem Biol 17:449–459. https://doi.org/10.1016/j.chembiol.2010.03.014

Poon WW, Marbois BN, Faull KF, Clarke CF (1995) 3-Hexaprenyl-4-hydroxybenzoic acid forms a predominant intermediate pool in ubiquinone biosynthesis in *Saccharomyces cerevisiae*. Arch Biochem Biophys 320:305–314. https://doi.org/10.1016/0003-9861(95)90014-4

Poon WW, Do TQ, Noelle Marbois B, Clarke CF (1997) Sensitivity to treatment with polyunsaturated fatty acids is a general characteristic of the ubiquinone-deficient yeast coq mutants. Mol Asp Med 18:121–127. https://doi.org/10.1016/S0098-2997(97)00004-6

Quinzii CM, Kattah AG, Naini A, Akman HO, Mootha VK, DiMauro S, Hirano M (2005) Coenzyme Q deficiency and cerebellar ataxia associated with an aprataxin mutation. Neurology 64:539–541. https://doi.org/10.1212/01.WNL.0000150588.75281.58

Rea SL, Graham BH, Nakamaru-Ogiso E, Kar A, Falk MJ (2010) Bacteria, yeast, worms, and flies: exploiting simple model organisms to investigate human mitochondrial diseases. Dev Disabil Res Rev 16:200–218. https://doi.org/10.1002/ddrr.114

Roche Y, Peretti P, Bernard S (2006) Influence of the chain length of ubiquinones on their interaction with DPPC in mixed monolayers. Biochim Biophys Acta Biomembr 1758:468–478. https://doi.org/10.1016/j.bbamem.2006.03.015

Rodriguez-Aguilera JC, Navarro F, Arroyo A, Villalba JM, Navas P (1993) Transplasma membrane redox system of HL-60 cells is controlled by cAMP. J Biol Chem 268:26346–26349

Rodríguez-Hernández A, Cordero MD, Salviati L, Artuch R, Pineda M, Briones P, Gómez Izquierdo L, Cotán D, Navas P, Sánchez-Alcázar JA (2009) Coenzyme Q deficiency triggers mitochondria degradation by mitophagy. Autophagy 5:19–32. https://doi.org/10.4161/auto.5.1.7174

Rowland AA, Voeltz GK (2012) Endoplasmic reticulum–mitochondria contacts: function of the junction. Nat Rev Mol Cell Biol 13:607–625. https://doi.org/10.1038/nrm3440

Rugarli EI, Langer T (2012) Mitochondrial quality control: a matter of life and death for neurons. EMBO J 31:1336–1349. https://doi.org/10.1038/emboj.2012.38

Runner VM, Brewster JL (2003) A genetic screen for yeast genes induced by sustained osmotic stress. Yeast 20:913–920. https://doi.org/10.1002/yea.1019

Rusiñol AE, Cui Z, Chen MH, Vance JE (1994) A unique mitochondria-associated membrane fraction from rat liver has a high capacity for lipid synthesis and contains pre-Golgi secretory proteins including nascent lipoproteins. J Biol Chem 269:27494–27502

Saiki R, Nagata A, Uchida N, Kainou T, Matsuda H, Kawamukai M (2003) Fission yeast decaprenyl diphosphate synthase consists of Dps1 and the newly characterized Dlp1 protein in a novel heterotetrameric structure. Eur J Biochem 270:4113–4121. https://doi.org/10.1046/j.1432-1033.2003.03804.x

Sakumoto N, Mukai Y, Uchida K, Kouchi T, Kuwajima J, Nakagawa Y, Sugioka S, Yamamoto E, Furuyama T, Mizubuchi H, Ohsugi N, Sakuno T, Kikuchi K, Matsuoka I, Ogawa N, Kaneko Y, Harashima S (1999) A series of protein phosphatase gene disruptants in*Saccharomyces cerevisiae*. Yeast 15:1669–1679. https://doi.org/10.1002/(SICI)1097-0061(199911)15:15<1669::AID-YEA480>3.0.CO;2-6

Salviati L, Sacconi S, Murer L, Zacchello G, Franceschini L, Laverda AM, Basso G, Quinzii C, Angelini C, Hirano M, Naini AB, Navas P, DiMauro S, Montini G (2005) Infantile encephalomyopathy and nephropathy with CoQ_{10} deficiency: a CoQ_{10}-responsive condition. Neurology 65:606–608. https://doi.org/10.1212/01.wnl.0000172859.55579.a7

Samori B, Lenaz G, Battino M, Marconi G, Domini I (1992) On coenzyme Q orientation in membranes: a linear dichroism study of ubiquinones in a model bilayer. J Membr Biol 128:193–203. https://doi.org/10.1007/BF00231812

Sampaio-Marques B, Burhans W, Ludovico P (2014) Longevity pathways and maintenance of the proteome: the role of autophagy and mitophagy during yeast ageing. Microb Cell 1:118–127. https://doi.org/10.15698/mic2014.04.136

Santangelo GM (2006) Glucose signaling in *Saccharomyces cerevisiae*. Microbiol Mol Biol Rev 70:253–282. https://doi.org/10.1128/MMBR.70.1.253

Santos-Ocaña C, Navas P, Crane FL, Córdoba F (1995) Extracellular ascorbate stabilization as a result of transplasma electron transfer in *Saccharomyces cerevisiae*. J Bioenerg Biomembr 27:597–603

Santos-Ocaña C, Villalba JM, Córdoba F, Padilla S, Crane FL, Clarke CF, Navas P (1998) Genetic evidence for coenzyme Q requirement in plasma membrane electron transport. J Bioenerg Biomembr 30:465–475

Santos-Ocaña C, Do TQ, Padilla S, Navas P, Clarke CF (2002) Uptake of exogenous coenzyme Q and transport to mitochondria is required for bc1 complex stability in yeast coq mutants. J Biol Chem 277:10973–10981. https://doi.org/10.1074/jbc.M112222200

Scalori V, Alessandrì MG, Mian M, Giovannini L, Bertelli AA (1988) Plasma and tissue concentrations of coenzyme Q10 in the rat after its oral administration. Int J Tissue React 10:95–97

Scharwey M, Tatsuta T, Langer T (2013) Mitochondrial lipid transport at a glance. J Cell Sci 126:5317–5323. https://doi.org/10.1242/jcs.134130

Schepens B, Tinton SA, Bruynooghe Y, Parthoens E, Haegman M, Beyaert R, Cornelis S (2007) A role for hnRNP C1/C2 and Unr in internal initiation of translation during mitosis. EMBO J 26:158–169. https://doi.org/10.1038/sj.emboj.7601468

Schultz JR, Clarke CF (1999) Characterization of *Saccharomyces cerevisiae* ubiquinone-deficient mutants. Biofactors 9:121–129. https://doi.org/10.1002/biof.5520090206

Serrano A, Cordoba F, Gonzalez-Reyes JA, Santos C, Navas P, Villalba JM (1995) NADH-specific dehydrogenase from onion root plasma membrane: purification and characterization. Protoplasma 184:133–139. https://doi.org/10.1007/BF01276910

Sharmin D, Sasano Y, Sugiyama M, Harashima S (2014) Effects of deletion of different PP2C protein phosphatase genes on stress responses in *Saccharomyces cerevisiae*. Yeast 31:393–409. https://doi.org/10.1002/yea.3032

Sippel CJ, Goewert RR, Slachman FN, Olson RE (1983) The regulation of ubiquinone-6 biosynthesis by *Saccharomyces cerevisiae*. J Biol Chem 258:1057–1061

Stefely JA, Pagliarini DJ (2017) Biochemistry of mitochondrial coenzyme Q biosynthesis. Trends Biochem Sci 42:824–843. https://doi.org/10.1016/j.tibs.2017.06.008

Stefely JA, Reidenbach AG, Ulbrich A, Oruganty K, Floyd BJ, Jochem A, Saunders JM, Johnson IE, Minogue CE, Wrobel RL, Barber GE, Lee D, Li S, Kannan N, Coon JJ, Bingman CA, Pagliarini DJ (2015) Mitochondrial ADCK3 employs an atypical protein kinase-like fold to enable coenzyme Q biosynthesis. Mol Cell 57:83–94. https://doi.org/10.1016/j.molcel.2014.11.002

Stefely JA, Kwiecien NW, Freiberger EC, Richards AL, Jochem A, Rush MJP, Ulbrich A, Robinson KP, Hutchins PD, Veling MT, Guo X, Kemmerer ZA, Connors KJ, Trujillo EA, Sokol J, Marx H, Westphall MS, Hebert AS, Pagliarini DJ, Coon JJ (2016a) Mitochondrial protein functions elucidated by multi-omic mass spectrometry profiling. Nat Biotechnol 34:1191–1197. https://doi.org/10.1038/nbt.3683

Stefely JA, Licitra F, Laredj L, Reidenbach AG, Kemmerer ZA, Grangeray A, Jaeg-Ehret T, Minogue CE, Ulbrich A, Hutchins PD, Wilkerson EM, Ruan Z, Aydin D, Hebert AS, Guo X, Freiberger EC, Reutenauer L, Jochem A, Chergova M, Johnson IE, Lohman DC, Rush MJP, Kwiecien NW, Singh PK, Schlagowski AI, Floyd BJ, Forsman U, Sindelar PJ, Westphall MS, Pierrel F, Zoll J, Dal Peraro M, Kannan N, Bingman CA, Coon JJ, Isope P, Puccio H, Pagliarini DJ (2016b) Cerebellar ataxia and coenzyme Q deficiency through loss of unorthodox kinase activity. Mol Cell 63:608–620. https://doi.org/10.1016/j.molcel.2016.06.030

Stenmark P, Grünler J, Mattsson J, Sindelar PJ, Nordlund P, Berthold DA (2001) A new member of the family of di-iron carboxylate proteins. Coq7 (clk-1), a membrane-bound hydroxylase involved in ubiquinone biosynthesis. J Biol Chem 276:33297–33300. https://doi.org/10.1074/jbc.C100346200

Sugiyama Y, Fujita T, Matsumoto M, Okamoto K, Imada I (1985) Effects of idebenone (CV-2619) and its metabolites on respiratory activity and lipid peroxidation in brain mitochondria from rats and dogs. J Pharmacobiodyn 8:1006–1017. https://doi.org/10.1248/bpb1978.8.1006

Takahashi T, Okamoto T, Mori K, Sayo H, Kishi T (1993) Distribution of ubiquinone and ubiquinol homologues in rat tissues and subcellular fractions. Lipids 28:803–809. https://doi.org/10.1007/BF02536234

Tauche A, Krause-Buchholz U, Rödel G, Rodel G (2008) Ubiquinone biosynthesis in *Saccharomyces cerevisiae*: the molecular organization of O-methylase Coq3p depends on Abc1p/Coq8p. FEMS Yeast Res 8:1263–1275. https://doi.org/10.1111/j.1567-1364.2008.00436.x

Teclebrhansp H, Olssonsti J, Swiezewskasii E, Dallner G (1993) Biosynthesis of the side chain of ubiquinone : trans-prenyltransferase in rat liver microsomes. J Biol Chem 268:23081–23086

Tomasetti M, Alleva R, Solenghi MD, Littarru GP (1999) Distribution of antioxidants among blood components and lipoproteins: significance of lipids/CoQ 10 ratio as a possible marker of increased risk for atherosclerosis. Biofactors 9:231–240. https://doi.org/10.1002/biof.5520090218

Tran UC, Marbois B, Gin P, Gulmezian M, Jonassen T, Clarke CF (2006) Complementation of *Saccharomyces cerevisiae* coq7 mutants by mitochondrial targeting of the Escherichia coli UbiF polypeptide: two functions of yeast Coq7 polypeptide in coenzyme Q biosynthesis. J Biol Chem 281:16401–16409. https://doi.org/10.1074/jbc.M513267200

Troyano A, Fernández C, Sancho P, de Blas E, Aller P (2001) Effect of glutathione depletion on antitumor drug toxicity (apoptosis and necrosis) in U-937 human promonocytic cells. J Biol Chem 276:47107–47115. https://doi.org/10.1074/jbc.M104516200

Turunen M, Appelkvist E-L, Sindelar P, Dallner G (1999a) Blood concentration of coenzyme Q10 increases in rats when esterified forms are administered. J Nutr 129:2113–2118

Turunen M, Sindelar P, Dallner G (1999b) Induction of endogenous coenzyme Q biosynthesis by administration of peroxisomal inducers. Biofactors 9:131–139. https://doi.org/10.1002/biof.5520090207

Ulrich EL, Girvin ME, Cramer WA, Markley JL (1985) Location and mobility of ubiquinones of different chain lengths in artificial membrane vesicles. Biochemistry 24:2501–2508. https://doi.org/10.1021/bi00331a016

Vazquez Fonseca L, Doimo M, Calderan C, Desbats MA, Acosta MJ, Cerqua C, Cassina M, Ashraf S, Hildebrandt F, Sartori G, Navas P, Trevisson E, Salviati L (2017) Mutations in COQ8B (ADCK4) found in patients with steroid resistant nephrotic syndrome alter COQ8B function. Hum Mutat 39:406–414. https://doi.org/10.1002/humu.23376

Veling MT, Reidenbach AG, Freiberger EC, Kwiecien NW, Hutchins PD, Drahnak MJ, Jochem A, Ulbrich A, Rush MJP, Russell JD, Coon JJ, Pagliarini DJ (2017) Multi-omic mitoprotease profiling defines a role for Oct1p in coenzyme Q production. Mol Cell 68:970–977. https://doi.org/10.1016/j.molcel.2017.11.023

Villalba JM, Navarro F, Córdoba F, Serrano A, Arroyo A, Crane FL, Navas P (1995) Coenzyme Q reductase from liver plasma membrane: purification and role in trans-plasma-membrane electron transport. Proc Natl Acad Sci U S A 92:4887–4891

Villalba JM, Navarro F, Gomez-Diaz C, Arroyo A, Bello RI, Navas P (1997) Role of cytochrome b5 reductase on the antioxidant function of coenzyme Q in the plasma membrane. Mol Asp Med 18:S7–S13

Willis RA, Folkers K, Tucker JL, Ye CQ, Xia LJ, Tamagawa H (1990) Lovastatin decreases coenzyme Q levels in rats. Proc Natl Acad Sci U S A 87:8928–8930

Wirth C, Brandt U, Hunte C, Zickermann V (2016) Structure and function of mitochondrial complex I. Biochim Biophys Acta Bioenerg 1857:902–914. https://doi.org/10.1016/j.bbabio.2016.02.013

Xia D, Yu C-A, Kim H, Xia J-Z, Kachurin AM, Zhang L, Yu L, Deisenhofer J (1997) Crystal structure of the cytochrome bc 1 complex from bovine heart mitochondria. Science 277:60–66. https://doi.org/10.1126/science.277.5322.60

Xie LX, Hsieh EJ, Watanabe S, Allan CM, Chen JY, Tran UC, Clarke CF (2011) Expression of the human atypical kinase ADCK3 rescues coenzyme Q biosynthesis and phosphorylation of Coq polypeptides in yeast coq8 mutants. Biochim Biophys Acta 1811:348–360. https://doi.org/10.1016/j.bbalip.2011.01.009. doi:S1388-1981(11)00019-9 [pii]

Xie LX, Ozeir M, Tang JY, Chen JY, Jaquinod S-K, Fontecave M, Clarke CF, Pierrel F, Kieffer-Jaquinod S (2012) Over-expression of the Coq8 kinase in *Saccharomyces cerevisiae* coq null mutants allows for accumulation of diagnostic intermediates of the coenzyme Q6 biosynthetic pathway. J Biol Chem 287:23571–23581. https://doi.org/10.1074/jbc.M112.360354

Yu CA, Yu L (1993) Mitochondrial ubiquinol-cytochrome c reductase complex: crystallization and protein: ubiquinone interaction. J Bioenerg Biomembr 25:259–273

Yu C, Gu L, Lin Y, Yu L (1985) Effect of alkyl side chain variation on the electron-transfer activity of ubiquinone derivatives. Biochemistry 24:3897–3902. https://doi.org/10.1021/bi00336a013

Yuzuriha T, Takada M, Katayama K (1983) Transport of [14C]coenzyme Q10 from the liver to other tissues after intravenous administration to guinea pigs. Biochim Biophys Acta Gen Subj 759:286–291. https://doi.org/10.1016/0304-4165(83)90325-2

Zhang H, Javor GT (2003) Regulation of the isofunctional genes ubiD and ubiX of the ubiquinone biosynthetic pathway of Escherichia coli. FEMS Microbiol Lett 223:67–72. https://doi.org/10.1016/S0378-1097(03)00343-4

Zhang Y, Aberg F, Appelkvist EL, Dallner G, Ernster L (1995) Uptake of dietary coenzyme Q supplement is limited in rats. J Nutr 125:446–453

Zhou L, Wang JY, Wang J, Poplawsky A, Lin S, Zhu B, Chang C, Zhou T, Zhang LH, He YW (2013) The diffusible factor synthase XanB2 is a bifunctional chorismatase that links the shikimate pathway to ubiquinone and xanthomonadins biosynthetic pathways. Mol Microbiol 87:80–93. https://doi.org/10.1111/mmi.12084

Zhou L, Li C, Gao L, Wang A (2015) High-density lipoprotein synthesis and metabolism (review). Mol Med Rep 12:4015–4021. https://doi.org/10.3892/mmr.2015.3930

Zhu QS, Beattie DS (1988) Direct interaction between yeast NADH-ubiquinone oxidoreductase, succinate-ubiquinone oxidoreductase, and ubiquinol-cytochrome c oxidoreductase in the reduction of exogenous quinones. J Biol Chem 263:193–199

Chapter 3
Coenzyme Q Function in Mitochondria

Maria Luisa Genova and Giorgio Lenaz

Abstract In this chapter we provide a review with a focus on the function of Coenzyme Q (CoQ, ubiquinone) in mitochondria. The notion of a mobile pool of CoQ in the lipid bilayer as the vehicle of electrons from respiratory complexes has somewhat changed with the discovery of respiratory supramolecular units, in particular the supercomplex comprising Complexes I and III; in such assembly the electron transfer is thought to be mediated by direct channelling, and we provide evidence for a kinetic advantage on the transfer based on random collisions. The CoQ pool, however, has a fundamental function in establishing a dissociation equilibrium with bound CoQ, besides being required for electron transfer from other dehydrogenases to Complex III. CoQ bound to Complex I and to Complex III is also involved in proton translocation; although the mechanism of the Q-cycle is well established for Complex III, the involvement of CoQ in proton translocation by Complex I is still debated. This review also briefly examines some additional roles of CoQ, such as the antioxidant effect of its reduced form and its postulated action at the transcriptional level.

Keywords Coenzyme Q · Respiratory chain · Supercomplexes · Channelling · Electron transport

3.1 Introduction: Physical and Chemical Properties of Coenzyme Q

This chapter is a review with a focus on the role of Coenzyme Q (CoQ) in mitochondrial bioenergetics, therefore other functions related to CoQ molecules engaged in extra-mitochondrial reactions or not directly related to mitochondrial energy conversion are not considered here in detail. To this purpose it is useful to briefly remind some properties of CoQ that are relevant to its function (Lenaz and Genova 2010,

M. L. Genova (✉) · G. Lenaz
Department of Biomedical and Neuromotor Sciences, Alma Mater Studiorum – Università di Bologna, Bologna, Italy
e-mail: marialuisa.genova@unibo.it; giorgio.lenaz@unibo.it

G. López Lluch (ed.), *Coenzyme Q in Aging*,
https://doi.org/10.1007/978-3-030-45642-9_3

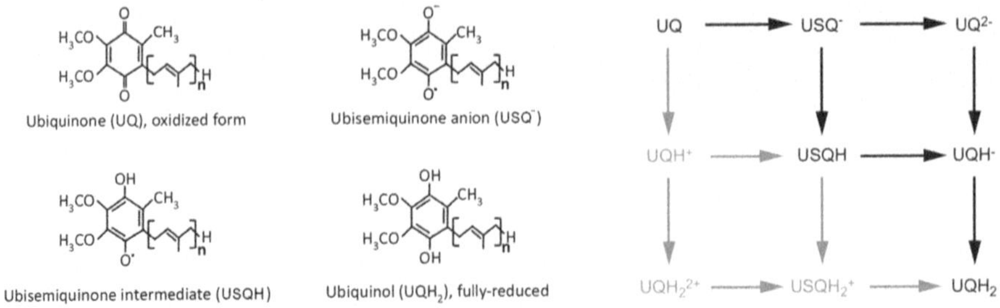

Fig. 3.1 Possible redox and protonation states of quinone molecules. The benzoquinone ring can exist in several redox states. The states that can be detected experimentally in solution or in catalytic sites of proteins are shown in black. Gray represents the states not present in biological environment. Vertical and horizontal arrows denote steps of protonation and reduction of the molecules, respectively. The transition between oxidized (UQ) and reduced (UQH_2) form involves formation of semiquinone intermediates which are unstable oxygen-centred free radicals (cf. Gunner et al. 2008; Sarewicz and Osyczka 2015).'n' indicates the number of isoprenoid units in the side-chain (e.g. n = 9 in mice and n = 10 in humans)

Genova and Lenaz 2011). CoQ exists in three redox states, fully oxidized (ubiquinone, Q), semiquinone (ubisemiquinone, SQ), and fully reduced (ubiquinol, QH_2). In addition, the existence of different possible levels of protonation of the quinone ring increases the possible redox forms although only six states can be experimentally detected in biological environments like in bulk lipid phases or in the catalytic site of proteins (Fig. 3.1). In fact, due to its extreme hydrophobicity, natural CoQ (CoQ_{10} in humans) can be present in three physical states: forming micellar aggregates, dissolved in lipid bilayers, and bound to proteins. Although the former state may be experimentally important (Fato et al. 1986), in the living cell CoQ is distributed among the other two states.

The extent to which CoQ is bound to mitochondrial proteins is an important parameter in its function. If we consider bound CoQ in a 1:1 stoichiometry with the mammalian mitochondrial respiratory complexes interacting with the quinone (i.e. Complex I, Complex II and Complex III), in beef heart mitochondria (BHM) we come up to about 0.35 nmol CoQ_{10}/mg protein. Since these enzymes may possess more than one bound CoQ molecule and other enzymes feeding electrons to the respiratory chain may also have bound CoQ molecules, this figure may become higher. However, the total content of CoQ in BHM (Capaldi 1982) is much higher than such estimation therefore we must assume that most CoQ is free in the membrane bilayer. Indeed, a direct study (Lass and Sohal 1999) of the amount of CoQ bound to mitochondrial proteins in five different mammalian species showed that the protein-bound aliquot ranges between 10% and 32% of total CoQ.

It has been assumed for long time that the shape of the CoQ molecule is linear, with some possibility of rotation allowed for the long isoprenoid tail. Bending of the molecule (Lenaz 1988) is confirmed by linear dichroism studies (Samorì et al. 1992) of its location in the hydrophobic mid-plane of the lipid bilayer, with the polar head and the proximal segment of the tail (about the third isoprene unit) oscillating between the mid-plane (i.e. wholly linear shape of the molecule) and the polar heads of the membrane phospholipids (i.e. maximal bending of 90°). This model allows for movement of the redox centre of CoQ, which is required for the interaction with the redox centres that are situated in the partner mitochondrial complexes at the level of the hydrophilic heads of the membrane phospholipids (Xia et al. 1997; Tocilescu et al. 2010).

A computer simulation of molecular dynamics of CoQ homologues in the vacuum showed that the lowest energy level for both oxidized and reduced quinones is characterized by a folded conformation, where the polar head is in tight contact with the last isoprenoid unit of the hydrophobic tail (Di Bernardo et al. 1998). We suggested that a folded conformation may also apply to CoQ in the hydrophobic interior of natural membranes; the suggestion is supported by the experimental demonstration by magnetic resonance techniques that ubisemiquinones are folded in organic solvents (Joela et al. 1997). A neutron diffraction study (Seelert et al. 2009) localized CoQ_{10} in the membrane midplane, although the actual shape could not be defined.

Galassi and Arantes (2015) have presented molecular dynamics simulations of several ubiquinone homologs with variable isoprenoid tail lengths complexed to single-component phosphatidylcholine bilayers. Free energy profiles for ubiquinone insertion in the lipid bilayer allow for the determination of the equilibrium location of ubiquinone in the membrane as well as for the validation of the simulation model by direct comparison with experimental partition coefficients. A detailed analysis of structural properties and interactions shows that the ubiquinone polar head group is localized at the water-bilayer interface at the same depth of the lipid glycerol groups and oriented normal to the membrane plane. Both the localization and orientation of ubiquinone head groups do not change significantly when increasing the number of isoprenoid units. The isoprenoid tail is extended and packed with the lipid acyl chains. For ubiquinones with long tails, the terminal isoprenoid units have high flexibility.

The results of Galassi and Arantes (2015) are in qualitative agreement with the results of atomistic molecular dynamics simulations performed by Kaurola et al. (2016) in order to assess the interactions of ubiquinone and its analogues with their immediate surroundings in a many-component lipid bilayer that mimics the inner mitochondrial membrane including cardiolipin (CL). Kaurola et al. (2016) showed that there is a strong tendency of both quinone and quinol molecules to localize in the vicinity of the lipid acyl groups, right under the lipid head group region and that there is also a location in the middle of the bilayer where the oxidized quinone molecules tend to stabilize and diffuse freely along the membrane normal direction of the lipid bilayer (Z-direction) up to a distance of ~1.5 nm, allowing the quinone headgroup to flip very fast (in 10–100 ns time scale) and to translocate from one side of the bilayer to the other. This characteristic is critical for the optimal turnover of the entire electron transport chain, since a quinone molecule formed upon oxidation at the P-side of the inner mitochondrial membrane (e.g. by activity at the Q_o-site of Complex III) will undergo a flip to reach the N-side and be reduced by Complex I or by the Q_i-site of Complex III. Ubiquinol was also found to translocate through the lipid bilayer, but with a time scale at least one order of magnitude

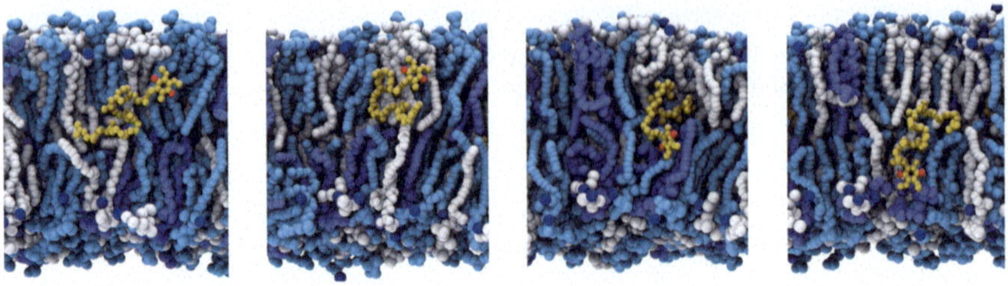

Fig. 3.2 Snapshots illustrating the translocation of quinol through a lipid bilayer. Quinol is shown in yellow, tetra-18:2-CL in white, DLPC in light blue, and DLPE in dark blue. (Reprinted from Kaurola et al. (2016), Copyright 2016, with permission from Elsevier)

slower. It is interesting to point out that the flipping of ubiquinol takes place in the neighbourhood of CL molecules (Fig. 3.2) whereas the simulation of Kaurola et al. (2016) showed practically no interactions of ubiquinone with di-18:2-PE (DLPE) and CL.

The cyclohexane/water partition coefficients of different quinones are a function of their hydrophobicities (Braun et al. 1986; Rich and Harper 1990). An additional consequence of the high hydrophobicity of ubiquinones, related to their partition coefficients, is their extent of solubility in monomeric state (Battino et al. 1986; Lenaz 1998); water insolubility is particularly serious a phenomenon for oxidized quinones, as in determination of Complex I activity. An important consideration concerning the use of ubiquinones as substrates is their reaction with the partner enzymes from within the lipid bilayer: a method to determine the true membrane *K*m (Michaelis constant, expressed as quinone concentration in the membrane) and the lipid/water partition coefficients was described by steady-state kinetic measurements at varying phospholipid fractional values in the assay medium (Fato et al. 1988).

The lateral diffusion of CoQ in lipid bilayers has received particular attention in the past in relation to its role as a small component of the mitochondrial respiratory chain that, according to the 'random collision model' of electron transfer proposed by Hackenbrock et al. (1986), undergoes considerably faster lateral diffusion compared to the macromolecular respiratory complexes and that, together with cytochrome *c*, assures electron transfer by random collisions with its partner enzymes (cf. also Sect. 3.2). In addition, Hackenbrock et al. (1986) suggested that CoQ diffusion in the mitochondrial membrane is the rate limiting step in the whole electron transfer process. However, our kinetic studies provided evidence against diffusion control of CoQ-linked reactions (Lenaz and Fato 1986).

A variety of techniques have been employed to measure the lateral diffusion of CoQ homologues in lipid bilayers and in natural membranes, yielding a broad range of values for the diffusion coefficients (D_l). Collision-dependent methods measuring lateral diffusion in the nm scale include excimer formation and fluorescence quenching, whereas methods based on the redistribution of probe molecules measure diffusion on a μm scale for which the preferred technique is fluorescence recovery after photobleaching (FRAP). Exploiting the FRAP technique with fluorescent labelled ubiquinone analogues, diffusion coefficients were calculated in mitochondrial membranes in the range of 10^{-9} cm²/s (Hackenbrock et al. 1986; Rajarathnam et al. 1989). It is worth noting that the FRAP technique requires chemical modification of the CoQ molecule by covalent binding of fluorescent reporter groups, thus increasing the size and hydrophilicity of CoQ. For that reason, some doubts arise on the very low values of D_l found in the FRAP experiments (Lenaz 1988).

On the other hand, exploiting collisional fluorescence quenching of membrane-partitioned fluorophores by genuine oxidized CoQ homologues, Fato et al. (1986) calculated diffusion coefficients in the range of 10^{-6}–10^{-7} cm²/s in both liposomes and mitochondrial membranes. Values of the order of 10^{-7} cm²/s were also calculated in the simulation study of Galassi and Arantes (2015) for different ubiquinones in a lipid bilayer.

3.2 Organization of the Respiratory Chain: Historical Outline

In the early '60s, Hatefi et al. (1962a) accomplished the systematic resolution and reconstitution of four respiratory complexes from mitochondria, i.e. NADH-Coenzyme Q reductase (Complex I, C_I), succinate-CoQ reductase (Complex II, C_{II}), ubiquinol-cytochrome c reductase (Complex III, C_{III}) and cytochrome c oxidase (Complex IV, C_{IV}) and led Green and Tzagoloff (1966) to postulate that the overall respiratory activity is the result of both intra-complex electron transfer between redox components having fixed steric relation and, in addition, of inter-complex electron transfer ensured by rapid diffusion of the mobile components acting as co-substrates, i.e. CoQ and cytochrome c (cyt. c). This proposal was substantially confirmed over the following years by a variety of kinetic and structural studies, leading Hackenbrock et al. (1986) to postulate the "Random Collision Model of Electron Transfer". The organization of the respiratory chain represented a major research subject in the 1970–1980s, culminating with acceptation of the random collision model by the majority of investigators in the field.

Recent experimental evidence obtained with newly developed techniques has replaced the random collision model of electron transfer with a model of supramolecular organisation based upon specific interactions between some of the respiratory enzymes, and leading to the acquisition of new properties (substrate channelling, supramolecular assembly, morphological organisation of the respiratory chain) that were unpredictable in the earlier reductionist approach.

However, a retrospective analysis of the literature reveals that the idea of supramolecular associations between respiratory enzymes has been present since the early times, dating from the original view of Chance and Williams (1955), who

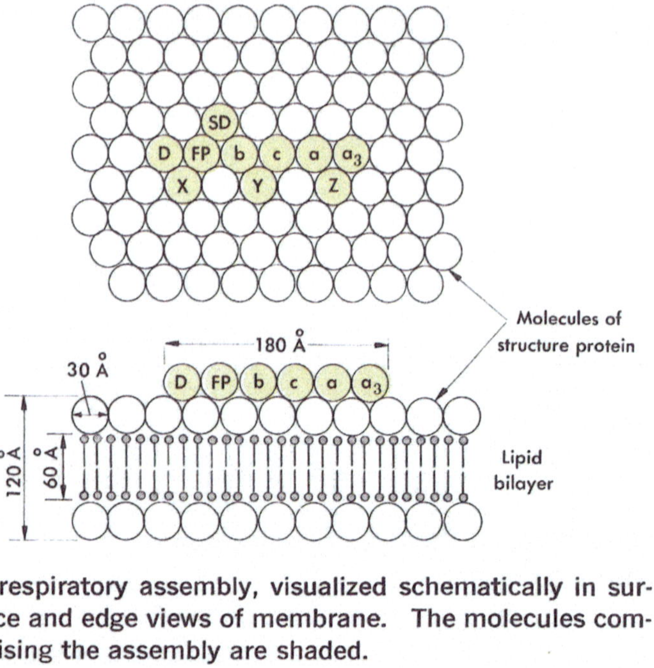

Fig. 3.3 Scheme of the respiratory assembly as originally drawn by Lehninger (1965) according to the early model of Chance and Williams (1955). *D* reduced diphosphopyridine nucleotide, *FP* flavoprotein, *SD* succinate dehydrogenase, *b, c, a, a3* designate the corresponding cytochromes, *X, Y, Z* designate hypothetical high-energy intermediates that were suggested to serve for donating high-energy phosphate groups to ATP generation

depicted the respiratory chain as a solid-state assembly of the prosthetic groups known up to then (i.e. only flavins and cytochromes) in a protein matrix (Fig. 3.3).

Contrary to common recollection, evidence against a random distribution of respiratory complexes also derived from the same early investigations of Hatefi et al. (1962b) that brought to the isolation of the respiratory complexes: in fact, during isolation of the C_I and C_{III} it was reported that units of C_I-plus- C_{III} were also present, indicating that those two respiratory complexes may be preferentially associated in the native membrane. The authors of these studies were well aware that such preferential structural associations are physiological and reveal the existence of functional supramolecular units of electron transfer. In their paper on the reconstitution of the electron transfer system from individual respiratory complexes, Hatefi et al. (1962a) showed that the four primary enzyme systems (i.e. C_I to C_{IV}) may be suitably combined to form supramolecular units capable of catalysing the sum of the reactions catalysed by the respective enzyme components. A careful examination of these pioneering papers allows one to conclude that specific supramolecular assemblies composed of respiratory complexes (i.e. supercomplexes, SCs), as well as the entire NADH oxidase (i.e. the respirasome, see later), had been isolated and reconstituted. Nevertheless the possible contradiction between the report of functional SCs and the model of Green and Tzagoloff (1966) of free complexes linked by mobile components was not reported or discussed at that time.

The publication of the fluid mosaic model of membranes (Singer and Nicolson 1972) strongly influenced the researchers involved in the study of mitochondrial membranes, and this is probably the reason why the random collision model of Hackenbrock et al. (1986) was so well and uncritically accepted.

There were, however, a few reports before the year 2000 on the possible presence of specific associations between respiratory complexes, either fixed (Ozawa et al. 1987) or dynamic (Hochman et al. 1985).

A drastic change occurred in 2000 when Schägger and Pfeiffer (2000) applied the previously introduced technique of polyacrylamide gel electrophoresis under non-denaturing conditions (Blue-Native Polyacrylamide Gel Electrophoresis, BN-PAGE) to digitonin-solubilized yeast and mammalian mitochondria. The newly discovered supramolecular associations were considered to represent the physiological state of the respiratory complexes. In the same paper the authors also described a dimeric state for the ATP-synthase complex.

Today the evidence of SCs association of the respiratory chain is well consolidated (Lenaz et al. 2016), although their possible relation with a random distribution of the individual complexes is not completely clarified (Acín-Peréz et al. 2008; Genova and Lenaz 2013, 2014; Enriquez and Lenaz 2014). This aspect will be widely discussed subsequently in this chapter.

3.3 An Aliquot of Coenzyme Q Is Segregated in Respiratory Supercomplexes

Employing BN-PAGE in digitonin-solubilized mitochondria of *Saccharomyces cerevisiae*, which possesses no C_I, Schägger and Pfeiffer (2000) revealed two bands with apparent masses of ~750 and 1000 kDa, each containing the subunits of both C_{III} and C_{IV}. Similarly, in bovine heart mitochondria, C_I–C_{III} interactions were apparent from the presence of C_I in the form of the SC I_1III_2, which was also found further assembled into larger assemblies (respirasomes, SC $I_1III_2IV_{1–4}$) comprising different copy numbers of C_{IV} (Schägger and Pfeiffer 2001). Only 14–16% of total C_I was found in free form suggesting that all C_I is bound to C_{III} in the absence of detergents. The initial criticism that SC may be an artefact observed because of mild detergents like digitonin used in the solubilisation procedure was largely abated by evidence that purified SCs are stable, catalytically active stoichiometric units (Schägger and Pfeiffer 2000; Acín-Peréz et al. 2008).

Some SCs have been purified and analysed by negative-stain electron microscopy (Schäfer et al. 2006) and single-particle cryo-EM (Althoff et al. 2011; Dudkina et al. 2011, Mileykovskaya et al. 2012). The 1.7 MDa bovine heart SC consists of one copy of C_I, one C_{III} dimer, and one C_{IV} monomer. X-ray structures of the component complexes were fitted to the 3D map to produce pseudo-atomic models of the bovine respirasome showing that C_{III} interacts with both C_I and C_{IV} (Fig. 3.4), the latter being absent in the smaller SC I_1III_2 of approx. 1.5 MDa.

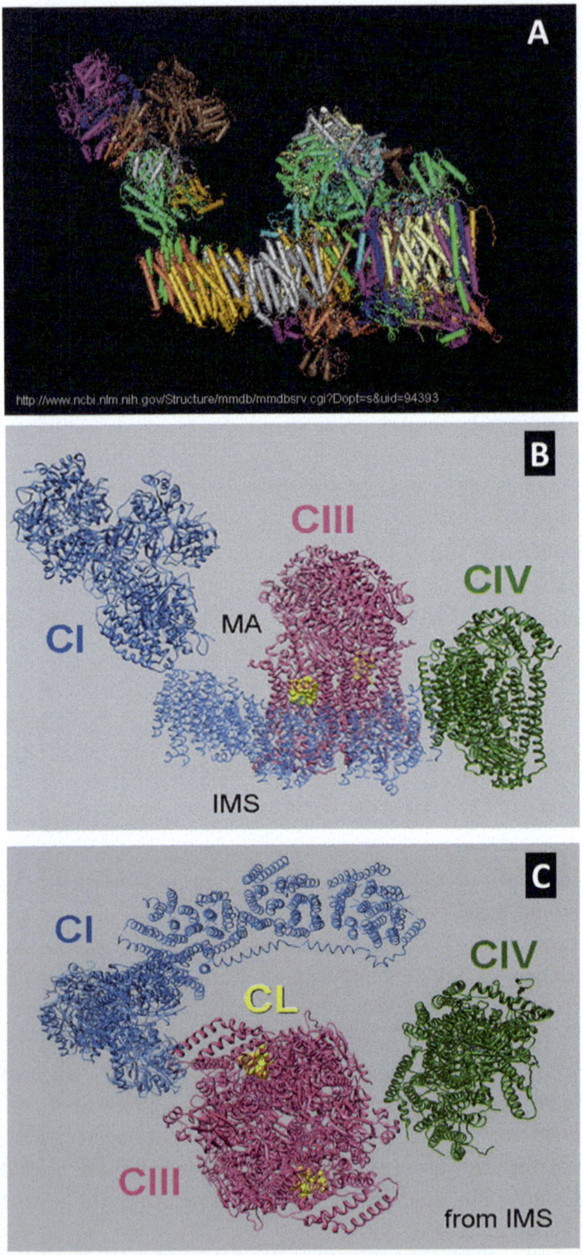

Fig. 3.4 Bovine mitochondrial SC $I_1III_2IV_1$. (**a**) Fitted model by single particle cryo-EM. (PDB ID: 2YBB; Althoff et al. 2011); (**b**) side view and (**c**) view from the intermembrane space (IMS) showing two CL molecules (in yellow) in the cavity of each monomer of Complex III formed by cytochromes c1 and b (Mileykovskaya and Dowhan 2014). *MA* denotes the mitochondrial matrix. (Reprinted from Lenaz et al. (2016), Copyright 2016, with permission from Elsevier)

Interactions between respiratory chain complexes within SCs depend on protein-cardiolipin (CL) interactions (Mileykovskaya and Dowhan 2014). SCs have 2–5 nm gaps at the transmembrane interfaces of the individual complexes (Fig. 3.4C). These gaps lie within the membrane imbedded domains of the SC and are therefore most likely filled with lipids. The mammalian SC has larger gaps than the yeast SC (Althoff et al. 2011; Dudkina et al. 2011; Mileykovskaya et al. 2012). In Barth syndrome patients, where cardiolipin remodelling is altered due to the mutation of the gene Tafazzin, SCs are unstable, leading to the mitochondrial functional impairment that underlies the disease (McKenzie et al. 2006). Direct involvement of CL in the formation of SC was demonstrated in genetically manipulated strains of *S. cerevisiae* in which the CL content can be regulated *in vivo* (Zhang et al. 2002). BN-PAGE of digitonin extracts of mitochondria revealed that yeast mutants completely lacking CL did not form the SC III_2IV_2 as observed in the wild type parental strain, whereas the total amount of individual respiratory complexes was not affected by the lack of CL (Zhang et al. 2002; Pfeiffer et al. 2003). However, no experimental evidences about C_I-containing SCs could be obtained in these studies, since *S. cerevisiae* lacks C_I.

In the case of bovine respirasome it was suggested that the lipid filled space between CI and CIII could serve as a diffusion microdomain, which restricts movement of CoQ facilitating its channelling inside the SC. This hypothesis is also consistent with the spatial arrangement of the individual complexes in the architecture of ovine (Letts et al. 2016) and porcine (Gu et al. 2016) supercomplexes as recently determined by near atomic cryo-electron microscopy.

The importance of lipid composition in the stabilization of C_I-containing SCs is supported by our results showing that both the stability of SC I_1III_2 from bovine heart and its NADH-cytochrome c oxidoreductase activity are hampered by reconstitution of CoQ_{10}-enriched proteoliposomes at high lipid to protein ratio (30:1 w:w) (Genova et al. 2008; Maranzana et al. 2013). On the contrary, SC I_1III_2 and efficient NADH-cytochrome c oxidoreductase activity may be preserved when similar high-lipid proteoliposomes are enriched with 20% CL (w:w), resembling the percent content of cardiolipin in the mitochondrial membrane (M. Kopuz, Y. Birinci, S. Nesci, G. Lenaz and M.L. Genova, unpublished data). A likely explanation is that dilution of native protein-bound CL in the excess exogenous lipids is counteracted by increasing the CL content in the proteoliposomes, thus shifting the equilibrium to CL binding to protein.

Recently SCs from bovine heart mitochondria were separated by sucrose density gradient centrifugation (Shinzawa-Itoh et al. 2016). The SC sample did not contain cytochrome c but did contain complexes I, III, and IV at a ratio of 1:2:1, 6 molecules of CoQ_{10}, and 623 atoms of phosphorus. When cytochrome c was added in the solution, the SC sample exhibited KCN-sensitive NADH oxidase activity, confirming both the previous studies of Acín-Peréz et al. (2008) on the respiratory activity of SCs isolated from BN-PAGE bands and the analogous studies of Stroh et al. (2004) on the respirasome from *P. denitrificans*.

Despite recent experimental advances in our capability to characterize and describe redox proteins in respiratory SCs, two particular challenges need to be met:

determine how the intermolecular redox reactions are coordinated with intramolecular redox reactions and how redox reactions are coordinated with other chemical events catalysed by SCs such as proton transfer or substrate binding and release (cf. Various authors (2015) and Letts et al. (2019) for further review).

3.4 Evidence for Coenzyme Q Channelling Between Complex I and Complex III

The question whether the respiratory SCs provide a functional advantage on respiration is not a trivial one, since it may have profound consequences on physiological regulation of metabolic fluxes and on pathological alterations of mitochondrial bioenergetics.

After the discovery of SCs it was proposed that the natural role of such assemblies is substrate channelling (Ovàdi 1991) in inter-complex electron transfer; this means direct transfer of electrons between the active sites of two enzymes catalysing consecutive reactions by successive reduction and reoxidation of the intermediate substrate without its diffusion in the bulk medium.

3.4.1 Structural Evidence

Some evidence for possible channelling comes from the 3D structure of the mitochondrial SC $I_1III_2IV_1$ (cf. Figure 3.1 in Sect. 3.3); a unique arrangement of the three component complexes indicates the pathways along which ubiquinone and cytochrome c can travel to shuttle electrons between their respective protein partners. The available data suggest that the binding sites for CoQ reduction in C_I (Baradaran et al. 2013) and for CoQ reoxidation in C_{III} (Zhang et al. 1998) face each other and are separated by a 13-nm gap, presumably filled with lipids, within the membrane core of the SC; CoQ is likely to run a trajectory through this gap (Althoff et al. 2011) and undergo a rapid flip to reach the two sides of the bilayer where the entry points for CoQ on the interacting proteins of SCs are positioned. Significant amounts of bound phospholipids are present in the purified SC from mammalian mitochondria with enrichment of CL compared with bovine heart total lipid (Althoff et al. 2011). Moreover, HPLC analysis of the lipid extracts indicated that each SC contains at least one molecule of ubiquinol (Althoff et al. 2011).

On the basis of kinetic analysis to be discussed later in this chapter, Blaza et al. (2014) claimed that SCs have no specific function and only derive from the necessity to maintain a high protein concentration in the inner membrane avoiding unspecific aggregation. It is certainly true that SC formation depends on the lipid concentration, being favoured at high protein to lipid ratio (Genova et al. 2008); on the other hand, if electron transfer by channelling actually does not occur, it is difficult to understand why the active sites of interacting complexes should be located in strict vicinity, in stereospecific non-random manner.

3.4.2 Evidence for Rate Advantage in the CoQ Region

If two membrane-bound redox enzymes are connected by a mobile redox carrier undergoing long-range diffusion in the medium, the overall reaction rate would be governed by the frequency of useful collisions between the mobile carrier and its two redox partners (pool behaviour, cf. Sect. 3.5 for insights). On the other hand, if the whole redox system is fixed in a solid state arrangement, the frequency of encounters will be dictated only by the steric proximity and fixed contacts between the redox enzymes and their intermediate substrate (channelling).

3.4.2.1 Fixed Assemblies: Stoichiometric Behaviour

Ragan and Heron et al. (1978) provided evidence that purified C_I and C_{III}, when mixed as concentrated solutions in detergent and then co-dialysed, combine reversibly in a 1:1 molar ratio to form a SC I-III unit (NADH-cytochrome c oxidoreductase); any extra molecules of CI or CIII that in principle might exchange electrons with such SC I-III unit in reality do not significantly contribute to the overall rate of cytochrome c reduction. In fact the results of Ragan and Heron et al. (1978) show that the reduction of cytochrome b by NADH is biphasic and the extents of the fast and slow phases of reduction are determined by the amount of CIII specifically associated with CI, the extent of the slow phase increasing only when CIII is in excess over the SC unit. These studies (Ragan and Heron 1978) were able for the first time to demonstrate the existence of an active SC formed by C_I and C_{III}, and that electron transfer within the SC I-III is fast, whereas electron transfer from the SC I-III to free CIII via the CoQ-pool is possible but extremely slow (cf. Sect. 3.5 for insights on the functional significance of a random distribution of mitochondrial complexes). Heron and co-workers (Heron et al. 1978) proposed that relative mobility is lost and complexes are frozen in their SC I-III assembly when phospholipid in excess of that needed to form an annulus is absent, thus favouring a stable orientation of the site of reduction of ubiquinone in CI with respect to the site of oxidation in CIII. However, full CoQ-pool behaviour could be restored and C_I and C_{III} could be made to operate independently of each other by raising the concentrations of phospholipid and ubiquinone (approx. a twofold and a sixfold increase, respectively) in the concentrated mixture (Ragan and Heron 1978).

Heron et al. (1978) also reported that endogenous CoQ_{10} leaks out of the SC I-III when extra phospholipid is present in the proteoliposomes, causing a decrease in NADH-cytochorome c oxidoreductase activity that could be alleviated by adding more ubiquinone. It is likely that the function of the large amount of ubiquinone in natural mitochondrial membranes may be, therefore, to maintain the CoQ_{10} content in the SC unit when it is formed.

Zhu et al. (1982) in studying the mechanism by which CoQ interacts with the energy conserving electron transfer chain of submitochondrial particles prepared from BHM concluded that both direct interaction between CoQ-loaded enzymes and diffusion of ubiquinone (and ubiquinol) are involved in the electron transfer from the dehydrogenases to C_{III}. In particular, they found that the rate of ubiquinol oxidase activity varied with the substrate used to reduce CoQ, a lower value being found with NADH than with succinate and this difference growing larger at low levels of the quinone (1% of the original amount as obtained by successive pentane extractions) whereas decreasing to zero on extrapolation to infinite CoQ content. They assumed that only bound quinone is present at very low residual levels of CoQ and therefore they favoured the view that electron transfer occurs via direct collision of the dehydrogenases with the bc1 complexes. Although the data of Zhu et al. (1982) cannot easily be interpreted in terms of SC I_1III_2 assemblies, they point out a partial pool function of free CoQ and a different behaviour of NADH and succinate oxidation (cf. also Sect. 3.5.1).

A more direct comparison of the effect of CoQ-channelling with respect to CoQ-pool behaviour was performed in our laboratory using proteoliposomes obtained by fusing a crude mitochondrial fraction (R4B) (Rieske 1967) enriched in C_I and C_{III} (besides residual C_{II}) with different amounts of phospholipids and CoQ_{10} (Lenaz et al. 1999). The comparison of the experimentally determined NADH-cytochrome c reductase activity with the values expected by theoretical calculation applying the pool equation (cf. Sect. 3.5) showed overlapping results at phospholipid dilutions (w/w) from 1:10 to 1:40 protein to lipid ratios. On the contrary, pool behaviour was not effective and the observed rates of NADH-cytochrome c reductase were higher than the theoretical values (Lenaz et al. 1999; Bianchi et al. 2003; Genova et al. 2008) at low protein:lipid dilution (1:1 w/w) resembling the mean nearest neighbour distance between respiratory complexes in mitochondria (Vanderkooi 1978; Schwerzmann et al. 1986; Lenaz 1988).

Moreover when the same proteoliposomes at 1:1 protein:lipid ratio were treated with n-dodecyl-β-D-maltoside (DDM) in the same amount as dissociation of SC I_1III_2 was detected by BN-PAGE, the NADH cytochrome c reductase activity fell dramatically, whereas both C_I and C_{III} individual activities were unchanged (Maranzana et al. 2013); an analogous behaviour was detected by treating bovine heart mitochondria with the same detergent. Our data can be interpreted as maintenance of CoQ channelling within the SC I_1III_2 allowing NADH oxidation to take place at high rate as long as the stability of the SC itself is not impaired (e.g. by high lipid dilution and by DDM *in vitro*).

This interpretation strongly favours our early hypothesis (Bianchi et al. 2003) that C_I-containing SCs physiologically exist in equilibrium with isolated respiratory complexes and that electron transfer in the respiratory chain would either follow specific CoQ-channelling or be governed by random collisions, depending on metabolic conditions of the cell.

3.4.2.2 Evidence for CoQ-Channelling by Metabolic Flux Control Analysis

The first demonstration that respiratory SCs are autonomous units carrying electron transfer by channelling was achieved in our laboratory (Bianchi et al. 2003, 2004) exploiting the flux control analysis and the principle that the sum of the flux control coefficients (FCC) of the individual enzymes in an integrated pathway must equal 1 unless these enzymes form supramolecular units and establish substrate channelling. In the latter case, these enzymes would be all equally rate-limiting and the sum of the control coefficients would be higher than 1 (Kholodenko and Westerhoff 1993).

Using this principle and specific inhibitors in order to define the extent of metabolic control exerted by each individual complex over the entire respiration, we found that both C_I and C_{III} have flux control coefficients approaching 1 in bovine heart mitochondria (Table 3.1), thus suggesting that the two complexes behave as a single enzymatic unit and that electron transfer through CoQ is accomplished by channelling between the two redox enzymes (Bianchi et al. 2004). Using the same method for succinate oxidation we found that CII is rate-limiting whereas CIII is not, supporting the notion that CII does not form SCs and that the oxidation of succinate follows pool behaviour. This approach is similar to that previously applied by Boumans et al. (1998), who had found that CoQ does not follow pool behaviour in yeast mitochondria unless they are treated with chaotropic agents. Also the inhibitor titration proposed by Kröger and Klingenberg (1973b) for evaluating pool behaviour obeys to the same principle.

It is worth noting that, in our flux control analysis using cyanide inhibition (Bianchi et al. 2004), C_{IV} appears to be randomly distributed, or in other words that a large excess of active enzyme exists in free form in the pathway from NADH to oxygen.

Very few other studies using metabolic control analysis were addressed to the functional aspects of SCs (Quarato et al. 2011; Kaambre et al. 2012, 2013); these studies confirmed that the respiratory chain, at least under certain conditions, is organized in functionally relevant supramolecular structures. In

Table 3.1 Flux control coefficients of respiratory complexes over NADH oxidase and succinate oxidase activity

Sample	Respiratory activity	Complex I	Complex II	Complex III	Sum of FCCs[a]	Ref.
BHM	NADH oxidase	1.06	n.a.	0.90	2	(b)
	Succinate oxidase	n.a.	0.88	0.34	1	(b)
R4B 1:1	NADH-cytochrome c oxidoreductase	0.93	n.a.	0.73	2	(c)
R4B 1:30	NADH-cytochrome c oxidoreductase	0.92	n.a.	0.15	1	(c)

BHM bovine heart mitochondria, *R4B* mitochondrial fraction enriched in Complex I and Complex III reconstituted in proteoliposomes at different protein:phospholipid ratios (w:w), *n.a.* not applicable in this sample. [a]Values rounded to the nearest unit; [b]Bianchi et al. 2004; [c]Lenaz et al. 2010 digitonin-permeabilized HepG2 cells, Quarato et al. (2011) observed that under conditions of high membrane potential (state-4 respiration) the sum of the FCCs calculated for Complexes I, III and IV activities exceeded 1, supporting the proposition that they are complexed in a supramolecular unit.

In saponin-permeabilized breast and cholorectal tumor samples Kaambre et al. (2012, 2013) observed flux control coefficients for mitochondrial oxidative phosphorylation activities whose sum approached 4 and interpreted the data as due to the presence of SC association.

Blaza et al. (2014) criticized the evidence for channelling in the respiratory chain deriving from flux control analysis (Bianchi et al. 2004): they reasoned that rotenone is competitive with CoQ and therefore the extent of its inhibition of C_I is affected by the additional presence of the exogenous quinone employed in CI assay but absent in the assay of NADH aerobic oxidation. Indeed, using rotenone and CoQ_1, Blaza et al. (2014) find FCC for CI that is actually not valid, since it exceeds 1. The discrepancy with (Bianchi et al. 2004) may however be ascribed to the different exogenous substrates used for CI assay in the two studies, i.e. CoQ_1 and decylubiquinone (DB) respectively, and also to the technical difficulty of calculating FCC from the initial slopes of the inhibition curves. DB has much lower affinity than CoQ_1 for CI (Fato et al. 1996), thus exerting lower competition with rotenone and lower influence on the inhibition efficiency. On the other hand, we note that the FCC values found by Blaza et al. (2014) using alternative inhibitors of CI not competitive with CoQ (e.g. piericidn and diphenyleneiodonium are oddly low; it is likely that in this case the low FCC for C_I over NADH oxidase activity is caused by the limiting amount of cytochrome c in their samples, which would necessarily shift the major control of the chain to the cytochrome c region. Significantly, in the study of Blaza et al. (2014), addition of cytochrome c to the mitochondrial membranes raised the above mentioned FCC of CI from 0.19 to 0.67.

We emphasize however that the major point discussed by us (Bianchi et al. 2004) to demonstrate the existence of a SC I+III was not so much the high FCC of CI but the concomitant high FCC of CIII in NADH oxidation (cf. Table 3.1; not measured, on the contrary, by Blaza et al. 2014), which is certainly incompatible with a CoQ-pool model postulating CI and CIII as independent molecules in the membrane. Such high FCC value of CIII, calculated by inhibitor titration with mucidin, is not an artefact because the corresponding FCC of CIII measured by using the same inhibitor in succinate oxidation is low (as expected from the lack of SC II+III). In other words, the FCC of CIII can be close to 1 if CIII is assembled together with the quinone reductase that precedes in the electron pathway and, in our hands, this condition is experimentally observed only for the NADH-dependent pathway involving CI whereas it does not occur in the case of CII. In addition, in the proteoliposome system described above, the FCC of CIII over NADH oxidation is drastically decreased after dissociation of the SC I_1III_2 by reconstitution in excess phospholipids (cf. R4B 1:1 vs. 1:30 in Table 3.1) or in peroxidised phospholipids (Lenaz and Genova 2009a).

3.5 Collision-Based Electron Transport: The "CoQ-Pool" Behaviour

The functional significance of a random distribution of mitochondrial complexes connected by CoQ was supported in the early times by the kinetic analysis of Kröger and Klingenberg (1973a); they showed that steady-state respiration in submitochondrial particles from beef heart, could be modelled as a simple two-enzyme system, the first causing reduction of ubiquinone (V_{red}) and the second causing oxidation of ubiquinol (V_{ox}), behaving kinetically as a homogeneous pool. According to this assumption the total CoQ molecules must effectively link any number of the dehydrogenase with any other number of the oxidase, and the activities of all the reducing and oxidizing enzymes determine the overall steady-state activity (V_{obs}) that is related to V_{red} and V_{ox} as indicated in Eq. 3.1.

$$V_{obs} = \left(V_{red} * V_{ox}\right) / \left(V_{red} + V_{ox}\right) \tag{3.1}$$

This expression (the "pool equation") was verified under a wide variety of input and output rates and establishes that CoQ distributes electrons randomly among the CoQ-reducing flavin dehydrogenases and the bc_1 complexes, behaving indeed as a laterally diffusing pool of molecules in a variety of systems (previously discussed by Lenaz and Genova 2007, 2009b).

Further evidence was provided by the characteristic effect of changing V_{red} or V_{ox} on inhibitor titration curves (Kröger and Klingenberg 1973b). In the case of the CoQ pool and titration of C_{III} by antimycin, V_{obs} is related to V_{red} and to V_{ox} modified by a factor *x*, the fraction of C_{III} inhibited by antimycin (Eq. 3.2).

$$V_{obs} = \left[V_{red} * V_{ox}\left(1-x\right)\right] / \left[V_{red} + V_{ox}\left(1-x\right)\right] \tag{3.2}$$

According to this concept, pool behaviour is characterized by a convex hyperbolic relationship between the integrated V_{obs} rate and the inhibitor concentration.

However, it is worth mentioning that most available data on CoQ pool concern succinate oxidation in submitochondrial particles, whereas fewer data are available for NADH oxidation mediated by C_I. On the other hand in mitochondrial systems the rate of C_I activity is strongly underestimated, due to the properties of CoQ analogues used as acceptors (Fato et al. 1996) so that the pool equation is not directly applicable. As a consequence of this observation any calculations based on absolute values of NADH-CoQ reductase activity are to be taken with extreme caution. A detailed analysis of the possible errors concerning the interpretation of the pool equation can be found in Lenaz and Genova (2007, 2009b). Deviations from pool behaviour of CoQ were also described in the past, raising some doubts on its universal validity (Ragan and Cottingham 1985; Gutman 1985). As a conclusion we may state with some certainty that, in beef heart mitochondria, succinate oxidation exhibits pool behaviour, indicating the presence of CoQ as a diffusible intermediate between C_{II} and C_{III}; on the other hand, the same statement for NADH oxidation is supported by less clear-cut evidence.

3.5.1 Interaction Between NADH and Succinate Pathways

Several reports have suggested that the CoQ pool is not homogeneous; even Kröger and Klingenberg (1973a) observed that a residual aliquot of CoQ cannot be reduced either by NADH or by succinate; later, Jørgensen et al. (1985) noticed that three pools of ubiquinone appeared to be present in heart mitochondria: a metabolically inactive pool consisting of reduced as well as oxidized ubiquinone, a pool coupled to oxidation of added (cytoplasmic) NADH, and the well-known pool coupled to citric acid cycle oxidations. The latter pool, however, could not distinguish NADH-dependent from succinate-dependent reduction of CoQ.

Benard et al. (2008) described the existence of three different pools of CoQ during succinate-dependent steady-state respiration in rat liver and muscle mitochondria: one pool is directly utilised, another (approx. 8% in muscle and 23% in liver) is mobilized as a reserve in case of a perturbation to maintain the energy fluxes at normal values (e.g. due to inhibition of the respiratory complexes or in case of mitochondrial diseases), and a third one (approx. 79% in muscle and 21% in liver) cannot be mobilized at all.

The effect on respiration of the simultaneous addition of NADH and succinate was examined in an early study (Gutman and Silman 1972) showing incomplete additivity and partial mutual competition between succinate oxidation and NADH oxidation; this effect was considered unlikely to be due to competition of the dehydrogenases for CoQ; also on the basis of other observations (Gutman et al. 1971), a compartmentalization of the CoQ pool was suggested, in which two sub-pools were able to partially interact through a spill-over diffusion-mediated mechanism.

The non-homogeneity of the ubiquinone pool with respect to succinate and NADH oxidation may be interpreted today in terms of compartmentalization of CoQ within SCs $I_1III_2IV_n$, besides the free CoQ pool used for connecting C_{II} and C_{III}.

Some investigations addressed the problem of the interaction between different substrates and their oxidation through the entire respiratory chain to oxygen, although the purpose was often not directed to understanding the state of homogeneity of the CoQ pool.

Early studies in Singer's laboratory (Ringler and Singer 1959), performed when the function of CoQ was not yet understood, examined the effect of the simultaneous addition of succinate and glycerol-3-phosphate on oxidation by either O_2 or cytochrome c in pig brain mitochondria. The activities were not at all additive, despite the fact that the respective dehydrogenases did not interfere with one another. The authors suggested that these activities followed the same path in the respiratory chain. Retrospective analysis indicates that succinate and glycerol-3-phosphate oxidation use the same CoQ pool. On the other hand a later study (Jackman and Willis 1996) showed that the combined addition of pyruvate plus malate and glycerol-3-phosphate induced an almost completely additive respiratory activity in mitochondria from rabbit gracilis skeletal muscle, suggesting that the NAD-linked pathway may be separate from the succinate pathway.

Lapuente-Brun et al. (2013) demonstrated that the physical assembly between C_I and C_{III} determines a preferential pathway for electrons mediated by a dedicated subset of CoQ molecules. According to their results, this compartmentalization prevents significant cross talk between NADH oxidation (C_I-dependent) and succinate oxidation (dependent on C_{II}) or other flavoenzyme-dependent oxidations. Those C_{III} molecules that physically interact with C_I in the formation of SCs are also exclusively dedicated to NADH oxidation while those C_{III} molecules that are not bound to C_I are mainly responsible for oxidation of succinate and other substrates using the free CoQ pool. Interestingly, when a partial loss of C_{III} occurs, preservation of the association between C_I and C_{III} is preferred to the free state of the complexes. In this situation NADH oxidation catalysed by the SC I+III is preferentially maintained despite the risk of compromising the oxidation of FAD-linked substrates (Lapuente-Brun et al. 2013).

The already quoted paper by Blaza et al. (2014) showed that the steady-state rates of aerobic NADH and succinate oxidation were not additive in bovine heart submitochondrial particles; moreover, the extents of cytochromes reduction (i.e. b_H, b_L, c and c_1) in the same cyanide-inhibited particles were similar if the reductant was either NADH or succinate or a mixture of the two substrates. Blaza et al. interpreted the results as a demonstration that a single homogeneous pool of CoQ molecules exists that receives electrons indifferently from C_I and C_{II} and that, consequently, free CIII and SC-bound CIII are able to equally receive electrons from the CoQ pool. We cannot however discriminate on the basis of these mere data whether this effect is due to the homogeneity of the CoQ-pool or to the homogeneity of the cytochrome c pool, since electron transfer to oxygen as under Blaza's experimental conditions comprises also the steps of cytochrome c and C_{IV} reduction and re-oxidation.

It is likely that cytochrome c be rate-limiting, thus provoking a bottleneck step both in NADH oxidation and in succinate oxidation, with electrons from both substrates competing for the same pool of cytochrome c and C_{IV} molecules in free form. Indeed most cytochrome c and C_{IV} are free in BHM, with only a small portion of C_{IV} forming the respirasome I_1III_2IV, and are able to receive electrons from any C_{III}, as demonstrated by our flux control analysis data (Bianchi et al. 2004) that showed no channelling in the cytochrome c region (see previous sections).

However, when we performed a study of succinate and NADH oxidation by exogenous cytochrome c as final electron acceptor under saturating conditions in KCN-inhibited BHM at steady-state, the electron transfer pathway for both oxidation reactions was shortened by including only CoQ and C_{III} as redox partners (Lenaz et al. 2016). Under such condition, the NADH and succinate pathways became clearly kinetically distinguishable because of their different CoQ compartments: we found that NADH- and succinate-cytochrome c oxidoreductase activity are additive and close to the theoretical summation of the two activities (Table 3.2).

The additivity was progressively lost by inhibitor titration with mucidin (data not shown), suggesting that the increased number of inactive C_{III} units forced CoQ reduced by succinate to access also the residual active units bound to CI in the SC (see also Sect. 3.6.1.1.).

Table 3.2 NADH and succinate oxidation by exogenous cytochrome c in cyanide-inhibited BHM

Substrate as electron donor	Rate of cytochrome c reduction μmoles/min/mg protein
NADH	0.356 ± 0.031 (5)
Succinate	0.591 ± 0.060 (5)
NADH+succinate	0.883 ± 0.071 (5)

Cytochrome c reductase activity was assayed spectrophotometrically in frozen and thawed BHM, as described in Lenaz et al. (2016). Data are mean values ± standard deviation of separate experiments, as indicated by the numbers in round brackets

3.5.2 *The Case of Reverse Electron Transfer*

Gutman (1985) investigated the properties of the NADH and succinate oxidation in submitochondrial particles in relation to the rates of energy-dependent reverse electron transfer from succinate to NAD^+ and of forward electron transfer from NADH to fumarate, concluding that "*the electron flux from succinate dehydrogenase to oxygen (forward electron transfer towards Complex III) or to NADH dehydrogenase (reverse electron transfer) employs the same carrier and is controlled by the same reaction*" whereas "*the electron transfer from NADH to oxygen does not share the same pathway through which electrons flow in the NADH-fumarate reductase*". In other words, C_I and C_{II} are linked by a different pathway with respect to C_I and C_{III}.

This observation poses the puzzling question (Lenaz and Genova 2009b) about how ubiquinol reduced in the membrane pool by C_{II} interacts with the CoQ binding site in C_I, since C_I is totally engaged in the SC. The same dilemma applies to the NADH-fumarate reductase activity that also involves interaction of C_I and C_{II}. In view of the recent progress in the knowledge of the detailed atomic structure of C_I (Efremov and Sazanov 2011; Baradaran et al. 2013), the previous view that two different routes may exist for forward and reverse electron transfer within C_I (Grivennikova et al. 2003) is no longer tenable as such, unless we consider two different conformations, of which the one present during reverse electron transfer makes the CoQ site more accessible to the pool.

It must be noted that the ATP-driven reverse electron transfer from succinate to NAD^+ occurs in the presence of a high mitochondrial transmembrane protonmotive force that, according to Piccoli et al. (2006), might be the physiological signal causing the structural reorganization of the respiratory complexes. The model hypothesis suggests that the SC $I_1III_2IV_n$ would dissociate its constituting complexes under high $\Delta\mu_{H+}$ condition, and this would no longer limit the access from the CoQ pool to the binding site in C_I.

This model is apparently incompatible with the observation reported by Gutman (1985) that NADH-fumarate reductase, that occurs at low membrane potential, shares the same pathway of the reverse reduction of NAD^+ by succinate. We have to keep in mind, however, that the rate of NADH-fumarate reductase is one order of magnitude lower than the rate of NADH-cytochrome c reductase and that of NAD reduction by succinate, therefore it might well be within the time range of a dynamic equilibrium between CoQ in the SC and CoQ in the membrane pool (cf. Sect. 3.6.1).

The most likely explanation, however, is that the pathways linking C_{II} and C_I during either direct or reverse electron transfer may require the interaction of the CoQ pool with C_I within the SC and that C_I in the SC is somehow accessible to the CoQ pool. This accessibility, however, would be no proof against the existence of channelling between C_I and C_{III}. Our conclusions are consistent with very recent observations by the group of Sazanov (Letts et al. 2016) showing accessibility of the CoQ-binding cavities of the mammalian respirasome that are open to the membrane and possibly allow free exchange with the CoQ pool.

3.6 What Is the Function of the Coenzyme Q Pool?

Given the considerations of the previous sections, there is no doubt that a mobile pool of CoQ in the inner mitochondrial membrane coexists with protein-bound CoQ. Is this pool just a reservoir of an excess of CoQ molecules without a specific function or is the pool necessary for functioning of the respiratory chain and/or for additional functions?

3.6.1 Dissociation Equilibrium of Bound CoQ

3.6.1.1 Free and Bound CoQ Equilibrate

As previously described, C_I is almost totally associated in a SC with C_{III}, and CoQ channelling is likely to occur in the lipid boundary comprised between the two complexes. However, this does not exclude that free CoQ in the pool is also necessary for proper channelling in the SC. In fact, the bound inter-complex quinone that allows electron flow directly from C_I to C_{III} is in dissociation equilibrium with the CoQ pool, so that the amount of bound CoQ, at steady state, would be dictated by the size of the pool: the existence of this equilibrium is suggested by the saturation kinetics for total ubiquinone exhibited by the integrated activity of C_I and C_{III} (Estornell et al. 1992) and by the decrease of respiratory activities in mitochondria fused with phospholipids with subsequent dilution of the CoQ pool (Schneider et al. 1982). To be in agreement with the experimental observations in favour of channelling, this proposition requires that the dissociation rate constants (k_{off}) of bound CoQ be considerably slower than the rates of inter-complex electron transfer via the same bound quinone molecules. Several observations reported in the previous sections indicate that such an equilibrium may be kinetically operative under some physiological conditions (e.g. concomitant NADH and succinate oxidation, energy-driven reverse electron transfer).

By this way, free CoQ acts as a reservoir for binding to the SC $I_1III_2IV_n$; in addition, free CoQ may be a reservoir for other functions believed to require CoQ binding to specific proteins, such as uncoupling proteins (Echtay et al. 2000) and the permeability transition pore (Walter et al. 2002).

3.6.1.2 May the CoQ Pool Operate in NADH Oxidation?

A different question is whether electron transfer between C_I and C_{III} can occur via the CoQ pool in absence of SC organization. Analysis of the literature does not offer clear-cut examples of electron flow between C_I and C_{III} in mitochondrial membranes mediated with certainty by the CoQ pool. Studies of respiration in pathological conditions (Van Raam et al. 2008; Rosca et al. 2008) showed that electron transfer in absence of SC organization is lost even if activity of the individual complexes is normal. Early reconstitution studies, however, had indicated that electron transfer is possible in both modes: the association of C_I with C_{III} (Ragan and Heron 1978; Heron et al. 1978) allows both channelling (electron transfer stoichiometric with the percentage amount of C_{III} associated to C_I) and CoQ pool behaviour (hyperbolic relation). In our reconstitution studies of C_I and C_{III} in phospholipid vesicles (Genova et al. 2008), NADH-cytochrome c reductase activity follows pool behaviour at protein dilutions with phospholipids higher than 1:10, whereas at lower dilution pool behaviour is not effective any more. In a proteoliposome system where C_I was reconstituted together with an alternative oxidase and CoQ_{10}, Jones et al. (2016) found high rates of NADH oxidation in absence of a C_I-containing SC, which demonstrates that CoQ_{10} was able to shuttle electrons from C_I by following pool behaviour.

3.6.2 Electron Transfer Between Individual Complexes Not Involved in Supercomplex Organization

The CoQ pool is required for electron transfer from C_{II} to C_{III}: indeed C_{II} kinetically follows pool behaviour after extraction and reconstitution (Kröger and Klingenberg 1973a, b) and in intact mitochondria (Stoner 1984) in accordance with the lack of SCs found by both BN-PAGE and flux control analysis. Since no clear association was demonstrated between C_I and C_{II} and most authors agree that C_{II} is not a significant part of SCs, also reverse electron transfer from succinate to NAD^+, involving sequential interaction of C_{II} and C_I by means of CoQ, must take place by collisional interactions in the CoQ pool. This observation consequently poses the puzzling question (Lenaz and Genova 2007) of whether and how ubiquinol produced by C_{II} can interact from the pool with the CoQ site in C_I at a rate compatible with the steady state kinetics of reverse electron transfer if all or most C_I units are associated with C_{III}, and the interaction of CoQ in the pool with the quinone-binding site in common between the latter two enzymes is necessarily slow. This question has been addressed in the previous sections of this chapter.

Furthermore, other enzymes such as glycerol-3-phosphate dehydrogenase, ETF dehydrogenase, dihydroorotate dehydrogenase, choline dehydrogenase, sulphide dehydrogenase, that are likely to be in minor amounts and strongly rate-limiting in integrated electron transfer, can probably feed electrons to the respiratory chain by interaction through the CoQ pool (Lenaz and Genova 2010). A study addressed to this problem (Rauchová et al. 1997) demonstrated that in brown adipose tissue (BAT) mitochondria the inhibition curve of glycerol phosphate-cytochrome c reductase is sigmoidal in the presence of myxothiazol and antimycin, suggesting the presence of a homogeneous CoQ pool between glycerol phosphate dehydrogenase (mtGPDH) and C_{III}. More recently, it was shown that the delivery of electrons from mtGPDH to C_{III} in human neutrophil mitochondria takes place in the absence of SC organisation and of NAD-linked respiration (Van Raam et al. 2008), in line with the notion that mtGPDH operates in mitochondria through the CoQ pool. Preliminary studies by BN-PAGE (M.L. Genova and H. Rauchova, unpublished) show that mtGPDH does not appear linked to any of the respiratory complexes. Accordingly, Mráček et al. (2014) demonstrated that mtGPDH associates into homooligomers (presumably as dimer, trimer and tetramer of this rather hydrophobic dehydrogenase) as well as high molecular weight SCs of more than 1000 kDa of yet unknown composition, better evidenced by clear-native-PAGE than by BN-PAGE, but none of them associated with C_I, C_{III} or C_{IV} (i.e. OXPHOS complexes that may share common electron transfer pathway with mGPDH and would therefore make kinetic sense). The mtGPDH homoligomers are endowed with in-gel activity; it is quite plausible that they represent native, though relatively labile (i.e. weak electrostatic interactions that are easily dissociated after Coomassie dye addition), forms of mtGPDH in the membrane of BAT mitochondria.

Schönfeld et al. (2010) observed that oxidation of carnitine esters of medium- and long-chain fatty acids by rat heart mitochondria is not accompanied by reverse electron transfer to NAD^+ through C_I although it produces the same or higher energization of mitochondria as compared to succinate oxidation by C_{II}. No association could be found by BN-PAGE analysis between C_I or other OXPHOS complexes and the electron transferring flavoprotein (ETF) that participates in fatty acid oxidation. It must be remarked, however, that membrane-bound ETF dehydrogenase was not included in Schönfeld's study. The scheme in Fig. 3.5 depicts the current view of the respiratory chain in terms of a mixed model of collisional interactions and channelling in SCs.

3.6.3 *Saturation Kinetics: Is CoQ Mitochondrial Concentration Saturating for Electron Transfer?*

In 1959, Crane et al. first demonstrated that succinate oxidation is lost by extracting bovine heart mitochondria with acetone, but the activity could be recovered by adding the newly discovered CoQ_{10} (Crane et al. 1957), disputing in such way the

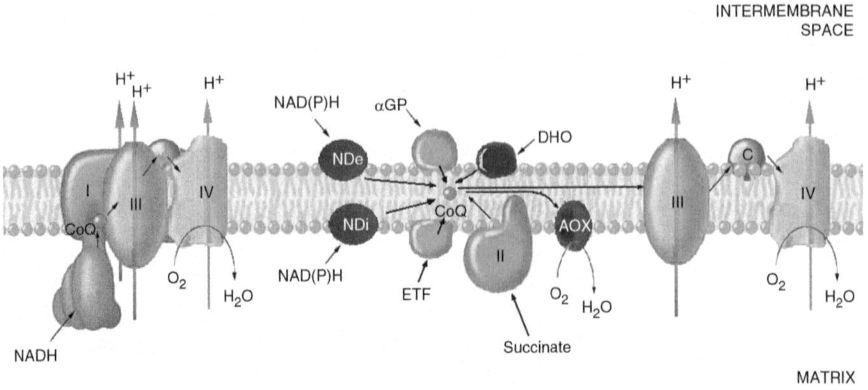

Fig. 3.5 A schematic drawing of the respiratory chain depicting the protein complexes and their substrates in relation with the inner mitochondrial membrane. Complex I is depicted as being totally part of a I-III2-IV supercomplex, whereas all other complexes are also shown in free form. *I* NADH:ubiquinone oxidoreductase, *II* succinate:ubiquinone oxidoreductase, *III* ubiquinol: cytochrome c oxidoreductase, *IV* cytochrome oxidase, *NDi and NDe* internal and external alternative NAD(P)H dehydrogenases, *AOX* alternative oxidase, *αGP* glycerol-3-phosphate, *ETF* electron transfer flavoprotein, *DHO* dihydroorotate, *CoQ* oenzyme Q, C cytochrome c

suggestion that CoQ is located in a side pathway with respect to the main stream of electron transfer (Redfearn and Pumphrey 1960). Unfortunately a role of CoQ in NADH oxidation could not be tested, because acetone treatment damages C_I activity. The problem was solved in Green's laboratory by Szarkowska (1966) who demonstrated in lyophilized and pentane-extracted mitochondria that both NADH and succinate oxidation are lost by CoQ extraction however can be largely restored (80% and 90% of the original activity, respectively) by addition of CoQ_{10}, mitochondrial phospholipids and cytochrome c (Fig. 3.6). Relatively large amounts of CoQ_{10} were required for maximal restoration of NADH oxidase activity as compared to the amounts of CoQ (about 5 nmoles/mg protein) normally present in mitochondria (see below for more data from Estornell et al. 1992). Notably, when added to lyophilized, but unextracted, mitochondria, CoQ_{10} and phospholipids had no stimulatory effect on the oxidation of either substrate.

In their early extraction–reconstitution studies, Lenaz et al. (1968) first reported that succinate oxidase is not very specific in the structural and steric requirements for the isoprenoid side chain, whereas NADH oxidase is rather specific, being reactivated only by CoQ homologs having long isoprenoid side chains (>6 units). Since CoQ reduced by either enzyme is re-oxidised by C_{III}, the specific requirements of NADH oxidation must be referred to properties in the C_I active site. Accordingly, CoQ homologs having short isoprenoid chains (like CoQ_2 and CoQ_3) inhibit NADH oxidation competitively with long-chain homologs (Landi et al. 1984; Fato et al. 1996). According to Degli Esposti (1998) CoQ_2 behaves as a type C inhibitor, i.e. an antagonist of the formation or release of the product ubiquinol.

The difference in specificity of the isoprenoid side chain of CoQ in restoring NADH oxidase and succinate oxidase activities of CoQ-depleted mitochondria led

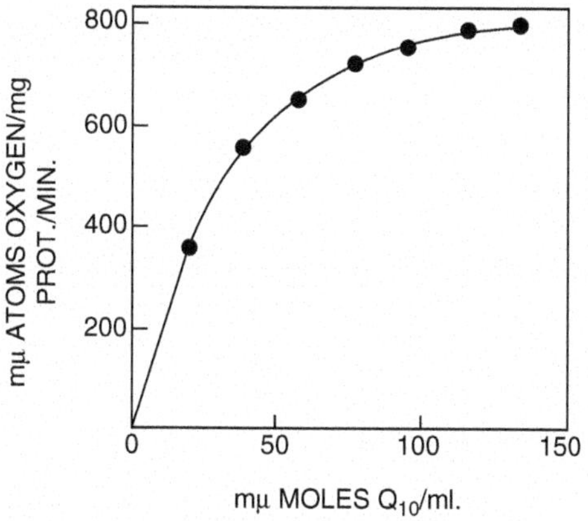

Fig. 3.6 NADH-oxidase activity as a function of CoQ_{10} concentration in a lyophilized, pentane-extracted preparation of BHM. The experimental conditions included mitochondria 0.25–0.28 mg protein/ml, cytochrome c 33 μg/ml, mitochondrial phospholipids 342 μg/ml and CoQ_{10} added as a solution in absolute ethanol 33 μl/ml. (Reprinted from Szarkowska (1966), Copyright 1966, with permission from Elsevier)

to the formulation of the existence of two sterically different sites for CoQ in NADH CoQ reductase and succinate CoQ reductase (Lenaz et al. 1968), having different sensitivities to the lipoidal and steric nature of the isoprenoid side chain, as also indicated by the different specificity of the two enzymes for the quinones as acceptors and by the different sensitivity to CoQ competitive inhibitors (Tan et al. 1993).

It was subsequently found that the saturated chain analog DB was as active as CoQ_{10} in restoring NADH oxidation (Lenaz et al. 1997), in agreement with its high acceptor activity (Fato et al. 1996). The hydroxydecyl analog idebenone, used in clinics to correct respiratory chain deficiencies (Mashima et al. 1992) and active in restoring succinate oxidation (Imada et al. 1989), is however a potent respiratory chain inhibitor at the level of C_I (Degli Esposti et al. 1996), making its therapeutic use of questionable efficacy.

In previous studies we have discussed the relation existing between rate of respiration and CoQ membrane concentration (Q_t) (Lenaz and Fato 1986); the rate is hyperbolically related to Q_t and maximal turnovers of electron transfer are attained only at Q_t saturating both rates of reduction and oxidation of the quinone pool. The "K_m" derived by the complex equation describing this composite system (Lenaz and Fato 1986; Ragan and Cottingham 1985) is a poised function of V_{max} and dissociation constants for CoQ of the complexes involved; this "K_m" can be therefore varying with rate changes of the complexes linked by the CoQ-pool, but is anyway an important parameter, in that it is operationally described as the total CoQ concentration (Q_t) yielding half-maximal velocity of integrated electron transfer.

The relation between electron transfer rate and CoQ concentration was seen for NADH and succinate oxidation in reconstituted systems and in phospholipid-enriched mitochondria (Estornell et al. 1992; Schneider et al. 1982; Parenti Castelli et al. 1987). Direct titrations of CoQ-depleted mitochondria reconstituted with different CoQ supplements yielded a "K_m" of NADH oxidation for Q_t in the range of 2–5 nmol/mg mitochondrial protein (Estornell et al. 1992), corresponding to a Q_t value of 4–10 mM in the lipid bilayer. A K_m value of similar order of magnitude for CoQ_{10} in NADH oxidation was reported by Jones et al. (2016) in a proteoliposome system in which C_I was totally rate-limiting and the CoQ pool fully oxidized. A puzzling observation is that the K_m for CoQ_{10} of NADH-cytochrome c reductase is much higher than that of succinate-cytochrome c reductase (Estornell et al. 1992). Analysis of the literature shows that the physiological CoQ content of several types of mitochondria (Battino et al. 1990) is in the range of the K_m for NADH oxidation, and therefore not saturating for this activity.

It has been demonstrated that incubation of beef heart submitochondrial particles in a CoQ_{10} solution leads to incorporation of CoQ_{10} in their membranes (Lenaz et al. 1994). The same authors found that kinetic saturation with CoQ_{10} could not be achieved because of the intrinsic insolubility of the molecule, thus concluding that the upper limit rate of electron transfer from NADH is a function of CoQ_{10} solubility in the membrane phospholipids. Nevertheless we could demonstrate that a soluble formulation of CoQ_{10} (Qter®) can enter mitochondria and enhance respiratory activity when administered to cultured cells (Bergamini et al. 2012). We measured the cellular and mitochondrial ubiquinone content in two cell lines (T67 and H9c2) after supplementation with Qter® and with native CoQ_{10}. Our results show that the water soluble formulation is more efficient in increasing ubiquinone levels in the cells' mitochondria. We have evaluated the bioenergetics effect of ubiquinone treatment, demonstrating that intracellular CoQ_{10} content after Qter® supplementation positively correlates with an improved mitochondrial functionality (increased oxygen consumption rate, transmembrane potential, ATP synthesis) and resistance to oxidative stress.

Attempts to raise CoQ concentration and respiratory activity in mitochondria by exogenous supplementation of the quinone in humans have been successful in some cases of mitochondrial genetic diseases characterized by CoQ deficiency, but the studies have shown substantial difficulties due to pharmacokinetic constraints preventing efficient uptake by cells and mitochondria with requirement of extremely high dosage (Quinzii and Hirano 2010). Rosenfeldt et al. (2005) used CoQ_{10} in cardiac surgery patients receiving 300 mg/day orally of CoQ_{10} dispensed in soy bean oil demonstrating a four-fold increase in serum concentration of CoQ_{10}, and a 2.5-fold increase in of CoQ_{10} in atrial myocardium which included a 2.4-fold increase of CoQ_{10} in atrial mitochondria. The coupled respiration was low in the mitochondria from CoQ_{10}-treated patients, however the ADP:O ratios were significantly higher than in untreated patients. It should be borne in mind that there are multiple rate-limiting steps in coupled respiration (Moreno-Sanchez et al. 1991), so that the extent of CoQ_{10} deficiency and its correction by exogenous quinone may not be apparent if a rate-limiting step involves a reaction where CoQ is not involved.

The manipulation of mouse mutants defective in CoQ biosynthesis provides a means of evaluating CoQ redox function in vivo, establishing a relation between residual CoQ content and respiratory activity (Wang and Hekimi 2016). Interestingly, this relation seems to be organ-dependent. Wang and Hekimi (2016) have been able to rescue CoQ deficiency in mice KO for the enzyme MCLK1/COQ7, which is responsible for CoQ hydroxylation in 6-position, by administering 2,4-dihydroxybenzoate (DHB) as biosynthetic precursor that is already hydroxylated in the becoming C6 position of the quinone ring. Contrary to that, supplementation with exogenous CoQ_{10} was almost ineffective.

3.7 Concluding Evidence About CoQ-Channelling

As we described in the previous sections of this paper, the major observations supporting the notion that SC association determines channelling in the CoQ region are the following: (a) rate advantage of NADH-cytochrome c reductase when C_I-containing SCs are present; (b) both C_I and C_{III} are rate-limiting as measured by flux control analysis; (c) evidence for two compartments of CoQ in experiments of competition of NADH and succinate oxidation.

One main point favouring controversy if not recognized is the dynamic character of CoQ bound within the SC, that is in dissociation equilibrium with the free pool of CoQ in the membrane. According to our hypothesis, some CoQ molecules are trapped in a lipid micro-domain within the SC and are channelled from C_I to C_{III} during electron transfer at steady state; however, when electron transfer is slow or blocked by an inhibitor, the relevance of CoQ dissociation from the SC to the pool becomes significant. This notion raises the puzzling question whether the mentioned low flux control coefficients found by Quarato et al. (2011) in state-4 (i.e. low-rate respiration) were due to the predominance of CoQ dissociation over CoQ channelling rather than to the physical disassembly of SC. To this respect, the plasticity model (Acín-Peréz and Enriquez 2014), would be a functional rather than structural feature of the respiratory chain, at least in the CoQ region. It is desirable that studies on the dissociation constants of CoQ from SC will give an answer to this scientific dilemma in the future. In addition, the rate of NADH-fumarate reductase (Gutman 1985; Genova and Lenaz 2014) may represent an indirect parameter of the dissociation turnover of CoQ from SC to the pool.

Does channelling occur under physiological conditions? A possible bottleneck around cytochrome c might induce interaction of the NADH pathway (SC) with the succinate pathway (pool); however, if a rate-limiting step is situated upstream, i.e. in or before the dehydrogenases, the reducing pressure of CoQ on SC may not be present and the two routes would take place independently. When oxidative metabolism proceeds mainly via the glycolytic pathway and the Krebs cycle, there is a prevalence of oxidation of NAD-linked substrates over succinate oxidation, allowing an undisturbed electron flux through SC I_1III_2. Strong oxidation of FAD-linked substrates as in fatty acid oxidation (where FAD-linked ETF dehydrogenase is most probably not forming a SC) might however induce strong interaction of the highly reduced CoQ pool with the SC. At the metabolic level these fast kinetic adjustments would be followed by gene expression changes of the level of individual complexes as postulated by Enriquez and collaborators (Lapuente-Brun et al. 2013).

3.8 Coenzyme Q and Proton Translocation

The involvement of CoQ in proton translocation to create the electrochemical potential used to drive ATP synthesis has been pioneered by Mitchell (1975) in its Q-cycle hypothesis. In the following years, the characteristics of energy-linked proton translocation by C_{III} have been documented by extensive experimentation (see Crofts 2004 and Cramer et al. 2011 for historical and structure perspectives). Involvement of CoQ in proton translocation by C_I was also postulated and several models, either cyclic or linear, were proposed, although only the recent crystallization of the complex has offered hints to elucidate the linkage of proton chemistry to electron transfer during NADH oxidation and quinone reduction.

3.8.1 *Complex III*

According to the Q-cycle model, recently refined by Crofts et al. (2013), proton-coupled oxidation of QH_2 at the outer positive site (called Q_o-site or site P) of the inner membrane occurs through a bifurcated reaction delivering the electrons to two different acceptor chains (Fig. 3.7C). The first electron reduces the high potential chain (Rieske iron-sulphur protein and hence cytochromes c_1 and c), and generates an intermediate semiquinone USQH in neutral form (see Fig. 3.1 for formulas). Removal of a H^+ from USQH yields the unstable anionic species, USQ^-, which is retained in the Q_o-site volume; the net result is release of two protons in the aqueous P-phase of the intermembrane space. The second electron, released from USQ^-, reduces the low potential chain consisting of hemes b_{566} (b_L) and b_{562} (b_H) of cyt b, which deliver the electron across the membrane to reduce ubiquinone (or semiquinone, see below) at the internal negative site (called site N or Q_i-site). The cycle is completed by oxidation of a second molecule of QH_2. In this second cycle the electron released by b_H reduces the semiquinone at the Q_o-site regenerating ubiquinol by taking up two protons from the matrix. Thus two turns of the catalytic cycle result in the release of four protons into the intermembrane space and consumption of two protons from the matrix side; the final stoichiometry of the Q-cycle is given by the following expressions:

$$2QH_2 + 2cyt.c\,Fe^{3+} + 2H^+_{(in)} \rightarrow Q + QH_2 + 2cyt.c\,Fe^{2+} + 4H^+_{(out)} \quad (3.3)$$

that is

$$QH_2 + 2cyt.c\,Fe^{3+} + 2H^+_{(in)} \rightarrow Q + 2cyt.c\,Fe^{2+} + 4H^+_{(out)} \quad (3.4)$$

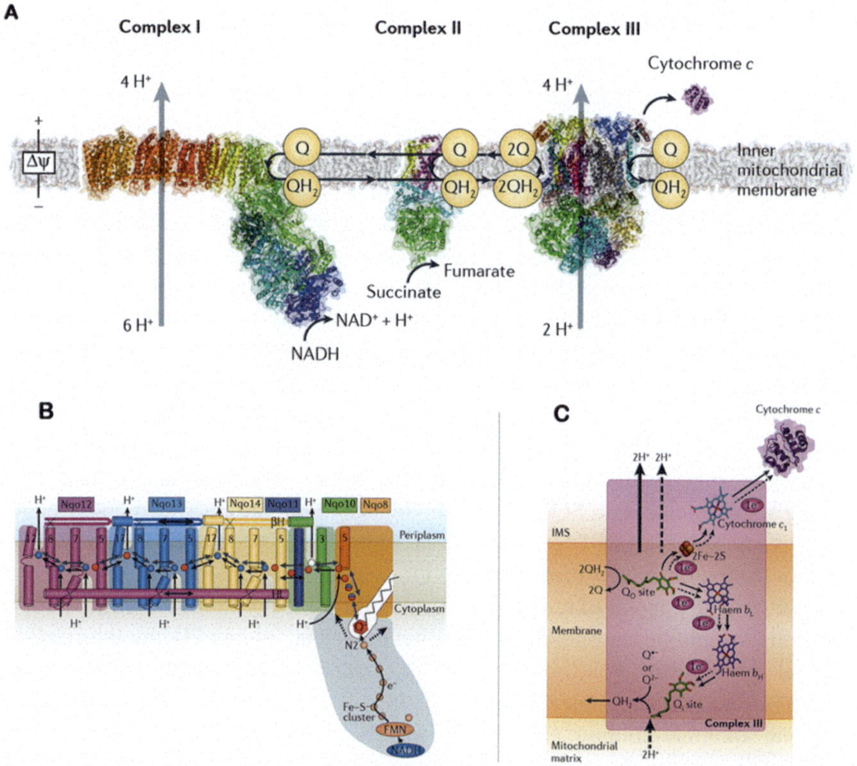

Fig. 3.7 Proton translocation in the CoQ region of the respiratory chain. (**a**) The structure of CoQ-dependent respiratory complexes is presented: Complex I from *Thermus thermophilus* (protein databank (PDB) identifier 4HEA; Baradaran et al. 2013), Complex II from *Sus scrofa* (PDB identifier 1ZOY; Sun et al. 2005), Complex III from *Bos taurus* (PDB identifier 1BGY; Iwata et al. 1998). (**b**) Proposed coupling mechanism of C_I; Upon electron transfer from the Fe–S cluster N2, negatively charged quinone (or charged residues nearby) initiates a cascade of conformational changes, propagating from the E-channel (at Nqo8, Nqo10 and Nqo11) to the antiporters via the central axis (indicated by grey arrows) comprising charged and polar residues that are located around flexible breaks in key transmembrane helices (TMHs). Cluster N2-driven shifts (dashed arrows) of Nqo4 and Nqo6 helices33 (not shown) are likely to assist overall conformational changes. Helix HL and the βH element help to coordinate conformational changes by linking discontinuous TMHs between the antiporters. Key charged residues can be protonated from the cytoplasm through several possible pathways, including inter-subunit transfer (indicated by black arrows). Following the reduction of CoQ and completion of conformational changes, Lys or GluTM12 in the antiporters and Glu32 from Nqo11 in the E-channel each eject a proton into the periplasm. TMHs are numbered and key charged residues are indicated by red and blue circles. (**c**) Mechanism of C_{III}; electron transfer in the first step of the Q-cycle is shown by solid arrows; dashed arrows indicate the same steps with a second ubiquinol. The oxidation of two ubiquinol molecules at the Qo-site releases four protons into the IMS. Two protons are taken up from the matrix as ubiquinol at the Q_i-site is reduced. FMN, flavin mononucleotide; $\Delta\Psi$, membrane potential. (Reprinted from Sazanov (2015) by permission from Macmillan Publishers Ltd.: Nature Reviews Molecular Cell Biology, Copyright (2015))

3.8.2 Complex I

The redox centre N2 is the direct electron donor to bound ubiquinone (Ohnishi 1998) and this step is linked to proton translocation (Yano and Ohnishi 2001; Flemming et al. 2005), although the mechanism has been strongly debated (Friedrich 2001; Vinogradov 2001; Brandt et al. 2003; Ohnishi and Salerno 2005; Sherwood and Hirst 2006). Most models implied at least partly CoQ redox chemistry in the mechanism of H^+ transport. Nevertheless, since all redox groups in the enzyme appear to be located in the hydrophilic arm or at least at the interface with the hydrophobic arm, direct coupling mechanisms appear unlikely to be solely responsible for H^+ movements; this implies that the driving force for proton translocation must be transduced over a considerable distance to the actual pumping process in the membrane arm via conformational coupling (Zickermann et al. 2008).

Ohnishi and collaborators (2010) reported that two distinct EPR-detectable semiquinone species also play important roles in C_I. They were called SQ_{Nf} (fast relaxing semiquinone) and SQ_{Ns} (slow relaxing semiquinone). It was proposed that Q_{Nf} plays a role in a "direct" redox-driven proton pump, while Q_{Ns} triggers an "indirect" conformation-driven proton pump. Q_{Nf} and Q_{Ns} together serve as ($1e^-/2e^-$) converter, for the transfer of reducing equivalent to the CoQ-pool. According to this model, $2H^+$ would be transported by the direct mechanism and $2H^+$ by the conformational mechanism. Note however that the controversy about the existence of more than one quinone molecule in the C_I is still open and the idea of a second functional quinone-binding site, which has been suggested to be located in antiporter-like subunit Nqo14 (ND2), is not in agreement with structural and mutagenesis data indicating that all three antiporter-like subunits have similar roles in proton translocation (cf. Sazanov 2015 for extensive review and discussion about notable difference between the mitochondrial and bacterial structure around the quinone-binding site of C_I).

The recent X-ray structure of C_I allows a better understanding of the mechanism of H^+ translocation (Sazanov 2015). The bacterial L-shaped assembly consists of the alpha-helical model for the membrane domain, with 63 transmembrane helices, and the known structure of the hydrophilic domain. Each symmetry-related set of five helices in the antiporter-like subunits (Nqo12, Nqo13 and Nqo14) forms an apparent half-channel for proton translocation (Fig. 3.7). The architecture of the complex provides strong clues about the coupling mechanism: the most widely accepted model currently is probably the 'one-stroke one-site' model, which proposes that all four protons are translocated at once, driven by the redox chemistry of one bound CoQ molecule, which takes into account the known redox potentials of quinone reduction intermediates and the reversibility of the overall C_I reaction (Efremov and Sazanov 2012). In brief, upon electron transfer from N2, the negatively charged ubiquinol (or charged residues nearby that control its protonation) can interact electrostatically with key negatively charged residues in the quinone cavity and drive conformational changes in the E-channel (at Nqo8, Nqo10 and Nqo11). Cluster N2 could also contribute, as helices that directly contact Nqo8 move following N2 reduction. Conformational changes in the E-channel can be transmitted to the nearest antiporter-like subunit Nqo14 and propagated to the tip of the membrane domain of C_I through interactions of charged residues in neighboring subunits. As a result, key charged residues would be protonated and de-protonated, and access to the cytoplasm and periplasm gated, as required for the H^+ pumping cycle.

It is worth noting that the key role of the CoQ redox cycle in driving conformational changes is consistent with the reversibility of the overall reaction catalysed by C_I. This complex functions close to an equilibrium *in vivo* and, under conditions of high proton motive force and in the presence of a highly reduced CoQ pool, the reaction can be driven in the reverse direction so that QH_2 reduces NAD^+. In this case, conformational changes driven by high $\Delta\Psi$ can result in high affinity for QH_2 and in a low redox potential of bound QH_2 so that electron transfer can proceed in reverse towards FMN (Efremov and Sazanov 2012).

3.9 Coenzyme Q as Mitochondrial Antioxidant and Anti-apoptotic Agent

Besides being a redox component of the respiratory chain, physiological CoQ_{10} behaves as a strong antioxidant in its reduced form. Ubiquinol can prevent the initiation of lipid peroxidation or also break the chain reaction by reacting with lipid peroxide radicals (LOO·), thus generating ubisemiquinone and a non-radical lipid hydroperoxide as indicated in eq. 3.5 (Ernster and Dallner 1995):

$$QH_2 + LOO\cdot \rightarrow Q\cdot + LOOH + H^+ \tag{3.5}$$

Physiological ubisemiquinones having long side chains do not react with oxygen, except in C_{III} under the very special conditions of the Q-cycle. On the other hand, short chain ubiquinones, as CoQ_1 and also hydroxydecyl-ubiquinone (idebenone), a compound having some clinical application, have a pro-oxidant effect in C_I. Idebenone also inhibits C_I (Degli Esposti et al. 1996); despite these effects, idebenone does not appear to be prooxidant in vivo (cf. McDaniel et al. 2005; Duveau et al. 2010), presumably because of prevalence of its reduced form in the cells (Imada et al. 2008).

In mitochondria the reduced antioxidant form of CoQ is regenerated by the respiratory chain (cf. previous sections). However, several other enzymes also catalyse CoQ reduction to achieve its antioxidant reduced state in eukaryotic cells. NADH-cytochrome b_5 reductase can reduce CoQ through a one-electron reaction mechanism (Arroyo et al. 1998). The soluble enzyme NAD(P)H-quinone oxidoreductase 1 (NQO1, DT-diaphorase) can reduce quinones by a two-electron reaction and maintains the reduced state of CoQ_{10} *in vitro* (Beyer et al. 1996). A distinct cytosolic NADPH-CoQ reductase different from NQO1 has been also described (Takahashi et al. 1995).

Murphy and coworkers (Kelso et al. 2001) developed a new antioxidant, named MitoQ, a ubiquinone derivative targeted to mitochondria by covalent attachment to a lipophilic triphenylphosphonium cation through an aliphatic carbon chain. Due to the large mitochondrial membrane potential, the cation is accumulated within mitochondria inside cells, where the ubiquinone moiety inserts into the lipid bilayer and is reduced by the respiratory chain. The ubiquinol derivative thus formed is an effective antioxidant that prevents lipid peroxidation and protects mitochondria from oxidative damage. After detoxifying a reactive oxygen species, the ubiquinol moiety is regenerated by the respiratory chain enabling its antioxidant activity to be recycled. In cell culture studies (Kelso et al. 2001), the mitochondrially localized antioxidant protects mammalian cells from hydrogen peroxide-induced apoptosis. On the other hand, it is worth noting that MitoQs of different alkyl chain lengths, although specifically directed to mitochondria, are not able to restore electron transfer in CoQ-deficient mitochondria (James et al. 2005) and do not always exhibit antioxidant properties (Plecitá-Hlavatá et al. 2009).

Other studies (Yamamura et al. 2001; Papucci et al. 2003; Naderi et al. 2006) have confirmed a protective role of CoQ_{10} against apoptosis by showing inhibition of cell death independently of its antioxidant effect, presumably by inhibition of opening of the permeability transition pore (PTP), a high conductance protein channel located in the inner mitochondrial membrane (Bernardi and Forte 2007) which depolarizes the mitochondrion and leads to the release in the cytoplasm of proteins contained in the space between the two mitochondrial membranes, such as cytochrome c and other factors that trigger the process of programmed cell death (apoptosis). In fact quinones have been shown to exert a direct effect on PTP. Walter et al. (2000) found that three functional classes of quinone analogues could be defined in relation to PTP: (i) PTP inhibitors, as CoQ_0, CoQ_2, decylubiquinone; (ii) PTP inducers, as idebenone (2,3-dimethoxy-5-methyl-6-(10-hydroxydecyl)-1,4-benzoquinone); (iii) PTP-inactive quinones, that counteract the effects of both inhibitors and inducers, such as CoQ_1. The quinones modulate the PTP through a common binding site rather than through oxidation-reduction reactions. Occupancy of this site can modulate the PTP open-closed transitions, possibly through secondary changes of the Ca^{2+} binding affinity for the pore. PTP opening by its inducers led the cells to apoptosis (Devun et al. 2010). In these studies the effect of hydrophobic long-chain quinones could not be investigated, however an indirect study by Li et al. (2005) suggests that CoQ_{10} may be a PTP inhibitor. These authors exposed SHSY5Y neuroblastoma cells to neurotoxic β-amyloid peptides (Aβ) and oxygen/glucose deprivation (OGD) to investigate the neuroprotective effect of 10 μM CoQ_{10}. Pore opening and superoxide anion concentration were increased in the group $A\beta^+/OGD^+$ relative to control, and were attenuated to the control level when CoQ_{10} was administered, indirectly demonstrating that CoQ_{10} inhibits the opening of the pore besides reducing the concentration of superoxide anion.

Durán-Prado et al. (2014) also demonstrated that pretreatment with CoQ at physiological concentrations in human plasma after oral CoQ supplementation protects HUVEC endothelial cells from β-amyloid-induced injury by preventing cell necrosis and apoptosis and restoring their ability to proliferate, migrate and form tube-like structures *in vitro*, which is mirrored by a restoration of the cell metabolic profile to control levels. Moreover, CoQ reduced the influx of extracellular Ca^{2+}, and Ca^{2+} release from mitochondria caused by PTP opening after β-amyloid administration, in addition to decreasing $O_2^{\cdot -}$ and H_2O_2 levels.

Similar studies indicated a protective effect of CoQ_{10} on PTP opening in amitriptyline toxicity (Cordero et al. 2009) and in ischemia and reperfusion in the heart (Sahach et al. 2007). It is not clear in these studies whether the effect of CoQ_{10} on the transition pore is a direct one or is mediated by the antioxidant effect.

3.9.1 Transcriptional Effects of CoQ

Several studies have revealed effects of CoQ_{10} on gene expression (Linnane et al. 2002; Gorelick et al. 2004; Groneberg et al. 2005; Nohl et al. 2005; Schmelzer et al. 2009; Sohet et al. 2009). In theory, these effects might be mediated directly by, for example, interactions with a transcription factor. However, reactive oxygen species (ROS) are also potent inducers of gene expression. H_2O_2 has been identified as an activator of the pro-inflammatory nuclear transcription factor NFκB (Kaltschmidt et al. 1999). In view of the antioxidant properties of the reduced form of CoQ_{10} and the effective enzymatic conversion of oxidized CoQ_{10} into its reduced form, CoQ_{10} might mediate anti-inflammatory effects via gene expression.

Triggering of cells with lipopolysaccharide LPS induces downstream signalling cascades of the transcription factor NFκB, which in turn leads to the induction of inflammatory genes; CoQ_{10} downregulates LPS-inducible genes in the monocytic cell line THP 1, presumably due to its antioxidant impact on gene expression (Schmelzer and Döring 2010).

In another study (Park et al. 2009) caloric restriction inhibited age-related expression of five genes in heart and cerebellum; among dietary antioxidants, lipoic acid and CoQ_{10} were as effective as caloric restriction in the cerebellum.

Tian et al. (2014) reported that dietary supplementation with ubiquinol-10 also prevents age-related decreases in the expression of sirtuin gene family members, which results in the activation of PGC-1α (peroxisome proliferator-activated receptor γ coactivator 1α), a major factor that controls mitochondrial biogenesis and respiration, as well as superoxide dismutase-2 (SOD2) and isocitrate dehydrogenase-2 (IDH2), which are major mitochondrial antioxidant enzymes and protect against the symptoms of age-related diseases. Furthermore, these authors demonstrate that ubiquinol-10 may activate Sirt1 and PGC-1α gene expression in SAMP1 mice by increasing cyclic adenosine monophosphate (cAMP) levels that, in turn, activate cAMP response element-binding protein (CREB) and AMP-activated protein kinase (AMPK).

Similar effects at the transcription level are promoted by the mitochondria-direct derivative MitoQ (Chacko et al. 2010). Thus it is likely that all effects of CoQ at the genetic level may be mediated by its antioxidant effect.

References

Acín-Peréz R, Enriquez JA (2014) The function of respiratory supercomplexes: the plasticity model. Biochim Biophys Acta 1837:444–450

Acín-Peréz R, Fernández-Silva P, Peleato ML et al (2008) Respiratory active mitochondrial supercomplexes. Mol Cell 32:529–539

Althoff T, Mills DJ, Popot JL et al (2011) Arrangement of electron transport chain components in bovine mitochondrial supercomplex I1III2IV1. EMBO J 30:4652–4664

Arroyo A, Navarro F, Navas P et al (1998) Ubiquinol regeneration by plasma membrane ubiquinone reductase. Protoplasma 205:107–113

Baradaran R, Berrisford JM, Minhas GS et al (2013) Crystal structure of the entire respiratory complex I. Nature 494:443–448

Battino M, Fahmy T, Lenaz G (1986) Determination of the critical micelle concentration of short-chain ubiquinones in model systems. Biochim Biophys Acta 851:377–384

Battino M, Ferri E, Villa RF et al (1990) Natural distribution and occurrence of coenzyme Q homologues. Membr Biochem 9:179–190

Benard G, Faustin B, Galinier A et al (2008) Functional dynamic compartmentalization of respiratory chain intermediate substrates: implications for the control of energy production and mitochondrial diseases. Int J Biochem Cell Biol 40:1543–1554

Bergamini C, Moruzzi N, Sblendido A et al (2012) A water soluble CoQ_{10} formulation improves intracellular distribution and promotes mitochondrial respiration in cultured cells. PLoS One 7:e33712

Bernardi P, Forte M (2007) The mitochondrial permeability transition pore. Novartis Found Symp 287:157–164; discussion 164-169

Beyer RE, Segura-Aguilar J, Di Bernardo S et al (1996) The role of DT-diaphorase in the maintenance of the reduced antioxidant form of coenzyme Q in membrane systems. Proc Natl Acad Sci U S A 93:2528–2532

Bianchi C, Fato R, Genova ML et al (2003) Structural and functional organization of Complex I in the mitochondrial respiratory chain. Biofactors 18:3–9

Bianchi C, Genova ML, Parenti Castelli G et al (2004) The mitochondrial respiratory chain is partially organized in a supercomplex assembly: kinetic evidence using flux control analysis. J Biol Chem 279:36562–36569

Blaza JN, Serrelli R, Jones AJ et al (2014) Kinetic evidence against partitioning of the ubiquinone pool and the catalytic relevance of respiratory chain supercomplexes. Proc Natl Acad Sci U S A 111:15735–15740

Boumans H, Grivell LA, Berden JA (1998) The respiratory chain in yeast behaves as a single functional unit. J Biol Chem 273:4872–4877

Brandt U, Kerscher S, Drose S et al (2003) Proton pumping by NADH:ubiquinone oxidoreductase. A redox driven conformational change mechanism? FEBS Lett 545:9–17

Braun BS, Benbow U, Lloyd-Williams P et al (1986) Determination of partition coefficients of quinones by high-performance liquid chromatography. Methods Enzymol 125:119–129

Capaldi RA (1982) Arrangement of proteins in the mitochondrial inner membrane. Biochim Biophys Acta 694:292–306

Chacko BK, Reily C, Srivastava A et al (2010) Prevention of diabetic nephropathy in Ins2(+/)-(AkitaJ) mice by the mitochondria-targeted therapy MitoQ. Biochem J 432:9–19

Chance B, Williams GR (1955) A method for the localization of sites for oxidative phosphorylation. Nature 176:250–254

Cordero MD, Moreno-Fernández AM, Gomez-Skarmeta JL et al (2009) Coenzyme Q10 and alpha-tocopherol protect against amitriptyline toxicity. Toxicol Appl Pharmacol 235:329–337

Cramer WA, Hasan SS, Yamashita E (2011) The Q cycle of cytochrome bc complexes: a structure perspective. Biochim Biophys Acta 1807:788–802

Crane FL, Hatefi Y, Lester RL et al (1957) Isolation of a quinone from beef heart mitochondria. Biochim Biophys Acta 25:220–221

Crane FL, Widmer C, Lester RL et al (1959) Studies on the electron transport system. XV. Coenzyme Q (Q275) and the succinoxidase activity of the electron transport particle. Biochim Biophys Acta 31:476–489

Crofts AR (2004) The Q-cycle, a personal perspective. Photosynth Res 80:223–243

Crofts AR, Hong S, Wilson C et al (2013) The mechanism of ubihydroquinone oxidation at the Qo-site of the cytochrome bc1 complex. Biochim Biophys Acta 1827:1362–1377

Degli Esposti M (1998) Inhibitors of NADH-ubiquinone reductase: an overview. Biochim Biophys Acta 1364:222–235

Degli Esposti M, Ngo A, Ghelli A et al (1996) The interaction of Q analogs, particularly hydroxydecyl benzoquinone (idebenone), with the respiratory complexes of heart mitochondria. Arch Biochem Biophys 330:395–400

Devun F, Walter L, Belliere J et al (2010) Ubiquinone analogs: a mitochondrial permeability transition pore-dependent pathway to selective cell death. PLoS One 5:e11792

Di Bernardo S, Fato R, Casadio R et al (1998) A high diffusion coefficient for coenzyme Q10 might be related to a folded structure. FEBS Lett 426:77–80

Dudkina NV, Kudryashev M, Stahlberg H et al (2011) Interaction of complexes I, III, and IV within the bovine respirasome by single particle cryoelectron tomography. Proc Natl Acad Sci U S A 108:15196–15200

Durán-Prado M, Frontiñán J, Santiago-Mora R et al (2014) Coenzyme Q10 protects human endothelial cells from β-amyloid uptake and oxidative stress-induced injury. PLoS One 9:e109223

Duveau DY, Arce PM, Schoenfeld RA et al (2010) Synthesis and characterization of mitoQ and idebenone analogues as mediators of oxygen consumption in mitochondria. Bioorg Med Chem 18:6429–6441

Echtay KS, Winkler E, Klingenberg M (2000) Coenzyme Q is an obligatory cofactor for uncoupling protein function. Nature 408:609–613

Efremov RG, Sazanov LA (2011) Structure of the membrane domain of respiratory complex I. Nature 476:414–420

Efremov RG, Sazanov LA (2012) The coupling mechanism of respiratory complex I - a structural and evolutionary perspective. Biochim Biophys Acta 1817:1785–1795

Enriquez JA, Lenaz G (2014) Coenzyme Q and the respiratory chain: coenzyme Q pool and mito chondrial supercomplexes. Mol Syndromol 5:119–140

Ernster L, Dallner G (1995) Biochemical, physiological and medical aspects of ubiquinone function. Biochim Biophys Acta 1271:195–204

Estornell E, Fato R, Castelluccio C et al (1992) Saturation kinetics of coenzyme Q in NADH and succinate oxidation in beef heart mitochondria. FEBS Lett 311:107–109

Fato R, Battino M, Degli Esposti M et al (1986) Determination of partition and lateral diffusion coefficients of ubiquinones by fluorescence quenching of n-(9-anthroyloxy)stearic acids in phospholipid vesicles and mitochondrial membranes. Biochemistry 25:3378–3390

Fato R, Castelluccio C, Palmer G et al (1988) A simple method for the determination of the kinetic constants of membrane enzymes utilizing hydrophobic substrates: Ubiquinol cytochrome c reductase. Biochim Biophys Acta 932:216–222

Fato R, Estornell E, Di Bernardo S et al (1996) Steady- state kinetics of the reduction of coenzyme Q analogs by complex I (NADH:ubiquinone oxidoreductase) in bovine heart mitochondria and submitochondrial particles. Biochemistry 35:2705–2716

Flemming D, Stolpe S, Schneider D et al (2005) A possible role for iron-sulfur cluster N2 in proton translocation by the NADH: ubiquinone oxidoreductase (complex I). J Mol Microbiol Biotechnol 10:208–222

Friedrich T (2001) Complex I: a chimaera of a redox and conformation-driven proton pump? J Bioenerg Biomembr 33:169–177

Galassi VV, Arantes GM (2015) Partition, orientation and mobility of ubiquinones in a lipid bilayer. Biochim Biophys Acta 1847:1560–1573

Genova ML, Lenaz G (2011) New developments on the functions of coenzyme Q in mitochondria. Biofactors 37:330–354

Genova ML, Lenaz G (2013) A critical appraisal of the role of respiratory supercomplexes in mitochondria. Biol Chem 394:631–639

Genova ML, Lenaz G (2014) Functional role of mitochondrial respiratory supercomplexes. Biochim Biophys Acta 1837:427–443

Genova ML, Baracca A, Biondi A et al (2008) Is supercomplex organization of the respiratory chain required for optimal electron transfer activity? Biochim Biophys Acta 1777:740–746

Gorelick C, Lopez-Jones M, Goldberg GL et al (2004) Coenzyme Q10 and lipid-related gene induction in HeLa cells. Am J Obstet Gynecol 190:1432–1434

Green DE, Tzagoloff A (1966) The mitochondrial electron transfer chain. Arch Biochem Biophys 116:293–304

Grivennikova VG, Roth R, Zakharova NV et al (2003) The mitochondrial and prokaryotic proton-translocating NADH:ubiquinone oxidoreductases: similarities and dissimilarities of the quinone-junction sites. Biochim Biophys Acta 1607:79–90

Groneberg DA, Kindermann B, Althammer M et al (2005) Coenzyme Q10 affects expression of genes involved in cell signalling, metabolism and transport in human CaCo-2 cells. Int J Biochem Cell Biol 37:1208–1218

Gu J, Wu M, Guo R et al (2016) The architecture of the mammalian respirasome. Nature 537:639–643

Gunner MR, Madeo J, Zhu Z (2008) Modification of quinone electrochemistry by the proteins in the biological electron transfer chains: examples from photosynthetic reaction centers. J Bioenerg Biomembr 40:509–519

Gutman M (1985) Kinetic analysis of electron flux through the quinones in the mitochondrial system. In: Lenaz G (ed) Coenzyme Q. Wiley, Chichester, pp 215–234

Gutman M, Silman N (1972) Mutual inhibition between NADH oxidase and succinoxidase activities in respiring submitochondrial particles. FEBS Lett 26:207–210

Gutman M, Coles CJ, Singer TP et al (1971) On the functional organization of the respiratory chain at the dehydrogenase-coenzyme Q junction. Biochemistry 10:2036–2043

Hackenbrock CR, Chazotte B, Gupte SS (1986) The random collision model and a critical assessment of diffusion and collision in mitochondrial electron transport. J Bioenerg Biomembr 18:331–368

Hatefi Y, Haavik AG, Fowler LR et al (1962a) Studies on the electron transfer system. XLII. Reconstitution of the electron transfer system. J Biol Chem 237:2661–2669

Hatefi Y, Haavik AG, Griffiths DE (1962b) Studies on the electron transfer system. XL. Preparation and properties of mitochondrial DPNH-Coenzyme Q reductase. J Biol Chem 237:1676–1680

Heron C, Ragan CI, Trumpower BL (1978) The interaction between mitochondrial NADH- ubiquinone oxidoreductase and ubiquinol-cytochrome c oxidoreductase. Restoration of ubiquinone-pool behavior. Biochem J 174:791–800

Hochman J, Ferguson-Miller S, Schindler M (1985) Mobility in the mitochondrial electron transport chain. Biochemistry 24:2509–2516

Imada I, Fujita T, Sugiyama Y et al (1989) Effects of idebenone and related compounds on respiratory activities of brain mitochondria, and on lipid peroxidation of their membranes. Arch Gerontol Geriatr 8:323–341

Imada I, Sato EF, Kira Y et al (2008) Effect of CoQ homologues on reactive oxygen generation by mitochondria. Biofactors 32:41–48

Iwata S, Lee JW, Okada K et al (1998) Complete structure of the 11-subunit bovine mitochondrial cytochrome bc1 complex. Science 281:64–71

Jackman MR, Willis WT (1996) Characteristics of mitochondria isolated from type I and type IIb skeletal muscle. Am J Physiol 270:C673–C678

James AM, Cochemé HM, Smith RA et al (2005) Interactions of mitochondria-targeted and untargeted ubiquinones with the mitochondrial respiratory chain and reactive oxygen species. Implications for the use of exogenous ubiquinones as therapies and experimental tools. J Biol Chem 280:21295–21312

Joela H, Kasa S, Lehtovuori P et al (1997) EPR, ENDOR and TRIPLE resonance and MO studies on ubiquinones (Q-n): comparison of radical anions and cations of coenzymes Q-10 and Q-6 with the model compounds Q-2 and Q-0. Acta Chem Scand 51:233–241
Jones AJ, Blaza JN, Bridges HR et al (2016) A self-assembled respiratory chain that catalyzes NADH oxidation by ubiquinone-10 cycling between complex I and the alternative oxidase. Angew Chem Int Ed Engl 55:728–731
Jørgensen BM, Rasmussen HN, Rasmussen UF (1985) Ubiquinone reduction pattern in pigeon heart mitochondria. Identification of three distinct ubiquinone pools. Biochem J 229:621–629
Kaambre T, Chekulayev V, Shevchuk I et al (2012) Metabolic control analysis of cellular respiration in situ in intraoperational samples of human breast cancer. J Bioenerg Biomembr 44:539–558
Kaambre T, Chekulayev V, Shevchuk I et al (2013) Metabolic control analysis of respiration in human cancer tissue. Front Physiol 4:151(1–6)
Kaltschmidt B, Sparna T, Kaltschmidt C (1999) Activation of NF-kappa B by reactive oxygen intermediates in the nervous system. Antioxid Redox Signal 1:129–144
Kaurola P, Sharma V, Vonk A et al (2016) Distribution and dynamics of quinones in the lipid bilayer mimicking the inner membrane of mitochondria. Biochim Biophys Acta 1858:2116–2122
Kelso GF, Porteous CM, Coulter CV et al (2001) Selective targeting of a redox-active ubiquinone to mitochondria within cells: antioxidant and antiapoptotic properties. J Biol Chem 276:4588–4596
Kholodenko NB, Westerhoff HV (1993) Metabolic channelling and control of the flux. FEBS Lett 320:71–74
Kröger A, Klingenberg M (1973a) The kinetics of the redox reactions of ubiquinone related to the electron-transport activity in the respiratory chain. Eur J Biochem 34:358–368
Kröger A, Klingenberg M (1973b) Further evidence of the pool function of ubiquinone as derived from the inhibition of the electron transport by antimycin. Eur J Biochem 39:313–323
Landi L, Pasquali P, Cabrini L et al (1984) On the mechanism of inhibition of NADH oxidase by ubiquinone-3. J Bioenerg Biomembr 16:153–166
Lapuente-Brun E, Moreno-Loshuertos R, Acín-Peréz R et al (2013) Supercomplex assembly determines electron flux in themitochondrial electron transport chain. Science 340:1567–1570
Lass A, Sohal RS (1999) Comparisons of coenzyme Q bound to mitochondrial membrane proteins among different mammalian species. Free Radic Biol Med 27:220–226
Lehninger AL (1965) Bionergetics: the molecular basis of biological energy transformations. WA Benjamin Inc, New York
Lenaz G (1988) Role of mobility of redox components in the inner mitochondrial membrane. J Membrane Biol 104:193–209
Lenaz G (1998) Quinone specificity of complex I. Biochim Biophys Acta 1364:207–221
Lenaz G, Fato R (1986) Is ubiquinone diffusion rate-limiting for electron transfer? J Bioenerg Biomembr 18:369–401
Lenaz G, Genova ML (2007) Kinetics of integrated electron transfer in the mitochondrial respiratory chain: random collisions vs solid state electron channeling. Am J Physiol Cell Physiol 292:C1221–C1239
Lenaz G, Genova ML (2009a) Structural and functional organization of the mitochondrial respiratory chain: a dynamic super-assembly. Int J Biochem Cell Biol 41:1750–1772
Lenaz G, Genova ML (2009b) Mobility and function of Coenzyme Q (ubiquinone) in the mitochondrial respiratory chain. Biochim Biophys Acta 1787:563–573
Lenaz G, Genova ML (2010) Structure and organization of mitochondrial respiratory complexes: a new understanding of an old subject. Antioxid Redox Signal 12:961–1008
Lenaz G, Daves GD, Folkers K (1968) Organic structural specificity and sites of coenzyme Q in succinoxidase and DPNH-oxidase systems. Arch Biochem Biophys 123:539–550
Lenaz G, Fato R, Castelluccio C et al (1994) An updating of the biochemical function of coenzyme Q in mitochondria. Mol Aspects Med 15(Suppl):S29–S36

Lenaz G, Bovina C, Castelluccio C et al (1997) Mitochondrial complex I defects in aging. Mol Cell Biochem 174:329–333
Lenaz G, Fato R, Di Bernardo S et al (1999) Localization and mobility of coenzyme Q in lipid bilayers and membranes. Biofactors 9:87–93
Lenaz G, Baracca A, Barbero G et al (2010) Mitochondrial respiratory chain super-complex I-III in physiology and pathology. Biochim Biophys Acta 1797:633–640
Lenaz G, Tioli G, Falasca AI et al (2016) Complex I function in mitochondrial supercomplexes. Biochim Biophys Acta 1857:991–1000
Letts JA, Fiedorczuk K, Sazanov LA (2016) The architecture of respiratory supercomplexes. Nature 537:644–648
Letts JA, Fiedorczuk K, Degliesposti G, Skehel M, Sazanov LA (2019) Structures of respiratory rupercomplex I+III_2 reveal functional and conformational crosstalk. Mol Cell 75:1131–1146.e6
Li G, Zou LY, Cao CM et al (2005) Coenzyme Q10 protects SHSY5Y neuronal cells from beta amyloid toxicity and oxygen-glucose deprivation by inhibiting the opening of the mitochondrial permeability transition pore. Biofactors 25:97–107
Linnane AW, Kopsidas G, Zhang C et al (2002) Cellular redox activity of coenzyme Q10: effect of CoQ_{10} supplementation on human skeletal muscle. Free Radic Res 36:445–453
Maranzana E, Barbero G, Falasca AI et al (2013) Mitochondrial respiratory supercomplex association limits production of reactive oxygen species from Complex I. Antioxid Redox Signal 19:1469–1480
Mashima Y, Hiida Y, Oguchi Y (1992) Remission of Leber's hereditary optic neuropathy with idebenone. Lancet 340:368–369
McDaniel DH, Neudecker BA, Di Nardo JC et al (2005) Idebenone: a new antioxidant – Part I. Relative assessment of oxidative stress protection capacity compared to commonly known antioxidants. J Cosmet Dermatol 4:10–17
McKenzie M, Lazarou M, Thorburn DR et al (2006) Mitochondrial respiratory chain supercomplexes are destabilized in Barth syndrome patients. J Mol Biol 361:462–469
Mileykovskaya E, Dowhan W (2014) Cardiolipin-dependent formation of mitochondrial respiratory supercomplexes. Chem Phys Lipids 17:942–948
Mileykovskaya E, Penczek PA, Fang J et al (2012) Arrangement of the respiratory chain complexes in Saccharomyces cerevisiae supercomplex III2IV2 revealed by single particle cryo-electron microscopy (EM). J Biol Chem 287:23095–23103
Mitchell P (1975) The protonmotive Q cycle: a general formulation. FEBS Lett 59:137–139
Moreno-Sanchez S, Devars S, Lopez-Gomez F et al (1991) Distribution of control of oxidative phosphorylation in mitochondria oxidizing NAD-linked substrates. Biochim Biophys Acta 1060:284–292
Mráček T, Holzerová E, Drahota Z et al (2014) ROS generation and multiple forms of mammalian mitochondrial glycerol-3-phosphate dehydrogenase. Biochim Biophys Acta 1837:98–111
Naderi J, Somayajulu-Nitu M, Mukerji A et al (2006) Water-soluble formulation of Coenzyme Q10 inhibits Bax-induced destabilization of mitochondria in mammalian cells. Apoptosis 11:1359–1369
Nohl H, Rohr-Udilova N, Gille L et al (2005) Suppression of tumour-promoting factors in fat-induced colon carcinogenesis by the antioxidants caroverine and ubiquinone. Anticancer Res 25:2793–2800
Ohnishi T (1998) Iron-sulfur clusters/semiquinones in complex I. Biochim Biophys Acta 1364:186–206
Ohnishi T, Salerno JC (2005) Conformation-driven and semiquinone-gated proton-pump mechanism in the NADH-ubiquinone oxidoreductase (complex I). FEBS Lett 579:4555–4561
Ohnishi ST, Salerno JC, Ohnishi T (2010) Possible roles of two quinone molecules in direct and indirect proton pumps of bovine heart NADH-quinone oxidoreductase (complex I). Biochim Biophys Acta 1797:1891–1893
Ovàdi J (1991) Physiological significance of metabolic channelling. J Theor Biol 152:135–141

Ozawa T, Nishikimi M, Suzuki H et al (1987) Structure and assembly of mitochondrial electron-transfer complexes. In: Ozawa T, Papa S (eds) Bioenergetics: structure and function of energy-transducing systems. Japan Sci Soc Press, Tokyo, pp 101–119
Papucci L, Schiavone N, Witort E et al (2003) Coenzyme Q10 prevents apoptosis by inhibiting mitochondrial depolarization independently of its free radical scavenging property. J Biol Chem 278:28220–28228
Parenti Castelli G, Fato R, Castelluccio C et al (1987) Kinetic studies on the pool function of ubiquinone in mitochondrial systems. Chem Scr 27:161–166
Park SK, Kim K, Page GP et al (2009) Gene expression profiling of aging in multiple mouse strains: identification of aging biomarkers and impact of dietary antioxidants. Aging Cell 8:484–495
Pfeiffer K, Gohil V, Stuart RA et al (2003) Cardiolipin stabilizes respiratory chain supercomplexes. J Biol Chem 278:52873–52880
Piccoli C, Scrima R, Boffoli D et al (2006) Control by cytochrome c oxidase of the cellular oxidative phosphorylation system depends on the mitochondrial energy state. Biochem J 396:573–583
Plecitá-Hlavatá L, Jezek J, Jezek P (2009) Pro-oxidant mitochondrial matrix-targeted ubiquinone MitoQ10 acts as anti-oxidant at retarded electron transport or proton pumping within Complex I. Int J Biochem Cell Biol 41:1697–1707
Quarato G, Piccoli C, Scrima R et al (2011) Variation of flux control coefficient of cytochrome c oxidase and of the other respiratory chain complexes at different values protonmotive force occurs by a threshold mechanism. Biochim Biophys Acta 1807:1114–1124
Quinzii CM, Hirano M (2010) Coenzyme Q and mitochondrial disease. Dev Disabil Res Rev 16:183–188
Ragan CI, Cottingham IR (1985) The kinetics of quinone pools in electron transport. Biochim Biophys Acta 811:13–31
Ragan CI, Heron C (1978) The interaction between mitochondrial NADH-ubiquinone oxidoreductase and ubiquinol-cytochrome c oxidoreductase – evidence for stoicheiometric association. Biochem J 174:783–790
Rajarathnam K, Hochman J, Schindler M et al (1989) Synthesis, location, and lateral mobility of fluorescently labeled ubiquinone 10 in mitochondrial and artificial membranes. Biochemistry 28:3168–3176
Rauchová H, Fato R, Drahota Z et al (1997) Steady-state kinetics of reduction of Coenzyme Q analogs by glycerol-3-phosphate dehydrogenase in brown adipose tissue mitochondria. Arch Biochem Biophys 344:235–241
Redfearn ER, Pumphrey AM (1960) The kinetics of ubiquinone reactions in heart-muscle preparations. Biochem J 76:64–71
Rich PR, Harper R (1990) Partition coefficients of quinones and hydroquinones and their relation to biochemical reactivity. FEBS Lett 269:139–144
Rieske JS (1967) Preparation and properties of reduced coenzyme Q-cytochrome c reductase (complex III of the respiratory chain). Methods Enzymol 10:239–245
Ringler RL, Singer TP (1959) Studies on the mitochondrial alpha-glycerophosphate dehydrogenase. I. Reaction of the dehydrogenase with electron acceptors and the respiratory chain. J Biol Chem 234:2211–2217
Rosca MG, Vazquez EJ, Kerner J et al (2008) Cardiac mitochondria in heart failure: decrease in respirasomes and oxidative phosphorylation. Cardiovasc Res 80:30–39
Rosenfeldt F, Marasco S, Lyon W et al (2005) Coenzyme Q10 therapy before cardiac surgery improves mitochondrial function and in vitro contractility of myocardial tissue. J Thorac Cardiovasc Surg 129:25–32
Sahach VF, Vavilova HL, Rudyk OV et al (2007) Inhibition of mitochondrial permeability transition pore is one of the mechanisms of cardioprotective effect of coenzyme Q10. Fiziol Zh 53:35–42
Samorì B, Lenaz G, Battino M et al (1992) On coenzyme Q orientation in membranes: a linear dichroism study of ubiquinones in a model bilayer. J Membr Biol 128:193–203

Sarewicz M, Osyczka A (2015) Electronic connection between the quinone and cytochrome c redox pools and its role in regulation of mitochondrial electron transport and redox signaling. Physiol Rev 95:219–243

Sazanov LA (2015) A giant molecular proton pump: structure and mechanism of respiratory complex I. Nat Rev Mol Cell Biol 16:375–388

Schäfer E, Seelert H, Reifschneider NH et al (2006) Architecture of active mammalian respiratory chain supercomplexes. J Biol Chem 281:15370–15375

Schägger H, Pfeiffer K (2000) Supercomplexes in the respiratory chains of yeast and mammalian mitochondria. EMBO J 19:1777–1783

Schägger H, Pfeiffer K (2001) The ratio of oxidative phosphorylation complexes I–V in bovine heart mitochondria and the composition of respiratory chain supercomplexes. J Biol Chem 276:37861–37867

Schmelzer C, Döring F (2010) Identification of LPS-inducible genes downregulated by ubiquinone in human THP-1 monocytes. Biofactors 36:222–228

Schmelzer C, Kitano M, Rimbach G et al (2009) Effects of ubiquinol-10 on microRNA-146a expression in vitro and in vivo. Mediators Inflamm 2009:415437

Schneider H, Lemasters JJ, Hackenbrock CR (1982) Lateral diffusion of ubiquinone during electron transfer in phospholipid- and ubiquinone-enriched mitochondrial membranes. J Biol Chem 257:10789–10793

Schönfeld P, Wieckowski MR, Lebiedzińska M et al (2010) Mitochondrial fatty acid oxidation and oxidative stress: lack of reverse electron transfer-associated production of reactive oxygen species. Biochim Biophys Acta 1797:929–938

Schwerzmann K, Cruz-Orive LM, Eggman R et al (1986) Molecular architecture of the inner membrane of mitochondria from rat liver: a combined biochemical and stereological study. J Cell Biol 102:97–103

Seelert H, Dani DN, Dante S et al (2009) From protons to OXPHOS supercomplexes and Alzheimer's disease: structure-dynamics-function relationships of energy-transducing membranes. Biochim Biophys Acta 1787:657–671

Sherwood S, Hirst J (2006) Investigation of the mechanism of proton trasnslocation by NADH:ubiquinone oxidoreductase (complex I) from bovine heart mitochondria: does the enzyme operate by a Q-cycle mechanism? Biochem J 400:541–550

Shinzawa-Itoh K, Shimomura H, Yanagisawa S et al (2016) Purification of active respiratory supercomplex from bovine heart mitochondria enables functional studies. J Biol Chem 291:4178–4184

Singer SJ, Nicolson GL (1972) The fluid mosaic model of the structure of cell membranes. Science 175:720–731

Sohet FM, Neyrinck AM, Pachikian BD et al (2009) Coenzyme Q10 supplementation lowers hepatic oxidative stress and inflammation associated with diet-induced obesity in mice. Biochem Pharmacol 78:1391–1400

Stoner CD (1984) Steady-state kinetics of the overall oxidative phosphorylation reaction in heart mitochondria. Determination of the coupling relationships between the respiratory reactions and miscellaneous observations concerning rate-limiting steps. J Bioenerg Biomembr 16:115–141

Stroh A, Anderka O, Pfeiffer K et al (2004) Assembly of respiratory complexes I, III, and IV into NADH oxidase supercomplex stabilizes complex I in Paracoccus denitrificans. J Biol Chem 279:5000–5007

Sun F, Huo X, Zhai Y et al (2005) Crystal structure of mitochondrial respiratory membrane protein complex II. Cell 121:1043–1057

Szarkowska L (1966) The restoration of DPNH oxidase activity by coenzyme Q (ubiquinone). Arch Biochem Biophys 113:519–525

Takahashi T, Yamaguchi T, Shitashige M et al (1995) Reduction of ubiquinone in membrane lipids by rat liver cytosol and its involvement in the cellular defence system against lipid peroxidation. Biochem J 309:883–890

Tan AK, Ramsay RR, Singer TP et al (1993) Comparison of the structures of the quinone-binding sites in beef heart mitochondria. J Biol Chem 268:19328–19333
Tian G, Sawashita J, Kubo H et al (2014) Ubiquinol-10 supplementation activates mitochondria functions to decelerate senescence in senescence-accelerated mice. Antioxid Redox Signal 20:2606–2620
Tocilescu MA, Zickermann V, Zwicker K et al (2010) Quinone binding and reduction by respiratory complex I. Biochim Biophys Acta 1797:1883–1890
Van Raam BJ, Sluiter W, de Wit E et al (2008) Mitochondrial membrane potential in human neutrophils is maintained by complex III activity in the absence of supercomplex organisation. PLoS One 3:e2013
Vanderkooi G (1978) Organization of protein and lipid components in membranes. In: Fleischer S, Hatefi Y, MacLennan D, Tzagoloff A (eds) Molecular biology of membranes. Plenum Publishing Corp, New York, pp 25–55
Various authors (2015) In: Louro RO, Diaz-Moreno I (eds) Redox proteins in supercomplexes and signalosomes. CRC Press, Boca Raton
Vinogradov AD (2001) Respiratory complex I: structure, redox components, and possible mechanisms of energy transduction. Biochemistry (Mosc) 66:1086–1097
Walter L, Nogueira V, Leverve X et al (2000) Three classes of ubiquinone analogs regulate the mitochondrial permeability transition pore through a common site. J Biol Chem 275:29521–29527
Walter L, Miyoshi H, Leverve X et al (2002) Regulation of the mitochondrial permeability transition pore by ubiquinone analogs. A progress report. Free Radic Res 36:405–412
Wang Y, Hekimi S (2016) Understanding ubiquinone. Trends Cell Biol 26:367–378
Xia D, Yu CA, Kim H et al (1997) Crystal structure of the cytochrome bc1 complex from bovine heart mitochondria. Science 277:60–66
Yamamura T, Otani H, Nakao Y et al (2001) Dual involvement of coenzyme Q10 in redox signaling and inhibition of death signaling in the rat heart mitochondria. Antioxid Redox Signal 3:103–112
Yano T, Ohnishi T (2001) The origin of cluster N2 of the energy-transducing NADH-quinone oxidoreductase: comparisons of phylogenetically related enzymes. J Bioenerg Biomembr 33:213–222
Zhang Z, Huang L, Shulmeister VM et al (1998) Electron transfer by domain movement in cytochrome bc1. Nature 392:677–684
Zhang M, Mileykovskaya E, Dowhan W (2002) Gluing the respiratory chain together Cardiolipin is required for supercomplex formation in the inner mitochondrial membrane. J Biol Chem 277:43553–43556
Zhu QS, Berden JA, De Vries S et al (1982) On the role of ubiquinone in the respiratory chain. Biochim Biophys Acta 680:69–79
Zickermann V, Dröse S, Tocilescu MA et al (2008) Challenges in elucidating structure and mechanism of proton pumping NADH:ubiquinone oxidoreductase (complex I). J Bioenerg Biomembr 40:475–483

Chapter 4
Extramitochondrial Coenzyme Q_{10} in Aging

Guillermo López-Lluch

Abstract Organisms maintain a complex relationship between the production of oxidative radicals and endogenous antioxidant levels and antioxidant enzymatic activities. Apart of its bioenergetics role in mitochondria, CoQ_{10} is also present in the rest of cell membranes and in plasma lipoproteins. In these structures, CoQ_{10} plays a key antioxidant role, preventing lipidic oxidative damage and regulating enzymatic and regulatory activities. Many years ago, two essential CoQ_{10}-dependent dehydrogenases were characterized, cytochrome b_5 reductase and NQO1. These enzymes are able to maintain a redox cycle of CoQ_{10} in membrane, preventing oxidative damage and maintaining other antioxidant systems depending on ascorbic acid and α-tocopherol. Recently, other CoQ_{10}-dependent enzymes such as a mitochondrial dehydrogenase (FSP1) and a dehydrogenase associated with the outer leaf of the plasma membrane have increased the importance of CoQ_{10} in the prevention of oxidative damage and the regulation of apoptosis. The role of these enzymes in aging remains to be clearly determined but CR, a known prolongevity procedure, increases the activity of CoQ_{10}-dependent dehydrogenases and prevents oxidative damage associated with aging. Then, maintenance of the activity of these extramitochondrial CoQ_{10}-dependent activities seems to be important for aging and longevity.

Keywords Coenzyme Q · Plasma-membrane redox system · Cytochrome B5 reductase · NQO1 · Apoptosis

---

G. López-Lluch (✉)
Centro Andaluz de Biología del Desarrollo, CABD-CSIC, CIBERER,
Instituto de Salud Carlos III, Universidad Pablo de Olavide, Sevilla, Spain
e-mail: glopllu@upo.es

G. López Lluch (ed.), *Coenzyme Q in Aging*,
https://doi.org/10.1007/978-3-030-45642-9_4

4.1 Introduction

Coenzyme Q (CoQ) is widely known to be located in the inner mitochondrial membrane where it transfer electrons in the mitochondrial electron transport chain (Lopez-Lluch 2017). In addition to its bioenergetics role in mitochondria, CoQ is also present in the rest of cell membranes where it acts as main lipidic antioxidant. In all the cell membranes, including mitochondrial membranes, the reduced form of CoQ, named ubiquinol, is maintained in a redox cycle by many oxidoreductases (Lopez-Lluch et al. 2010). In mitochondrial membranes, electron transport chain complex I and II and Acyl-CoQ reductase are able to transfer electrons from matrix NADH, succinate or fatty acids to the oxidized form, ubiquinone, bound to the inner mitochondrial membrane. Further, ubiquinol transfers these electrons to Complex III being, then, essential for the respiratory activity of mitochondria (Crane et al. 1957). Mitochondrial CoQ is also a substrate for receiving electrons from other enzymes associated to mitochondria. Dihydroorotate dehydrogenase is involved in the novo synthesis of pyrimidine nucleotides and uses ubiquinone as electron acceptor (Hey-Mogensen et al. 2014; Loffler et al. 1997). Deficiency in this enzyme causes severe diseases by leading to mitochondrial dysfunction (Fang et al. 2013). Premature aging has been associated with a rise in the ratio of ubiquinol/ubiquinone associated with a defect in the synthesis of pyrimidine nucleotides and the capacity of reparation of DNA and cell division (Olgun 2009). Mitochondrial glycerol-3-phosphate dehydrogenase is a known enzyme involved in lipid metabolism in mitochondria and receives electrons from cytosolic redox intermediates (Alcázar-Fabra et al. 2016). Other enzyme is sulfide:quinone oxidoreductase involved in the oxidation pathway of sulfides that can affect the metabolism in glutathione, glutamate, serotonin and catecholamines (Quinzii et al. 2017; Ziosi et al. 2017).

In addition to the essential role of CoQ in the above indicated metabolic activities, another key function of CoQ is a main lipidic antioxidant. CoQ protects phospholipids against oxidative damage directly by disrupting the lipid peroxidation chain or indirectly through maintaining α-tocopherol or ascorbic acid in their respective activated state (Fernandez-Ayala et al. 2000). In plasma membrane (PM), maintenance of CoQ levels in its active form, ubiquinol, depends on the activity of different oxidoreductases (Ernster and Dallner 1995) that accept electrons from cytosolic NAD(P)H and donate them to intramembranous CoQ. Once in its reduced form, ubiquinol transfers electrons to membrane or exogenous electron acceptors in a system known as plasma membrane redox system (PMRS) (Villalba et al. 1995). NAD(P)H-dehydrogenases include a NADH-cytochrome b_5 reductase (Cyb_5R_3) that reduces CoQ by using a one-electron mechanism (Arroyo et al. 2000; Navarro et al. 1995) and NQO1 (NAD(P)H:quinone oxidoreductase 1) that reduces CoQ by a two electron mechanism (Beyer et al. 1996).

PMRS has been also associated with many regulatory activities such as the prevention of the activation of the neutral sphingomyelinase activity and then, the release of ceramide and induction of cell death (Fig. 4.1) or the activation of adenylyl-cyclase and the modulation of the release of cAMP (Lopez-Lluch et al.

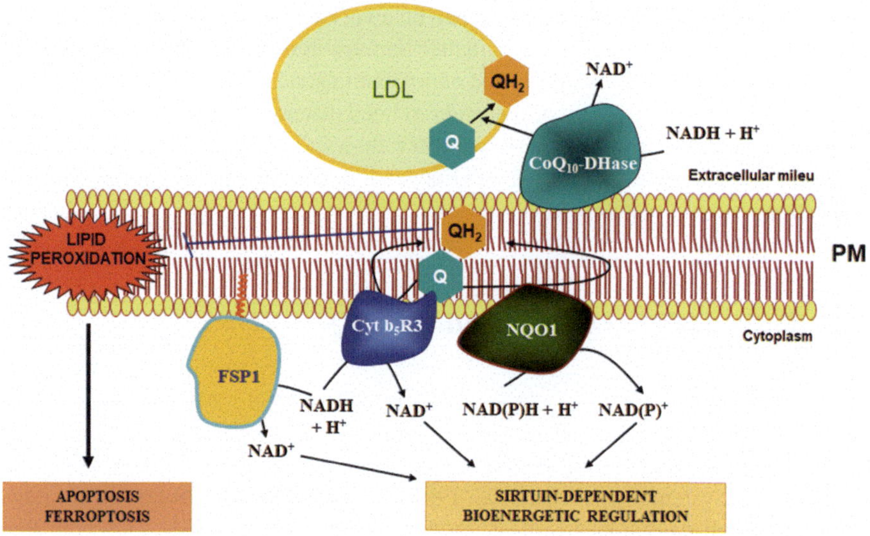

Fig. 4.1 Essential role of CoQ-dependent plasma membrane oxidoreductases. Maintenance of the redox cycle of CoQ_{10} in plasma membrane and other membrane prevents lipid peroxidation and the release of proapoptotic and proferroptotic compounds. For this role, the presence and activity of Cyb_5R_3, located at the plasma membrane, NQO1, translocated from cytoplasm, and FSP1, translocated from mitochondria, play a key role. Further, the activity of these enzymes is also associated with the local regulation of redox-associated signaling pathways and the bioenergetics regulation of the cell through Sirtuin and other enzymes. Further, a recent study indicates the presence of an enzyme located at the outer side of the plasma membrane of hepatocytes able to maintain $CoQ_{10}H_2$ in plasma lipoproteins

1998; Lopez-Lluch et al. 2005a) or the maintenance of growth in mitochondrial deficiency conditions.

Protection of lipids against oxidative damage in plasma lipoproteins is also a key extramitochondrial function of CoQ_{10}. Blood plasma CoQ_{10} is associated with lipoproteins, mainly with LDL (Johansen et al. 1991). Levels of CoQ_{10} in LDL are directly related with the resistance of these particles to oxidative damage (Mohr et al. 1992) whereas, low CoQ_{10} levels in plasma have been associated with a higher degree of lipid peroxidation increasing the risk for cardiovascular disease (Stocker et al. 1991). As in cell membranes, CoQ_{10} maintains also α-tocopherol levels in plasma and thus, it can be considered the main lipophilic antioxidant preventing lipid peroxidation (Stocker et al. 1991). Further, plasma CoQ_{10} can also suppress the oxidative damage in endothelial cells by regulating oxidative stress-dependent response (Tsai et al. 2011). The importance of plasma CoQ_{10} in the prevention of oxidative damage in cardiovascular diseases has been highlighted by the inverse relationship between CoQ_{10} levels and the risk of mortality in acute phase patients (Shimizu et al. 2017)

CoQ_{10} plays essential functions in cells from bioenergetics to antioxidant protection in mitochondria, cell membranes and blood plasma lipoproteins. The putative decrease of CoQ_{10} levels during aging can be a very important risk factor for mitochondrial dysfunction but also for oxidative damage in cell membranes and in lipoproteins. Further, a dysfunction of the antioxidant capacity of CoQ_{10} during aging can aggravate the progression of many age-related diseases that depends on plasma membrane signaling processes. This chapter shows the importance of extramitochondrial functions of CoQ_{10} and CoQ-dependent activities during aging and its regulation with prolongevity effectors.

4.2 Origen of Extramitochondrial CoQ

Location of the synthesis of CoQ is a controversial subject for debate since it has been associated with the endoplasmic reticulum (ER)-Golgi system or with the matrix side of the inner mitochondrial membrane. If the synthesis is associated with the ER-Golgi system (Teclebrhan et al. 1995), the distribution to the rest of extramitochondrial cell membranes is clear through the secretory pathway. However, the location of the synthesis of CoQ_{10} in the inner mitochondrial membrane indicates the existence of a CoQ_{10} transfer mechanism from mitochondria to ER-Golgi system. Recently, it has been suggested that UBIAD1, a prenyltransferase located in the Golgi membrane is responsible for the synthesis of CoQ_{10} since its depletion reduces the cytosolic pool of CoQ_{10} and increases lipid peroxidation in endothelial cells (Mugoni et al. 2013). However, depletion of this protein also affects the cholesterol metabolism and its distribution into mitochondria (Morales et al. 2014) indicating a main function in the metabolism of the CoQ_{10}-isoprene side-chain than in the modification of the benzene ring that clearly occurs in mitochondria (Stefely and Pagliarini 2017).

Cell culture experiments performed by our group demonstrate that in human cells *de novo* synthesis of CoQ_{10} in human cells occurs in mitochondria (Fernandez-Ayala et al. 2005a). By using a radioactive precursor of the CoQ_{10} benzene ring, [14C]-parahydroxy-benzoate, we found that CoQ_{10}-associated radioactivity was initially found in mitochondria and later in mitochondria-associated ER membranes (MAMs), ER, Golgi and plasma membrane (Fernandez-Ayala et al. 2005a). Exogenously added CoQ also reach mitochondrial membranes and can affect the mitochondrial electron transport chain (Fernandez-Ayala et al. 2005b). This means that a flux between the ER-Golgi system and mitochondria exists through MAMs that link mitochondria to ER (Sharma et al. 2019). The important role of these membranes in mitochondrial dynamics, biogenesis and transport in cells suggests the transport of CoQ_{10} to all the cell membranes can be associated with the function of MAMs. Interestingly, mitochondrial-ER contacts decrease in aged cells (Janikiewicz et al. 2018) probably by the decrease in the levels of mitofusin 2, a key protein in mitochondrial fusion and dynamics (de Brito and Scorrano 2008). Interestingly, mitofusin 2 is required to maintain CoQ levels in mitochondria indicating the importance of mitochondrial fusion and/or interaction with ER through MAMs in mitochondrial synthesis and traffic (Mourier et al. 2015). The synthesis of CoQ_{10}

requires the coordinated activity of different enzymes and regulatory proteins (Chap. 2 of this book). These proteins are known as COQ proteins and to date more than 12 members have been described (Stefely and Pagliarini 2017). As a two senses road, the incorporation of the isoprene side-chain synthesized by COQ1 complex and its binding to pHB through COQ2 can occurs in these MAMs whereas the modification of the benzene ring occurs in the inner mitochondrial membrane. Then, the decrease in MAMs-mitochondria contacts during aging can affects both, de novo synthesis of CoQ_{10} in mitochondria and its transfer to the rest of membranes and also the incorporation of exogenous CoQ_{10} to mitochondria.

4.3 Extramitochondrial CoQ Oxidoreductases

The redox cycle of CoQ in extramitochondrial membranes is maintained by several oxidoreductases that transfer electrons from cytosolic electron donors to ubiquinone. This function is mainly played by two main oxidoreductases, cytochrome b_5 reductase, (Cyb_5Rase) and NAD(P)H:quinone oxidoreductase 1, (NQO1) (Ross and Siegel 2017). Although with minor capacity, other enzymes such a cytosolic NAD(P)H-ubiquinone reductase (Takahashi et al. 1992), lipoamide dehydrogenase (Olsson et al. 1999), thioredoxin reductase and even glutathione reductase can also transfer electrons to ubiquinone (Nordman et al. 2003). Recently, other mitochondria-linked enzymes have emerged as CoQ-dependent oxidoreductases affecting extramitochondrial functions of CoQ. These enzymes are ferroptosis suppressor protein 1 (FSP1) (Bersuker et al. 2019) and CDGSH iron sulfur domain 1 protein (mitoNEET) (Wang et al. 2017). Further, an extracellular NADH-dependent CoQ_{10} reductase seems to be responsible of the maintenance of blood plasma ubiquinol levels (Takahashi et al. 2019). All these enzymes offers a complete system for the extramitochondrial maintenance of ubiquinol levels in cell membranes and to control many physiological activities in cells and tissues.

4.3.1 *Cyb_5Rase*

Cyb_5Rase (E.C. 1.6.5.5.), was discovered as a phospholipid-dependent NADH-CoQ reductase associated with plasma membrane (Navarro et al. 1995). Known also as ascorbate free radical reductase (Villalba et al. 1995) it is not only associated with the protection against oxidative stress but also in the regulation of metabolic homeostasis (Siendones et al. 2014). In fact, deficiency of membrane bound Cyb_5Rase causes recessive hereditary methaemoglobinaemia in humans. This incurable rare disease causes, among other symptoms, severe neurological disorders (Siendones et al. 2018). In Cyb_5R_3-deficient cells, a decrease in the NAD^+/NADH ratio is found, indicating the importance of this enzyme in metabolic regulation. Further, this enzyme has been also associated with mitochondrial activity since its deficiency affects mitochondrial respiration rate, activity of mitochondrial electron transport chain complexes and ATP production (Martin-Montalvo et al. 2016).

Cyb_5R_3 is regulated by erythroid-derived 2-like2 (NRF2) transcription factor, indicating a redox stress-dependent regulation (Siendones et al. 2014). However, experiments performed in cultured cells demonstrated that high levels of ROS produce a negative effect in the levels of Cyb_5R_3 (Bello et al. 2003). It seems that Cyb_5Rase activity and levels associated to plasma membrane are affected by aging and also modulated by prolongevity effectors as caloric restriction and resveratrol indicating a role in aging progression (Rodriguez-Bies et al. 2016; Rodriguez-Bies et al. 2015).

4.3.2 NQO1

NQO1, also known as DT-diaphorase (E.C. 1.6.99.2.), is the most known quinone oxidoreductases in cells. NQO1 reduces quinones by a two-electron mechanism. It is highly inducible and seems to play several roles in adaptive and stress cell responses (Joseph et al. 1994; Prochaska et al. 1992) by their induction through the activation of the NRF2 transcription factor or the Aryl carbon (Ah) receptor (Jaiswal 2000; Vasiliou et al. 1994). The activity of NQO1 in chemoprotection and catabolism of quinones has been extensively studied (Ernster et al. 1986; Ernster et al. 1972; Ross et al. 1990). Other functions of the enzyme are associated with stabilization or proteins such as p53, p63, ornithine decarboxylase and PGC1α (Ross and Siegel 2017) or in the control of the stability of mRNA of serpins and others (Di Francesco et al. 2016; Ross and Siegel 2017).

NQO1 levels increase rapidly under stress conditions indicating their essential role as member of the antioxidant system in cells. As a target of NRF2, NQO1 is induced under oxidative challenge and by inducers of oxidative damage response such as polyphenols (Hsieh et al. 2006). As antioxidant enzyme, NQO1 acts a CoQ-reductase transferring two electrons from NADPH to ubiquinone in cell membranes. In fact, the protective activity of CoQ_{10} against membrane damage could be blocked by inhibition of NQO1 (Beyer et al. 1996).

4.3.3 Lipoamide Reductase, Glutathione Oxidoreductase and Thioredoxin Reductase

Lipoamide dehydrogenase, a member of the family of pyridine nucleotide disulfide dehydrogenases was proposed to reduce ubiquinone in cell membranes (Xia et al. 2001). This enzyme belongs to the same family than glutathione and thioredoxin reductases, is present both, inside and outside mitochondria and reduces ubiquinone (Olsson et al. 1999), lipoic acid and lipoamide (Williams et al. 1967) and has been proposed to play an important role in the protection of biological membranes from lipid peroxidation (Xia et al. 2001). The activity of this enzyme has been associated with the protection of dopaminergic cells against neurotoxicity in Parkinson's disease (Dhanasekaran et al. 2008).

The selenoprotein thioredoxin reductase (TrxR) and thioredoxin (Trx) is an ubiquitous antioxidant system also playing regulatory roles of many physiological functions in cells (Nordberg and Arner 2001). TrxR1 reduces ubiquinone in a selenium-dependent manner and this relationship has been used to stablish the association of CoQ_{10} levels and selenium (Xia et al. 2003). TrxR and glutathione reductase has been strongly associated with the maintenance of the levels of ubiquinol in cells (Nordman et al. 2003).

4.3.4 FSP1

Ferroptosis suppressor protein 1 (FSP1), previously known as apoptosis-inducing factor mitochondrial 2 (AIFM2, AIF2), has been very recently shown as a CoQ_{10}-dependent oxidoreductase linked to plasma membrane by a myristoyl chain (Bersuker et al. 2019). As in the case of Cyb_5R_3 and NQO1, FSP1 association with plasma membrane permits to maintain CoQ_{10} in its reduced form, ubiquinol, by using cytosolic NAD(P)H as electron donor (Doll et al. 2019). Interestingly, the function of FSP1 is essential in the protection of cells against ferroptosis in Glutathione Peroxidase 4 knock out cells indicating a putative synergistic effect of FSP1 with this peroxidase in the prevention of phospholipid peroxidation (Doll et al. 2019).

The importance of this protein in aging and cell survival remains to be stablished. However, it has been associated with the protection of neurons against ferroptosis associated with the progression of age-related neurodegenerative diseases (Santoro 2020). Considered in the beginning as a brain specific form of AIF, this redox-active mitochondrial enzyme was considered to be strongly anchored to the inner membrane of mitochondria facing the intermembrane space (Hangen et al. 2010). Downregulation of FSP1/AIF2 has been also associated with mitochondrial dysfunction indicating also a role in the maintenance of mitochondrial activity (Chen et al. 2019) probably by regulating the NAD^+/NADH ratio through its activity at the plasma membrane (Doll et al. 2019). Its study in aging and longevity field would indicate its importance in the prevention of age-related diseases.

4.3.5 mitoNEET

MitoNEET (CISD1, CDGSH iron sulfur domain 1 protein) is a small 2Fe-2S containing protein located at the outer mitochondrial membrane, able to regulate oxidative capacity (Wiley et al. 2007). It has been recently considered as a therapeutic target for type II diabetes and other age-associated diseases such as neurodegeneration or cancer (Lipper et al. 2019). MitoNEET is associated with mitochondrial homeostasis by regulating mitochondrial pore (VDAC1) opening in a redox-dependent manner (Lipper et al. 2019) and with the control of energy metabolism, iron homeostasis and ROS production (Li et al. 2018). Interestingly, long time ago, regulation of this pore in mitochondria was also suggested through an ubiquinone-binding site (Fontaine et al. 1998). As in the case of FSP1, mitoNEET has been also linked to the protection against ferroptosis by protecting mitochondria from lipid peroxidation (Yuan et al. 2016). Interestingly, two recent studies have suggested that ubiquinone can be a natural acceptor of electrons from this protein regulating its redox status and many of their functions (Li et al. 2018; Wiley et al. 2007). CoQ_{10} has been proposed to accept electrons and maintain this protein in its oxidized status. Taken into consideration that the oxidized form of this protein is the active form that can transfer [2Fe-2S] clusters to cytosolic and mitochondrial proteins, ubiquinone can play an important regulatory role in the homeostasis of mitochondrial activity through this protein (Wang et al. 2017).

4.4 CoQ_{10} in Blood Plasma

The presence of CoQ_{10} in blood plasma is associated with the lipoproteins that distribute cholesterol among different organs (Stocker et al. 1991). Interestingly, CoQ_{10} in lipoproteins is maintained in its reduced form. Even after supplementation with ubiquinone, the ratio ubiquinol/ubiquinone in plasma remains unchanged and near the 100% (Okamoto et al. 1989) indicating the existence of a system that maintains CoQ_{10} in its reduced form in plasma. The putative reductase involved in maintenance of CoQ_{10} in its reduced form has been elusive although recently it has been shown that a liver cancer cell line, HepG2, can reduce extracellular CoQ_{10} in a mechanism independent on the reduction of intracellular CoQ_{10} or the leakage of intracellularly reduced CoQ_{10} (Takahashi et al. 2019). This capacity was already found some years ago with HepG2 cells and red blood cells membranes although in some way discarded since reduction of short chain CoQ forms was much faster than LDL-associated CoQ_{10} (Stocker and Suarna 1993). This new finding opens the possibility of the presence of members of the PMRS able to transfer electrons from intracellular reductants to oxidized extracellular CoQ_{10} or the presence of a reductase in the outer side of the plasma membrane able to use extracellular NAD(P)H to reduce ubiquinone (Fukuwatari and Shibata 2009) both involved in the maintenance of the high levels of blood plasma ubiquinol.

Decrease in the ratio ubiquinol/ubiquinone in blood plasma has been considered as a biomarker of aging in European adults (Niklowitz et al. 2016). Further, this ratio has been considered as a marker of oxidative stress in blood plasma (Yamamoto and Yamashita 1997). Levels of CoQ_{10} in plasma seems to respond to physical activity of individuals but affecting in a different manner depending of the age of the individuals. Whereas high activity is associated with lower levels of CoQ_{10} in plasma in young individuals, this relationship is positive in older people indicating an age-dependent regulation (Del Pozo-Cruz et al. 2014a). Sedentary lifestyle and obesity show an inverse relationship with CoQ_{10} levels in plasma increasing the risk for cardiovascular disease by reducing lipid peroxidation protection and showing a gender difference since it affects more to women than men (Del Pozo-Cruz et al. 2014b). The discovery of an extracellular CoQ_{10} reductase in liver can associate age-dependent changes in plasma ubiquinol/ubiquinone ratio with changes in the expression or activity of this enzyme (Takahashi et al. 2019). Further research must be performed in order to clarify the importance of this enzyme in the progression of cardiovascular disease and plasma oxidation during aging.

4.5 Antioxidant Role of CoQ in Cell Membranes

Phospholipids, sphingolipids and cholesterol are the main components of cell membranes. Cell membranes also contains other lipids involved in the antioxidant protection against oxidative damage such as α-tocopherol and CoQ. In cell membranes, α-tocopherol protects unsaturated fatty acids from peroxidation (Niki 1987). On the other hand, ubiquinol is responsible for the maintenance of α-tocopherol and also prevention of lipid peroxidation (Lass and Sohal 1998). A great body of evidence have demonstrated that ubiquinol is the main component in cell membranes for preventing lipid peroxidation and blocking oxidative damage cascade. We stablished the existence of an inverse relationship between the levels of CoQ in plasma membrane and the rate of lipid peroxidation (Lopez-Lluch et al. 1999). We clearly demonstrated that the presence of CoQ in plasma membrane blocks lipid peroxidation cascade directly but also indirectly by regenerating other main antioxidants such as extracellular ascorbate (Gomez-Diaz et al. 1997a; Santos-Ocana et al. 1998) and α-tocopherol (Bello et al. 2003; Crane and Navas 1997). This protective activity of ubiquinol against oxidation in cell membranes has been associated with the prevention of apoptosis induced by chemical compounds but not by receptor-dependent apoptotic signaling (Alleva et al. 2001).

Deficiency in α-tocopherol and selenium is compensated by the induction of the enzymatic system involved in the maintenance of CoQ in its reduced form in plasma membrane. This induction not only affected the amount of CoQ but also increased the levels of CoQ-dependent reductases such as Cyb_5Rase and NQO1 that accumulated in plasma membrane (Navarro et al. 1998). Maintenance of the system involved in CoQ reduction in PM is key to keep its antioxidant function. High rates of lipid peroxidation in PM in aged animals are associated with low levels of CoQ reductases (De Cabo et al. 2004). Moreover, overexpression of NQO1 and Cyb_5Rase increased the resistance of neuronal cells to oxidative insults whereas cells depleted of these enzymes were more vulnerable to oxidative stress (Hyun et al. 2012; Hyun and Lee 2015). All these results indicate that the maintenance of a redox recycling system for CoQ in membranes is essential to prevent lipid peroxidation cascade and then, to reduce cell damage and death.

The activity of CoQ as antioxidant makes it as an essential factor in the protection of apoptosis induced by oxidative stress. This activity blocks the apoptotic program induced by oxidative stress induced by serum deprivation in human cells (Fernandez-Ayala et al. 2000; Rodriguez-Aguilera et al. 2000). Plasma membrane CoQ was very effective blocking the activation of plasma membrane bound Mg^{2+}-dependent neutral sphingomyelinase (Mg^{2+}-nSMase) and the subsequent activation of ceramide-dependent caspase activation in serum deprived cells (Navas et al. 2002; Villalba and Navas 2000). The same effect was found in aged rats fed with different nutritional oils (Bello et al. 2006). In comparison with animals fed with sunflower oil, animals fed with olive oil showed less lipoperoxidation and higher levels of CoQ in plasma membrane resulting in lower activation of Mg^{2+}-nSMase in the liver of aged rats (Bello et al. 2006). The same inhibitory effect was found when rats fed with a PUFA-rich diet were supplemented with CoQ_{10} indicating the importance of CoQ levels in plasma membrane to inhibit the apoptotic program dependent of oxidative damage (Bello et al. 2005).

4.6 Regulatory Role of CoQ in Plasma Membrane

PM is also the barrier that delimits cells but also the structure by which the cell interacts with the environment receiving and releasing signals to the extracellular milieu. The importance of the physiological role of CoQ in plasma membrane was clarified in by its essential role in the maintenance of cell growth and survival in serum-free animal (Sun et al. 1990) and plant cells (Crane et al. 1991). CoQ-dependent oxidoreductases are also regulated by growth factors, insulin and pituitary extracts indicating a modulatory role in cell signaling associated with the response to hormones (Brightman et al. 1992). The promotion of cell growth by the activation of PMRS was associated with many functions including the above indicated membrane antioxidant protection but also with intracellular redox homeostasis maintenance, modulation of intracellular signals, opening of membrane pore or regulation of intracellular NAD^+/NADH levels (Sun et al. 1992a, b; Lopez-Lluch et al. 2010). The role of this system in the promotion of cell growth and survival was highlighted by the effect of ferrycianide, as extracellular electron acceptor of the PMRS, and CoQ_{10} in the promotion of growth in mitochondrial electron transport chain defective ρ^0 cells (Martinus et al. 1993). Then, it seems clear that, PMRS can affect cell growth and survival by controlling NAD^+/NADH ratio in cells when mitochondrial activity is defective.

Furthermore, the activity of the PMRS has been proposed to play an important role in development (Crane and Navas 1997). The activity of CoQ-dependent oxidoreductases in PM is modulated during erythrocyte differentiation (Gomez-Diaz et al. 1997a). Moreover, the activation of this system in myeloid cells enhances the differentiation to monocytes induced by 1α,25-dihydroxyvitamin D_3 (Lopez-Lluch et al. 1998; Quesada et al. 1996). In this effect, we found that PMRS activation modulates second messengers such as cAMP that regulate the activity of transcription factors involved in differentiation (Lopez-Lluch et al. 1998, 2001; Quesada et al. 1996).

The redox state of CoQ in plasma membrane has been also associated with the regulation of the activity of tyrosine kinases through the local generation of H_2O_2 (Crane et al. 1994). The inhibitory effect of H_2O_2 on protein phosphatases could increase the activated status of many tyrosine kinases associated with growth receptors (Brightman et al. 1992). It has been proposed that in this role, Cyb_5R_3 would play an important function since this enzyme reduces CoQ by an one-electron mechanism and, thus, can produce semiquinone forms that can act as prooxidant (Villalba et al. 1995).

Another important function of CoQ-dependent reductases in the PMRS is based on the regulation of the electron donors in cytoplasm. The control of intracellular NAD^+/NADH ratio can regulate the activity of NAD^+-dependent Sirtuins and, then, modulate the activity of enzymes regulated by acetylation/deacetylation, many of them involved in cell homeostasis and metabolism (Lopez-Lluch et al. 2010). Sirtuins are key factors in aging process and its activation has shown effects on cell metabolism, physiology and lifespan (Baur et al. 2006). As it has been mentioned before, the regulation of the NAD^+/NADH ratio affects cell bioenergetics (Larm et al. 1994) and up-regulation of PMRS activity maintains growth in cells lacking functional mitochondria (Gomez-Diaz et al. 1997b; Hyun and Lee 2015; Larm et al. 1994). For this reason, PMRS activity has been considered essential in the maintenance of cell bioenergetics when the activity of mitochondria decreases as happends in aging and many age-related diseases (de Grey 2001).

4.7 Maintenance of CoQ-Dependent Extramitochondrial Activities Can Be Associated with Aging Progression

Aging, many chronic diseases, and other age-related diseases are associated with mitochondrial dysfunction (Haas 2019). When mitochondrial dysfunction occurs, cells can adapt to maintain ATP production through metabolic adaptive mechanisms such as lactate fermentation. For this reason, mitochondrial DNA-depleted cells, ρ^0 cells, unable to maintain the activity of mitochondrial electron transport chain, can survive by enhancing glycolysis and also to increase the activity of CoQ-dependent PMRS that oxidizes the excess of NADH (Gomez-Diaz et al. 1997b; Piechota et al. 2006). In these ρ^0 cells, production of ROS and the activity of plasma membrane redox enzymes are induced to maintain redox homeostasis (Hyun et al. 2007). The same compensating response can be attributed to the increase of the activity of plasma membrane redox enzymes found in insulin-dependent human patients showing deficiency in mitochondrial activity (Lenaz et al. 2002). Activation of PMRS has been associated with the control of mitochondrial function. In yeast, the activation of NQR1, a plasma membrane associated CoQ reductase, increases when cells shift to respiratory metabolism (Jimenez-Hidalgo et al. 2009). Then, the capacity to increase PMRS activity under caloric restriction or with polyphenols such as resveratrol helps to the adaptation of senescent cells to mitochondrial dysfunction associated with aging (Hyun et al. 2006; Kim et al. 2013; Rodriguez-Bies et al. 2016; Tung et al. 2014).

Overexpression of $CytB_5Rase$ in neuronal cells protects them against oxidative, metabolic, energetic and proteotoxic stress, many of them associated with aging progression (Hyun and Lee 2015). Similarly, overexpression of NQO1 also protects cells against metabolic and proteotoxic stress through maintaining bioenergetics and increasing mitochondrial function (Hyun et al. 2012; Kim et al. 2013). On the other hand, dysfunction of these enzymes have been associated with the progression of aging and age-related diseases. The expression of NQO1 decreases in a mouse model of Alzheimer's disease (Torres-Lista et al. 2014) and alterations of NQO1 activity have been also found in Alzheimer's Disease patients showing increase in the staining in astrocytes and neurites surrounding senile plaques (SantaCruz et al. 2004).

The activation of PMRS can be also associated with the modifications of mitochondrial physiology after nutritional challenging. Caloric restriction increases mitochondrial fusion by inhibition of a protein involved in fission, DRP1, through the increase of cAMP and activation of PKA (Jheng et al. 2012). We demonstrated that both, $CytB_5Rase$ and NQO1 are induced under caloric restriction in liver of both, rats (De Cabo et al. 2004) and mice (Lopez-Lluch et al. 2005b). Activation of PMRS is also associated with the enhancement of monocyte differentiation through induction of cAMP (Lopez-Lluch et al. 2005a), an effect that was also associated with changes in mitochondrial physiology (Lopez-Lluch et al. 2001). Furthermore, activity of PMRS has been associated with the maintenance of mitochondrial function and the delay in senescence progression in a senescence-accelerated mice model (Tian et al. 2014), in a process associated with a rise of cAMP and the activation of Sirt1 and PGC1-α. Then, the maintenance of mitochondrial physiology during aging can be associated with the activity of CoQ_{10}-dependent plasma membrane associated activities through modulation of cAMP levels and activation of PKA that can affect both mitochondrial structure (Jheng et al. 2012) and transport (Wang et al. 2015). All these evidence indicate that activation of PMRS and CoQ-dependent oxidoreductases is important in the maintenance of cell activity and in the prevention of mitochondrial dysfunction during aging.

As has been above indicated, PMRS can also regulate SMase-dependent signaling. Activation of Mg^{2+}-nSMase releases ceramides from PM to the cytosol and is involved in cell signaling and apoptosis (Chatterjee 1999). Activity of this enzyme increases in brain and liver during maturation and aging (Petkova et al. 1988; Spence and Burgess 1978; Venable et al. 1995). Its activity is also enhanced in cerebral cortex in accelerated senescence mice models (Kim et al. 1997). For this reason, the role of Mg^{2+}-nSMase in aging has been suggested (Venable et al. 1995) and its chronic activation and the release of ceramide has been associated with aging progression and aging-related diseases (Cutler and Mattson 2001; Lightle et al. 2000). We demonstrated that the activity of PMRS inhibits nSMase and prevents ceramide-dependent apoptosis in human cells (Fernandez-Ayala et al. 2000; Navas et al.

2002). For this reason, the maintenance of CoQ levels and CoQ-dependent activities at the plasma membrane during aging can be considered important factor to prevent senescence and cell death during aging.

4.8 Conclusion

It seems clear that activation of PMRS in aging can be considered one important prolongevity strategy (Saraswat and Rizvi 2017). Extramitochondrial CoQ-dependent oxidoreductase such as Cyb_5Rase and NQO1 are induced by prolongevity interventions such as caloric restriction, nutritional bioactive compounds such as resveratrol and physical activity. Activation of these oxidoreductases and other CoQ-dependent enzymes increase the resistance of cell membranes to oxidative stress and the protection against oxidative damage-related senescence and apoptosis. Further, activation of PMRS, maintains redox homeostasis in cells, preventing or delaying mitochondrial dysfunction and maintaining redox conditions to compensate mitochondrial deficiency. This process is regulated by NAD^+-dependent deacetylases such as sirtuins and by second messengers such as cAMP induced by the activation of PMRS.

Acknowledgements This work was supported by grants from the Instituto de Salud Carlos III.

References

Alcázar-Fabra M, Navas P, Brea-Calvo G (2016) Coenzyme Q biosynthesis and its role in the respiratory chain structure. Biochimica et Biophysica Acta (BBA) – Bioenergetics 1857(8):1073–1078. https://doi.org/10.1016/j.bbabio.2016.03.010

Alleva R, Tomasetti M, Andera L, Gellert N, Borghi B, Weber C, Murphy MP, Neuzil J (2001) Coenzyme Q blocks biochemical but not receptor-mediated apoptosis by increasing mitochondrial antioxidant protection. FEBS Lett 503(1):46–50

Arroyo A, Kagan VE, Tyurin VA, Burgess JR, de Cabo R, Navas P, Villalba JM (2000) NADH and NADPH-dependent reduction of coenzyme Q at the plasma membrane. Antioxid Redox Signal 2(2):251–262. https://doi.org/10.1089/ars.2000.2.2-251

Baur JA, Pearson KJ, Price NL, Jamieson HA, Lerin C, Kalra A, Prabhu VV, Allard JS, Lopez-Lluch G, Lewis K, Pistell PJ, Poosala S, Becker KG, Boss O, Gwinn D, Wang M, Ramaswamy S, Fishbein KW, Spencer RG, Lakatta EG, Le Couteur D, Shaw RJ, Navas P, Puigserver P, Ingram DK, de Cabo R, Sinclair DA (2006) Resveratrol improves health and survival of mice on a high-calorie diet. Nature 444(7117):337–342. https://doi.org/10.1038/nature05354

Bello RI, Alcain FJ, Gomez-Diaz C, Lopez-Lluch G, Navas P, Villalba JM (2003) Hydrogen peroxide- and cell-density-regulated expression of NADH-cytochrome b5 reductase in HeLa cells. J Bioenerg Biomembr 35(2):169–179

Bello RI, Gomez-Diaz C, Buron MI, Alcain FJ, Navas P, Villalba JM (2005) Enhanced anti-oxidant protection of liver membranes in long-lived rats fed on a coenzyme Q10-supplemented diet. Exp Gerontol 40(8–9):694–706. https://doi.org/10.1016/j.exger.2005.07.003

Bello RI, Gomez-Diaz C, Buron MI, Navas P, Villalba JM (2006) Differential regulation of hepatic apoptotic pathways by dietary olive and sunflower oils in the aging rat. Exp Gerontol 41(11):1174–1184. https://doi.org/10.1016/j.exger.2006.08.012

Bersuker K, Hendricks J, Li Z, Magtanong L, Ford B, Tang PH, Roberts MA, Tong B, Maimone TJ, Zoncu R, Bassik MC, Nomura DK, Dixon SJ, Olzmann JA (2019) The CoQ oxidoreductase FSP1 acts parallel to GPX4 to inhibit ferroptosis. Nature. https://doi.org/10.1038/s41586-019-1705-2

Beyer RE, Segura-Aguilar J, Di Bernardo S, Cavazzoni M, Fato R, Fiorentini D, Galli MC, Setti M, Landi L, Lenaz G (1996) The role of DT-diaphorase in the maintenance of the reduced antioxidant form of coenzyme Q in membrane systems. Proc Natl Acad Sci U S A 93(6):2528–2532. https://doi.org/10.1073/pnas.93.6.2528

Brightman AO, Wang J, Miu RK, Sun IL, Barr R, Crane FL, Morre DJ (1992) A growth factor- and hormone-stimulated NADH oxidase from rat liver plasma membrane. Biochim Biophys Acta 1105(1):109–117. https://doi.org/10.1016/0005-2736(92)90168-l

Chatterjee S (1999) Neutral sphingomyelinase: past, present and future. Chem Phys Lipids 102(1–2):79–96. https://doi.org/10.1016/s0009-3084(99)00077-8

Chen W, Liu H, Wang T, Bao G, Wang N, Li RC (2019) Downregulation of AIF-2 inhibits proliferation, migration, and invasion of human glioma cells via mitochondrial dysfunction. J Mol Neurosc 68(2):304–310. https://doi.org/10.1007/s12031-019-01306-y

Crane FL, Navas P (1997) The diversity of coenzyme Q function. Mol Asp Med 18(Suppl):S1–S6

Crane FL, Hatefi Y, Lester RL, Widmer C (1957) Isolation of a quinone from beef heart mitochondria. Biochim Biophys Acta 25(1):220–221

Crane FL, Sun IL, Barr R, Low H (1991) Electron and proton transport across the plasma membrane. J Bioenerg Biomembr 23(5):773–803. https://doi.org/10.1007/bf00786001

Crane FL, Sun IL, Crowe RA, Alcain FJ, Low H (1994) Coenzyme Q10, plasma membrane oxidase and growth control. Mol Asp Med 15(Suppl):S1–11

Cutler RG, Mattson MP (2001) Sphingomyelin and ceramide as regulators of development and lifespan. Mech Ageing Dev 122(9):895–908. https://doi.org/10.1016/s0047-6374(01)00246-9

de Brito OM, Scorrano L (2008) Mitofusin 2: a mitochondria-shaping protein with signaling roles beyond fusion. Antioxid Redox Signal 10(3):621–633. https://doi.org/10.1089/ars.2007.1934

De Cabo R, Cabello R, Rios M, Lopez-Lluch G, Ingram DK, Lane MA, Navas P (2004) Calorie restriction attenuates age-related alterations in the plasma membrane antioxidant system in rat liver. Exp Gerontol 39(3):297–304. https://doi.org/10.1016/j.exger.2003.12.003

de Grey AD (2001) A proposed mechanism for the lowering of mitochondrial electron leak by caloric restriction. Mitochondrion 1(2):129–139. https://doi.org/10.1016/s1567-7249(01)00008-3

Del Pozo-Cruz J, Rodriguez-Bies E, Ballesteros-Simarro M, Navas-Enamorado I, Tung BT, Navas P, Lopez-Lluch G (2014a) Physical activity affects plasma coenzyme Q10 levels differently in young and old humans. Biogerontology 15(2):199–211. https://doi.org/10.1007/s10522-013-9491-y

Del Pozo-Cruz J, Rodriguez-Bies E, Navas-Enamorado I, Del Pozo-Cruz B, Navas P, Lopez-Lluch G (2014b) Relationship between functional capacity and body mass index with plasma coenzyme Q10 and oxidative damage in community-dwelling elderly-people. Exp Gerontol 52:46–54. https://doi.org/10.1016/j.exger.2014.01.026

Dhanasekaran M, Albano CB, Pellet L, Karuppagounder SS, Uthayathas S, Suppiramaniam V, Brown-Borg H, Ebadi M (2008) Role of lipoamide dehydrogenase and metallothionein on 1-methyl-4-phenyl-1,2,3,6- tetrahydropyridine-induced neurotoxicity. Neurochem Res 33(6):980–984. https://doi.org/10.1007/s11064-007-9468-9

Di Francesco A, Di Germanio C, Panda AC, Huynh P, Peaden R, Navas-Enamorado I, Bastian P, Lehrmann E, Diaz-Ruiz A, Ross D, Siegel D, Martindale JL, Bernier M, Gorospe M, Abdelmohsen K, de Cabo R (2016) Novel RNA-binding activity of NQO1 promotes SERPINA1 mRNA translation. Free Radic Biol Med 99:225–233. https://doi.org/10.1016/j.freeradbiomed.2016.08.005

Doll S, Freitas FP, Shah R, Aldrovandi M, da Silva MC, Ingold I, Grocin AG, Xavier da Silva TN, Panzilius E, Scheel C, Mourao A, Buday K, Sato M, Wanninger J, Vignane T, Mohana V, Rehberg M, Flatley A, Schepers A, Kurz A, White D, Sauer M, Sattler M, Tate EW, Schmitz W, Schulze A, O'Donnel V, Proneth B, Popowicz GM, Pratt D, Angeli JPF, Conrad M (2019) FSP1 is a glutathione-independent ferroptosis suppressor. Nature. https://doi.org/10.1038/s41586-019-1707-0

Ernster L, Dallner G (1995) Biochemical, physiological and medical aspects of ubiquinone function. Biochim Biophys Acta 1271(1):195–204. https://doi.org/10.1016/0925-4439(95)00028-3

Ernster L, Lind C, Rase B (1972) A study of the DT-diaphorase activity of warfarin-resistant rats. Eur J Biochem 25(1):198–206. https://doi.org/10.1111/j.1432-1033.1972.tb01685.x

Ernster L, Atallah AS, Hochstein P (1986) DT diaphorase and the cytotoxicity and mutagenicity of quinone-derived oxygen radicals. Prog Clin Biol Res 209A:353–363

Fang J, Uchiumi T, Yagi M, Matsumoto S, Amamoto R, Takazaki S, Yamaza H, Nonaka K, Kang D (2013) Dihydro-orotate dehydrogenase is physically associated with the respiratory complex and its loss leads to mitochondrial dysfunction. Biosci Rep 33(2):e00021. https://doi.org/10.1042/bsr20120097

Fernandez-Ayala DJ, Martin SF, Barroso MP, Gomez-Diaz C, Villalba JM, Rodriguez-Aguilera JC, Lopez-Lluch G, Navas P (2000) Coenzyme Q protects cells against serum withdrawal-induced apoptosis by inhibition of ceramide release and caspase-3 activation. Antioxid Redox Signal 2(2):263–275. https://doi.org/10.1089/ars.2000.2.2-263

Fernandez-Ayala DJ, Brea-Calvo G, Lopez-Lluch G, Navas P (2005a) Coenzyme Q distribution in HL-60 human cells depends on the endomembrane system. Biochim Biophys Acta 1713(2):129–137. https://doi.org/10.1016/j.bbamem.2005.05.010

Fernandez-Ayala DJ, Lopez-Lluch G, Garcia-Valdes M, Arroyo A, Navas P (2005b) Specificity of coenzyme Q10 for a balanced function of respiratory chain and endogenous ubiquinone biosynthesis in human cells. Biochim Biophys Acta 1706(1–2):174–183. https://doi.org/10.1016/j.bbabio.2004.10.009

Fontaine E, Ichas F, Bernardi P (1998) A ubiquinone-binding site regulates the mitochondrial permeability transition pore. J Biol Chem 273(40):25734–25740. https://doi.org/10.1074/jbc.273.40.25734

Fukuwatari T, Shibata K (2009) Consideration of diurnal variations in human blood NAD and NADP concentrations. J Nutr Sci Vitaminol 55(3):279–281. https://doi.org/10.3177/jnsv.55.279

Gomez-Diaz C, Rodriguez-Aguilera JC, Barroso MP, Villalba JM, Navarro F, Crane FL, Navas P (1997a) Antioxidant ascorbate is stabilized by NADH-coenzyme Q10 reductase in the plasma membrane. J Bioenerg Biomembr 29(3):251–257. https://doi.org/10.1023/a:1022410127104

Gomez-Diaz C, Villalba JM, Perez-Vicente R, Crane FL, Navas P (1997b) Ascorbate stabilization is stimulated in rho(0)HL-60 cells by CoQ_{10} increase at the plasma membrane. Biochem Biophys Res Commun 234(1):79–81. https://doi.org/10.1006/bbrc.1997.6582

Haas RH (2019) Mitochondrial dysfunction in aging and diseases of aging. Biology (Basel) 8(2). https://doi.org/10.3390/biology8020048

Hangen E, De Zio D, Bordi M, Zhu C, Dessen P, Caffin F, Lachkar S, Perfettini JL, Lazar V, Benard J, Fimia GM, Piacentini M, Harper F, Pierron G, Vicencio JM, Benit P, de Andrade A, Hoglinger G, Culmsee C, Rustin P, Blomgren K, Cecconi F, Kroemer G, Modjtahedi N (2010) A brain-specific isoform of mitochondrial apoptosis-inducing factor: AIF2. Cell Death Differ 17(7):1155–1166. https://doi.org/10.1038/cdd.2009.211

Hey-Mogensen M, Goncalves RL, Orr AL, Brand MD (2014) Production of superoxide/H2O2 by dihydroorotate dehydrogenase in rat skeletal muscle mitochondria. Free Radic Biol Med 72:149–155. https://doi.org/10.1016/j.freeradbiomed.2014.04.007

Hsieh TC, Lu X, Wang Z, Wu JM (2006) Induction of quinone reductase NQO1 by resveratrol in human K562 cells involves the antioxidant response element ARE and is accompanied by nuclear translocation of transcription factor Nrf2. Med Chem 2(3):275–285. https://doi.org/10.2174/157340606776930709

Hyun DH, Lee GH (2015) Cytochrome b5 reductase, a plasma membrane redox enzyme, protects neuronal cells against metabolic and oxidative stress through maintaining redox state and bioenergetics. Age (Dordr) 37(6):122. https://doi.org/10.1007/s11357-015-9859-9

Hyun DH, Emerson SS, Jo DG, Mattson MP, de Cabo R (2006) Calorie restriction up-regulates the plasma membrane redox system in brain cells and suppresses oxidative stress during aging. Proc Natl Acad Sci U S A 103(52):19908–19912. https://doi.org/10.1073/pnas.0608008103

Hyun DH, Hunt ND, Emerson SS, Hernandez JO, Mattson MP, de Cabo R (2007) Up-regulation of plasma membrane-associated redox activities in neuronal cells lacking functional mitochondria. J Neurochem 100(5):1364–1374. https://doi.org/10.1111/j.1471-4159.2006.04411.x

Hyun DH, Kim J, Moon C, Lim CJ, de Cabo R, Mattson MP (2012) The plasma membrane redox enzyme NQO1 sustains cellular energetics and protects human neuroblastoma cells against metabolic and proteotoxic stress. Age (Dordr) 34(2):359–370. https://doi.org/10.1007/s11357-011-9245-1

Jaiswal AK (2000) Regulation of genes encoding NAD(P)H:quinone oxidoreductases. Free Radic Biol Med 29(3–4):254–262

Janikiewicz J, Szymanski J, Malinska D, Patalas-Krawczyk P, Michalska B, Duszynski J, Giorgi C, Bonora M, Dobrzyn A, Wieckowski MR (2018) Mitochondria-associated membranes in aging and senescence: structure, function, and dynamics. Cell Death Dis 9(3):332. https://doi.org/10.1038/s41419-017-0105-5

Jheng HF, Tsai PJ, Guo SM, Kuo LH, Chang CS, Su IJ, Chang CR, Tsai YS (2012) Mitochondrial fission contributes to mitochondrial dysfunction and insulin resistance in skeletal muscle. Mol Cell Biol 32(2):309–319. https://doi.org/10.1128/mcb.05603-11

Jimenez-Hidalgo M, Santos-Ocana C, Padilla S, Villalba JM, Lopez-Lluch G, Martin-Montalvo A, Minor RK, Sinclair DA, de Cabo R, Navas P (2009) NQR1 controls lifespan by regulating the promotion of respiratory metabolism in yeast. Aging Cell 8(2):140–151. https://doi.org/10.1111/j.1474-9726.2009.00461.x

Johansen K, Theorell H, Karlsson J, Diamant B, Folkers K (1991) Coenzyme Q10, alpha-tocopherol and free cholesterol in HDL and LDL fractions. Ann Med 23(6):649–656. https://doi.org/10.3109/07853899109148098

Joseph P, Xie T, Xu Y, Jaiswal AK (1994) NAD(P)H:quinone oxidoreductase1 (DT-diaphorase): expression, regulation, and role in cancer. Oncol Res 6(10–11):525–532

Kim SS, Kang MS, Choi YM, Suh YH, Kim DK (1997) Sphingomyelinase activity is enhanced in cerebral cortex of senescence-accelerated mouse-P/10 with advancing age. Biochem Biophys Res Commun 237(3):583–587. https://doi.org/10.1006/bbrc.1997.7133

Kim J, Kim SK, Kim HK, Mattson MP, Hyun DH (2013) Mitochondrial function in human neuroblastoma cells is up-regulated and protected by NQO1, a plasma membrane redox enzyme. PLoS One 8(7):e69030. https://doi.org/10.1371/journal.pone.0069030

Larm JA, Vaillant F, Linnane AW, Lawen A (1994) Up-regulation of the plasma membrane oxidoreductase as a prerequisite for the viability of human Namalwa rho 0 cells. J Biol Chem 269(48):30097–30100

Lass A, Sohal RS (1998) Electron transport-linked ubiquinone-dependent recycling of alpha-tocopherol inhibits autooxidation of mitochondrial membranes. Arch Biochem Biophys 352(2):229–236. https://doi.org/10.1006/abbi.1997.0606

Lenaz G, Paolucci U, Fato R, D'Aurelio M, Parenti Castelli G, Sgarbi G, Biagini G, Ragni L, Salardi S, Cacciari E (2002) Enhanced activity of the plasma membrane oxidoreductase in circulating lymphocytes from insulin-dependent diabetes mellitus patients. Biochem Biophys Res Commun 290(5):1589–1592. https://doi.org/10.1006/bbrc.2002.6392

Li X, Wang Y, Tan G, Lyu J, Ding H (2018) Electron transfer kinetics of the mitochondrial outer membrane protein mitoNEET. Free Radic Biol Med 121:98–104. https://doi.org/10.1016/j.freeradbiomed.2018.04.569

Lightle SA, Oakley JI, Nikolova-Karakashian MN (2000) Activation of sphingolipid turnover and chronic generation of ceramide and sphingosine in liver during aging. Mech Ageing Dev 120(1–3):111–125. https://doi.org/10.1016/s0047-6374(00)00191-3

Lipper CH, Stofleth JT, Bai F, Sohn YS, Roy S, Mittler R, Nechushtai R, Onuchic JN, Jennings PA (2019) Redox-dependent gating of VDAC by mitoNEET. Proc Natl Acad Sci U S A 116(40):19924–19929. https://doi.org/10.1073/pnas.1908271116

Loffler M, Jockel J, Schuster G, Becker C (1997) Dihydroorotat-ubiquinone oxidoreductase links mitochondria in the biosynthesis of pyrimidine nucleotides. Mol Cell Biochem 174(1–2):125–129

Lopez-Lluch G (2017) Mitochondrial activity and dynamics changes regarding metabolism in ageing and obesity. Mech Ageing Dev 162:108–121. https://doi.org/10.1016/j.mad.2016.12.005

Lopez-Lluch G, Buron MI, Alcain FJ, Quesada JM, Navas P (1998) Redox regulation of cAMP levels by ascorbate in 1,25-dihydroxy- vitamin D3-induced differentiation of HL-60 cells. Biochem J 331(Pt 1):21–27

Lopez-Lluch G, Barroso MP, Martin SF, Fernandez-Ayala DJ, Gomez-Diaz C, Villalba JM, Navas P (1999) Role of plasma membrane coenzyme Q on the regulation of apoptosis. Biofactors 9(2–4):171–177

Lopez-Lluch G, Blazquez MV, Perez-Vicente R, Macho A, Buron MI, Alcain FJ, Munoz E, Navas P (2001) Cellular redox state and activating protein-1 are involved in ascorbate effect on calcitriol-induced differentiation. Protoplasma 217(1–3):129–136

Lopez-Lluch G, Fernandez-Ayala DJ, Alcain FJ, Buron MI, Quesada JM, Navas P (2005a) Inhibition of COX activity by NSAIDs or ascorbate increases cAMP levels and enhances differentiation in 1alpha,25-dihydroxyvitamin D3-induced HL-60 cells. Arch Biochem Biophys 436(1):32–39. https://doi.org/10.1016/j.abb.2004.12.031

Lopez-Lluch G, Rios M, Lane MA, Navas P, de Cabo R (2005b) Mouse liver plasma membrane redox system activity is altered by aging and modulated by calorie restriction. Age (Dordr) 27(2):153–160. https://doi.org/10.1007/s11357-005-2726-3

Lopez-Lluch G, Rodriguez-Aguilera JC, Santos-Ocana C, Navas P (2010) Is coenzyme Q a key factor in aging? Mech Ageing Dev 131(4):225–235. https://doi.org/10.1016/j.mad.2010.02.003

Martin-Montalvo A, Sun Y, Diaz-Ruiz A, Ali A, Gutierrez V, Palacios HH, Curtis J, Siendones E, Ariza J, Abulwerdi GA, Sun X, Wang AX, Pearson KJ, Fishbein KW, Spencer RG, Wang M, Han X, Scheibye-Knudsen M, Baur JA, Shertzer HG, Navas P, Villalba JM, Zou S, Bernier M, de Cabo R (2016) Cytochrome b5 reductase and the control of lipid metabolism and healthspan. NPJ Aging Mech Dis 2:16006. https://doi.org/10.1038/npjamd.2016.6

Martinus RD, Linnane AW, Nagley P (1993) Growth of rho 0 human Namalwa cells lacking oxidative phosphorylation can be sustained by redox compounds potassium ferricyanide or coenzyme Q10 putatively acting through the plasma membrane oxidase. Biochem Mol Biol Int 31(6):997–1005

Mohr D, Bowry VW, Stocker R (1992) Dietary supplementation with coenzyme Q10 results in increased levels of ubiquinol-10 within circulating lipoproteins and increased resistance of human low-density lipoprotein to the initiation of lipid peroxidation. Biochim Biophys Acta 1126(3):247–254. https://doi.org/10.1016/0005-2760(92)90237-p

Morales CR, Grigoryeva LS, Pan X, Bruno L, Hickson G, Ngo MH, McMaster CR, Samuels ME, Pshezhetsky AV (2014) Mitochondrial damage and cholesterol storage in human hepatocellular carcinoma cells with silencing of UBIAD1 gene expression. Mol Genet Metab Rep 1:407–411. https://doi.org/10.1016/j.ymgmr.2014.09.001

Mourier A, Motori E, Brandt T, Lagouge M, Atanassov I, Galinier A, Rappl G, Brodesser S, Hultenby K, Dieterich C, Larsson NG (2015) Mitofusin 2 is required to maintain mitochondrial coenzyme Q levels. J Cell Biol 208(4):429–442. https://doi.org/10.1083/jcb.201411100

Mugoni V, Postel R, Catanzaro V, De Luca E, Turco E, Digilio G, Silengo L, Murphy MP, Medana C, Stainier DY, Bakkers J, Santoro MM (2013) Ubiad1 is an antioxidant enzyme that regulates eNOS activity by CoQ_{10} synthesis. Cell 152(3):504–518. https://doi.org/10.1016/j.cell.2013.01.013

Navarro F, Villalba JM, Crane FL, Mackellar WC, Navas P (1995) A phospholipid-dependent NADH-coenzyme Q reductase from liver plasma membrane. Biochem Biophys Res Commun 212(1):138–143. https://doi.org/10.1006/bbrc.1995.1947

Navarro F, Navas P, Burgess JR, Bello RI, De Cabo R, Arroyo A, Villalba JM (1998) Vitamin E and selenium deficiency induces expression of the ubiquinone-dependent antioxidant system at the plasma membrane. FASEB J 12(15):1665–1673. https://doi.org/10.1096/fasebj.12.15.1665
Navas P, Fernandez-Ayala DM, Martin SF, Lopez-Lluch G, De Caboa R, Rodriguez-Aguilera JC, Villalba JM (2002) Ceramide-dependent caspase 3 activation is prevented by coenzyme Q from plasma membrane in serum-deprived cells. Free Radic Res 36(4):369–374. https://doi.org/10.1080/10715760290021207
Niki E (1987) Antioxidants in relation to lipid peroxidation. Chem Phys Lipids 44(2–4):227–253
Niklowitz P, Onur S, Fischer A, Laudes M, Palussen M, Menke T, Doring F (2016) Coenzyme Q10 serum concentration and redox status in European adults: influence of age, sex, and lipoprotein concentration. J Clin Biochem Nutr 58(3):240–245. https://doi.org/10.3164/jcbn.15-73
Nordberg J, Arner ES (2001) Reactive oxygen species, antioxidants, and the mammalian thioredoxin system. Free Radic Biol Med 31(11):1287–1312. https://doi.org/10.1016/s0891-5849(01)00724-9
Nordman T, Xia L, Bjorkhem-Bergman L, Damdimopoulos A, Nalvarte I, Arner ES, Spyrou G, Eriksson LC, Bjornstedt M, Olsson JM (2003) Regeneration of the antioxidant ubiquinol by lipoamide dehydrogenase, thioredoxin reductase and glutathione reductase. Biofactors 18(1–4):45–50. https://doi.org/10.1002/biof.5520180206
Okamoto T, Matsuya T, Fukunaga Y, Kishi T, Yamagami T (1989) Human serum ubiquinol-10 levels and relationship to serum lipids. Int J Vitam Nutr Res 59(3):288–292
Olgun A (2009) Converting NADH to NAD+ by nicotinamide nucleotide transhydrogenase as a novel strategy against mitochondrial pathologies during aging. Biogerontology 10(4):531–534. https://doi.org/10.1007/s10522-008-9190-2
Olsson JM, Xia L, Eriksson LC, Bjornstedt M (1999) Ubiquinone is reduced by lipoamide dehydrogenase and this reaction is potently stimulated by zinc. FEBS Lett 448(1):190–192. https://doi.org/10.1016/s0014-5793(99)00363-4
Petkova DH, Momchilova-Pankova AB, Markovska TT, Koumanov KS (1988) Age-related changes in rat liver plasma membrane sphingomyelinase activity. Exp Gerontol 23(1):19–24. https://doi.org/10.1016/0531-5565(88)90016-2
Piechota J, Szczesny R, Wolanin K, Chlebowski A, Bartnik E (2006) Nuclear and mitochondrial genome responses in HeLa cells treated with inhibitors of mitochondrial DNA expression. Acta Biochim Pol 53(3):485–495
Prochaska HJ, Santamaria AB, Talalay P (1992) Rapid detection of inducers of enzymes that protect against carcinogens. Proc Natl Acad Sci U S A 89(6):2394–2398
Quesada JM, Lopez LG, Buron MI, Alcain FJ, Borrego F, Velde JP, Blanco I, Bouillon R, Navas P (1996) Ascorbate increases the 1,25 dihydroxyvitamin D3-induced monocytic differentiation of HL-60 cells. Calcif Tissue Int 59(4):277–282
Quinzii CM, Luna-Sanchez M, Ziosi M, Hidalgo-Gutierrez A, Kleiner G, Lopez LC (2017) The role of Sulfide oxidation impairment in the pathogenesis of primary CoQ deficiency. Front Physiol 8:525. https://doi.org/10.3389/fphys.2017.00525
Rodriguez-Aguilera JC, Lopez-Lluch G, Santos-Ocana C, Villalba JM, Gomez-Diaz C, Navas P (2000) Plasma membrane redox system protects cells against oxidative stress. Redox Rep 5(2–3):148–150. https://doi.org/10.1179/135100000101535528
Rodriguez-Bies E, Navas P, Lopez-Lluch G (2015) Age-dependent effect of every-other-day feeding and aerobic exercise in ubiquinone levels and related antioxidant activities in mice muscle. J Gerontol A Biol Sci Med Sci 70(1):33–43. https://doi.org/10.1093/gerona/glu002
Rodriguez-Bies E, Tung BT, Navas P, Lopez-Lluch G (2016) Resveratrol primes the effects of physical activity in old mice. Br J Nutr 116(6):979–988. https://doi.org/10.1017/S0007114516002920
Ross D, Siegel D (2017) Functions of NQO1 in cellular protection and CoQ_{10} metabolism and its potential role as a redox sensitive molecular switch. Front Physiol 8:595. https://doi.org/10.3389/fphys.2017.00595

Ross D, Siegel D, Gibson NW, Pacheco D, Thomas DJ, Reasor M, Wierda D (1990) Activation and deactivation of quinones catalyzed by DT-diaphorase. Evidence for bioreductive activation of diaziquone (AZQ) in human tumor cells and detoxification of benzene metabolites in bone marrow stroma. Free Radic Res Commun 8(4–6):373–381. https://doi.org/10.3109/10715769009053371

SantaCruz KS, Yazlovitskaya E, Collins J, Johnson J, DeCarli C (2004) Regional NAD(P)H:quinone oxidoreductase activity in Alzheimer's disease. Neurobiol Aging 25(1):63–69. https://doi.org/10.1016/s0197-4580(03)00117-9

Santoro MM (2020) The antioxidant role of non-mitochondrial CoQ_{10}: mystery solved! Cell Metab 31(1):13–15. https://doi.org/10.1016/j.cmet.2019.12.007

Santos-Ocana C, Villalba JM, Cordoba F, Padilla S, Crane FL, Clarke CF, Navas P (1998) Genetic evidence for coenzyme Q requirement in plasma membrane electron transport. J Bioenerg Biomembr 30(5):465–475. https://doi.org/10.1023/a:1020542230308

Saraswat K, Rizvi SI (2017) Novel strategies for anti-aging drug discovery. Expert Opin Drug Discov 12(9):955–966. https://doi.org/10.1080/17460441.2017.1349750

Sharma A, Smith HJ, Yao P, Mair WB (2019) Causal roles of mitochondrial dynamics in longevity and healthy aging. EMBO Rep:e48395. https://doi.org/10.15252/embr.201948395

Shimizu M, Miyazaki T, Takagi A, Sugita Y, Yatsu S, Murata A, Kato T, Suda S, Ouchi S, Aikawa T, Hiki M, Takahashi S, Hiki M, Hayashi H, Kasai T, Shimada K, Miyauchi K, Daida H (2017) Low circulating coenzyme Q10 during acute phase is associated with inflammation, malnutrition, and in-hospital mortality in patients admitted to the coronary care unit. Heart Vessel 32(6):668–673. https://doi.org/10.1007/s00380-016-0923-x

Siendones E, SantaCruz-Calvo S, Martin-Montalvo A, Cascajo MV, Ariza J, Lopez-Lluch G, Villalba JM, Acquaviva-Bourdain C, Roze E, Bernier M, de Cabo R, Navas P (2014) Membrane-bound CYB5R3 is a common effector of nutritional and oxidative stress response through FOXO3a and Nrf2. Antioxid Redox Signal 21(12):1708–1725. https://doi.org/10.1089/ars.2013.5479

Siendones E, Ballesteros M, Navas P (2018) Cellular and molecular mechanisms of recessive hereditary methaemoglobinaemia type II. J Clin Med 7(10). https://doi.org/10.3390/jcm7100341

Spence MW, Burgess JK (1978) Acid and neutral sphingomyelinases of rat brain. Activity in developing brain and regional distribution in adult brain. J Neurochem 30(4):917–919. https://doi.org/10.1111/j.1471-4159.1978.tb10804.x

Stefely JA, Pagliarini DJ (2017) Biochemistry of mitochondrial coenzyme Q biosynthesis. Trends Biochem Sci 42(10):824–843. https://doi.org/10.1016/j.tibs.2017.06.008

Stocker R, Suarna C (1993) Extrecellular reduction of ubiquinone-1 and -10 by human Hep G2 and blood cells. Biochim Biophys Acta Gen Subj 1158(1):15–22. https://doi.org/10.1016/0304-4165(93)90090-U

Stocker R, Bowry VW, Frei B (1991) Ubiquinol-10 protects human low density lipoprotein more efficiently against lipid peroxidation than does alpha-tocopherol. Proc Natl Acad Sci U S A 88(5):1646–1650. https://doi.org/10.1073/pnas.88.5.1646

Sun IL, Sun EE, Crane FL, Morré DJ (1990) Evidence for coenzyme Q function in transplasma membrane electron transport. Biochem Biophys Res Commun 172(3):979–984. https://doi.org/10.1016/0006-291X(90)91542-Z

Sun IL, Sun EE, Crane FL (1992a) Stimulation of serum-free cell proliferation by Coenzyme Q. Biochem Biophys Res Commun 189(1):8–13. https://doi.org/10.1016/0006-291X(92)91517-T

Sun IL, Sun EE, Crane FL, Morre DJ, Lindgren A, Low H (1992b) Requirement for coenzyme Q in plasma membrane electron transport. Proc Natl Acad Sci U S A 89(23):11126–11130. https://doi.org/10.1073/pnas.89.23.11126

Takahashi T, Shitashige M, Okamoto T, Kishi T, Goshima K (1992) A novel ubiquinone reductase activity in rat cytosol. FEBS Lett 314(3):331–334. https://doi.org/10.1016/0014-5793(92)81499-c

Takahashi T, Mine Y, Okamoto T (2019) Extracellular coenzyme Q10 (CoQ_{10}) is reduced to ubiquinol-10 by intact Hep G2 cells independent of intracellular CoQ_{10} reduction. Arch Biochem Biophys 672:108067. https://doi.org/10.1016/j.abb.2019.108067

Teclebrhan H, Jakobsson-Borin A, Brunk U, Dallner G (1995) Relationship between the endoplasmic reticulum-Golgi membrane system and ubiquinone biosynthesis. Biochim Biophys Acta 1256(2):157–165. https://doi.org/10.1016/0005-2760(95)00016-6

Tian G, Sawashita J, Kubo H, Nishio SY, Hashimoto S, Suzuki N, Yoshimura H, Tsuruoka M, Wang Y, Liu Y, Luo H, Xu Z, Mori M, Kitano M, Hosoe K, Takeda T, Usami S, Higuchi K (2014) Ubiquinol-10 supplementation activates mitochondria functions to decelerate senescence in senescence-accelerated mice. Antioxid Redox Signal 20(16):2606–2620. https://doi.org/10.1089/ars.2013.5406

Torres-Lista V, Parrado-Fernandez C, Alvarez-Monton I, Frontinan-Rubio J, Duran-Prado M, Peinado JR, Johansson B, Alcain FJ, Gimenez-Llort L (2014) Neophobia, NQO1 and SIRT1 as premorbid and prodromal indicators of AD in 3xTg-AD mice. Behav Brain Res 271:140–146. https://doi.org/10.1016/j.bbr.2014.04.055

Tsai KL, Chen LH, Chiou SH, Chiou GY, Chen YC, Chou HY, Chen LK, Chen HY, Chiu TH, Tsai CS, Ou HC, Kao CL (2011) Coenzyme Q10 suppresses oxLDL-induced endothelial oxidative injuries by the modulation of LOX-1-mediated ROS generation via the AMPK/PKC/NADPH oxidase signaling pathway. Mol Nutr Food Res 55(Suppl 2):S227–S240. https://doi.org/10.1002/mnfr.201100147

Tung BT, Rodriguez-Bies E, Ballesteros-Simarro M, Motilva V, Navas P, Lopez-Lluch G (2014) Modulation of endogenous antioxidant activity by resveratrol and exercise in mouse liver is age dependent. J Gerontol A Biol Sci Med Sci 69(4):398–409. https://doi.org/10.1093/gerona/glt102

Vasiliou V, Theurer MJ, Puga A, Reuter SF, Nebert DW (1994) Mouse dioxin-inducible NAD(P)H: menadione oxidoreductase: NMO1 cDNA sequence and genetic differences in mRNA levels. Pharmacogenetics 4(6):341–348

Venable ME, Lee JY, Smyth MJ, Bielawska A, Obeid LM (1995) Role of ceramide in cellular senescence. J Biol Chem 270(51):30701–30708. https://doi.org/10.1074/jbc.270.51.30701

Villalba JM, Navas P (2000) Plasma membrane redox system in the control of stress-induced apoptosis. Antioxid Redox Signal 2(2):213–230. https://doi.org/10.1089/ars.2000.2.2-213

Villalba JM, Navarro F, Cordoba F, Serrano A, Arroyo A, Crane FL, Navas P (1995) Coenzyme Q reductase from liver plasma membrane: purification and role in trans-plasma-membrane electron transport. Proc Natl Acad Sci U S A 92(11):4887–4891

Wang Z, Zhang L, Liang Y, Zhang C, Xu Z, Zhang L, Fuji R, Mu W, Li L, Jiang J, Ju Y, Wang Z (2015) Cyclic AMP mimics the anti-ageing effects of calorie restriction by up-regulating sirtuin. Sci Rep 5:12012. https://doi.org/10.1038/srep12012

Wang Y, Landry AP, Ding H (2017) The mitochondrial outer membrane protein mitoNEET is a redox enzyme catalyzing electron transfer from FMNH2 to oxygen or ubiquinone. J Biol Chem 292(24):10061–10067. https://doi.org/10.1074/jbc.M117.789800

Wiley SE, Murphy AN, Ross SA, van der Geer P, Dixon JE (2007) MitoNEET is an iron-containing outer mitochondrial membrane protein that regulates oxidative capacity. Proc Natl Acad Sci U S A 104(13):5318–5323. https://doi.org/10.1073/pnas.0701078104

Williams CH Jr, Zanetti G, Arscott LD, McAllister JK (1967) Lipoamide dehydrogenase, glutathione reductase, thioredoxin reductase, and thioredoxin. J Biol Chem 242(22):5226–5231

Xia L, Bjornstedt M, Nordman T, Eriksson LC, Olsson JM (2001) Reduction of ubiquinone by lipoamide dehydrogenase. An antioxidant regenerating pathway. Eur J Biochem 268(5):1486–1490. https://doi.org/10.1046/j.1432-1327.2001.02013.x

Xia L, Nordman T, Olsson JM, Damdimopoulos A, Bjorkhem-Bergman L, Nalvarte I, Eriksson LC, Arner ES, Spyrou G, Bjornstedt M (2003) The mammalian cytosolic selenoenzyme thioredoxin reductase reduces ubiquinone. A novel mechanism for defense against oxidative stress. J Biol Chem 278(4):2141–2146. https://doi.org/10.1074/jbc.M210456200

Yamamoto Y, Yamashita S (1997) Plasma ratio of ubiquinol and ubiquinone as a marker of oxidative stress. Mol Asp Med 18(Suppl):S79–S84. https://doi.org/10.1016/s0098-2997(97)00007-1

Yuan H, Li X, Zhang X, Kang R, Tang D (2016) CISD1 inhibits ferroptosis by protection against mitochondrial lipid peroxidation. Biochem Biophys Res Commun 478(2):838–844. https://doi.org/10.1016/j.bbrc.2016.08.034

Ziosi M, Di Meo I, Kleiner G, Gao XH, Barca E, Sanchez-Quintero MJ, Tadesse S, Jiang H, Qiao C, Rodenburg RJ, Scalais E, Schuelke M, Willard B, Hatzoglou M, Tiranti V, Quinzii CM (2017) Coenzyme Q deficiency causes impairment of the sulfide oxidation pathway. EMBO Mol Med 9(1):96–111. https://doi.org/10.15252/emmm.201606356

Chapter 5
Regulation of Synthesis of Coenzyme Q_{10}

María Victoria Cascajo-Almenara and Guillermo López-Lluch

Abstract The regulation of gene expression needs many mechanisms of control which are multiple and diverse, encompassing from DNA structure to synthesis, modification and location of proteins. These processes, in addition to being tissue-specific, change during embryonic development and throughout the life of the organism. CoQ_{10} is essential to maintain cell homeostasis and to increase life-span. CoQ_{10} biosynthesis pathway is strictly regulated and this program involves different levels of regulation (transcriptional, post-transcriptional and post-translational). Here we have focused on processes of post-transcriptional regulation governed by RNA-binding proteins (RBPs) and review the role of these RBPs in senescence and aging. Finally, we report post-transcriptional and post-translational events involved in the regulation of several *COQs* genes involved in the CoQ_{10} synthesis complex assembly. The knowledge of processes that control CoQ_{10} levels in cells is key to develop therapies to stimulate the endogenous synthesis of CoQ_{10}.

Keywords Coenzyme Q · Protein stability · Protein regulation · HuR · RNA binding proteins (RBPs)

5.1 Introduction

The expression of genes changes enormously along the life of organisms. The set of genes characterizing each cell and tissue in different stage of the development are transcriptionally regulated by different transcription factors that bind to regulatory elements located normally near the promotor or each gene. However, this is not the only level of regulation that affects the transit from DNA to protein, several post-transcriptional regulatory processes are also involved in the control of the levels of the proteins codified by these genes. This post-transcriptional regulation involves

M. V. Cascajo-Almenara (✉) · G. López-Lluch
Centro Andaluz de Biología del Desarrollo, CABD-CSIC, CIBERER,
Instituto de Salud Carlos III, Universidad Pablo de Olavide, Sevilla, Spain
e-mail: mvcasalm@upo.es

G. López Lluch (ed.), *Coenzyme Q in Aging*,
https://doi.org/10.1007/978-3-030-45642-9_5

processes related with mRNA metabolism, such as splicing, nuclear export, cytoplasm localization, stability and translation (Glisovic et al. 2008; Mata et al. 2005; Zhao et al. 2017).

Many authors defend the existence of common expression patterns occurring in a coordinated way. Thus, cells are able to adapt and respond coordinately to different alterations or stimuli. This adaptation also occurs during aging. Many of mRNA encoding proteins related with aging are associated to RNA-binding proteins (RBPs) which regulate protein production, at least in part, by modification of the mRNA stability and translation level (Masuda et al. 2012). These post-transcriptional regulation processes involve cis-acting elements in the RNA molecule, such as AU-rich elements (AREs) generally located at the 5′ and 3′mRNA untranslated regions (UTR), and poliA tail. AREs play a critical role in the regulation of gene expression during cell growth, differentiation, immune response and cancer (Chen and Shyu 1995).

AREs are responsible for regulating the stability of transcripts (Franks and Lykke-Andersen 2008; Pesole et al. 2001). The regulation of these elements is important during aging since many senescence-associated genes contain ARE sequences (Borbolis and Syntichaki 2015). These motifs specifically interact with trans-acting factors, microRNA and RNA-binding proteins, forming ribonucleoprotein complexes (RNPs) (Krecic and Swanson 1999). Theses interactions are very dynamic and RNPs composition depends on cell type, growth conditions, cell compartment and target mRNA (Masuda et al. 2009). In this chapter, we will focus in RBPs and its function in cellular senescence, aging and control of the members of the CoQ synthesis complex.

5.2 Molecular Mechanisms of mRNA Stability Regulation

The molecular mechanisms by which RBPs regulate mRNA stability are still not completely clarified. Among RBPs, several of them, decrease the mRNA stability whereas others stabilize mRNA and even regulate translational rate. RBPs associate mRNA with cellular structures such as exosome, processing P-bodies and stress granules involved in the mRNA turnover (Adjibade and Mazroui 2014; Chen et al. 2001; Kedersha and Anderson 2002; Sheth and Parker 2003; von Roretz et al. 2011). Some RBPs increase half-life of mRNA through a binding competition preventing the association of destabilizing RBPs (Kuwano et al. 2008). It has been also described that several RBPs act at the translational level, specifically during initiation, elongation and/or termination regulating the translation rate of transcripts with which they interact (Kawai et al. 2006; Mazan-Mamczarz et al. 2006).

Different RBPs include some proteins that show affinity for the same mRNA. Consequently, these proteins probably associate and function in a cooperative or competitive manner on the regulation of the same target mRNA (Lal et al. 2004). Among the RBPs identified involved in senescence and aging, we can find stabilizing proteins such as HuR, a RBP ubiquitously expressed belonging to ELAV (embryonic lethal abnormal visual)/Hu protein family (Herman and Autieri 2017) and QKI, Quaking RBP (Darbelli and Richard 2016). AUF1, ARE /poly(U)-binding/degradation factor 1 (Gratacos and Brewer 2010), CUGBP1, a member of the CUG-BP- and ETR-3-like factors family (Jones et al. 2012), KSRP, KH-type splicing regulatory protein are involved in alternative splicing, destabilization and translation enhancement (Apponi et al. 2011). CRT, calreticulin (Timchenko et al. 2002), and TTP, tristetraprolin (Apponi et al. 2011) regulate the destabilization of mRNA. On the other hand, other RBPs show dual activity, for example TIA-1, T-cell restricted intracellular antigen 1, that functions as a translational repressor of mitochondrial proteins such as cytochrome c (Kawai et al. 2006) whereas promotes mitochondrial fragmentation by enhancing mitochondrial fission factor translation (Tak et al. 2017). In general, all these proteins have been associate with the regulation of transcripts and translation during development and aging (Campos-Melo et al. 2014; Kai 2016; Kim et al. 2017). Among the proteins regulated by the RBPs, we can fing the cyclin kinase inhibitors p16, p21 and p27 and cyclins A and B (Cho et al. 2010; Kullmann et al. 2002; Wang et al. 2000, 2005). Levels of these proteins change with age, and they are involved in stress response, proliferation and cell cycle control.

5.3 RBPs-Dependent Regulation of mRNA Stability in Cell Growth and Senescence

Cellular senescence is the process by cells remain retained in cell cycle, after a finite number of divisions, but they are viable and metabolically active for a long time. HuR is one of the most studied RBP and has been associated with cell senescence. This family also includes to HuB, HuC y HuD, which are expressed predominantly in neurons. HuR has three RNA-recognition motives (RRM) through which it binds to AU and U-rich elements mainly located in 3′UTR of the transcripts (Lopez de Silanes et al. 2004; Ma et al. 1996; Peng et al. 1998). Although, HuR is found primarily in the nucleus, its function on target mRNA depends of its cytoplasmic localization (Atasoy et al. 1998). In fact, HuR can be shuttled between nucleus and cytoplasm through its RNS sequence (shuttling sequence) located between second and third RRM (Fan and Steitz 1998).

Cell cycle progression depends on HuR cytoplasmic level since HuR regulates mRNA encoding cyclins A and B1 (Wang et al. 2000). The stability of both cyclins increases by HuR interaction. Conversely, reduction of HuR level destabilizes the mRNA of cyclin A, B1 and c-fos mRNA, decreasing its respective protein level and thereby inducing cell senescence (Wang et al. 2000). The effect of HuR is controlled by a methyltransferase (CARM1) that enhances the regulatory capacity of HuR on these mRNAs (Pang et al. 2013) Further, it's been shown that this RBP binds to the cycling kinase inhibitor $p27^{kip1}$ mRNA at its 5′UTR inhibiting its translation (Kullmann et al. 2002). Since $p27^{kip1}$ binds to cyclin E/Cdk2 or cyclin D/Cdk 2 inhibiting the progression through G_1 phase of the cell cycle, its inhibition by HuR will reduce the possibility of cell cycle blocking (Polyak et al. 1994). On the other hand, interaction of HuR to p16/INK4 3′UTR induces a decrease of mRNA stability, by recruitment of another RBP, AUF1 (Chang et al. 2010).

HuR also regulates stress and immune response proteins, such as, c-myc, vascular endothelial growth factor (VEGF), hypoxia-inducible factor (HIF)-1, prothymosin, cyclooxygenase (COX)-2, tumor necrosis factor-alpha (TNF-α), and several interleukins, indicating a putative role in the control of the immune system and in immunosenescence (Casolaro et al. 2008; Dean et al. 2001; Galban et al. 2008; Lal et al. 2005; Levy et al. 1998; Liu et al. 2009; Sengupta et al. 2003).

Expression of three RBPs, HuR, AUF1 and TIA-1, were analyzed in a model of cellular aging, such as human diploid fibroblasts (HDF) in culture (Wang et al. 2001). In these cells, the interaction of RBPs to target transcripts decreased during replicative senescence, so this situation induced a reduction of mRNA stability and the level of protein regulated by HuR. On the other hand, HuR, AUF1 and TIA-1 level studied from biopsies of different human tissues and different age (fetal, young, adult and old) remains unchanged, or even a slight increase in lung and intestine, suggesting that HuR is functional during tissue aging (Kim et al. 2017; Masuda et al. 2009; Pesole et al. 2001). The reduction of HuR protein expression in senescence seems to be due to a lower translation rate, because mRNA levels remain unchanged. Additionally, HuR cellular distribution changes during senescence (Yi et al. 2010). In a positive feedback, HuR can associate to 3′UTR of HuR mRNA regulating the nuclear export and consequently increasing the HuR translation rate. This mechanism of regulation might explain the progressive loss of HuR protein during senescence and increase in cancer model.

The discrepancies between these studies raises the differences between in vivo and HDF in culture senescence. Scientific community agrees that senescence of HDF cells recapitulates some features of elderly cells although not all of them in the whole organism. Senescent cells accumulate with age and they contribute to process of aging by accelerating the tissue regeneration loss. However, the exact connection between cellular senescence and aging is still not clear. All these works show the need to deepen in study of senescence models used and the influence of RBPs in this process.

5.4 Deregulation of the Stability of mRNA Is Associated with Aging and Age Related Diseases

Aging is accompanied by multiple levels of deregulation of several systems affecting cells and tissue in different ways. In the case of mRNA stability, the role of RBPs in the decline of functionality of organs and tissues is not completely clear. However, although the information about the functions of these proteins in the response of organs such as muscle to disuse, aging or exercise, is limited, it has been recently proposed that the dynamic changes in RBPs can play a key role in the response of muscle to muscle loss in aged people (Van Pelt et al. 2019).

Differences in the profile of transcripts during aging are clear although the mechanisms involved have not been clarified to date. In studies carried out in human cohorts, the profile of N6-methyladenosine (m6A) decreases and this decrease negatively affects the levels of several mRNAs (Min et al. 2018). Dysregulation of mRNA stability has been associated with aged tissues and importantly with the progression of age-associated diseases like atherosclerosis and inflammatory profile (Herman and Autieri 2018; Masuda et al. 2009). However, the information about the dynamic of RBPs during aging and immunosenescence is not clear and needs further research (Herman and Autieri 2018).

In the case of mitochondrial dysfunction as hallmark of aging, RBPs can be involved in a vicious cycle. The mechanism of response to mitochondrial protein stress, named mitochondrial unfolded protein response (mtUPR), is essential to maintain mitochondrial turnover (Yi et al. 2018). However, in some cases this mechanism can be blocked by mRNA instability that affects aging or the progression of aging-related diseases. The response to mtUPR seems to depend on a mitochondrial transporter known as ABCB10. Deletion of this protein causes increase of ROS levels and cell death indicating an important role in mitochondrial activity. Huntingtin, the protein involved in Hungtinton's disease, decreases the levels of ABCB10 protein by reducing the stability of its mRNA and decreases its translation (Fu et al. 2019).

5.5 Post-transcriptional Regulation of of CoQ_{10}

Regulation of the half life of mRNA transcripts and translation rate of the components of the CoQ_{10} synthesis machinery is not clear. We have found organ-dependent changes in the levels of the transcripts of many of these genes along mice life showing a puzzling regulation. In many of the organs, especially in those showing mitotic capacity, the mRNA levels of the components of the synthesis increase during maturity and decrease to young levels or less in the case of aged animals (Campos-Silva et al. 2017). This behavior of the COQs transcripts resembles the pattern of expression found in humans in which the levels of expression of miRNAs in centenarians are more similar to the pattern in young people than to the pattern in octogenarians (Serna et al. 2012). In fact, centenarians seem to maintain the capacity to produce microRNAs whereas this pathway is impaired in octogenarians (Borras et al. 2017) and their transcriptome is distinct from septuagenarians, probably indicating the induction of regulatory pathways involved in survival at very advanced ages resembling the regulation of young people (Borras et al. 2016).

As it has been previously indicated in Chap. 2 of this book that Coenzyme Q (CoQ_{10} in human and CoQ_6 in yeast) is a lipid endogenously synthesized in mitochondria, and it is present in all the cell membranes. CoQ biosynthesis pathway is a complex process that involved at least the product of 13 genes in mammals *(COQs and ADCKs)*. The products of *COQs* genes form a multi-enzyme complex, anchored in the inner mitochondrial membrane. PDSS1, PDSS2, COQ2, COQ3, COQ5, COQ6 and COQ7 proteins play an enzymatic function. The role of several other proteins is still unknown, such as COQ4, COQ8A, COQ8B, COQ9, COQ10 and COQ11 and some authors hypothesized that they play regulatory functions (Stefely and Pagliarini 2017).

Homeostasis of ubiquinone is essential to maintain mitochondrial function, and therefore, to keep appropriate physiological activities in the cell. CoQ content and *COQs* gene regulation change in function of tissue, nutritional, physiological and pathological conditions and it's known that CoQ is essential in the longevity of different animal models (SPAcosta et al. 2016; Linnane et al. 2002; Navas et al. 2005; Parrado-Fernandez et al. 2011). We have found that gene expression patterns change enormously throughout the life of an organism (Campos-Silva et al. 2017). All these evidences suggest that biosynthetic pathway must be tightly regulated, including several levels of regulation (transcriptional and post-translational), to balance oxidative phosphorylation and adapting to changes in nutritional and/or physiological conditions and mitochondrial activity.

In humans, deficit of CoQ_{10} induces heterogeneous and very severe symptoms, such as nephrotic syndrome, encephalomyopathy and ataxia (Quinzii et al. 2007). CoQ_{10} deficiency is considered a rare disease caused by either mutation in any gene involved in its biosynthesis (primary deficiency) or by other mitochondrial dysfunctions producing CoQ_{10} deficiency in which CoQ-synthome is not be directly affected (secondary deficiency) (Doimo et al. 2014; Kuhl et al. 2017; Yubero et al. 2016).

Regulatory mechanisms of CoQ biosynthesis include transcriptional regulation of enzymes in the mevalonate pathway, which is shared with other lipids such as dolichol and cholesterol, via the transcriptional factor PPARα (Bentinger et al. 2008; Bentinger et al. 2003). The transcriptional induction of *COQ7* expression has been associated to the activity of NF-κB in response to oxidative stress and probably in inflammation responses (Brea-Calvo et al. 2009). Interestingly, CoQ_{10} has been associated with the regulation of NF-E2 related factor 2 (NRF2)/antioxidant response element (ARE) in many systems (Choi et al. 2009; Li et al. 2016), but, despite its important role as antioxidant (Chap. 3 in this book), this regulatory factor has not been directly related with the control of *COQs* gene expression.

Several findings have demonstrated that, in many cases, there is not a direct relationship between mRNA and protein levels. In theses cases, increase at the mRNA levels are not accompanied by high protein amount indicating the presence of post-translational mechanisms of regulation affecting mRNA stability and/or translational rate (Lohman et al. 2014; Luna-Sanchez et al. 2015). Recent literature highlights the importance of post-transcriptional regulation of *COQ* genes and the structure of the complex of CoQ synthesis in the modulation of ubiquinone levels. At post-transcriptional level, it has been described that mRNA-binding proteins (RBPs) regulate CoQ_{10} biosynthesis. Specifically, HuR and hnRNP C1/C2 bind to 3'UTR of COQ7 mRNA modulating the stability and translation of the transcript of COQ7. HuR and hnRNP C1/C2 play opposite roles. Whereas HuR stabilizes COQ7 mRNA, maintaining the physiological level of protein, hnRNP C1/C2 induces

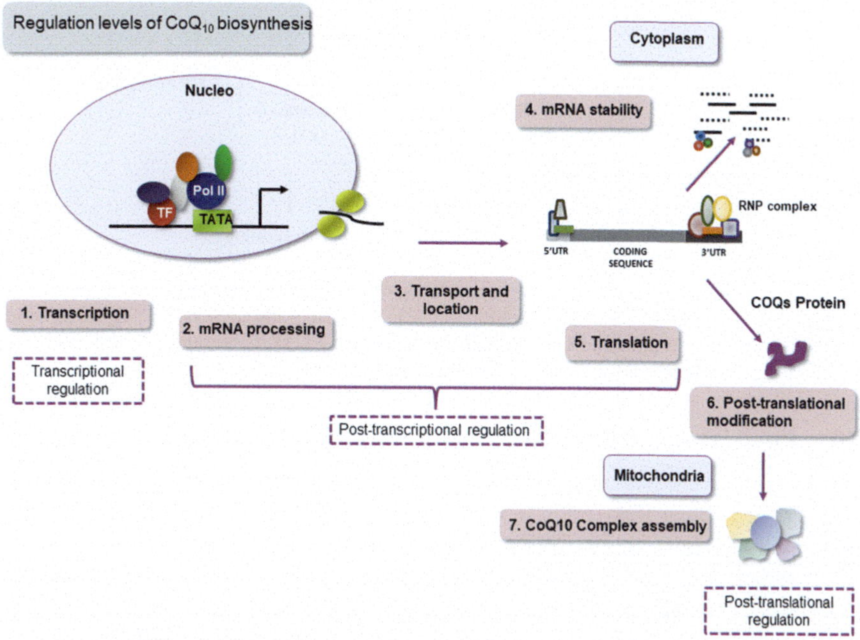

Fig. 5.1 Regulation levels of CoQ_{10} biosynthesis. Synthesis of CoQ_{10} depends on several levels of regulation. 1. Transcriptional regulation is not completely known since many of the probable transcriptional factors involved in the expression of the members of the Q-synthome are not known yet. 2–5. Regulation of the mRNA processing, such as transport, location, stability and translation control depends on RBPs and these processes seem to be key points in regulation of the synthesis of CoQ_{10}. 6. Post-translational modifications involve putative phosphorylation points and proteolysis into mitochondria. 7. Proteins are finally assembled in a multicomplex and it is also an essential regulation point at the end of the process

COQ7 mRNA destabilization (Cascajo et al. 2016). Through these factors, the dynamic of COQ7 mRNA would depend on the specific tissue and on metabolic and stress conditions.

In the case of COQ3 and COQ6 transcripts, it seems that they could be possible targets of the Clu1/CluA homologue (CLUH) (Gao et al. 2014). CLUH binds to several mRNA of nuclear-encoded mitochondrial proteins and also induces mitochondrial clustering, probably linking the biogenesis of mitochondrial proteins with the distribution of the organelle. Although these authors did not demonstrate specifically the function of this RBP in *COQs* genes, they showed that CLUH depletion reduces the translation of several of its target proteins. Another member of the CoQ-synthome, COQ5, has been shown as a target of Puf3 in yeast (Lapointe et al. 2018). In fact, this RBP regulates CoQ6 biosynthesis in yeast (Garcia-Rodriguez et al. 2007; Lapointe et al. 2018). All these results support the idea that post-transcriptional regulation through RBPs regulating mRNA half life and translational rate is a key process in control of CoQ homeostasis (Figs. 5.1 and 5.2).

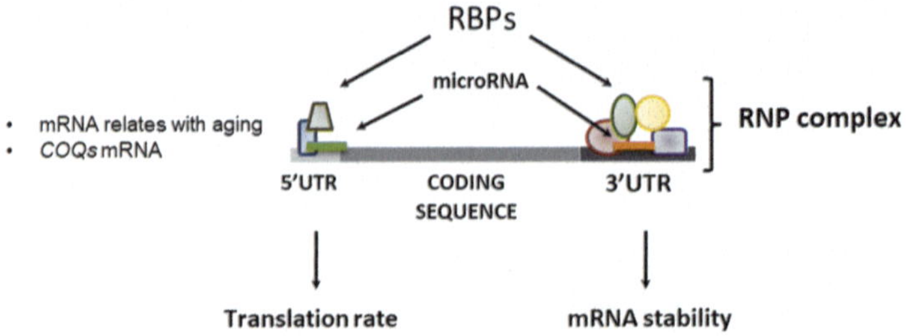

Fig. 5.2 mRNA regulatory processes of members of the Q-synthome. RBPs bind to mRNA at the 5′UTR or the 3′UTR ends. 5′UTR binding proteins regulate translation rate of the mRNA and can control the level of protein. Probably RBPs bound to the 3′UTR control mRNA stability, thus keeping the level of mRNA available for quick synthesis of proteins.

5.6 Post-translational Regulation of CoQ_{10} Synthesis

Although it is clear that CoQ_{10} synthesis is driven by a complex of proteins, the interactions between these proteins and the regulation of the activity of the synthome remains elusive. Post-translational modifications of COQs proteins such as phosphorylations and proteolysis also regulate the biosynthesis of CoQ. PPTC7, a human mitochondrial phosphatase, modulates CoQ_{10} content by controlling COQ7 phosphorylation (Gonzalez-Mariscal et al. 2018; Niemi et al. 2019; Stefely et al. 2015; Xie et al. 2011). No further regulation by phosphorylation has been suggested although many of the members of the CoQ-synthome show putative phosphorylation sites.

Regarding proteolytic processes, COQ5 has been suggested to be a target of the mitoprotease Oct1p. Proteolytic modification of COQ5 improves protein stability and in consequence CoQ_6 synthesis in yeast (Veling et al. 2017). On the other hand, we have recently found that PARL, a human protease found in mitochondria, is involved in CoQ_{10} biosynthesis and cell respiration by regulating COQ4 and UQCRFS1, a respiratory Complex III subunit (Spinazzi et al. 2019).

Other regulatory mechanisms of CoQ synthesis have been recently described. Some of the COQ proteins, with yet unknown function (COQ4, COQ9 and COQ8), would coordinate the association between CoQ precursors and enzymatic proteins to regulate the assembly of the CoQ synthesis complex (Kamzalov and Sohal 2004; Marbois et al. 2009; Zampol et al. 2010). Specifically, COQ9 protein interacts with COQ7 affecting both level and enzymatic function of COQ7 protein in mice (Lohman et al. 2014). This result is also supported by the fact that patients with *COQ9* mutations show reduced levels of COQ7 protein (Danhauser et al. 2016). These authors showed that COQ9 stabilizes other COQs proteins and they suggest that this regulatory process is tissue-specific. In addition, it has been described that COQ9 possess lipid-binding activity (Lohman et al. 2014). Both CoQ and its precursors are extremely hydrophobic due to isoprenoid tail, which keep it embedded within of mitochondrial inner membrane. This tail is attached to polar head that is sequentially modified by enzymatic COQ proteins to produce the mature CoQ molecule. In addition to binding and stabilizing COQs proteins, COQ9 is capable of interacting with different lipid intermediates of CoQ synthesis pathway. In fact, other non-enzymatic proteins of CoQ pathway, such as COQ8/ADCK3 or ADCK4, also show lipid-binding activity (Reidenbach et al. 2018).

On the other hand, other processes no directly involved in CoQ_{10} synthesis can influence the structure of the complex. It has been demonstrated that head-modifying enzymes are specifically organized into domains inside mitochondria, both in yeast and human cells. Concentration and distribution of these domains depend on functional status of CoQ biosynthesis pathway. ER-mitochondrial contacts are enriched sites in lipids and enzymatic proteins that could play an important role in this regulatory process (Subramanian et al. 2019). In fact, mitofusin 2 protein, a key factor in mitochondrial fusion/fission dynamics, is enriched in ER-mitochondrial contacts and is required to maintain CoQ levels in cells (Mourier et al. 2015).

According to our recent studies, ADCK2 emerges as other protein with regulatory function in CoQ synthesis (Vazquez-Fonseca et al. 2019). ADCKs (ADCK1–5) are domain-containing mitochondrial protein kinases, but their functions need to be studied in more details (Vazquez-Fonseca et al. 2019). Recently, we have found that lipid metabolism and CoQ biosynthesis are affected by ADCK2 mutations in mice and humans. In fact, it has been proposed that ADCK2 could be involved in transport of lipids inside the mitochondria. More experimental approaches are needed in order to demonstrate if ADCK2 have a role in CoQ domains and ER-mitochondrial contacts.

All these elements are required to maintain the concentration and spatial distribution of lipid intermediates, adequate protein levels for the synthesis of CoQ_{10}, and, in consequence, an efficient CoQ_{10} homeostasis within the mitochondria and other cell membranes. All these processes of regulation, from *COQ* gene expression (transcriptional control), to mRNA metabolism, protein synthesis (post-transcriptional control) and protein complex assembly (post-translational control), might be perfectly coordinated. It seems that they can be tissue-specific and they are probably associated with mitochondrial turnover and dynamic, being for this reason modulated throughout development and life of an organism. Because alterations in any of these mechanisms could cause a CoQ deficit, their knowledge would help to develop specific therapies to modulate CoQ content.

References

Adjibade P, Mazroui R (2014) Control of mRNA turnover: implication of cytoplasmic RNA granules. Semin Cell Dev Biol 34:15–23. https://doi.org/10.1016/j.semcdb.2014.05.013

Apponi LH, Corbett AH, Pavlath GK (2011) RNA-binding proteins and gene regulation in myogenesis. Trends Pharmacol Sci 32(11):652–658. https://doi.org/10.1016/j.tips.2011.06.004

Atasoy U, Watson J, Patel D, Keene JD (1998) ELAV protein HuA (HuR) can redistribute between nucleus and cytoplasm and is upregulated during serum stimulation and T cell activation. J Cell Sci 111(Pt 21):3145–3156

Bentinger M, Turunen M, Zhang XX, Wan YJ, Dallner G (2003) Involvement of retinoid X receptor alpha in coenzyme Q metabolism. J Mol Biol 326(3):795–803. https://doi.org/10.1016/s0022-2836(02)01447-x

Bentinger M, Tekle M, Brismar K, Chojnacki T, Swiezewska E, Dallner G (2008) Stimulation of coenzyme Q synthesis. Biofactors 32(1–4):99–111. https://doi.org/10.1002/biof.5520320112

Borbolis F, Syntichaki P (2015) Cytoplasmic mRNA turnover and ageing. Mech Ageing Dev 152:32–42. https://doi.org/10.1016/j.mad.2015.09.006

Borras C, Abdelaziz KM, Gambini J, Serna E, Ingles M, de la Fuente M, Garcia I, Matheu A, Sanchis P, Belenguer A, Errigo A, Avellana JA, Barettino A, Lloret-Fernandez C, Flames N, Pes G, Rodriguez-Manas L, Vina J (2016) Human exceptional longevity: transcriptome from centenarians is distinct from septuagenarians and reveals a role of Bcl-xL in successful aging. Aging (Albany NY) 8(12):3185–3208. https://doi.org/10.18632/aging.101078

Borras C, Serna E, Gambini J, Ingles M, Vina J (2017) Centenarians maintain miRNA biogenesis pathway while it is impaired in octogenarians. Mech Ageing Dev 168:54–57. https://doi.org/10.1016/j.mad.2017.07.003

Brea-Calvo G, Siendones E, Sanchez-Alcazar JA, de Cabo R, Navas P (2009) Cell survival from chemotherapy depends on NF-kappaB transcriptional up-regulation of coenzyme Q biosynthesis. PLoS One 4(4):e5301. https://doi.org/10.1371/journal.pone.0005301

Campos-Melo D, Droppelmann CA, Volkening K, Strong MJ (2014) RNA-binding proteins as molecular links between cancer and neurodegeneration. Biogerontology 15(6):587–610. https://doi.org/10.1007/s10522-014-9531-2

Campos-Silva C, Reyes-Torres I, Rivera M, Meza-Torres C, Hernandez-Camacho JD, Rodriguez-Bies E, Navas P, Lopez-Lluch G (2017) Regulation of the expression of coenzyme Q-synthesis complex during ageing. Rev Esp Geriatr Gerontol 52(6):307–312. https://doi.org/10.1016/j.regg.2017.03.003

Cascajo MV, Abdelmohsen K, Noh JH, Fernandez-Ayala DJ, Willers IM, Brea G, Lopez-Lluch G, Valenzuela-Villatoro M, Cuezva JM, Gorospe M, Siendones E, Navas P (2016) RNA-binding proteins regulate cell respiration and coenzyme Q biosynthesis by post-transcriptional regulation of COQ7. RNA Biol 13(7):622–634. https://doi.org/10.1080/15476286.2015.1119366

Casolaro V, Fang X, Tancowny B, Fan J, Wu F, Srikantan S, Asaki SY, De Fanis U, Huang SK, Gorospe M, Atasoy UX, Stellato C (2008) Posttranscriptional regulation of IL-13 in T cells: role of the RNA-binding protein HuR. J Allergy Clin Immunol 121(4):853–859.e854. https://doi.org/10.1016/j.jaci.2007.12.1166

Chang N, Yi J, Guo G, Liu X, Shang Y, Tong T, Cui Q, Zhan M, Gorospe M, Wang W (2010) HuR uses AUF1 as a cofactor to promote p16INK4 mRNA decay. Mol Cell Biol 30(15):3875–3886. https://doi.org/10.1128/mcb.00169-10

Chen CY, Shyu AB (1995) AU-rich elements: characterization and importance in mRNA degradation. Trends Biochem Sci 20(11):465–470. https://doi.org/10.1016/s0968-0004(00)89102-1

Chen CY, Gherzi R, Ong SE, Chan EL, Raijmakers R, Pruijn GJ, Stoecklin G, Moroni C, Mann M, Karin M (2001) AU binding proteins recruit the exosome to degrade ARE-containing mRNAs. Cell 107(4):451–464. https://doi.org/10.1016/s0092-8674(01)00578-5

Cho SJ, Zhang J, Chen X (2010) RNPC1 modulates the RNA-binding activity of, and cooperates with, HuR to regulate p21 mRNA stability. Nucleic Acids Res 38(7):2256–2267. https://doi.org/10.1093/nar/gkp1229

Choi HK, Pokharel YR, Lim SC, Han HK, Ryu CS, Kim SK, Kwak MK, Kang KW (2009) Inhibition of liver fibrosis by solubilized coenzyme Q10: role of Nrf2 activation in inhibiting transforming growth factor-beta1 expression. Toxicol Appl Pharmacol 240(3):377–384. https://doi.org/10.1016/j.taap.2009.07.030

Danhauser K, Herebian D, Haack TB, Rodenburg RJ, Strom TM, Meitinger T, Klee D, Mayatepek E, Prokisch H, Distelmaier F (2016) Fatal neonatal encephalopathy and lactic acidosis caused by a homozygous loss-of-function variant in COQ9. Eur J Hum Genet 24(3):450–454. https://doi.org/10.1038/ejhg.2015.133

Darbelli L, Richard S (2016) Emerging functions of the Quaking RNA-binding proteins and link to human diseases. Wiley Interdiscip Rev RNA 7(3):399–412. https://doi.org/10.1002/wrna.1344
Dean JL, Wait R, Mahtani KR, Sully G, Clark AR, Saklatvala J (2001) The 3′ untranslated region of tumor necrosis factor alpha mRNA is a target of the mRNA-stabilizing factor HuR. Mol Cell Biol 21(3):721–730. https://doi.org/10.1128/mcb.21.3.721-730.2001
Doimo M, Desbats MA, Cerqua C, Cassina M, Trevisson E, Salviati L (2014) Genetics of coenzyme q10 deficiency. Mol Syndromol 5(3–4):156–162. https://doi.org/10.1159/000362826
Fan XC, Steitz JA (1998) HNS, a nuclear-cytoplasmic shuttling sequence in HuR. Proc Natl Acad Sci U S A 95(26):15293–15298. https://doi.org/10.1073/pnas.95.26.15293
Franks TM, Lykke-Andersen J (2008) The control of mRNA decapping and P-body formation. Mol Cell 32(5):605–615. https://doi.org/10.1016/j.molcel.2008.11.001
Fu Z, Liu F, Liu C, Jin B, Jiang Y, Tang M, Qi X, Guo X (2019) Mutant huntingtin inhibits the mitochondrial unfolded protein response by impairing ABCB10 mRNA stability. Biochim Biophys Acta Mol basis Dis 1865(6):1428–1435. https://doi.org/10.1016/j.bbadis.2019.02.015
Galban S, Kuwano Y, Pullmann R Jr, Martindale JL, Kim HH, Lal A, Abdelmohsen K, Yang X, Dang Y, Liu JO, Lewis SM, Holcik M, Gorospe M (2008) RNA-binding proteins HuR and PTB promote the translation of hypoxia-inducible factor 1alpha. Mol Cell Biol 28(1):93–107. https://doi.org/10.1128/mcb.00973-07
Gao J, Schatton D, Martinelli P, Hansen H, Pla-Martin D, Barth E, Becker C, Altmueller J, Frommolt P, Sardiello M, Rugarli EI (2014) CLUH regulates mitochondrial biogenesis by binding mRNAs of nuclear-encoded mitochondrial proteins. J Cell Biol 207(2):213–223. https://doi.org/10.1083/jcb.201403129
Garcia-Rodriguez LJ, Gay AC, Pon LA (2007) Puf3p, a Pumilio family RNA binding protein, localizes to mitochondria and regulates mitochondrial biogenesis and motility in budding yeast. J Cell Biol 176(2):197–207. https://doi.org/10.1083/jcb.200606054
Glisovic T, Bachorik JL, Yong J, Dreyfuss G (2008) RNA-binding proteins and post-transcriptional gene regulation. FEBS Lett 582(14):1977–1986. https://doi.org/10.1016/j.febslet.2008.03.004
Gonzalez-Mariscal I, Martin-Montalvo A, Vazquez-Fonseca L, Pomares-Viciana T, Sanchez-Cuesta A, Fernandez-Ayala DJ, Navas P, Santos-Ocana C (2018) The mitochondrial phosphatase PPTC7 orchestrates mitochondrial metabolism regulating coenzyme Q10 biosynthesis. Biochim Biophys Acta Bioenerg 1859(11):1235–1248. https://doi.org/10.1016/j.bbabio.2018.09.369
Gratacos FM, Brewer G (2010) The role of AUF1 in regulated mRNA decay. Wiley Interdiscip Rev RNA 1(3):457–473. https://doi.org/10.1002/wrna.26
Herman AB, Autieri MV (2017) Inflammation-regulated mRNA stability and the progression of vascular inflammatory diseases. Clin Sci (Lond) 131(22):2687–2699. https://doi.org/10.1042/cs20171373
Herman AB, Autieri MV (2018) Cardiovascular disease, inflammation, and mRNA stability. Aging (Albany NY) 10(11):3046–3047. https://doi.org/10.18632/aging.101619
Kai M (2016) Roles of RNA-binding proteins in DNA damage response. Int J Mol Sci 17(3):310. https://doi.org/10.3390/ijms17030310
Kamzalov S, Sohal RS (2004) Effect of age and caloric restriction on coenzyme Q and alpha-tocopherol levels in the rat. Exp Gerontol 39(8):1199–1205. https://doi.org/10.1016/j.exger.2004.04.007
Kawai T, Lal A, Yang X, Galban S, Mazan-Mamczarz K, Gorospe M (2006) Translational control of cytochrome c by RNA-binding proteins TIA-1 and HuR. Mol Cell Biol 26(8):3295–3307. https://doi.org/10.1128/mcb.26.8.3295-3307.2006
Kedersha N, Anderson P (2002) Stress granules: sites of mRNA triage that regulate mRNA stability and translatability. Biochem Soc Trans 30(Pt 6):963–969. https://doi.org/10.1042/bst0300963
Kim C, Kang D, Lee EK, Lee JS (2017) Long noncoding RNAs and RNA-binding proteins in oxidative stress, cellular senescence, and age-related diseases. Oxidative Med Cell Longev 2017:2062384. https://doi.org/10.1155/2017/2062384

Krecic AM, Swanson MS (1999) hnRNP complexes: composition, structure, and function. Curr Opin Cell Biol 11(3):363–371. https://doi.org/10.1016/s0955-0674(99)80051-9
Kuhl I, Miranda M, Atanassov I, Kuznetsova I, Hinze Y, Mourier A, Filipovska A, Larsson NG (2017) Transcriptomic and proteomic landscape of mitochondrial dysfunction reveals secondary coenzyme Q deficiency in mammals. elife 6. https://doi.org/10.7554/eLife.30952
Kullmann M, Gopfert U, Siewe B, Hengst L (2002) ELAV/Hu proteins inhibit p27 translation via an IRES element in the p27 5′UTR. Genes Dev 16(23):3087–3099. https://doi.org/10.1101/gad.248902
Kuwano Y, Kim HH, Abdelmohsen K, Pullmann R Jr, Martindale JL, Yang X, Gorospe M (2008) MKP-1 mRNA stabilization and translational control by RNA-binding proteins HuR and NF90. Mol Cell Biol 28(14):4562–4575. https://doi.org/10.1128/mcb.00165-08
Lal A, Mazan-Mamczarz K, Kawai T, Yang X, Martindale JL, Gorospe M (2004) Concurrent versus individual binding of HuR and AUF1 to common labile target mRNAs. EMBO J 23(15):3092–3102. https://doi.org/10.1038/sj.emboj.7600305
Lal A, Kawai T, Yang X, Mazan-Mamczarz K, Gorospe M (2005) Antiapoptotic function of RNA-binding protein HuR effected through prothymosin alpha. EMBO J 24(10):1852–1862. https://doi.org/10.1038/sj.emboj.7600661
Lapointe CP, Stefely JA, Jochem A, Hutchins PD, Wilson GM, Kwiecien NW, Coon JJ, Wickens M, Pagliarini DJ (2018) Multi-omics reveal specific targets of the RNA-binding protein Puf3p and its orchestration of mitochondrial biogenesis. Cell Syst 6(1):125–135.e126. https://doi.org/10.1016/j.cels.2017.11.012
Levy NS, Chung S, Furneaux H, Levy AP (1998) Hypoxic stabilization of vascular endothelial growth factor mRNA by the RNA-binding protein HuR. J Biol Chem 273(11):6417–6423. https://doi.org/10.1074/jbc.273.11.6417
Li L, Du J, Lian Y, Zhang Y, Li X, Liu Y, Zou L, Wu T (2016) Protective effects of coenzyme Q10 against hydrogen peroxide-induced oxidative stress in PC12 cell: the role of Nrf2 and antioxidant enzymes. Cell Mol Neurobiol 36(1):103–111. https://doi.org/10.1007/s10571-015-0224-4
Linnane AW, Zhang C, Yarovaya N, Kopsidas G, Kovalenko S, Papakostopoulos P, Eastwood H, Graves S, Richardson M (2002) Human aging and global function of coenzyme Q10. Ann N Y Acad Sci 959:396–411; discussion 463-395. https://doi.org/10.1111/j.1749-6632.2002.tb02110.x
Liu L, Rao JN, Zou T, Xiao L, Wang PY, Turner DJ, Gorospe M, Wang JY (2009) Polyamines regulate c-Myc translation through Chk2-dependent HuR phosphorylation. Mol Biol Cell 20(23):4885–4898. https://doi.org/10.1091/mbc.E09-07-0550
Lohman DC, Forouhar F, Beebe ET, Stefely MS, Minogue CE, Ulbrich A, Stefely JA, Sukumar S, Luna-Sanchez M, Jochem A, Lew S, Seetharaman J, Xiao R, Wang H, Westphall MS, Wrobel RL, Everett JK, Mitchell JC, Lopez LC, Coon JJ, Tong L, Pagliarini DJ (2014) Mitochondrial COQ9 is a lipid-binding protein that associates with COQ7 to enable coenzyme Q biosynthesis. Proc Natl Acad Sci U S A 111(44):E4697–E4705. https://doi.org/10.1073/pnas.1413128111
Lopez de Silanes I, Zhan M, Lal A, Yang X, Gorospe M (2004) Identification of a target RNA motif for RNA-binding protein HuR. Proc Natl Acad Sci U S A 101(9):2987–2992. https://doi.org/10.1073/pnas.0306453101
Luna-Sanchez M, Diaz-Casado E, Barca E, Tejada MA, Montilla-Garcia A, Cobos EJ, Escames G, Acuna-Castroviejo D, Quinzii CM, Lopez LC (2015) The clinical heterogeneity of coenzyme Q10 deficiency results from genotypic differences in the Coq9 gene. EMBO Mol Med, 7 (5):670–687. https://doi.org/10.15252/emmm.201404632
Ma WJ, Cheng S, Campbell C, Wright A, Furneaux H (1996) Cloning and characterization of HuR, a ubiquitously expressed Elav-like protein. J Biol Chem 271(14):8144–8151. https://doi.org/10.1074/jbc.271.14.8144
Marbois B, Gin P, Gulmezian M, Clarke CF (2009) The yeast Coq4 polypeptide organizes a mitochondrial protein complex essential for coenzyme Q biosynthesis. Biochim Biophys Acta 1791(1):69–75. https://doi.org/10.1016/j.bbalip.2008.10.006

Masuda K, Marasa B, Martindale JL, Halushka MK, Gorospe M (2009) Tissue- and age-dependent expression of RNA-binding proteins that influence mRNA turnover and translation. Aging (Albany NY) 1(8):681–698. https://doi.org/10.18632/aging.100073

Masuda K, Kuwano Y, Nishida K, Rokutan K (2012) General RBP expression in human tissues as a function of age. Ageing Res Rev 11(4):423–431. https://doi.org/10.1016/j.arr.2012.01.005

Mata J, Marguerat S, Bahler J (2005) Post-transcriptional control of gene expression: a genome-wide perspective. Trends Biochem Sci 30(9):506–514. https://doi.org/10.1016/j.tibs.2005.07.005

Mazan-Mamczarz K, Lal A, Martindale JL, Kawai T, Gorospe M (2006) Translational repression by RNA-binding protein TIAR. Mol Cell Biol 26(7):2716–2727. https://doi.org/10.1128/mcb.26.7.2716-2727.2006

Min KW, Zealy RW, Davila S, Fomin M, Cummings JC, Makowsky D, McDowell CH, Thigpen H, Hafner M, Kwon SH, Georgescu C, Wren JD, Yoon JH (2018) Profiling of m6A RNA modifications identified an age-associated regulation of AGO2 mRNA stability. Aging Cell 17(3):e12753. https://doi.org/10.1111/acel.12753

Mourier A, Motori E, Brandt T, Lagouge M, Atanassov I, Galinier A, Rappl G, Brodesser S, Hultenby K, Dieterich C, Larsson NG (2015) Mitofusin 2 is required to maintain mitochondrial coenzyme Q levels. J Cell Biol 208(4):429–442. https://doi.org/10.1083/jcb.201411100

Navas P, Villalba JM, Lenaz G (2005) Coenzyme Q-dependent functions of plasma membrane in the aging process. Age (Dordr) 27(2):139–146. https://doi.org/10.1007/s11357-005-1632-z

Niemi NM, Wilson GM, Overmyer KA, Vogtle FN, Myketin L, Lohman DC, Schueler KL, Attie AD, Meisinger C, Coon JJ, Pagliarini DJ (2019) Pptc7 is an essential phosphatase for promoting mammalian mitochondrial metabolism and biogenesis. Nat Commun 10(1):3197. https://doi.org/10.1038/s41467-019-11047-6

Pang L, Tian H, Chang N, Yi J, Xue L, Jiang B, Gorospe M, Zhang X, Wang W (2013) Loss of CARM1 is linked to reduced HuR function in replicative senescence. BMC Mol Biol 14:15. https://doi.org/10.1186/1471-2199-14-15

Parrado-Fernandez C, Lopez-Lluch G, Rodriguez-Bies E, Santa-Cruz S, Navas P, Ramsey JJ, Villalba JM (2011) Calorie restriction modifies ubiquinone and COQ transcript levels in mouse tissues. Free Radic Biol Med 50(12):1728–1736. https://doi.org/10.1016/j.freeradbiomed.2011.03.024

Peng SS, Chen CY, Xu N, Shyu AB (1998) RNA stabilization by the AU-rich element binding protein, HuR, an ELAV protein. EMBO J 17(12):3461–3470. https://doi.org/10.1093/emboj/17.12.3461

Pesole G, Mignone F, Gissi C, Grillo G, Licciulli F, Liuni S (2001) Structural and functional features of eukaryotic mRNA untranslated regions. Gene 276(1–2):73–81. https://doi.org/10.1016/s0378-1119(01)00674-6

Polyak K, Lee MH, Erdjument-Bromage H, Koff A, Roberts JM, Tempst P, Massague J (1994) Cloning of p27Kip1, a cyclin-dependent kinase inhibitor and a potential mediator of extracellular antimitogenic signals. Cell 78(1):59–66. https://doi.org/10.1016/0092-8674(94)90572-x

Quinzii CM, DiMauro S, Hirano M (2007) Human coenzyme Q10 deficiency. Neurochem Res 32(4–5):723–727. https://doi.org/10.1007/s11064-006-9190-z

Reidenbach AG, Kemmerer ZA, Aydin D, Jochem A, McDevitt MT, Hutchins PD, Stark JL, Stefely JA, Reddy T, Hebert AS, Wilkerson EM, Johnson IE, Bingman CA, Markley JL, Coon JJ, Dal Peraro M, Pagliarini DJ (2018) Conserved lipid and small-molecule modulation of COQ8 reveals regulation of the ancient kinase-like UbiB family. Cell Chem Biol 25(2):154-165.e111. https://doi.org/10.1016/j.chembiol.2017.11.001

Sengupta S, Jang BC, Wu MT, Paik JH, Furneaux H, Hla T (2003) The RNA-binding protein HuR regulates the expression of cyclooxygenase-2. J Biol Chem 278(27):25227–25233. https://doi.org/10.1074/jbc.M301813200

Serna E, Gambini J, Borras C, Abdelaziz KM, Belenguer A, Sanchis P, Avellana JA, Rodriguez-Manas L, Vina J (2012) Centenarians, but not octogenarians, up-regulate the expression of microRNAs. Sci Rep 2:961. https://doi.org/10.1038/srep00961

Sheth U, Parker R (2003) Decapping and decay of messenger RNA occur in cytoplasmic processing bodies. Science 300(5620):805–808. https://doi.org/10.1126/science.1082320

SPAcosta MJ, Vazquez Fonseca L, Desbats MA, Cerqua C, Zordan R, Trevisson E, Salviati L (2016) Coenzyme Q biosynthesis in health and disease. Biochim Biophys Acta 1857(8):1079–1085. https://doi.org/10.1016/j.bbabio.2016.03.036

Spinazzi M, Radaelli E, Horre K, Arranz AM, Gounko NV, Agostinis P, Maia TM, Impens F, Morais VA, Lopez-Lluch G, Serneels L, Navas P, De Strooper B (2019) PARL deficiency in mouse causes complex III defects, coenzyme Q depletion, and Leigh-like syndrome. Proc Natl Acad Sci U S A 116(1):277–286. https://doi.org/10.1073/pnas.1811938116

Stefely JA, Pagliarini DJ (2017) Biochemistry of mitochondrial coenzyme Q biosynthesis. Trends Biochem Sci 42(10):824–843. https://doi.org/10.1016/j.tibs.2017.06.008

Stefely JA, Reidenbach AG, Ulbrich A, Oruganty K, Floyd BJ, Jochem A, Saunders JM, Johnson IE, Minogue CE, Wrobel RL, Barber GE, Lee D, Li S, Kannan N, Coon JJ, Bingman CA, Pagliarini DJ (2015) Mitochondrial ADCK3 employs an atypical protein kinase-like fold to enable coenzyme Q biosynthesis. Mol Cell 57(1):83–94. https://doi.org/10.1016/j.molcel.2014.11.002

Subramanian K, Jochem A, Le Vasseur M, Lewis S, Paulson BR, Reddy TR, Russell JD, Coon JJ, Pagliarini DJ, Nunnari J (2019) Coenzyme Q biosynthetic proteins assemble in a substrate-dependent manner into domains at ER-mitochondria contacts. J Cell Biol 218(4):1353–1369. https://doi.org/10.1083/jcb.201808044

Tak H, Eun JW, Kim J, Park SJ, Kim C, Ji E, Lee H, Kang H, Cho DH, Lee K, Kim W, Nam SW, Lee EK (2017) T-cell-restricted intracellular antigen 1 facilitates mitochondrial fragmentation by enhancing the expression of mitochondrial fission factor. Cell Death Differ 24(1):49–58. https://doi.org/10.1038/cdd.2016.90

Timchenko LT, Iakova P, Welm AL, Cai ZJ, Timchenko NA (2002) Calreticulin interacts with C/EBPalpha and C/EBPbeta mRNAs and represses translation of C/EBP proteins. Mol Cell Biol 22(20):7242–7257. https://doi.org/10.1128/mcb.22.20.7242-7257.2002

Van Pelt DW, Hettinger ZR, Vanderklish PW (2019) RNA-binding proteins: the next step in translating skeletal muscle adaptations? J Appl Physiol (1985) 127(2):654–660. https://doi.org/10.1152/japplphysiol.00076.2019

Vazquez-Fonseca L, Schaefer J, Navas-Enamorado I, Santos-Ocana C, Hernandez-Camacho JD, Guerra I, Cascajo MV, Sanchez-Cuesta A, Horvath Z, Siendones E, Jou C, Casado M, Gutierrez P, Brea-Calvo G, Lopez-Lluch G, Fernandez-Ayala DJM, Cortes-Rodriguez AB, Rodriguez-Aguilera JC, Matte C, Ribes A, Prieto-Soler SY, Dominguez-Del-Toro E, Francesco AD, Aon MA, Bernier M, Salviati L, Artuch R, Cabo R, Jackson S, Navas P (2019) ADCK2 Haploinsufficiency reduces mitochondrial lipid oxidation and causes myopathy associated with CoQ deficiency. J Clin Med 8(9). https://doi.org/10.3390/jcm8091374

Veling MT, Reidenbach AG, Freiberger EC, Kwiecien NW, Hutchins PD, Drahnak MJ, Jochem A, Ulbrich A, Rush MJP, Russell JD, Coon JJ, Pagliarini DJ (2017) Multi-omic Mitoprotease profiling defines a role for Oct1p in coenzyme Q production. Mol Cell 68(5):970–977.e911. https://doi.org/10.1016/j.molcel.2017.11.023

von Roretz C, Di Marco S, Mazroui R, Gallouzi IE (2011) Turnover of AU-rich-containing mRNAs during stress: a matter of survival. Wiley Interdiscip Rev RNA 2(3):336–347. https://doi.org/10.1002/wrna.55

Wang W, Caldwell MC, Lin S, Furneaux H, Gorospe M (2000) HuR regulates cyclin a and cyclin B1 mRNA stability during cell proliferation. EMBO J 19(10):2340–2350. https://doi.org/10.1093/emboj/19.10.2340

Wang W, Yang X, Cristofalo VJ, Holbrook NJ, Gorospe M (2001) Loss of HuR is linked to reduced expression of proliferative genes during replicative senescence. Mol Cell Biol 21(17):5889–5898. https://doi.org/10.1128/mcb.21.17.5889-5898.2001

Wang W, Martindale JL, Yang X, Chrest FJ, Gorospe M (2005) Increased stability of the p16 mRNA with replicative senescence. EMBO Rep 6(2):158–164. https://doi.org/10.1038/sj.embor.7400346

Xie LX, Hsieh EJ, Watanabe S, Allan CM, Chen JY, Tran UC, Clarke CF (2011) Expression of the human atypical kinase ADCK3 rescues coenzyme Q biosynthesis and phosphorylation of coq polypeptides in yeast coq8 mutants. Biochim Biophys Acta 1811(5):348–360. https://doi.org/10.1016/j.bbalip.2011.01.009

Yi J, Chang N, Liu X, Guo G, Xue L, Tong T, Gorospe M, Wang W (2010) Reduced nuclear export of HuR mRNA by HuR is linked to the loss of HuR in replicative senescence. Nucleic Acids Res 38(5):1547–1558. https://doi.org/10.1093/nar/gkp1114

Yi HS, Chang JY, Shong M (2018) The mitochondrial unfolded protein response and mitohormesis: a perspective on metabolic diseases. J Mol Endocrinol 61(3):R91–r105. https://doi.org/10.1530/jme-18-0005

Yubero D, Montero R, Martin MA, Montoya J, Ribes A, Grazina M, Trevisson E, Rodriguez-Aguilera JC, Hargreaves IP, Salviati L, Navas P, Artuch R, Jou C, Jimenez-Mallebrera C, Nascimento A, Perez-Duenas B, Ortez C, Ramos F, Colomer J, O'Callaghan M, Pineda M, Garcia-Cazorla A, Espinos C, Ruiz A, Macaya A, Marce-Grau A, Garcia-Villoria J, Arias A, Emperador S, Ruiz-Pesini E, Lopez-Gallardo E, Neergheen V, Simoes M, Diogo L, Blazquez A, Gonzalez-Quintana A, Delmiro A, Dominguez-Gonzalez C, Arenas J, Garcia-Silva MT, Martin E, Quijada P, Hernandez-Lain A, Moran M, Rivas Infante E, Avila Polo R, Paradas Lope C, Bautista Lorite J, Martinez Fernandez EM, Cortes AB, Sanchez-Cuesta A, Cascajo MV, Alcazar M, Brea-Calvo G (2016) Secondary coenzyme Q10 deficiencies in oxidative phosphorylation (OXPHOS) and non-OXPHOS disorders. Mitochondrion 30:51–58. https://doi.org/10.1016/j.mito.2016.06.007

Zampol MA, Busso C, Gomes F, Ferreira-Junior JR, Tzagoloff A, Barros MH (2010) Over-expression of CoQ_{10} in Saccharomyces cerevisiae inhibits mitochondrial respiration. Biochem Biophys Res Commun 402(1):82–87. https://doi.org/10.1016/j.bbrc.2010.09.118

Zhao BS, Roundtree IA, He C (2017) Post-transcriptional gene regulation by mRNA modifications. Nat Rev Mol Cell Biol 18(1):31–42. https://doi.org/10.1038/nrm.2016.132

Part II
Coenzyme Q in Aging Studies

Chapter 6
Coenzyme Q and Aging in *C. elegans*

Claudio Asencio

Abstract The free-living roundworm *Caenorhabditis elegans* has been extensively used in aging research because its short life span, its inexpensive laboratory growth and its easy genetic manipulation permitting an enormous amount of mutants. Aging studies using this organism as model have demonstrated that many environmental factors, including food availability, temperature, population density and drugs affect *C. elegans* life span. One of these factors is coenzyme Q, that has been associated with aging in this organism through its activity in the control of mitochondrial metabolism. The use of mutants in the synthesis of coenzyme Q in this organism associated a key role of clk-1 in coenzyme Q synthesis and longevity.

Keywords *C. elegans* · *clk-1* · Coenzyme Q · Worm · Longevity

6.1 Introduction

The roundworm *Caenorhabditis elegans* is a free-living nematode that has been extensively used in aging research (Gems and Partridge 2013). Several characteristics, like its short life span of about 3 weeks, its lack of strong heterosis (Johnson and Hutchinson 1993), its inexpensive laboratory growth and its easy genetic manipulation, make this organism especially suitable for longevity studies (Kenyon 2005). Many environmental factors, including food availability, temperature, population density and drugs are known to affect *C. elegans* life span (Kenyon 1997). However, the discovery more than 30 years ago of *age-1,* the first gene that when mutated resulted in increased life span (Friedman and Johnson 1988), suggested that the aging process might respond to a genetic program and therefore be potentially modulated. This extraordinary discovery triggered an intense research effort, which widely expanded our understanding of the aging process and its related diseases. Although several physiological processes, such as endocrine signaling, stress

C. Asencio (✉)
Pablo de Olavide University, Sevilla, Spain and European Molecular Biology Laboratory (EMBL), Heidelberg, Germany
e-mail: claudio.asencio@embl.de

G. López Lluch (ed.), *Coenzyme Q in Aging*,
https://doi.org/10.1007/978-3-030-45642-9_6

response, telomeres shortening and metabolism have been related to longevity in *C. elegans* (Kenyon 2005), in this chapter we are going to focus on coenzyme Q metabolism and its role in *C. elegans* aging.

6.2 The Synthesis of Coenzyme Q Affects Longevity in *C. elegans*

About two decades ago, the research on the role of coenzyme Q in *C. elegans* aging started unintentionally. Wong and colleagues used unbiased genetic screens, one of the most powerful genetic approaches to analyze physiological traits in *C. elegans*, to identify maternal-effect mutations. These experiments isolated mutations that slowed the timing of several physiological processes of the worms, such as growth, defecation rhythm, swimming and pharyngeal pumping. Interestingly, these mutations also slowed the rhythm of embryo and postembryonic development and extended the life span of the worms. Three of these mutations resulted in allelic variants, *e2519, qm51* and *qm30,* of the same gene, which was named *clk-1,* as these worms showed alterations in their physiological clock. In concordance with the design of the genetic screens, these three allelic mutations could be maternally rescued and the homozygous worms were viable (Wong et al. 1995; Hekimi et al. 1995). A detailed analysis of these *clk-1* alleles showed that *qm51* and *qm30* were null mutations and *e2519* was a missense mutation (Ewbank et al. 1997).

The *clk-1* gene was later found to encode for a protein with high similarity to yeast Coq7p (Jonassen et al. 1996) and to a still uncharacterized human protein (Ewbank et al. 1997). Felkai and colleagues observed that in addition to its mitochondrial expression in all somatic cells, the mutations in the *clk-1* gene affected mitochondrial function *in vivo*, resulting in a mild reduction in respiration. Furthermore, overexpression of the CLK-1 protein resulted in an increase in the rate of aging. Therefore, the CLK-1 protein was proposed to control respiration, behavior and aging in *C. elegans* (Felkai et al. 1999).

Subsequent work proved that the CLK-1/COQ7 protein is highly conserved in eukaryotes and that it is involved in coenzyme Q biosynthesis in *C. elegans*. In this way, complementation experiments showed that the human COQ7 expressed in a multicopy vector could restore the respiratory growth of COQ7 knock out yeast cells (Vajo et al. 1999). Instead, it was later shown that *clk-1* worms displayed alterations in quinone biosynthesis (Miyadera et al. 2001). Miyadera and colleagues employed reverse phase HPLC coupled to mass spectrometry analyses to show that *C. elegans* produces coenzyme Q_9. However, *clk-1* mutants did not produce Q_9 and were found to accumulate its immediate precursor, 5-demethoxyubiquinone 9 (DMQ_9), at a similar level to Q_9 in the WT worms (Levavasseur et al. 2001). These results proved that the CLK-1 protein is necessary for the step converting DMQ_9 to Q_9 in *C. elegans*, as was suggested by computational analyses (Rea 2001).

The striking connections between coenzyme Q and aging brought a remarkable interest in the ubiquinone biosynthesis pathway in *C. elegans*. The straightforward silencing of *C. elegans* genes by RNA interference (RNAi) (Fire et al. 1998; Timmons and Fire 1998) and the high homology of different genomic regions with the known yeast COQ genes allowed the identification of the novel genes *coq-1, coq-2, coq-4, coq-5, coq-6* and *coq-8* (Asencio et al. 2003). Silencing of these genes, and the previously described *coq-3* (Hihi et al. 2002) and *clk-1/coq-7* genes, resulted in a significant reduction in coenzyme Q content in *C. elegans* (Asencio et al. 2003). The identification of these genes was later confirmed by complementation experiments. Yeast deletion mutants for the genes COQ7 and COQ5 recovered growth in non-fermentable carbon sources when complemented with the *C. elegans clk-1* and *coq-5* genes respectively (Rodriguez-Aguilera et al. 2003). Silencing of *coq* genes resulted not only in reduced coenzyme Q content but also in a longer life span, partially resembling the longevity of *clk-1* mutants. Interestingly, silenced worms did not reproduce the rest of the slowed phenotypes of *clk-1* mutants. Normal I + III and II + III mitochondrial activities and a reduction of at least 30% in superoxide anions production was observed in silenced worms (Asencio et al. 2003). These results suggested that the content of coenzyme Q in silenced animals was sufficient to maintain respiration but low enough to significantly reduce the production of damaging superoxide anions, which led to increased longevity.

As coenzyme Q is essential for respiration in eukaryotes, researchers wanted to understand the unexpectedly almost normal respiration levels and survival rates of the homozygous *clk-1* mutants. Jonassen and colleagues detected coenzyme Q_8 in samples of both wild type worms and *clk-1* mutants (Jonassen et al. 2001, 2002). Coenzyme Q_8 is the endogenous ubiquinone isoform synthesized by *E. coli* OP50, the standard *C. elegans* laboratory food. These results indicated that *C. elegans* can readily incorporate coenzyme Q_8 from the diet. Likewise, when WT and *clk-1* mutants were fed with the Q-depleted *E. coli* GD1 strain (*ubiG KO*) the WT worms showed unaltered development and fertility, while the *clk-1* mutants arrested development at the second larval stage. It is remarkable that this developmental arrest was not observed with other long lived mutants, such as *daf-2*, suggesting that the phenotypes of *clk-1* mutants derived from a novel longevity pathway. In fact, the growth of the *clk-1* mutants, the accumulation of DMQ_9 and the presence of Q_8 were restored when the animals were fed a rescued *E. coli* strain expressing the depleted *ubiG* gene. This led Jonassen and colleagues to conclude that the uptake of coenzyme Q_8 from diet can partially rescue the otherwise lethal mutations of the *clk-1* gene, which was therefore confirmed to be involved in coenzyme Q_9 biosynthesis in *C. elegans* (Jonassen et al. 2001).

At that point, it was still unclear whether the survival of *clk-1* worms when fed with the Q-replete *E. coli* OP50 strain depended on the endogenous production of DMQ_9, the ingested coenzyme Q_8 or both. Hihi and colleagues observed that *clk-1* mutants did not grow when fed on *E. coli* strains carrying mutations in different *ubi-* genes. These genes are involved in coenzyme Q_8 biosynthesis in bacteria and as a consequence, these strains were unable to synthesize coenzyme Q_8. It is remarkable the lack of survival of the *clk-1* mutants when fed an *E.coli ubiF* mutant. This gene is the homologue of *clk-1* in *E. coli* and therefore this bacterial strain does not produce coenzyme Q_8 and accumulates DMQ_8. These experiments confirmed that the survival of *clk-1* mutants was not supported by DMQ_8 and solely depended on the presence of dietary coenzyme Q_8 (Hihi et al. 2002). Interestingly, this survival was not observed with other homozygous *coq* mutants, which did not produce either Q_9 nor DMQ_9, such as *coq-1* and *coq-2* (Gavilan et al. 2005) or *coq-3* (Hihi et al. 2002). Indeed, Arroyo and colleagues detected a small amount of coenzyme Q_9 in purified mitochondria from *clk-1* animals (Arroyo et al. 2006). Together, this indicated that the presence of DMQ_9 or maybe a small amount of coenzyme Q_9 was essential for the coenzyme Q_8-dependent survival of *clk*-1 mutants (Hihi et al. 2002). This observation resembles the maternal rescue of homozygous *clk-1* mutants derived from heterozygous mothers fed the Q-depleted *E. coli* GD1 bacteria. In this way, the amount of CLK-1 protein provided by the mothers to the homozygous embryos is enough to promote larval and reproductive development. However, this maternal rescue is lost in the next generation, and the homozygous *clk-1* mutants are not viable when kept on Q-depleted bacteria (Burgess et al. 2003). It was evident that the main difference between *clk-1* animals and other *coq* mutants resided in the presence of DMQ_9 in mitochondria. However, it was unclear whether the increased longevity of *clk-1* animals derived from the presence of DMQ_9, the effect of the dietary Q_8, the mild reduction in respiration, additional functions of the CLK-1 protein or a combination of some or all of these factors.

To investigate these questions, Branicky and colleagues carried out genetic screens to identify mutations that could suppress both the slow development of *clk-1* worms when fed the Q-replete *E. coli* OP50 bacteria and the developmental arrest when fed on the Q-deficient DM123 bacterial strain. Two and seven suppressors were found respectively for the *clk-1 e2519* mutant and none for *qm30*. Importantly, all the suppressor mutations were mapped to genes encoding for tRNAGlu synthetases. As the *e2519* mutation results in a Glu to Lys change in the CLK-1 protein, the suppression effects of these mutations were directly involved in the restoration, at least partially, of the endogenous synthesis of coenzyme Q_9 (Branicky et al. 2006). Suppressors still accumulated DMQ_9 and produced a very small amount of coenzyme Q_9. However, this small coenzyme Q_9 content was apparently sufficient for maintaining growth and fertility, since all the suppressors were fully fertile when fed in Q-depleted bacteria. Remarkably, it was observed that some of the *clk-1* phenotypes could be uncoupled. The otherwise still long-living suppressors showed neither delayed embryonic development nor slow behaviors. Indeed, the rhythm of pharyngeal pumping was restored to wild type levels, which suggested that the extended longevity of the original *clk-1* mutants did not simply derive from caloric restriction (Branicky et al. 2006). As the majority of the *clk-1* phenotypes were rescued in the suppressors but they still accumulated DMQ_9, Branicky and colleagues finally proposed that these *clk-1* phenotypes were not caused by the presence of this compound.

6.3 Antioxidant Role of Coenzyme Q and Longevity

As a well-known antioxidant, coenzyme Q is widely used as a nutritional supplement for humans. The survival and extension of life span for *clk-1* mutants when fed a diet containing Q_8 were in agreement with the beneficial effects of dietary supplementation with coenzyme Q. Surprisingly, Larsen and colleagues proved that dietary Q_8 reduces the life span of wild type worms (Larsen and Clarke 2002). Particularly, wild type animals and conditional fertility *fer-15* mutants fed on a Q-depleted *E. coli* strain lived 59% longer than when fed with the standard Q-replete *E. coli* OP50. As this shortening in life span appeared to be solely dependent on the Q_8 content of *E. coli*, the researchers aimed to study the effect of supplementing *C. elegans* with different isoforms of coenzyme Q varying in their side chain length. While the fertility of wild type animals and *daf*-2, *clk*-2 and *isp-1* long living mutants was not affected by the different isoforms of coenzyme Q, the *clk-1 qm30* and *qm51* mutants were only fertile when fed on Q_9 or Q_8 producing *E. coli*. This phenotype is less severe in the case of the missense *clk-1* mutant strain *e2519*, which was fertile when supplemented with dietary Q_7, Q_8 and Q_9 and for one generation when fed Q_6 (Hihi et al. 2003). Similar results were obtained by Jonassen and colleagues and in addition they showed that *e2519* and *qm30* mutants transport Q_8 to mitochondria more effectively than Q_7 (Jonassen et al. 2003). Only the exogenous supplementation with coenzyme Q_{10} was found to extend the life span of wild type worms and to reduce the production of superoxide anions (Ishii et al. 2004). These data were confirmed by Yang and colleagues who also proved that the mitochondrial oxidative stress and H_2O_2 production in *clk-1* mutants were significantly reduced when fed with coenzyme Q_{10} (Yang et al. 2009). As a consequence, the role of Coenzyme Q in *C. elegans* physiology and aging is more complex than originally expected. On the one side, coenzyme Q is essential for respiration and in this way, dietary Q_8 is required for the survival of *clk-1* mutants. However, on the other side, the supplementation of coenzyme Q to wild type worms modulates its life span according to the length of its side chain. Feeding wild type animals with coenzyme Q_8 reduces its life span (Larsen and Clarke 2002), while feeding them with coenzyme Q_{10} reduced oxidative stress and increased longevity (Ishii et al. 2004). Together, these results suggest that the balance between mitochondrial respiration and the generation of oxidative damage should be finely regulated to maintain cellular homeostasis and promote longevity.

In this context, several laboratories investigated the mitochondrial activities of *clk-1* mutants. Kayser and colleagues found normal complex I, complex II and complex II + III dependent activities in *clk-1* mitochondria. However, the complex I + III activity was found to be severely reduced in these animals, which suggested that complexes I and II may have different affinities for coenzyme Q (Kayser et al. 2004). Although the role of accumulated DMQ_9 in these mutants remained unclear, the authors proposed that this compound, together with rhodoquinone 9 (RQ_9) may support respiration in adult *clk-1* when fed on a Q-less diet. However, later data did not support this idea. Extraction of all quinones of *clk-1* and wild type mitochondria and later replenishment with coenzyme Q_9 resulted in similar I + III and II + III mitochondrial activities. Furthermore, no differences in the assembly of the respiratory supercomplexes were found between *clk-1* and wild type mitochondria as analyzed by blue native electrophoresis (Yang et al. 2011). To investigate the specific effect of DMQ $_9$ in mitochondria, *clk-1* pentane extracts were added to coenzyme Q_9 to enrich it with DMQ_9. The wild type quinone-depleted mitochondria were later replenished with coenzyme Q_9 with and without DMQ_9. Yang and colleagues found a significant reduction in complex I + III activities upon addition of DMQ_9 to replenished wild type mitochondria. As the complex II + III activity was not affected, and even though the presence of other inhibitory metabolites in the *clk*-1 extracts could not be excluded, it was proposed that DMQ_9 specifically inhibits either the transfer of electrons from complex I to Q or from Q to complex III (Yang et al. 2011).

6.4 Coenzyme Q Levels and Generation of ROS in Mitochondria

As mitochondria is the main source of oxidative damage in the cell, the generation of ROS and oxidative damage in *clk-1* mutants was studied by different laboratories. From these studies, ROS (Yang et al. 2009) and superoxide production (Braeckman et al. 2002) in *clk-1* mitochondria were found to be higher than in a wild type. Strikingly, despite the increased ROS production, *clk-1* mutants had normal or decreased levels of oxidative damage (Kayser et al. 2004; Yang et al. 2007, 2009; Yang and Hekimi 2010). A possible explanation for these counterintuitive results was the upregulation of *sod-2* and *sod-3* mRNAs observed in *clk-1* worms (Braeckman et al. 2002; Van Raamsdonk et al. 2010). This implies that the increased ROS production in *clk-1* mutants may trigger a defensive response leading to a net reduction in oxidative damage and therefore longer life span. In fact, the stimulation of mitochondrial respiration in *C. elegans* by the glycolysis inhibitor 2-deoxy-D-glucose (DOG), triggers the formation of ROS and a subsequent induction of stress resistance resulting in extended longevity (Schulz et al. 2007). Yang and colleagues later proposed that superoxide is indeed a protective signal that may initiate changes in gene expression affecting the life span of *C. elegans* (Yang and Hekimi 2010). In this way, gene expression profiling experiments of *clk-1* mutants showed important changes in gene expression compared to the wild type. This included upregulation of genes involved in glycolysis, gluconeogenesis, detoxification and collagen synthesis (Cristina et al. 2009; Fischer et al. 2014) and downregulation of genes involved in growth and reproduction (Fischer et al. 2014). The observed changes in gene expression in *clk-1* mutants were accompanied with an increase in mitochondrial DNA content that together resembled the retrograde response of yeast *petite* mitochondrial mutants, which are also long lived (Cristina et al. 2009). This response is called retrograde since the information flows in the reverse direction, in this case, from the mitochondria to the nucleus (Kirchman et al. 1999). One of the genes with increased gene expression in *clk*-1 mutants is *fstr-1*. The silencing of both *fstr-1* and its homologue *fstr-2* restored the behavioral rates and life span of *clk-1* animals to wild type levels. Importantly, these worms still accumulated DMQ_9 and did not produced coenzyme Q_9. In other words, these results implied that the *fstr-1/2 genes* slow the biological rhythms and extend the life span of the *clk-1* mutants in response to coenzyme Q levels (Cristina et al. 2009). Furthermore, the expression profile of other mutants with impaired respiration was found to be similar to that observed in *clk-1* animals and also resembled the yeast mitochondrial retrograde response (Cristina et al. 2009). Therefore, the retrograde response is probably necessary for the extended life span observed in other *C. elegans* respiratory mutants. However, the physical connection, in the form of a molecule or protein, linking the altered mitochondrial metabolism to changes in nuclear gene expression remained elusive.

Strikingly, Monaghan and colleagues demonstrated that, apart from its traditional mitochondrial localization, the CLK-1 protein can also localize to the nucleus in both *C. elegans* and human cells (Monaghan et al. 2015). Indeed, the CLK-1 protein contains two localization signals, an N-terminal mitochondrial transfer signal (MTS) that is degraded upon import into mitochondria, and a second nuclear transfer signal (NTS) downstream of the mitochondrial one. Both localization signals are so close to one another that they are degraded when CLK-1 is imported into mitochondria. Only uncleaved COQ7/CLK-1 proteins will keep the NTS and therefore be imported into the nucleus, where they were found to bind to chromatin and regulate the expression of a subset of genes. This includes genes involved in the mitochondrial unfolded protein response (UPR^{mt}) and other genes involved in the regulation of the cellular redox balance like GLNA-1 glutaminase and WWOX oxidoreductase. Monaghan and colleagues proved that nuclear content of COQ7/CLK-1 inversely correlated with the presence of ROS. Thus, the exposure of cells and worms to the antioxidant N-acetyl-L-cysteine (NAC) decreased the amount on COQ-7/CLK-1 in the nucleus, while an increase in ROS resulted in a higher proportion of uncleaved COQ7/CLK-1 and increased nuclear localization. Interestingly, the nuclear form of COQ7/CLK-1 functions independently of the mitochondrial one and was found to potentially regulate metabolic pathways that alter ROS production independently of coenzyme Q. Monaghan and colleagues proposed the nuclear CLK-1/COQ7 protein to work as a rheostat to maintain ROS homeostasis. In this way, during normal mitochondrial respiration, a basal production of ROS directs a certain amount of CLK-1 to the nucleus where it alters gene expression. As some of these regulated genes are involved in ROS metabolism, the nuclear CLK-1 could mediate a reduction in ROS production, which in turns leads CLK-1 to be mainly localized in mitochondria. CLK-1-dependent gene regulation is therefore decreased, basal ROS production is resumed and homeostasis is maintained in the cell (Monaghan et al. 2015)

These results strongly support the hormesis theory of aging (Rattan 2001), which implies that short non-lethal stressors trigger a response mechanism that not only mitigates the initial stress but also promotes overall health and increased life span. This theory was already proposed by Shulz and colleagues to explain the extended life span observed in worms upon glycolysis inhibition, resulting in the induction of mitochondrial respiration and increased oxidative stress (Schulz et al. 2007). In agreement with this theory, Sanchez-Blanco and colleagues also proposed that the persistent supplementation of dietary coenzyme Q_8 to wild type worms results in a lower cellular oxidation level that could impair the mechanisms maintaining the REDOX homeostasis in the cells resulting in a shorter life span (Sanchez-Blanco et al. 2016). Indeed, Schulz and colleagues advised against the widespread use of antioxidants as food supplements, as this might lead to a systemically impairment of the cellular hermetic response mechanism causing a net increase in oxidative damage and premature aging (Schulz et al. 2007).

6.5 Conclusions

In conclusion, the last two decades of research widely expanded our understanding of the role of coenzyme Q metabolism in the aging process of *C. elegans*. These efforts led to the identification of the first respiratory enzyme, CLK-1, that is targeted to two different cellular compartments, the mitochondria and the nucleus, were it can affect the aging process at different levels. In mitochondria, CLK-1 directly promotes respiration, whereas in the nucleus, CLK-1 senses metabolic activity and responds by modulating gene expression. More research will be required to increase our understanding of the exact mechanism by which CLK-1 modifies transcription of specific loci. Likewise, it remains unclear whether other mitochondrial enzymes also use the retrograde response of CLK-1. Research on this particular topic may lead to the discovery of a novel paradigm for maintaining homeostasis that promotes health and increased longevity.

References

Arroyo A, Santos-Ocana C, Ruiz-Ferrer M, Padilla S, Gavilan A, Rodriguez-Aguilera JC, Navas P (2006) Coenzyme Q is irreplaceable by demethoxy-coenzyme Q in plasma membrane of Caenorhabditis elegans. FEBS Lett 580(7):1740–1746. https://doi.org/10.1016/j.febslet.2006.02.025

Asencio C, Rodriguez-Aguilera JC, Ruiz-Ferrer M, Vela J, Navas P (2003) Silencing of ubiquinone biosynthesis genes extends life span in Caenorhabditis elegans. FASEB J 17(9):1135–1137. https://doi.org/10.1096/fj.02-1022fje

Braeckman BP, Houthoofd K, Brys K, Lenaerts I, De Vreese A, Van Eygen S, Raes H, Vanfleteren JR (2002) No reduction of energy metabolism in Clk mutants. Mech Ageing Dev 123(11):1447–1456

Branicky R, Nguyen PA, Hekimi S (2006) Uncoupling the pleiotropic phenotypes of clk-1 with tRNA missense suppressors in Caenorhabditis elegans. Mol Cell Biol 26(10):3976–3985. https://doi.org/10.1128/mcb.26.10.3976-3985.2006

Burgess J, Hihi AK, Benard CY, Branicky R, Hekimi S (2003) Molecular mechanism of maternal rescue in the clk-1 mutants of Caenorhabditis elegans. J Biol Chem 278(49):49555–49562. https://doi.org/10.1074/jbc.M308507200

Cristina D, Cary M, Lunceford A, Clarke C, Kenyon C (2009) A regulated response to impaired respiration slows behavioral rates and increases lifespan in Caenorhabditis elegans. PLoS Genet 5(4):e1000450. https://doi.org/10.1371/journal.pgen.1000450
Ewbank JJ, Barnes TM, Lakowski B, Lussier M, Bussey H, Hekimi S (1997) Structural and functional conservation of the Caenorhabditis elegans timing gene clk-1. Science (New York, NY) 275(5302):980–983
Felkai S, Ewbank JJ, Lemieux J, Labbe JC, Brown GG, Hekimi S (1999) CLK-1 controls respiration, behavior and aging in the nematode Caenorhabditis elegans. EMBO J 18(7):1783–1792. https://doi.org/10.1093/emboj/18.7.1783
Fire A, Xu S, Montgomery MK, Kostas SA, Driver SE, Mello CC (1998) Potent and specific genetic interference by double-stranded RNA in Caenorhabditis elegans. Nature 391(6669):806–811. https://doi.org/10.1038/35888
Fischer A, Niklowitz P, Menke T, Doring F (2014) Promotion of growth by Coenzyme Q10 is linked to gene expression in C. elegans. Biochem Biophys Res Commun 452(4):920–927. https://doi.org/10.1016/j.bbrc.2014.09.016
Friedman DB, Johnson TE (1988) Three mutants that extend both mean and maximum life span of the nematode, Caenorhabditis elegans, define the age-1 gene. J Gerontol 43(4):B102–B109
Gavilan A, Asencio C, Cabello J, Rodriguez-Aguilera JC, Schnabel R, Navas P (2005) C. elegans knockouts in ubiquinone biosynthesis genes result in different phenotypes during larval development. Biofactors 25(1–4):21–29
Gems D, Partridge L (2013) Genetics of longevity in model organisms: debates and paradigm shifts. Annu Rev Physiol 75:621–644. https://doi.org/10.1146/annurev-physiol-030212-183712
Hekimi S, Boutis P, Lakowski B (1995) Viable maternal-effect mutations that affect the development of the nematode Caenorhabditis elegans. Genetics 141(4):1351–1364
Hihi AK, Gao Y, Hekimi S (2002) Ubiquinone is necessary for Caenorhabditis elegans development at mitochondrial and non-mitochondrial sites. J Biol Chem 277(3):2202–2206. https://doi.org/10.1074/jbc.M109034200
Hihi AK, Kebir H, Hekimi S (2003) Sensitivity of Caenorhabditis elegans clk-1 mutants to ubiquinone side-chain length reveals multiple ubiquinone-dependent processes. J Biol Chem 278(42):41013–41018. https://doi.org/10.1074/jbc.M305034200
Ishii N, Senoo-Matsuda N, Miyake K, Yasuda K, Ishii T, Hartman PS, Furukawa S (2004) Coenzyme Q10 can prolong C. elegans lifespan by lowering oxidative stress. Mech Ageing Dev 125(1):41–46
Johnson TE, Hutchinson EW (1993) Absence of strong heterosis for life span and other life history traits in Caenorhabditis elegans. Genetics 134(2):465–474
Jonassen T, Marbois BN, Kim L, Chin A, Xia YR, Lusis AJ, Clarke CF (1996) Isolation and sequencing of the rat Coq7 gene and the mapping of mouse Coq7 to chromosome 7. Arch Biochem Biophys 330(2):285–289. https://doi.org/10.1006/abbi.1996.0255
Jonassen T, Larsen PL, Clarke CF (2001) A dietary source of coenzyme Q is essential for growth of long-lived Caenorhabditis elegans clk-1 mutants. Proc Natl Acad Sci U S A 98(2):421–426. https://doi.org/10.1073/pnas.021337498
Jonassen T, Marbois BN, Faull KF, Clarke CF, Larsen PL (2002) Development and fertility in Caenorhabditis elegans clk-1 mutants depend upon transport of dietary coenzyme Q8 to mitochondria. J Biol Chem 277(47):45020–45027. https://doi.org/10.1074/jbc.M204758200
Jonassen T, Davis DE, Larsen PL, Clarke CF (2003) Reproductive fitness and quinone content of Caenorhabditis elegans clk-1 mutants fed coenzyme Q isoforms of varying length. J Biol Chem 278(51):51735–51742. https://doi.org/10.1074/jbc.M308760200
Kayser EB, Sedensky MM, Morgan PG, Hoppel CL (2004) Mitochondrial oxidative phosphorylation is defective in the long-lived mutant clk-1. J Biol Chem 279(52):54479–54486. https://doi.org/10.1074/jbc.M403066200
Kenyon C (1997) Aging in C. elegans, vol 33. C. elegans, 2nd edn. Cold Spring Harbor Laboratory Press, New York

Kenyon C (2005) The plasticity of aging: insights from long-lived mutants. Cell 120(4):449–460. https://doi.org/10.1016/j.cell.2005.02.002

Kirchman PA, Kim S, Lai CY, Jazwinski SM (1999) Interorganelle signaling is a determinant of longevity in Saccharomyces cerevisiae. Genetics 152(1):179–190

Larsen PL, Clarke CF (2002) Extension of life-span in Caenorhabditis elegans by a diet lacking coenzyme Q. Science (New York, NY) 295(5552):120–123. https://doi.org/10.1126/science.1064653

Levavasseur F, Miyadera H, Sirois J, Tremblay ML, Kita K, Shoubridge E, Hekimi S (2001) Ubiquinone is necessary for mouse embryonic development but is not essential for mitochondrial respiration. J Biol Chem 276(49):46160–46164. https://doi.org/10.1074/jbc.M108980200

Miyadera H, Amino H, Hiraishi A, Taka H, Murayama K, Miyoshi H, Sakamoto K, Ishii N, Hekimi S, Kita K (2001) Altered quinone biosynthesis in the long-lived clk-1 mutants of Caenorhabditis elegans. J Biol Chem 276(11):7713–7716. https://doi.org/10.1074/jbc.C000889200

Monaghan RM, Barnes RG, Fisher K, Andreou T, Rooney N, Poulin GB, Whitmarsh AJ (2015) A nuclear role for the respiratory enzyme CLK-1 in regulating mitochondrial stress responses and longevity. Nat Cell Biol 17(6):782–792. https://doi.org/10.1038/ncb3170

Rattan SI (2001) Applying hormesis in aging research and therapy. Hum Exp Toxicol 20(6):281–285. discussion 293-284

Rea S (2001) CLK-1/Coq7p is a DMQ mono-oxygenase and a new member of the di-iron carboxylate protein family. FEBS Lett 509(3):389–394

Rodriguez-Aguilera JC, Asencio C, Ruiz-Ferrer M, Vela J, Navas P (2003) Caenorhabditis elegans ubiquinone biosynthesis genes. Biofactors 18(1–4):237–244

Sanchez-Blanco A, Rodriguez-Matellan A, Gonzalez-Paramas A, Gonzalez-Manzano S, Kim SK, Mollinedo F (2016) Dietary and microbiome factors determine longevity in Caenorhabditis elegans. Aging 8(7):1513–1539. https://doi.org/10.18632/aging.101008

Schulz TJ, Zarse K, Voigt A, Urban N, Birringer M, Ristow M (2007) Glucose restriction extends Caenorhabditis elegans life span by inducing mitochondrial respiration and increasing oxidative stress. Cell Metab 6(4):280–293. https://doi.org/10.1016/j.cmet.2007.08.011

Timmons L, Fire A (1998) Specific interference by ingested dsRNA. Nature 395(6705):854. https://doi.org/10.1038/27579

Vajo Z, King LM, Jonassen T, Wilkin DJ, Ho N, Munnich A, Clarke CF, Francomano CA (1999) Conservation of the Caenorhabditis elegans timing gene clk-1 from yeast to human: a gene required for ubiquinone biosynthesis with potential implications for aging. Mamm Genome 10(10):1000–1004

Van Raamsdonk JM, Meng Y, Camp D, Yang W, Jia X, Benard C, Hekimi S (2010) Decreased energy metabolism extends life span in Caenorhabditis elegans without reducing oxidative damage. Genetics 185(2):559–571. https://doi.org/10.1534/genetics.110.115378

Wong A, Boutis P, Hekimi S (1995) Mutations in the clk-1 gene of Caenorhabditis elegans affect developmental and behavioral timing. Genetics 139(3):1247–1259

Yang W, Hekimi S (2010) A mitochondrial superoxide signal triggers increased longevity in Caenorhabditis elegans. PLoS Biol 8(12):e1000556. https://doi.org/10.1371/journal.pbio.1000556

Yang W, Li J, Hekimi S (2007) A measurable increase in oxidative damage due to reduction in superoxide detoxification fails to shorten the life span of long-lived mitochondrial mutants of Caenorhabditis elegans. Genetics 177(4):2063–2074. https://doi.org/10.1534/genetics.107.080788

Yang YY, Gangoiti JA, Sedensky MM, Morgan PG (2009) The effect of different ubiquinones on lifespan in Caenorhabditis elegans. Mech Ageing Dev 130(6):370–376. https://doi.org/10.1016/j.mad.2009.03.003

Yang YY, Vasta V, Hahn S, Gangoiti JA, Opheim E, Sedensky MM, Morgan PG (2011) The role of DMQ(9) in the long-lived mutant clk-1. Mech Ageing Dev 132(6–7):331–339. https://doi.org/10.1016/j.mad.2011.06.009

Chapter 7
Coenzyme Q and Aging in the Fruit Fly *Drosophila melanogaster*

Daniel J. M. Fernández-Ayala and Alberto Sanz

Abstract Aging is the consequence of the gradual accumulation of molecular and cellular damage during life. Oxidative damage due to mitochondrial malfunction seems to be the main contributor to aging. Although, recently it has been propose that reverse electron transport participate in signalling more than in damage the cell by ROS production. Other molecules has been described to take part in the aging process, as they are NAD, antioxidants and several microRNAs, as well new pathways that regulate the progression of aging. In addition, gene regulation due to epigenetic modification seems to be the responsible of providing a protective or permissive environment to age faster or slower. In this chapter, we review these things using the fruit fly as a model organism.

Keywords Aging · *D. melanogaster* · Coenzyme Q · NAD · Longevity · Reactive oxygen species

7.1 Introduction

It is generally accepted that aging is a consequence of the gradual accumulation of molecular and cellular damage during life. The 'Free Radical Theory of Ageing' (Harman 1956) updated to the "Mitochondrial Free Radical Theory of Ageing" (Harman 1972) (MFRTA) proposed that Reactive Oxygen Species (ROS) produced as by-products during normal metabolism cause the accumulation of oxidative damage. Accumulation of oxidative damage, such as oxidized DNA and proteins, or

D. J. M. Fernández-Ayala (✉)
CABD-CSIC and CIBERER, Universidad Pablo Olavide, Sevilla, Spain
e-mail: dmorfer@upo.es

A. Sanz
Institute of Molecular, Cell and Systems Biology, University of Glasgow, Glasgow, UK

G. López Lluch (ed.), *Coenzyme Q in Aging*,
https://doi.org/10.1007/978-3-030-45642-9_7

even DNA breaks, may cause the fall of biological systems. Thus, oxidative damage can lead to higher mitochondrial ROS production through a vicious circle mechanism. Although MFRTA is the most popular damage theory, there are many other theories proposing alternative mechanisms of damage e.g. telomere attrition, accumulation of cellular debris, epigenetics changes, etc. (Lopez-Otin et al. 2013). Recently, a new paradigm has been postulated where damage is not a cause but a consequence of a genetic program or "quasi-program" driven by continuous activation of pro-growth pathways such as nutrient sensing pathways controlled by Target of Rapamycin (Blagosklonny 2006).

Although, the proximal cause of aging remains unknown, it is clear that manipulation of metabolism alter the rate of aging. Three different metabolic pathways have been shown to affect the rate of aging: (1) the activity level of the electron transport chain in mitochondria, (2) the Insulin-like growth factor (IIS) signalling pathway, and (3) the Sirtuin pathway responsive to caloric restriction. Although these pathways affect aging separately, they have additive increases in lifespan (Taylor and Dillin 2011).

Aging is the greatest known risk factor for most non-communicable human diseases. Apart of the obvious ethical reasons, the use of humans in aging research is complicated due to our extreme long natural life span and to several other environmental and social factors that would make impossible to control lifespan studies properly. Cellular models of human disease provide important information about the underlying mechanisms; however, they have limitations that need to be overcome by *in vivo* models. Therefore, the use of animal models in aging research is instrumental to understand the proximal cause of aging and find new therapies to delay, prevent or reverse age-related diseases (Mitchell et al. 2015).

Here, we review some of the most recent findings on aging research using the fruit fly *Drosophila melanogaster* as animal model. In addition, we report the latest interventions that extend life span and health span in *Drosophila*.

7.2 Aging in Flies

Drosophila melanogaster has been widely used for biological research in studies of genetics, development and organogenesis, physiology, pathogenesis, pharmacology, degenerative diseases, cardiomyopathies and inflammatory diseases, cancer and aging (Fernandez-Ayala et al. 2014). *Drosophila* has also been used to model mitochondrial disease in order to understand the underlying genetic and physiological mechanism and to try new therapies including gene therapy, or even for mitochondrial gene therapy (Fernandez-Ayala et al. 2014). *Drosophila* is perfect for aging studies because it is easy to culture, has a short lifespan and life cycle, produces abundant progeny and has a powerful genetics that allows spatiotemporal manipulation of gene expression.

Here, we will review some of the most exciting research recently published using *Drosophila melanogaster* in the aging field. We will briefly describe those pathways that seem to regulate aging rate, how age-related pathologies are studied in flies and how epigenetic modifications affect gene regulation in old individuals. For the sake of space, we will focus on the role of mitochondria and ROS, how manipulations during development affect adult lifespan, and how supplements in the diet can extend healthy lifespan. For a full description of a gene, function or the biological process where is involved, the reader is advised to visit FlyBase (http://flybase.org). The annotation symbol ("CG") of a gene is shown next to the name of the gene.

7.2.1 The Role of Mitochondria in Drosophila's Aging

Mitochondrial dysfunction is one of the few universal hallmarks of aging (Lopez-Otin et al. 2013). Aged individuals are characterized by reduced bioenergetics capacity, increased ROS production, more oxidative damage in mitochondrial DNA and RNA, and accumulation of mutations in mtDNA. Although, mutations in superoxide dismutase 1 and 2 increase superoxide levels and shorten lifespan (Phillips et al. 1989). The role of mitochondrial ROS in *Drosophila*'s aging remains controversial (see below).

Reduction in ATP generation caused by reduced mitochondrial function can compromise cellular viability triggering cell death. Accordingly, interruption of mitochondrial protein translation in *tko25t* mutants, due to a point mutation in the nuclear gene that encodes the mitoribosomal protein S12 (*CG7925*), causes a severe phenotype characterized by reduction in mitochondrial respiration, development delay, and a dramatic shortening in lifespan, reproducing many of the symptoms observed in human mitochondrial disorders (Toivonen et al. 2001). A subsequent transcriptomic analysis revealed systematic and compensatory changes in the expression of genes involved in metabolism, including up-regulation of lactate dehydrogenase and many other genes related to the catabolism of fat and proteins, the TCA cycle and various anaplerotic pathways, as well as several pathways involved in gut transport and absorption of lipids and proteins. Overall, these results indicate that OXPHOS dysfunction is perceived physiologically as a starvation for particular biomolecules, inducing mitochondria to trigger a general programmed response similar to those produced in response to caloric restriction (Fernandez-Ayala et al. 2010). However, caloric restriction or mild mitochondrial dysfunction (e.g. mild knock-down of components of the electron transport chain) extends lifespan (Partridge et al. 2005; Copeland et al. 2009) because their much more moderate effect on mitochondrial function.

Aging has been associated with a progressive decline of mitochondrial function accompanied with a reduction of physical activity in *Drosophila melanogaster* (Peleg et al. 2016). Accordingly, increasing mitochondrial turnover of damaged mitochondria by over-expressing the master regulator of mitophagy: *parkin* (CG10523) extends lifespan (Rana et al. 2013). Similarly, boosting mitochondrial function by over-expressing *spargel* (CG9809) in the intestine also increases longevity (Rera et al. 2011). However, aging can be much more complex than anticipated with unknown players participating in determining aging rate. For example, there is a growing interest in the role of metabolic intermediates (e.g. Krebs's cycle intermediates) in triggering cancer and other pathological processes (Yang et al. 2013). Interestingly, cellular metabolism during early aging showed an increment of both several mitochondrial metabolic substrates, like Acetyl-CoA, citrate and isocitrate, and their related-enzymatic activities, which are accompanied with an increment in oxygen consumption at middle age (Peleg et al. 2016). Acetyl-CoA is a key metabolite in the central metabolism, and also a cofactor for the acetylation of lysine residues. Protein-acetylation of specific aminoacids is a key posttranslational modification for many metabolic enzymes as well as nuclear regulators of gene expression, mainly histone proteins (Graff and Tsai 2013). Changes in acetyl-coA levels could trigger alterations in metabolic function that would explain some of the phenotypes observed during aging. However, it is still unclear if the aging process leads to metabolic alterations, or if changes in the metabolic state of an organism trigger the process of aging.

Recent research suggests that is possible to reverse mitochondrial decay with dietary supplements that increase cellular levels of NAD (nicotinamide adenine dinucleotide). NAD is a central coenzyme in the Kreb's cycle, but also participates in other metabolic redox reactions, being a precursor for several cellular signalling molecules, and a substrate for posttranslational modifications to proteins. *CG9940* encodes the NAD+ synthase protein that catalyses the final step in de novo NAD+ biosynthesis, and transfers the amide from either ammonia or glutamine to nicotinic acid adenine di-nucleotide (NaAD). The over-expression of *CG9940* enhances cardiac output and reduces heart failure in aged flies, delays age-related mobility decline, and prolongs lifespan (Wen et al. 2016). Conversely, knock-down of *CG9940* by RNA-interference has negative effects on these parameters. Also, expression levels of the NAD+ synthase regulates the adaptation to exercise in aging (Wen et al. 2016). Over-expression of *CG9940* boosts the response to exercise increasing cardiac function, mobility, and lifespan in old flies, whereas a low expression of the gene reduces the capacity of flies to respond to the stress caused by exercise.

7.2.2 The Role of ROS in Longevity in Drosophila

The role of ROS in determining Drosophila's lifespan is unclear with different studies reporting opposite results (Sanz 2016). As predicted by MFRTA, depletion of both SOD1 and SOD2 either in muscle or in nervous system produces ROS accumulation, impairment of locomotive activity and shortens lifespan (Oka et al. 2015). However, other studies have shown results that do not support MFRTA. For example, although mitochondrial ROS production has been shown to negatively correlate with lifespan in wild type strains (Sanz et al. 2010a), experimental reduction of ROS levels by ectopic expression of an alternative oxidase from *Ciona intestinalis* (AOX) does not extend lifespan in flies (Sanz et al. 2010a). AOX expression is able to complement the semilethality of partial knockdown of both *cyclope* (*CG14028*, COXVIc) and the complex IV assembly factor *Surf1 (CG9943)* (Fernandez-Ayala et al. 2009). Moreover, it also rescues the locomotor defect and the excess of mitochondrial ROS production caused by mutations in *dj-1beta*, a *Drosophila* homolog of the human Parkinson's disease gene DJ-1. These results indicate that ectopic expression of this ubiquinol oxidase can restore the electron flow in the mitochondrial respiratory chain when it is blocked, thereby decreasing the high level of ubiquinol and restoring the physiological properties of Coenzyme Q, and subsequently inhibiting the deleterious effects that mitochondrial defects produce on health and lifespan. However, these modifications do not produce higher life span in wild type animals besides their effect decreasing mtROS (Sanz et al. 2010a). Furthermore, mutations in *dj-1β* (*CG1349*) increase mitochondrial ROS levels and oxidative damage reducing fly activity, but do not short lifespan (Stefanatos et al. 2012). These results support that other mitochondrial-related factors apart of ROS are affecting longevity in *Drosophila* and/or that ROS act via signalling and not modulating oxidative stress.

ROS are essential cellular messengers (Reczek and Chandel 2014) and disruption of ROS signalling reduces cellular viability (Hamalainen et al. 2015). Interestingly, boosting ROS levels can extend lifespan by a mechanism that has been called mitohormesis (Ristow and Zarse 2010). Mitochondrial complex I (CI) has been proposed as a crucial regulator of animal longevity (Stefanatos and Sanz 2011). Mutations in CI are frequently associated with multiple pathologies in humans and flies, and compensation of CI function by an alternative NADH internal dehydrogenase (NDI1) from *Saccharomyces cerevisiae* extends lifespan (Sanz et al. 2010b). Paradoxically, mild knock-down of the complex I (CI) subunit *ND-75* (*CG2286*) increases ROS levels and extends lifespan in a ROS-dependent manner (Owusu-Ansah et al. 2013). Disruption of CI causes the activation of the mitochondrial unfolded protein response that represses IIS inducing lifespan-extension. Similarly, allotopic expression of NDI1 increases the generation of ROS via reverse electron transport at respiratory CI and extends lifespan (Scialo et al. 2016). In theory, mitohormesis extends lifespan by increasing endogenous antioxidant defences (Ristow and Zarse 2010). However, overexpression of antioxidant does not increase lifespan (Sanz 2016), and extension in lifespan conferred by increasing ROS at CI is suppressed by antioxidants (Owusu-Ansah et al. 2013; Scialo et al. 2016). This supports that ROS act via signalling and not by reducing oxidative stress through boosting antioxidant defences.

7.2.3 Links Between Aging and Development in Drosophila

Age-specific mortality rate reach a plateau in late life following the exponential increase in mortality rate that is characteristic of aging. However, depending on how a population develops, this is how aging is. Populations selected for fast development (9–10 day generation cycles) during several generations entered the late-life phase at an earlier age than flies selected for slow development (28 day generation cycles) (Shahrestani et al. 2016). Populations showed changes in age-specific rates, including differences in starvation resistance, desiccation resistance, time-in-motion and geotaxis. Age at which physiological decline occurs matches with the age at which the transition from exponential mortality to late life occurs (Shahrestani et al. 2016). This result suggests that late-life is physiologically distinct from aging.

Some microRNAs are known to affect lifespan by post-transcriptionally silencing mRNAs that play critical, beneficial roles at early stages of the life cycle but are deleterious when expressed inappropriately at later stages (Chawla et al. 2016). In adult flies, miR-34 promotes longevity and maintains neuronal homeostasis by repressing *Eip74EF (CG32180),* a transcription factor required for progression through early development that regulates oogenesis and autophagic cell death (Liu et al. 2012). Similarly, other microRNAs such as let-7, miR-125 and mi-100 regulate the transcription factor *chinmo (CG31666),* which is involved in imaginal disc-derived wing morphogenesis, mushroom body development and dendrite morphogenesis (Chawla et al. 2016), during nervous system formation. The coordinate upregulation of the three former miRNAs in old flies induces brain degeneration and shortens life span, although it seems that they operate in two different stages (development and adulthood) instead of acting at the same time. For example, let-7 is predominantly responsible for regulating *chinmo* during nervous system formation, and miR-125 acts during adulthood. In contrast, loss of miR-125 induces ectopic *chinmo* expression only in adult brains (Chawla et al. 2016). Other microRNA that has been involved in aging is miR-1000, whose loss leads to shorten lifespan (Verma et al. 2015).

7.2.4 Pathways Related to Aging in Drosophila

Extensive research has identified several orthologous genes that modulate a few specific conserved genetic pathways that affect longevity, metabolism and development from yeast to humans. These pathways are conserved in *Drosophila* and have been extensively investigated in this model organism (Bitto et al. 2015).

mTOR is a serine/threonine protein kinase that promotes cellular growth and cell division in response to nutrients and growth factors. The mTOR protein acts in at least two complexes, mTOR complex 1 (mTORC1) and mTOR complex 2 (mTORC2). While mTORC1 operates as a central regulator of longevity, by repressing autophagy and modulating mitochondrial metabolism, mTORC2 regulates the activity of several substrates involved in cytoskeleton reorganization and cell polarity (Bitto et al. 2015).

Interestingly, manipulation of TOR signalling via overexpression of TOR repressors extends *Drosophila's* lifespan by interrupting nutrient sensing and mimicking dietary restriction (Kapahi et al. 2004). The intracellular signalling regulated by the protein kinase TOR modulates several biological processes inside the cell, such as ribosome biogenesis, translation, autophagy, cell proliferation and cytoskeletal changes required for cell growth. The primary tissues in which these pathways operate are the fat body and the nervous system. Additionally, both the expression of a dominant-negative version of TOR and the inhibition of TOR-complex 1 by rapamycin also extend lifespan (Kapahi et al. 2004). Moreover, TOR downregulation together with insulin signalling improves locomotion and cardiac function and confers neuroprotection against neurodegenerative diseases such as Parkinson's and Alzheimer's disease (Hirth 2010; Partridge et al. 2011). Similarly, lifespan is extended in response to reduced levels of insulin-like peptides (Gronke et al. 2010) or by loss of function mutations in different components of the insulin signalling pathway, such as the insulin receptor (Tatar et al. 2001) or downstream substrates such as *chico* (*CG5686*) (Clancy et al. 2001; Slack et al. 2010).

There is a close connection between mitochondria and IIS. For example, IIS regulates mitochondrial respiration, while mitochondrial ROS fine-tune insulin receptor activation in neurons (Ramalingam and Kim 2015). The age-related reduction in mitochondrial function, including respiration and ATP production, can disrupt the insulin receptor activation in neurons leading to the development of cerebral insulin resistance in old age. Another important metabolic regulator is the AMP-activated protein kinase AMPK (*CG3051*). In *Drosophila,* activation of AMPK in neurons or in the gut increases life span and slows aging in a non-cell-autonomous manner, regulating both the uptake and oxidation of glucose, as well as the oxidation of fatty acids in response to changes in the relative abundance of AMP and ATP (Ulgherait et al. 2014). AMP:ATP ratio is strongly influenced by mitochondrial function. Paradoxically, metformin, a complex I inhibitor, increases the ratio AMP:ATP and activates AMPK, but does not extend lifespan in fruit flies (Slack et al. 2012) as it does in worms or mice (De Haes et al. 2014; Martin-Montalvo et al. 2013).

Repeated or chronic perturbations of circadian rhythms are strongly suspected to be detrimental to healthspan, affecting aging, locomotor capacity and longevity in both humans and flies. At the molecular level, circadian clocks are based on negative feedback loops (Allada and Chung 2010). The transcription factors Clock *(Clk, CG7391)* and Cycle *(Cyc, CG8727)* form a heterodimer that promotes the transcription of two negative elements, period *(per, CG2647)* and timeless *(tim, CG3234),* which form a heterodimer that inhibits the activity of Clk/Cyc. The inhibition of Clk/Cyc results in per/tim degradation and releases of Clk/Cyc from repression, allowing the start of a new cycle. Mutations in *per* gene displays robust short 16-h rhythms, reduced longevity and decreased startle-induced locomotion in aged flies (Vaccaro et al. 2016). Interestingly, the adjustment of environmental light-dark cycles to the endogenous rhythms of 16 h fully suppresses the acceleration of the age-related decline in per mutants, while accelerates it in wild-type flies.

It has been found that in the brain of old flies, sleep fragmentation induces ROS production, expression of insulin-like peptides and the insulin receptor, which are maintained even after 4 days of recovering normal sleep patterns (Williams et al. 2016). In this model, ROS production causes the expression of ARE genes in an attempt to reduce neuronal ROS levels, suggesting that the increment in ROS may damage mitochondria and induce the ER-UPR response, a cellular stress response related to the endoplasmic reticulum that is triggered by unfolded proteins (Williams et al. 2016). The ER-UPR response interrupts protein translation, induces the degradation of misfolded proteins, and activates the production of molecular chaperones involved in protein folding as an immediate response, however if stress is extended ER-UPR induces apoptosis. It has been proposed that higher ROS levels would improve insulin signalling in neurons, inhibiting cytochrome c release, and protecting neurons from ER-UPR induced apoptosis (Williams et al. 2016).

The ratio protein:carbohydrate in the fly diet is instrumental in determining the effect of dietary restriction on lifespan, with low protein diets extending lifespan (Mair et al. 2005; Partridge et al. 2005). Tryptophan hydroxylase *(Trh, CG9122),* the serotonin receptor 2A *(5-HT2A, CG1056),* and the solute carrier 7-family amino acid transporter *(JhI-21, CG12317)* are required for the transient feeding preference for dietary protein after starvation through their role in establishing protein value (Ro et al. 2016). Interestingly, disruption of any of these genes increases lifespan independently of food intake, suggesting that the perceived value of dietary protein modulates lifespan via serotonin signalling. These results suggest that serotonin acts through an intracellular pathway mediated by the 5-HT2A receptor, in which the amino acid transporter, JhI-21, is required for serotonergic evaluation of the dietary protein content, which subsequently affects lifespan (Ro et al. 2016).

7.2.5 Age-Related Pathologies That Affect Drosophila

As we reported above, aging is characterized by an increase in ROS production, which has been proposed to cause neuronal insulin resistance, induces antioxidant systems that protect against oxidative damage and activates the expression of neuroprotective endoplasmic reticulum molecular chaperones in dopaminergic neurons of middle-aged flies to prevent the induction of apoptosis (Williams et al. 2016). Paradoxically, the elevated activity of antioxidants systems in Parkinson's and Alzheimer's disease may contribute to dysfunctional insulin receptor activation and central insulin resistance, which would lead to a decline in synapses and synaptic function (Freiherr et al. 2013). Related to that, inhibition of *dj-1alpha (CG6646)* gene, a homologue of the familial Parkinson's disease gene DJ-1, leads to oxidative stress, mitochondrial dysfunction, and dopaminergic neuron loss in *Drosophila* (Faust et al. 2009). Feeding flies with drugs that combine antioxidant and anti-inflammatory properties shows protective effects and improving many of the deleterious phenotypes associated with dj1-alpha mutation.

Progeria recapitulates many of the characteristics associated with aging such as neurodegeneration. It can be caused by defects in RNA metabolism and transport through the nuclear envelope. Lamin C (*CG10119*) is an intermediate filament with structural function that is responsible for nuclear pore distribution and mitotic nuclear envelope assembly. Mutations in *Lamin C* alter RNA export through the nuclear envelope and is used to model progeria in *Drosophila* (Li et al. 2016). The phenotype of Lamin C mutants includes progressive jumping and flight defects that are exacerbated with age. Interestingly, most of RNA affected by mutations in *Lamin C* are related with genes required for maintaining mitochondrial integrity and function, e.g. Marf (CG3869) single fly orthologue for mitofusins 1 and 2. These studies connect defects in RNA export through nuclear envelop to progressive loss of mitochondrial integrity and premature aging.

7.2.6 Gene Regulation During Aging

Considerable natural variation for lifespan exists within human and animal populations, including *Drosophila melanogaster.* Highfill and collaborators studied this genetic variation to elucidate the pathways and genes involved in aging (Highfill et al. 2016). The comparative genetic analysis of more than 800 lines, which grew under strict inbred conditions for several generations, mapped five quantitative trait loci that contribute to explain lifespan variation between the different inbred lines. RNA sequencing analysis comparing young *versus* old flies showed pathways affecting lifespan, and identified genes within the previous mapped quantitative trait loci (Highfill et al. 2016). Old flies increased the expression of genes involved in antimicrobial defence, such as *Relish* (CG11992), and decreased the expression of those related to mitochondrial metabolism, including the electron transport chain *(UQCR-14, CG3560)* and the coenzyme Q biosynthesis *(coq2, CG9613).* This genetic variation could help to elucidate which pathways and genes are involved in aging, as well as genetic mechanisms underlying risk for age-related diseases.

It has been shown in several animal models that genetic differences and somatic mutations underlie longevity and that some environmental factors correlate with lifespan extension, such as a nutrient availability that can activate the stress response, reduce the basal metabolic rate or decrease fertility under dietary restriction situations (Sen et al. 2016). These observations support the idea of epigenetic mechanisms in modulating longevity pathways. Methylation of histones, particularly in histone 3, as for example the trimetylation in lysine 4 (H3K4me3) or in lysine 27 (H3K27me3) are epigenetic modifications that activate or repress transcription respectively, which had been directly linked to lifespan regulation in many organisms.

In flies, mutations in any of the subunits of Polycomb Repressive Complex 2 (PRC2), one of the H3K27 methyltransferases, reduce global levels of H3K27me3 and extend lifespan of male flies by derepressing target genes *Abd-B* (*CG11648*) and *Odc1* (*CG8721*). *Abd-B* belongs to the Hox gene family and encodes a homeodomain transcription factor that controls cell identity and contributes to the differentiation of posterior thorax and abdominal segment of the fly; *Odb1* is an Ornithine Decarboxylase that also plays a role in regulation of cell cycle. On the other hand, TRX antagonizes silencing mediated by PRC2 promoting the acetylation of H3K27. It had been described that mutations in *trx* (*CG8651*) increase H3K27me3 levels and suppress mutations of PRC2 subunits and their longevity phenotype. . Since acetylation and methylation at the same site are mutually exclusive, increased acetylation results in both decreased methylation and loss of silencing at PRC2 target sites (Siebold et al. 2010). In addition, overexpression of *Lid* (*CG9088*), the histone demethylase homolog to the worm rbr-2, reduces the level of H3K4me3 and extends lifespan in *Drosophila,* while its knockdown shortens lifespan (Li et al. 2010).

Similarly, there is evidence that histone modifications affect lifespan of flies, altering the expression of specific target genes involved in stress tolerance or modulation of nutrient sensing pathways (Sen et al. 2016; Wood et al. 2016). Old flies showed a loss of activating epigenetic marks like H3K4me3 and H3K36me3 and a gain of repressive epigenetic marks like H3K9me3 compared to young animals (Wood et al. 2010).

Acetylation of histones, another epigenetic modification that modifies gene activation, can affect fly longevity. The protein deacetylase Sirt1 (*CG5216*) maintains gene silencing in heterochromatin regions by epigenetic regulation of histone deacetylation. Overexpression of Sirt1 in *Drosophila* repressed the normal age-related loss of gene silencing in heterochromatin regions, as well as the loss of gene silencing undergoing genotoxic stress (Wood et al. 2016). Furthermore, it has been shown that spermidine inhibits the enzyme histone-acetyltransferase and generates chromatin states with a low acetylation grade. Because spermidine levels decline with age in several organisms, including *Drosophila,* the supplementation with spermidine increases lifespan (Eisenberg et al. 2009). The phenotype shown by these animals include upregulation of autophagy and improved stress response by gene activation due to histone acetylation. Alterations in the histone acetylation pattern, which alter the transcriptome, are shown in middle aged flies (Peleg et al. 2016). This can cause a general deterioration of the chromatin organization that underlies control of transcription during aging causing an increment in transcriptional noise and aberrant maturation of RNAs. These age-dependent changes in gene expression can be attenuated by environmental conditions such as dietary restriction, by the overexpression of heterochromatin components or by mutations in epigenetic regulators, such as histone modifying enzymes (Siebold et al. 2010).

In addition to histone modifications, epigenetic changes include chemical modifications of DNA, such as DNA methylation. DNA methylation is correlated with transcriptional repression and is important in genome imprinting, X-chromosome silencing, and repression of both centromeric and repetitive sequences (Sen et al. 2016). Although *Drosophila* has limited DNA methylation, overexpression of *Mt2 (FlyBase CG10692),* the *Drosophila* C-5 cytosine methyltransferase, increases lifespan, whereas *Mt2* mutants are short lived (Lin et al. 2005). Because both 5mC (5-methyl-cytosine) and 6mA (6-methyl-adenine) have been localized to transposons, DNA methylation in flies could function to mark and repress transposition (Zhang et al. 2015). Furthermore, it has been described that transposons are de-repressed in both senescent human cells (Sen et al. 2016) and *Drosophila*'s tissues (Chen et al. 2016). Overexpression of transposable elements is responsible for cytotoxic phenotypes (Li et al. 2013), and inhibition of transposition prevents cytotoxicity and senescence (Wood et al. 2016). Therefore, the deregulation of DNA methyltransferases may activate specific transcriptional programs in parallel with histone modification changes, being all together responsible for specific changes in gene expression found in aged flies.

Finally, dietary restriction represses expression of transposable elements as well as activation of genes located in heterochromatin during aging through changes in epigenetic factors (Wood et al. 2016). In *Drosophila*, the proteins Dicer-2 (*CG6493*) and Argonaute 2 (AGO2, *CG7439*), recruits the histone H3K9 methyltransferase Su(var)3–9 (*CG43664*) to catalyse formation of repressive heterochromatin at transposable elements sites. In order to do this, they use the small interfering RNA (siRNA) pathway. Accordingly, overexpression of Sir2, Su(var)3–9, and Dicer-2 mitigated the expression of transposable elements during aging and led to an increase in lifespan (Wood et al. 2016). The same effect was observed by decreasing the expression of the RNA-editing enzyme ADAR (adenosine deaminase acting on RNA, *CG12598*), which controls the efficiency of siRNA-directed silencing of transposable elements (Wood et al. 2016). These data support the retrotransposon theory of aging, which is based on the deleterious activation of epigenetically silenced transposons. According to this theory, maintenance of repressive heterochromatin preserves transposable elements silencing, prevents the damage caused by its activation and extends healthy life span.

7.3 Coenzyme Q Supplementation and Aging in *Drosophila*

Little is known about the effect of Coenzyme Q supplementation on *Drosophila*'s aging. One of the main problems with Coenzyme Q supplementation is verified that is incorporated into the tissues and is able to participate in physiological processes such as electron transport. For example, coenzyme Q supplementation does not rescue phenotypes associated with mutation in *dj-1alpha* (Faust et al. 2009), whose human homologue DJ-1 causes Parkinson's disease in humans, although other antioxidants and anti-inflammatory compounds rescues most of the phenotypes in *dj-1alpha* mutants. Interestingly, antioxidant supplementation do not extend *Drosophila*'s lifespan (Sanz 2016), whereas the role of anti-inflammatory compounds is not clear yet, although there is some evidence that they can have positive effects (He et al. 2014).

On the other hand, it has been shown that elimination of coenzyme Q from the diet shortens life span and accelerates the aging process (Palmer and Sackton 2003). Since Drosophila's laboratory diet usually includes yeast, flies fed Q-less yeast live shorter. However, feeding flies with a yeast deficient in the mitochondrial respiratory complex III showed similar results, suggesting that is mitochondrial functionality of the yeast, and not the absence of coenzyme Q the responsible for lifespan shortening (Palmer and Sackton 2003).

We finally conclude that new analyses should be done to clarify the role of coenzyme in *Drosophila* aging.

References

Allada R, Chung BY (2010) Circadian organization of behavior and physiology in Drosophila. Annu Rev Physiol 72:605–624. https://doi.org/10.1146/annurev-physiol-021909-135815

Bitto A, Wang AM, Bennett CF, Kaeberlein M (2015) Biochemical genetic pathways that modulate aging in multiple species. Cold Spring Harb Perspect Med 5(11). https://doi.org/10.1101/cshperspect.a025114

Blagosklonny MV (2006) Aging and immortality: quasi-programmed senescence and its pharmacologic inhibition. Cell Cycle 5(18):2087–2102

Chawla G, Deosthale P, Childress S, Wu YC, Sokol NS (2016) A let-7-to-miR-125 MicroRNA switch regulates neuronal integrity and lifespan in Drosophila. PLoS Genet 12(8):e1006247. https://doi.org/10.1371/journal.pgen.1006247

Chen H, Zheng X, Xiao D, Zheng Y (2016) Age-associated de-repression of retrotransposons in the Drosophila fat body, its potential cause and consequence. Aging Cell 15(3):542–552. https://doi.org/10.1111/acel.12465

Clancy DJ, Gems D, Harshman LG, Oldham S, Stocker H, Hafen E, Leevers SJ, Partridge L (2001) Extension of life-span by loss of CHICO, a Drosophila insulin receptor substrate protein. Science 292(5514):104–106. https://doi.org/10.1126/science.1057991

Copeland JM, Cho J, Lo T Jr, Hur JH, Bahadorani S, Arabyan T, Rabie J, Soh J, Walker DW (2009) Extension of Drosophila life span by RNAi of the mitochondrial respiratory chain. Curr Biol 19(19):1591–1598. https://doi.org/10.1016/j.cub.2009.08.016

De Haes W, Frooninckx L, Van Assche R, Smolders A, Depuydt G, Billen J, Braeckman BP, Schoofs L, Temmerman L (2014) Metformin promotes lifespan through mitohormesis via the peroxiredoxin PRDX-2. Proc Natl Acad Sci U S A 111(24):E2501–E2509. https://doi.org/10.1073/pnas.1321776111

Eisenberg T, Knauer H, Schauer A, Buttner S, Ruckenstuhl C, Carmona-Gutierrez D, Ring J, Schroeder S, Magnes C, Antonacci L, Fussi H, Deszcz L, Hartl R, Schraml E, Criollo A, Megalou E, Weiskopf D, Laun P, Heeren G, Breitenbach M, Grubeck-Loebenstein B, Herker E, Fahrenkrog B, Frohlich KU, Sinner F, Tavernarakis N, Minois N, Kroemer G, Madeo F (2009) Induction of autophagy by spermidine promotes longevity. Nat Cell Biol 11(11):1305–1314. https://doi.org/10.1038/ncb1975

Faust K, Gehrke S, Yang Y, Yang L, Beal MF, Lu B (2009) Neuroprotective effects of compounds with antioxidant and anti-inflammatory properties in a Drosophila model of Parkinson's disease. BMC Neurosci 10:109. https://doi.org/10.1186/1471-2202-10-109

Fernandez-Ayala DJ, Sanz A, Vartiainen S, Kemppainen KK, Babusiak M, Mustalahti E, Costa R, Tuomela T, Zeviani M, Chung J, O'Dell KM, Rustin P, Jacobs HT (2009) Expression of the Ciona intestinalis alternative oxidase (AOX) in Drosophila complements defects in mitochondrial oxidative phosphorylation. Cell Metab 9(5):449–460. https://doi.org/10.1016/j.cmet.2009.03.004

Fernandez-Ayala DJ, Chen S, Kemppainen E, O'Dell KM, Jacobs HT (2010) Gene expression in a Drosophila model of mitochondrial disease. PLoS One 5(1):e8549. https://doi.org/10.1371/journal.pone.0008549

Fernandez-Ayala DJ, Jimenez-Gancedo S, Guerra I, Navas P (2014) Invertebrate models for coenzyme q10 deficiency. Mol Syndromol 5(3-4):170–179. https://doi.org/10.1159/000362751

Freiherr J, Hallschmid M, Frey WH 2nd, Brunner YF, Chapman CD, Holscher C, Craft S, De Felice FG, Benedict C (2013) Intranasal insulin as a treatment for Alzheimer's disease: a review of basic research and clinical evidence. CNS Drugs 27(7):505–514. https://doi.org/10.1007/s40263-013-0076-8

Graff J, Tsai LH (2013) Histone acetylation: molecular mnemonics on the chromatin. Nat Rev Neurosci 14(2):97–111. https://doi.org/10.1038/nrn3427

Gronke S, Clarke DF, Broughton S, Andrews TD, Partridge L (2010) Molecular evolution and functional characterization of Drosophila insulin-like peptides. PLoS Genet 6(2):e1000857. https://doi.org/10.1371/journal.pgen.1000857

Hamalainen RH, Ahlqvist KJ, Ellonen P, Lepisto M, Logan A, Otonkoski T, Murphy MP, Suomalainen A (2015) mtDNA mutagenesis disrupts pluripotent stem cell function by altering redox signaling. Cell Rep 11(10):1614–1624. https://doi.org/10.1016/j.celrep.2015.05.009
Harman D (1956) Aging: a theory based on free radical and radiation chemistry. J Gerontol 11(3):298–300
Harman D (1972) The biologic clock: the mitochondria? J Am Geriatr Soc 20(4):145–147
He C, Tsuchiyama SK, Nguyen QT, Plyusnina EN, Terrill SR, Sahibzada S, Patel B, Faulkner AR, Shaposhnikov MV, Tian R, Tsuchiya M, Kaeberlein M, Moskalev AA, Kennedy BK, Polymenis M (2014) Enhanced longevity by ibuprofen, conserved in multiple species, occurs in yeast through inhibition of tryptophan import. PLoS Genet 10(12):e1004860. https://doi.org/10.1371/journal.pgen.1004860
Highfill CA, Reeves GA, Macdonald SJ (2016) Genetic analysis of variation in lifespan using a multiparental advanced intercross Drosophila mapping population. BMC Genet 17:113. https://doi.org/10.1186/s12863-016-0419-9
Hirth F (2010) Drosophila melanogaster in the study of human neurodegeneration. CNS Neurol Disord Drug Targets 9(4):504–523
Kapahi P, Zid BM, Harper T, Koslover D, Sapin V, Benzer S (2004) Regulation of lifespan in Drosophila by modulation of genes in the TOR signaling pathway. Curr Biol 14(10):885–890. https://doi.org/10.1016/j.cub.2004.03.059
Li L, Greer C, Eisenman RN, Secombe J (2010) Essential functions of the histone demethylase lid. PLoS Genet 6(11):e1001221. https://doi.org/10.1371/journal.pgen.1001221
Li W, Prazak L, Chatterjee N, Gruninger S, Krug L, Theodorou D, Dubnau J (2013) Activation of transposable elements during aging and neuronal decline in Drosophila. Nat Neurosci 16(5):529–531. https://doi.org/10.1038/nn.3368
Li Y, Hassinger L, Thomson T et al (2016) Lamin mutations accelerate aging via defective export of mitochondrial mRNAs through nuclear envelope budding. Curr Biol 26(15):2052–2059. https://doi.org/10.1016/j.cub.2016.06.007. Epub 2016 Jul 21. PMID: 27451905
Lin MJ, Tang LY, Reddy MN, Shen CK (2005) DNA methyltransferase gene dDnmt2 and longevity of Drosophila. J Biol Chem 280(2):861–864. https://doi.org/10.1074/jbc.C400477200
Liu N, Landreh M, Cao K, Abe M, Hendriks GJ, Kennerdell JR, Zhu Y, Wang LS, Bonini NM (2012) The microRNA miR-34 modulates ageing and neurodegeneration in Drosophila. Nature 482(7386):519–523. https://doi.org/10.1038/nature10810
Lopez-Otin C, Blasco MA, Partridge L, Serrano M, Kroemer G (2013) The hallmarks of aging. Cell 153(6):1194–1217. https://doi.org/10.1016/j.cell.2013.05.039
Mair W, Piper MD, Partridge L (2005) Calories do not explain extension of life span by dietary restriction in Drosophila. PLoS Biol 3(7):e223. https://doi.org/10.1371/journal.pbio.0030223
Martin-Montalvo A, Mercken EM, Mitchell SJ, Palacios HH, Mote PL, Scheibye-Knudsen M, Gomes AP, Ward TM, Minor RK, Blouin MJ, Schwab M, Pollak M, Zhang Y, Yu Y, Becker KG, Bohr VA, Ingram DK, Sinclair DA, Wolf NS, Spindler SR, Bernier M, de Cabo R (2013) Metformin improves healthspan and lifespan in mice. Nat Commun 4:2192. https://doi.org/10.1038/ncomms3192
Mitchell SJ, Scheibye-Knudsen M, Longo DL, de Cabo R (2015) Animal models of aging research: implications for human aging and age-related diseases. Annu Rev Anim Biosci 3:283–303. https://doi.org/10.1146/annurev-animal-022114-110829
Oka S, Hirai J, Yasukawa T, Nakahara Y, Inoue YH (2015) A correlation of reactive oxygen species accumulation by depletion of superoxide dismutases with age-dependent impairment in the nervous system and muscles of Drosophila adults. Biogerontology 16(4):485–501. https://doi.org/10.1007/s10522-015-9570-3
Owusu-Ansah E, Song W, Perrimon N (2013) Muscle mitohormesis promotes longevity via systemic repression of insulin signaling. Cell 155(3):699–712. https://doi.org/10.1016/j.cell.2013.09.021
Palmer MR, Sackton TB (2003) The effects of dietary coenzyme Q on Drosophila life span. Aging Cell 2(6):335–339

Partridge L, Piper MD, Mair W (2005) Dietary restriction in Drosophila. Mech Ageing Dev 126(9):938–950. https://doi.org/10.1016/j.mad.2005.03.023

Partridge L, Alic N, Bjedov I, Piper MD (2011) Ageing in Drosophila: the role of the insulin/Igf and TOR signalling network. Exp Gerontol 46(5):376–381. https://doi.org/10.1016/j.exger.2010.09.003

Peleg S, Feller C, Forne I, Schiller E, Sevin DC, Schauer T, Regnard C, Straub T, Prestel M, Klima C, Schmitt Nogueira M, Becker L, Klopstock T, Sauer U, Becker PB, Imhof A, Ladurner AG (2016) Life span extension by targeting a link between metabolism and histone acetylation in Drosophila. EMBO Rep 17(3):455–469. https://doi.org/10.15252/embr.201541132

Phillips JP, Campbell SD, Michaud D, Charbonneau M, Hilliker AJ (1989) Null mutation of copper/zinc superoxide dismutase in Drosophila confers hypersensitivity to paraquat and reduced longevity. Proc Natl Acad Sci U S A 86(8):2761–2765

Ramalingam M, Kim SJ (2015) Insulin exerts neuroprotective effects via Akt/Bcl-2 signaling pathways in differentiated SH-SY5Y cells. J Recept Signal Transduct Res 35(1):1–7. https://doi.org/10.3109/10799893.2014.922576

Rana A, Rera M, Walker DW (2013) Parkin overexpression during aging reduces proteotoxicity, alters mitochondrial dynamics, and extends lifespan. Proc Natl Acad Sci U S A 110(21):8638–8643. https://doi.org/10.1073/pnas.1216197110

Reczek CR, Chandel NS (2014) ROS-dependent signal transduction. Curr Opin Cell Biol 33C:8–13. https://doi.org/10.1016/j.ceb.2014.09.010

Rera M, Bahadorani S, Cho J, Koehler CL, Ulgherait M, Hur JH, Ansari WS, Lo T Jr, Jones DL, Walker DW (2011) Modulation of longevity and tissue homeostasis by the Drosophila PGC-1 homolog. Cell Metab 14(5):623–634. https://doi.org/10.1016/j.cmet.2011.09.013

Ristow M, Zarse K (2010) How increased oxidative stress promotes longevity and metabolic health: the concept of mitochondrial hormesis (mitohormesis). Exp Gerontol 45(6):410–418. https://doi.org/10.1016/j.exger.2010.03.014

Ro J, Pak G, Malec PA, Lyu Y, Allison DB, Kennedy RT, Pletcher SD (2016) Serotonin signaling mediates protein valuation and aging. eLife 5. doi:https://doi.org/10.7554/eLife.16843

Sanz A (2016) Mitochondrial reactive oxygen species: do they extend or shorten animal lifespan? Biochim Biophys Acta 1857(8):1116–1126. https://doi.org/10.1016/j.bbabio.2016.03.018

Sanz A, Fernandez-Ayala DJ, Stefanatos RK, Jacobs HT (2010a) Mitochondrial ROS production correlates with, but does not directly regulate lifespan in Drosophila. Aging 2(4):200–223. https://doi.org/10.18632/aging.100137

Sanz A, Soikkeli M, Portero-Otin M, Wilson A, Kemppainen E, McIlroy G, Ellila S, Kemppainen KK, Tuomela T, Lakanmaa M, Kiviranta E, Stefanatos R, Dufour E, Hutz B, Naudi A, Jove M, Zeb A, Vartiainen S, Matsuno-Yagi A, Yagi T, Rustin P, Pamplona R, Jacobs HT (2010b) Expression of the yeast NADH dehydrogenase Ndi1 in Drosophila confers increased lifespan independently of dietary restriction. Proc Natl Acad Sci U S A 107(20):9105–9110. https://doi.org/10.1073/pnas.0911539107

Scialo F, Sriram A, Fernandez-Ayala D, Gubina N, Lohmus M, Nelson G, Logan A, Cooper HM, Navas P, Enriquez JA, Murphy MP, Sanz A (2016) Mitochondrial ROS produced via reverse electron transport extend animal lifespan. Cell Metab 23(4):725–734. https://doi.org/10.1016/j.cmet.2016.03.009

Sen P, Shah PP, Nativio R, Berger SL (2016) Epigenetic mechanisms of longevity and aging. Cell 166(4):822–839. https://doi.org/10.1016/j.cell.2016.07.050

Shahrestani P, Wilson JB, Mueller LD, Rose MR (2016) Patterns of physiological decline due to age and selection in Drosophila melanogaster. Evolution. https://doi.org/10.1111/evo.13065

Siebold AP, Banerjee R, Tie F, Kiss DL, Moskowitz J, Harte PJ (2010) Polycomb Repressive Complex 2 and Trithorax modulate Drosophila longevity and stress resistance. Proc Natl Acad Sci U S A 107(1):169–174. https://doi.org/10.1073/pnas.0907739107

Slack C, Werz C, Wieser D, Alic N, Foley A, Stocker H, Withers DJ, Thornton JM, Hafen E, Partridge L (2010) Regulation of lifespan, metabolism, and stress responses by the Drosophila SH2B protein, Lnk. PLoS Genet 6(3):e1000881. https://doi.org/10.1371/journal.pgen.1000881

Slack C, Foley A, Partridge L (2012) Activation of AMPK by the putative dietary restriction mimetic metformin is insufficient to extend lifespan in Drosophila. PLoS One 7(10):e47699. https://doi.org/10.1371/journal.pone.0047699

Stefanatos R, Sanz A (2011) Mitochondrial complex I: a central regulator of the aging process. Cell Cycle 10(10):1528–1532

Stefanatos R, Sriram A, Kiviranta E, Mohan A, Ayala V, Jacobs HT, Pamplona R, Sanz A (2012) dj-1beta regulates oxidative stress, insulin-like signaling and development in Drosophila melanogaster. Cell Cycle 11(20):3876–3886. https://doi.org/10.4161/cc.22073

Tatar M, Kopelman A, Epstein D, Tu MP, Yin CM, Garofalo RS (2001) A mutant Drosophila insulin receptor homolog that extends life-span and impairs neuroendocrine function. Science 292(5514):107–110. https://doi.org/10.1126/science.1057987

Taylor RC, Dillin A (2011) Aging as an event of proteostasis collapse. Cold Spring Harb Perspect Biol 3(5). https://doi.org/10.1101/cshperspect.a004440

Toivonen JM, O'Dell KM, Petit N, Irvine SC, Knight GK, Lehtonen M, Longmuir M, Luoto K, Touraille S, Wang Z, Alziari S, Shah ZH, Jacobs HT (2001) Technical knockout, a Drosophila model of mitochondrial deafness. Genetics 159(1):241–254

Ulgherait M, Rana A, Rera M, Graniel J, Walker DW (2014) AMPK modulates tissue and organismal aging in a non-cell-autonomous manner. Cell Rep 8(6):1767–1780. https://doi.org/10.1016/j.celrep.2014.08.006

Vaccaro A, Birman S, Klarsfeld A (2016) Chronic jet lag impairs startle-induced locomotion in Drosophila. Exp Gerontol 85:24–27. https://doi.org/10.1016/j.exger.2016.09.012

Verma P, Augustine GJ, Ammar MR, Tashiro A, Cohen SM (2015) A neuroprotective role for microRNA miR-1000 mediated by limiting glutamate excitotoxicity. Nat Neurosci 18(3):379–385. https://doi.org/10.1038/nn.3935

Wen DT, Zheng L, Ni L, Wang H, Feng Y, Zhang M (2016) The expression of CG9940 affects the adaptation of cardiac function, mobility, and lifespan to exercise in aging Drosophila. Exp Gerontol 83:6–14. https://doi.org/10.1016/j.exger.2016.07.006

Williams MJ, Perland E, Eriksson MM, Carlsson J, Erlandsson D, Laan L, Mahebali T, Potter E, Frediksson R, Benedict C, Schioth HB (2016) Recurrent sleep fragmentation induces insulin and neuroprotective mechanisms in middle-aged flies. Front Aging Neurosci 8:180. https://doi.org/10.3389/fnagi.2016.00180

Wood JG, Hillenmeyer S, Lawrence C, Chang C, Hosier S, Lightfoot W, Mukherjee E, Jiang N, Schorl C, Brodsky AS, Neretti N, Helfand SL (2010) Chromatin remodeling in the aging genome of Drosophila. Aging Cell 9(6):971–978. https://doi.org/10.1111/j.1474-9726.2010.00624.x

Wood JG, Jones BC, Jiang N, Chang C, Hosier S, Wickremesinghe P, Garcia M, Hartnett DA, Burhenn L, Neretti N, Helfand SL (2016) Chromatin-modifying genetic interventions suppress age-associated transposable element activation and extend life span in Drosophila. Proc Natl Acad Sci U S A 113(40):11277–11282. https://doi.org/10.1073/pnas.1604621113

Yang M, Soga T, Pollard PJ (2013) Oncometabolites: linking altered metabolism with cancer. J Clin Invest 123(9):3652–3658. https://doi.org/10.1172/JCI67228

Zhang G, Huang H, Liu D, Cheng Y, Liu X, Zhang W, Yin R, Zhang D, Zhang P, Liu J, Li C, Liu B, Luo Y, Zhu Y, Zhang N, He S, He C, Wang H, Chen D (2015) N6-methyladenine DNA modification in Drosophila. Cell 161(4):893–906. https://doi.org/10.1016/j.cell.2015.04.018

Chapter 8
The Aging Process and Coenzyme Q: Clk-1 Mouse Models

Mayumi Takahashi, Kazuhide Takahashi, and Takuji Shirasawa

Abstract Mitochondria are causally linked to lifespan regulation and the aging process. Coenzyme Q (CoQ), an electron transporter in the mitochondrial respiratory chain, is a key molecule in the regulation of mitochondrial function. Loss of function of the *clk-1* gene, which encodes a biosynthetic enzyme of CoQ, results in lifespan extension and slowed biological rhythms in *Caenorhabditis elegans*. The structure and function of the *clk-1* gene are evolutionarily conserved from yeast to humans, however, *clk-1*$^{-/-}$ mice that lack CoQ are embryonic lethal. *Clk-1* mutant nematodes with no dietary supply of CoQ also exhibit developmental arrest and larval death. Taken together, these results indicate that CoQ is critical for survival in the early development of both nematodes and mice, and play an important role for lifespan extension in nematodes. This review provides an overview of the role of *clk-1* and CoQ in the regulation of lifespan and biological rhythms of mice.

Keywords Clk-1 · Mouse · Aging · Mitochondria · Lifespan

8.1 Introduction

Many mutants related to longevity were isolated from the nematode *Caenorhabditis elegans* (*C. elegans*) in the late 1900s and the genes responsible for these mutants were subsequently identified. Thus, the presence of gerontogenes that are involved in regulating the aging process and lifespan was confirmed (Johnson and Lithgow 1992). The *clock-1* (*clk-1*) gene was identified as a longevity gene in a *C. elegans* mutant that had a lifespan of approximately 50% longer than that of wild-type *C. elegans* (Wong et al. 1995). Because another important phenotype of the *clk-1*

M. Takahashi (✉) · K. Takahashi
Biological Process of Aging, Tokyo Metropolitan Institute of Gerontology, Itabashi-ku, Tokyo, Japan
e-mail: mmtaka@tmig.or.jp; ktaka11@tmig.or.jp

T. Shirasawa
Shirasawa Anti-Aging Medical Institute, Bunkyo-ku, Tokyo, Japan
e-mail: shirasawa@shirasawa-acl.net

G. López Lluch (ed.), *Coenzyme Q in Aging*,
https://doi.org/10.1007/978-3-030-45642-9_8

mutant is the slowness of biological rhythms such as developmental speed, embryonic cell cycle, pharyngeal pumping rate, swimming rate, and defecation cycle, the responsible gene was named *clk-1,* as it regulates biological clocks (Wong et al. 1995). As *clk-1* mutant nematodes have a loss of function mutation in the *clk-1* gene, this gene has been thought to control both lifespan and biological rhythms.

The *clk-1* gene, a homologue of *COQ7* in yeast, is a nuclear gene and encodes the CLK-1 protein, which has a molecular mass of approximately 24 kDa and consists of 217 amino acids with a mitochondrial targeting signal at its N-terminus and localizes to the mitochondrial inner membrane (Asaumi et al. 1999; Jonassen et al. 1996; Stenmark et al. 2001). CLK-1 is a hydroxylase that converts demethoxyubiquinone (DMQ) to 5-hydroxy-ubiqinone in the penultimate step of the biosynthesis of coenzyme Q (CoQ) (Fig. 8.1) (Marbois and Clarke 1996; Miyadera et al. 2001; Jonassen et al. 1998). Therefore, *clk-1* mutant nematodes do not generate CoQ but accumulate DMQ instead (Miyadera et al. 2001). The biogenesis of CoQ is mediated by at least six known enzymes (Kawamukai 2009; Turunen et al. 2004), and silencing CoQ biosynthetic genes other than *clk-1* also results in lifespan extension in *C. elegans* (Asencio et al. 2003).

The *clk-1* structure is evolutionarily conserved among eukaryotes from yeast to humans (Ewbank et al. 1997; Asaumi et al. 1999; Vajo et al. 1999; Jonassen et al. 1996; Marbois and Clarke 1996) and in some bacteria (Stenmark et al. 2001; Andersson et al. 1998). Transgenic expression of the human or mouse*clk-1* gene in *clk-1* mutant nematodes restored the extended lifespan and rescued the slowed rhythmic behaviors to levels comparable with those of wild-type nematodes (Takahashi et al. 2001). This finding indicates that the function of *clk-1* is evolutionarily conserved from nematodes to humans. Therefore, to investigate whether *clk-1* and CoQ regulate lifespan and aging in mammals, *clk-1*-deficient (*clk-1*$^{-/-}$) mice were generated by genetic engineering. This review provides an overview of experimental findings using these mice, and discusses the possible role of *clk-1* and CoQ in the regulation of lifespan and aging in mammals.

8.2 *clk-1* Is Essential for Aerobic Respiration Via CoQ

In an effort to elucidate the possible role of *clk-1* in lifespan regulation in mammals, *clk-1*$^{-/-}$mice were generated by targeted disruption of the *clk-1* gene, followed by intercrossing heterozygous (*clk-1*$^{+/-}$) female and male mice (Nakai et al. 2001; Levavasseur et al. 2001). However, *clk-1*$^{-/-}$mice are embryonic lethal; that is, all *clk-1*$^{-/-}$mice die by embryonic day (E) 11.5 (Nakai et al. 2001) or E13.5 (Levavasseur et al. 2001). The *clk-1*$^{-/-}$mice we generated exhibited small-sized bodies and developmental delay by at least 48 h at E10.5 as assessed by neural tube development evaluation (Fig. 8.2). The cerebral wall of *clk-1*$^{-/-}$mice exhibited an aberrant stratification with round neuroepithelial cells. These cells had abnormally enlarged mitochondria with vesicular cristae and enlarged lysosomes filled with disrupted

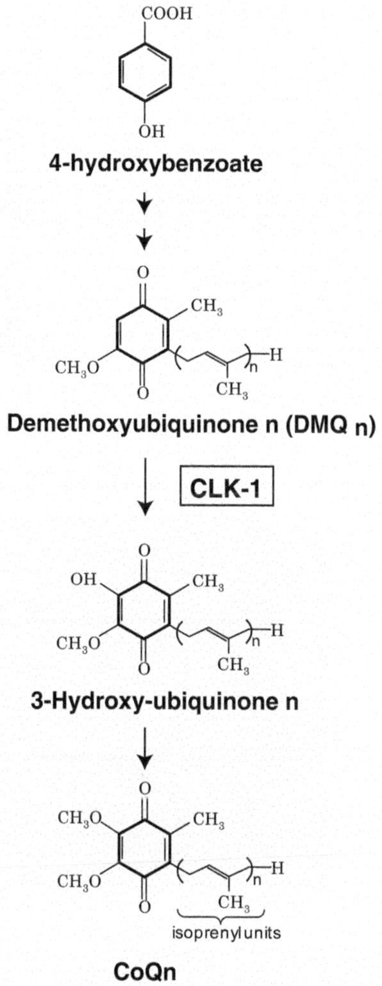

Fig. 8.1 The CoQ biosynthetic pathway. CLK-1 protein hydroxylates demethoxyubiquinone (DMQ) at the 5-position and converts it to 5-hydroxy-ubiquinone

membranes, which indicated mitochondrial dysfunction and was consistent with the phenotype found in mitochondrial encephalopathy (Nakai et al. 2001).

In *C. elegans*, *clk-1* mutants were initially identified as longevity mutants (Wong et al. 1995); however, they exhibited extended lifespan only when they were fed wild-type *Escherichia coli* (*E. coli*), which have endogenous CoQ (Jonassen et al. 2001). When the *clk-1* mutant nematodes were fed CoQ-depleted mutant *E.coli*, they showed developmental arrest at the L2 larval stage. Therefore, CoQ may be essential for early development and survival for both *C.elegans* and mice.

clk-1$^{-/-}$ mouse embryos failed to synthesize CoQ but had DMQ, the precursor of CoQ (Nakai et al. 2001; Levavasseur et al. 2001) (Fig. 8.1). This result was also the case for *coq7*Δ mutant yeast (Marbois and Clarke 1996) and *clk-1* mutant nematodes (Miyadera et al. 2001). Despite a complete loss of CoQ, embryonic cells that

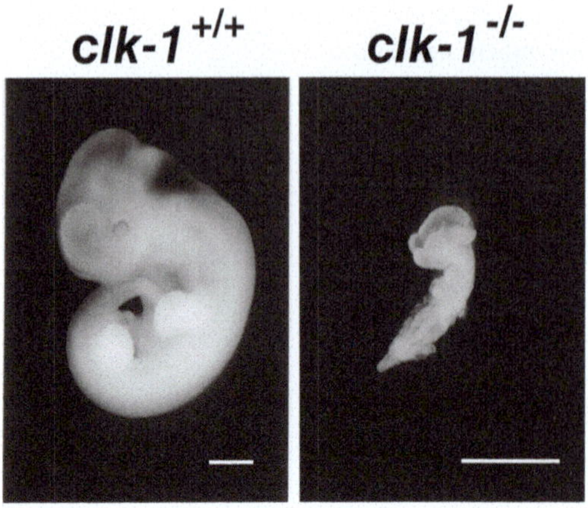

Fig. 8.2 Developmental delay in *clk-1*-deficient mice. Embryos of wild-type (*clk-1*$^{+/+}$) and *clk-1*-deficient (*clk-1*$^{-/-}$) mice at embryonic day 10.5. *clk-1*$^{-/-}$ mice are extremely small compared with wild-type mice and exhibit a developmental delay by at least 48 h. Scale bars: 1 mm. (From Takahashi et al. 2008)

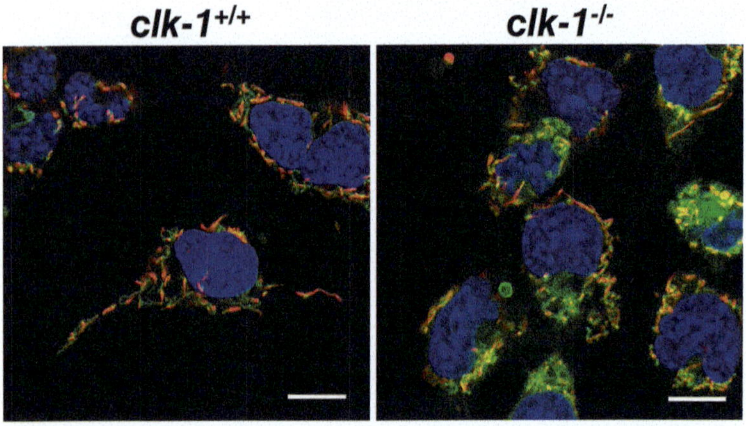

Fig. 8.3 Embryonic cells from *clk-1*$^{-/-}$mice partly exhibit high membrane potential. Embryonic cells dissociated from *clk-1*$^{+/+}$ and *clk-1*$^{-/-}$ mouse embryos at E10.5 were stained with JC-1 and Hoechst33342. Mitochondria with high and low mitochondrial membrane potentials are indicated with red and green, respectively. Scale bars: 10 μm. (From Takahashi et al. 2008)

were dissociated from *clk-1*$^{-/-}$mice at E10.5 had partially conserved mitochondrial membrane potential (Fig. 8.3). In agreement with this result, Levavasseur and colleagues reported that the respiratory enzyme activities and oxygen consumption in embryonic stem (ES) cells that were generated from *clk-1*$^{-/-}$mice were partially maintained but significantly reduced compared with those in *clk-1*$^{+/+}$ ES cells (Levavasseur et al. 2001). These results suggest that the respiration in ES cells or embryonic cells from *clk-1*$^{-/-}$mice is partially driven by DMQ instead of CoQ, but insufficient for the survival of *clk-1*$^{-/-}$ mouse embryos beyond E10.5.

When *clk-1*$^{+/-}$mice were intercrossed to obtain *clk-1*$^{-/-}$mice, the genotypes of *clk-1*$^{+/+}$, *clk-1*$^{+/-}$, and *clk-1*$^{-/-}$ mouse embryos were at the expected Mendelian ratio until E8.5. The number of mouse embryos with the *clk-1*$^{-/-}$ genotype gradually decreased, and they died by E11.5. Heart development and angiogenesis in the yolk sac were induced in mouse embryos at E8.5–10.5 (Kaufman and Navarataman 1981; Kaufman 1991), which supply oxygen to the embryos. Additionally, the mitochondrial inner membrane matures in rat embryos at E10.5–12.5 (Shepard et al. 1998), which corresponds to E8.5–10.5 in mouse embryos. These studies indicate that mitochondrial respiration changes from anaerobic to aerobic at E8.5–10.5 in mouse embryos. A small amount of ATP is generated in the glycolytic pathway under anaerobic conditions, but a large amount of ATP through oxidative phosphorylation is needed to meet the demand for embryonic growth under the aerobic conditions after mitochondrial maturation. Therefore, the quantity of ATP provided by glycolysis and DMQ-mediated oxidative phosphorylation might be insufficient for embryonic development from E8.5 onward in *clk-1*$^{-/-}$mice. Mice in which oxidative phosphorylation was abolished died at approximately E9.5, including mice with a disrupted gene encoding cytochrome *c* (Kang Li et al. 2000) or mitochondrial Tfam (Larsson et al. 1998). These reports suggest that sufficient amounts of CoQ, which support active production of ATP through normal mitochondrial function, are crucial for successive growth and development of mouse embryos after E9.5.

In *clk-1*$^{-/-}$mouse embryos, many condensed and fragmented nuclei and TUNEL-positive cells, all of which are characteristic of apoptosis, were frequently observed in *clk-1*$^{-/-}$ mouse embryos at E10.5 (Fig. 8.4) (Takahashi et al. 2008). Other hallmarks of apoptosis, such as caspase-3 activation and externalization of phosphatidylserine on the plasma membrane, were detected in dissociated cells from the *clk-1*$^{-/-}$ mouse embryos. Furthermore, *clk-1*-deficient cells exhibited release of cytochrome *c* from mitochondria into the cytoplasm, which is the first sign of mitochondrial-mediated apoptosis, and reduction in mitochondrial membrane potential and intracellular ATP levels. On the other hand, exogenous administration of a solubilized form of CoQ (sCoQ) to embryonic cells from *clk-1*$^{-/-}$mice rescued

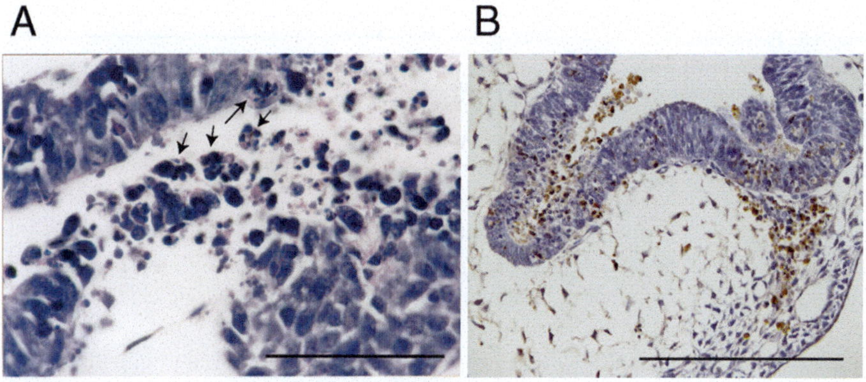

Fig. 8.4 Apoptosis is induced in *clk-1*$^{-/-}$mouse embryos at E10.5. (**A**) Condensed and fragmental nuclei (arrows) characteristic of apoptosis were observed in *clk-1*$^{-/-}$ mouse embryos at E10.5 with hematoxylin-eosin staining. (**B**) Many apoptotic cells, which are stained in brown, are present in *clk-1*$^{-/-}$ mice. Scale bars: 50 μm. (From Takahashi et al. 2008)

the reduced mitochondrial membrane potential and ATP levels, and suppressed apoptosis (Takahashi et al. 2008). Taken together, these results suggest that the embryonic lethality of *clk-1*$^{-/-}$mice is due to CoQ deficiency, and in turn, mitochondrial dysfunction and mitochondria-mediated apoptosis in the whole body.

8.3 Ultradian Rhythms Are Regulated by CoQ

Circadian rhythms that span 24 h are the most popular and well-investigated biological rhythms at the molecular level (Yu and Weaver 2011). By contrast, ultradian rhythms that have a period shorter than 24 h are not well known, but are involved in many important biological phenomena (Isomura and Kageyama 2014; Lloyd and Rossi 1992). There are many different ultradian rhythms such as hormonal circulation which cycle over several hours, rapid-eye-movement (REM) and non-REM sleep cycles that last approximately 90 min each, heartbeat, which cycles over tens of milliseconds, and cell division cycles. An interesting correlation was observed between lifespan and ultradian rhythms. Heart rate, which is an ultradian rhythm, is inversely correlated with mammalian lifespans (Fig. 8.5) (Levine 1997; Williams and Pleitropy 1957), indicating that slower heart beating results in longer living. Under experimental conditions, the long-lived *clk-1* mutant nematodes exhibited slower ultradian rhythms including the mitosis, pharyngeal pumping, swimming rates, and defecation cycle (Wong et al. 1995). This pioneering study suggests that biological rhythms are inversely correlated with a lifespan in *C. elegans* and that some ultradian rhythms and lifespan are regulated by a similar mechanism. Furthermore, exogenous administration of sCoQ completely rescued the slowed rhythmic behaviors and lifespan extension in *clk-1* mutant nematodes (Takahashi

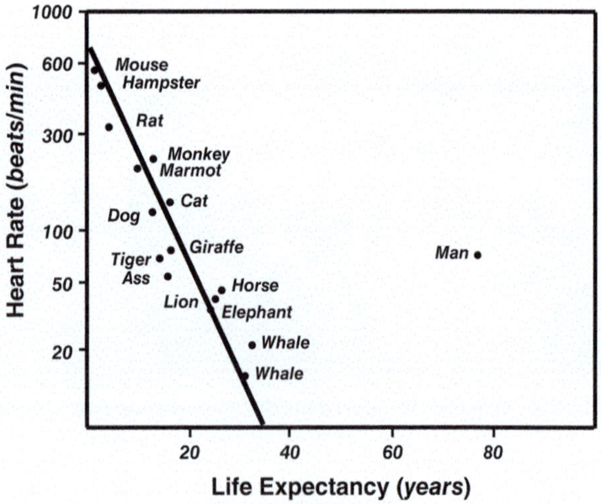

Fig. 8.5 Heart rate is inversely correlated with lifespan in mammals. (From Levine 1997)

et al. 2012a). This result suggests the possibility that CoQ regulates some ultradian rhythms and lifespan in *C. elegans*.

Cells from *clk-1*$^{-/-}$mouse embryos at E10.5 collected just before the embryonic death survived for at least a week in a culture medium supplemented with fetal bovine serum (Takahashi et al. 2012b). The DNA synthesis rate of embryonic cells from *clk-1*$^{-/-}$mice was less than half of that of *clk-1*$^{+/+}$ or *clk-1*$^{+/-}$ cells. The heart rates of whole hearts isolated from *clk-1*$^{-/-}$ mouse embryos in organ culture were very slow compared with those from *clk-1*$^{+/+}$ or *clk-1*$^{+/-}$ mouse embryos (Takahashi et al. 2012b). When cardiomyocytes dissociated from the hearts of *clk-1*$^{-/-}$ mouse embryos were cultivated in a micro-mass culture system, they exhibited intrinsic and synchronous beating. The beating rates were also very slow and approximately one-third of those from *clk-1*$^{+/+}$ and *clk-1*$^{+/-}$ cardiomyocytes. Moreover, the isolated hearts and embryonic cells from *clk-1*$^{-/-}$ mouse embryos showed reduced mitochondrial function, including reduced membrane potential and ATP production compared with those from *clk-1*$^{+/+}$ or *clk-1*$^{+/-}$ mouse embryos. These findings support the possible involvement of *clk-1* in the regulation of ultradian rhythms in mice.

The mechanisms underlying the regulation of ultradian rhythms by *clk-1* have yet to be fully elucidated. Nevertheless, exogenous administration of sCoQ completely restored the slowed heartbeat and the slowed DNA synthesis phenotype in embryonic cells from *clk-1*$^{-/-}$mice (Takahashi et al. 2012b). *clk-1* encodes a CoQ biosynthetic enzyme and the major role of CoQ is as an electron transporter in the mitochondrial respiratory chain to facilitate ATP generation. Therefore, *clk-1* is thought to regulate ultradian rhythms through the mitochondrial function driven by CoQ in mice.

8.4 The Role of *clk-1* and CoQ in the Regulation of Lifespan and the Aging Process

clk-1-deficient mice were generated to elucidate the participation of the *clk-1* gene in lifespan extension in mammals; however these mice were embryonic lethal, as described previously (Nakai et al. 2001; Levavasseur et al. 2001). To overcome the embryonic lethality of *clk-1*-deficient mice, we generated transgenic mice (Tg96) by transgenic expression of mouse *clk-1* into *clk-1*$^{-/-}$mice. The Tg96 mice were rescued from embryonic lethality and had decreased CoQ levels (Nakai et al. 2004). To increase litter size for lifespan measurement, Tg96 (C57BL/6NCr) mice were crossed with the prolific ICR strain mice, resulting in the generation of inbred progeny (Tg96/I mice). The Tg96/I mice exhibited extended lifespan (Fig. 8.6), smaller bodies and leaner leg skeletal muscles (Takahashi et al. 2014). Additionally, the Tg96/I mice had reduced whole-body oxygen consumption (VO_2) rates during the dark period and reduced mitochondrial VO_2 and ATP content in leg skeletal muscles compared with wild-type mice. These findings indicate a close relationship between lifespan extension and decreased mitochondrial function. Nevertheless, the

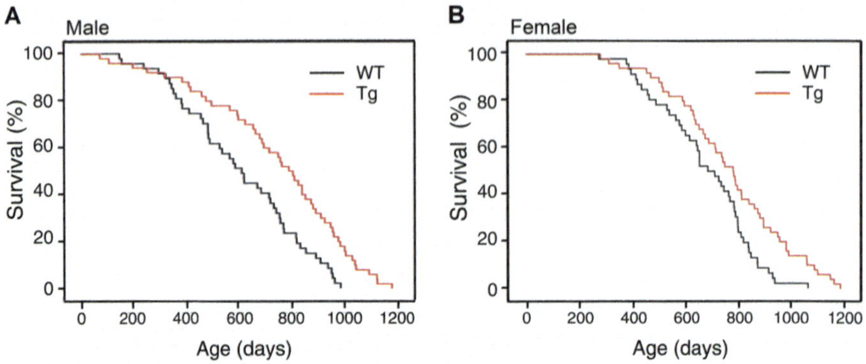

Fig. 8.6 Increased lifespan of Tg96/I mice. (**a**) Male Tg96/I mice (n = 50) lived longer than male wild-type mice (n = 47; $P < 0.0005$; Kaplan-Meier curves analyzed by the Cox-Mantel log-rank test). (**b**) Similarly, female Tg96/I mice (n = 50) lived longer than female wild-type mice (n = 46; $P < 0.005$; Cox-Mantel log-rank test). (From Takahashi et al. 2014)

mitochondrial CoQ levels in Tg96/I mice were unexpectedly reverted to levels comparable with the wild-type mice. The reason for the restoration of the CoQ levels in Tg96/I mice remains unknown, but might have been due to the intercrossing of Tg96 mice with ICR mice.

Another study contained a similar report of extended lifespan with no remarkable decrease in mitochondrial CoQ levels in *clk-1*$^{+/-}$mice (Lapointe and Hekimi 2008). These *clk-1*$^{+/-}$mice had half the amount of CLK-1 protein and comparable levels of CoQ, and they exhibited extended lifespan compared with wild-type mice. Additionally, mitochondrial functions such as VO_2 and ATP synthesis were decreased in the *clk-1*$^{+/-}$mice. The relationship between moderate reduction in mitochondrial function and extended lifespan was reported in various species (Chin et al. 2014; Copeland et al. 2009; Dell'agnello et al. 2007; Dillin et al. 2002; Kayser et al. 2004). However, the molecular and cellular mechanisms by which reduction of mitochondrial function causes lifespan extension remains to be fully explained for any species. In *clk-1 C. elegans* mutant, CoQ reduction may contribute to the extension in lifespan through decreased mitochondrial function. In a recent study evaluating long lived *clk-1*$^{+/-}$mice, CoQ was decreased in the mitochondrial inner membrane but increased in the outer membrane, maintaining an equivalent total CoQ content in the whole mitochondria to that in wild-type mice (Lapointe et al. 2012). The possible changes in CoQ distribution within the mitochondria is a novel observation and may explain the extended lifespan of Tg96/I mice, which have CoQ levels comparable to wild-type mice. Further studies are required to clarify the causal relationship between CoQ levels and lifespan in mice using an appropriate mouse model.

CoQ is a redox active molecule, i.e. it is a major source of reactive oxygen species (ROS) generation and possesses antioxidant activity, which can scavenge self-generated ROS (Turunen et al. 2004). "The free radical theory of aging" has been the major theory of aging during the past 60 years (Harman 1956, 1972). ROS were thought to damage macromolecules including protein, lipid, and DNA, thereby leading to the accumulation of cellular and tissue damage during aging. However, recent reports suggest that moderate ROS levels are considered to have rather beneficial effects on aging and lifespan through activating cell-protective stress responses (Gems and Doonan 2009; Yang and Hekimi 2010; Sanz et al. 2010; Lee et al. 2010). In Tg96 mice, decreased CoQ levels positively correlated with the activity of mitochondrial respiratory enzymes and ROS (Nakai et al. 2004). Embryonic cells from *clk-1*$^{-/-}$mice exhibited less H_2O_2 generation than wild-type cells (our unpublished data). Moreover, adult-onset *clk-1* knockout mice, which are *clk-1*-deficient in adulthood, lack CoQ and exhibit mitochondrial dysfunctions and reduced H_2O_2 generation compared with wild-type mice (Wang et al. 2015). These studies suggest that low levels or loss of CoQ causes decreased mitochondrial functions and ROS generation in mice.

By contrast, the long-lived *clk-1*$^{+/-}$mice had comparable amounts of mitochondrial CoQ, decreased mitochondrial functions (Lapointe et al. 2012), and increased H_2O_2 levels compared with wild-type mice (Lapointe and Hekimi 2008). The reason for this phenotype discrepancy among these mice that were generated by different genetic engineering techniques remains unclear. In the long-lived *clk-1*$^{+/-}$mice, it was suggested that elevated ROS induces hypoxia-inducible factor-1α (HIF-1α) expression, which in turn activates the immune response (Zheng et al. 2010), as reported in *clk-1* mutant nematodes (Lee et al. 2010). This finding may help decipher the mechanism by which moderately elevated levels of ROS leads to lifespan extension. More extensive studies are required to elucidate the role of ROS in aging and lifespan in relation to CoQ levels.

8.5 Concluding Remarks

A loss of function mutation of the *clk-1* gene results in slowed ultradian rhythms and extended lifespan in *C.elegans* (Wong et al. 1995). The *clk-1* gene encodes the CLK-1 protein, which is a hydroxylase involved in the CoQ biosynthetic pathway (Marbois and Clarke 1996; Miyadera et al. 2001; Jonassen et al. 1998). Therefore, *clk-1* mutant nematodes have no endogenous CoQ (Jonassen et al. 2001; Miyadera et al. 2001). To examine the role of *clk-1* and CoQ in the regulation of ultradian rhythms and lifespan in mammals, *clk-1*-deficient mice were generated (Nakai et al. 2001; Levavasseur et al. 2001). Unexpectedly, the *clk-1*-deficient mice were embryonic lethal. Similarly, the *clk-1* mutation in nematodes was demonstrated to be fatal when they were fed CoQ-deficient *E.coli* (Jonassen et al. 2001). Under the conditions in which the *clk-1* mutant nematodes were supplied with exogenous CoQ from *E.coli* as bait, they exhibited slowed ultradian rhythms and extended lifespan. Tg96/I mice were subsequently generated to overcome the embryonic lethality of the *clk-1*-deficient mice, and they had extended lifespans. However, the Tg96/I mice had a comparable amount of mitochondrial CoQ to that of wild-type mice (Takahashi et al. 2014). Similarly, *clk-1* heterozygous mice also exhibited extended lifespans but not decreased levels of mitochondrial CoQ (Lapointe and Hekimi 2008). Therefore, the causal relationship between decreased CoQ levels and lifespan extension remains to be fully elucidated in mice.

Meanwhile, exogenous administration of sCoQ completely rescued the slowed ultradian rhythms and extended lifespan in *clk-1* mutant nematodes (Takahashi et al. 2012a), and slowed ultradian rhythms in the hearts or cells isolated from *clk-1*$^{-/-}$ mouse embryos at E10.5 (Takahashi et al. 2012b). These results suggest that CoQ certainly regulates ultradian rhythms and lifespan in nematodes and ultradian rhythms in mice. Mitochondrial function is considered to play an important role in aging and lifespan regulation, and CoQ is a key molecule that modulates mitochondrial function. Further research using with appropriate mouse models that have lower amounts of CoQ is necessary to understand the role of CoQ in aging and lifespan in mammals.

References

Andersson SG, Zomorodipour A, Andersson JO, Sicheritz-Ponten T, Alsmark UC, Podowski RM, Naslund AK, Eriksson AS, Winkler HH, Kurland CG (1998) The genome sequence of Rickettsia prowazekii and the origin of mitochondria. Nature 396(6707):133–140

Asaumi S, Kuroyanagi H, Seki N, Shirasawa T (1999) Orthologues of the Caenorhabditis elegans longevity gene clk-1 in mouse and human. Genomics 58(3):293–301

Asencio C, Rodriguez-Aguilera JC, Ruiz-Ferrer M, Vela J, Navas P (2003) Silencing of ubiquinone biosynthesis genes extends life span in Caenorhabditis elegans. FASEB J 17(9):1135–1137. https://doi.org/10.1096/fj.02-1022fje

Chin RM, Fu X, Pai MY, Vergnes L, Hwang H, Deng G, Diep S, Lomenick B, Meli VS, Monsalve GC, Hu E, Whelan SA, Wang JX, Jung G, Solis GM, Fazlollahi F, Kaweeteerawat C, Quach A, Nili M, Krall AS, Godwin HA, Chang HR, Faull KF, Guo F, Jiang M, Trauger SA, Saghatelian A, Braas D, Christofk HR, Clarke CF, Teitell MA, Petrascheck M, Reue K, Jung ME, Frand AR, Huang J (2014) The metabolite alpha-ketoglutarate extends lifespan by inhibiting ATP synthase and TOR. Nature 510(7505):397–401. https://doi.org/10.1038/nature13264

Copeland JM, Cho J, Lo T Jr, Hur JH, Bahadorani S, Arabyan T, Rabie J, Soh J, Walker DW (2009) Extension of Drosophila life span by RNAi of the mitochondrial respiratory chain. Curr Bio 19(19):1591–1598. https://doi.org/10.1016/j.cub.2009.08.016

Dell'agnello C, Leo S, Agostino A, Szabadkai G, Tiveron C, Zulian A, Prelle A, Roubertoux P, Rizzuto R, Zeviani M (2007) Increased longevity and refractoriness to Ca(2+)-dependent neurodegeneration in Surf1 knockout mice. Hum Mol Genet 16(4):431–444. https://doi.org/10.1093/hmg/ddl477

Dillin A, Hsu AL, Arantes-Oliveira N, Lehrer-Graiwer J, Hsin H, Fraser AG, Kamath RS, Ahringer J, Kenyon C (2002) Rates of behavior and aging specified by mitochondrial function during development. Science 298(5602):2398–2401. https://doi.org/10.1126/science.1077780

Ewbank JJ, Barnes TM, Lakowski B, Lussier M, Bussey H, Hekimi S (1997) Structural and functional conservation of the Caenorhabditis elegans timing gene clk-1. Science 275(5302):980–983

Gems D, Doonan R (2009) Antioxidant defense and aging in C. elegans: is the oxidative damage theory of aging wrong? Cell Cycle 8(11):1681–1687. https://doi.org/10.4161/cc.8.11.8595

Harman D (1956) Aging: a theory based on free radical and radiation chemistry. J Gerontol 11(3):298–300

Harman D (1972) The biologic clock: the mitochondria? J Am Geriatr Soc 20(4):145–147

Isomura A, Kageyama R (2014) Ultradian oscillations and pulses: coordinating cellular responses and cell fate decisions. Development 141(19):3627–3636. https://doi.org/10.1242/dev.104497

Johnson TE, Lithgow GJ (1992) The search for the genetic basis of aging: the identification of gerontogenes in the nematode Caenorhabditis elegans. J Am Geriatr Soc 40(9):936–945

Jonassen T, Larsen PL, Clarke CF (2001) A dietary source of coenzyme Q is essential for growth of long-lived Caenorhabditis elegans clk-1 mutants. Proc Natl Acad Sci U S A 98(2):421–426

Jonassen T, Marbois BN, Kim L, Chin A, Xia YR, Lusis AJ, Clarke CF (1996) Isolation and sequencing of the rat Coq7 gene and the mapping of mouse Coq7 to chromosome 7. Arch Biochem Biophys 330(2):285–289

Jonassen T, Proft M, Randez-Gil F, Schultz JR, Marbois BN, Entian KD, Clarke CF (1998) Yeast Clk-1 homologue (Coq7/Cat5) is a mitochondrial protein in coenzyme Q synthesis. J Biol Chem 273(6):3351–3357

Kang Li YL, Shelton JM, Richardson JA, Spencer E, Chen ZJ, Wang X, Williams RS (2000) Cytochrome c deficiency causes embryonic lethality and attenuates stress-induced apoptosis. Cell 101:389–399

Kaufman MH (1991) Histochemical identification of primordial germ cells and differentiation of the gonads in homozygous tetraploid mouse embryos. J Anat 179:169–181

Kaufman MH, Navaratnam V (1981) Early differentiation of the heart in mouse embryos. J Anat 133:235–246

Kawamukai M (2009) Biosynthesis and bioproduction of coenzyme Q10 by yeasts and other organisms. Biotechnol Appl Biochem 53(Pt 4):217–226. https://doi.org/10.1042/BA20090035

Kayser EB, Sedensky MM, Morgan PG, Hoppel CL (2004) Mitochondrial oxidative phosphorylation is defective in the long-lived mutant clk-1. J Biol Chem 279(52):54479–54486

Lapointe J, Hekimi S (2008) Early mitochondrial dysfunction in long-lived Mclk1+/− mice. J Biol Chem 283(38):26217–26227

Lapointe J, Wang Y, Bigras E, Hekimi S (2012) The submitochondrial distribution of ubiquinone affects respiration in long-lived Mclk1+/− mice. J Cell Biol 199(2):215–224. https://doi.org/10.1083/jcb.201203090

Larsson NG, Wang J, Wilhelmsson H, Oldfors A, Rustin P, Lewandoski M, Barsh GS, Clayton DA (1998) Mitochondrial transcription factor A is necessary for mtDNA maintenance and embryogenesis in mice. Nat Genet 18(3):231–236

Lee SJ, Hwang AB, Kenyon C (2010) Inhibition of respiration extends C. elegans life span via reactive oxygen species that increase HIF-1 activity. Curr Biol 20(23):2131–2136. https://doi.org/10.1016/j.cub.2010.10.057

Levavasseur F, Miyadera H, Sirois J, Tremblay ML, Kita K, Shoubridge E, Hekimi S (2001) Ubiquinone is necessary for mouse embryonic development but is not essential for mitochondrial respiration. J Biol Chem 276(49):46160–46164

Levine HJ (1997) Rest heart rate and life expectancy. J Am Coll Cardiol 30(4):1104–1106

Lloyd D, Rossi E (1992) Ultradian rhythms in life processes: a fundamental inquiry into chronobiology and psychobiology. Springer, New York, pp 1–406

Marbois BN, Clarke CF (1996) The COQ7 gene encodes a protein in saccharomyces cerevisiae necessary for ubiquinone biosynthesis. J Biol Chem 6:2995–3004

Miyadera H, Amino H, Hiraishi A, Taka H, Murayama K, Miyoshi H, Sakamoto K, Ishii N, Hekimi S, Kita K (2001) Altered quinone biosynthesis in the long-lived clk-1 mutants of Caenorhabditis elegans. J Biol Chem 276(11):7713–7716

Nakai D, Shimizu T, Nojiri H, Uchiyama S, Koike H, Takahashi M, Hirokawa K, Shirasawa T (2004) coq7/clk-1 regulates mitochondrial respiration and the generation of reactive oxygen species via coenzyme Q. Aging Cell 3(5):273–281

Nakai D, Yuasa S, Takahashi M, Shimizu T, Asaumi S, Isono K, Takao T, Suzuki Y, Kuroyanagi H, Hirokawa K, Koseki H, Shirasawa T (2001) Mouse homologue of coq7/clk-1, longevity gene in Caenorhabditis elegans, is essential for coenzyme Q synthesis, maintenance of mitochondrial integrity, and neurogenesis. Biochem Biophys Res Commun 289(2):463–471

Sanz A, Fernandez-Ayala DJ, Stefanatos RK, Jacobs HT (2010) Mitochondrial ROS production correlates with, but does not directly regulate lifespan in Drosophila. Aging 2(4):200–223
Shepard TH, Muffley LA, Smith LT (1998) Ultrastructural study of mitochondria and their cristae in embryonic rats and primate (N. nemistrina). Anat Rec 252(3):383–392
Stenmark P, Grunler J, Mattsson J, Sindelar PJ, Nordlund P, Berthold DA (2001) A new member of the family of di-iron carboxylate proteins. Coq7 (clk-1), a membrane-bound hydroxylase involved in ubiquinone biosynthesis. J Biol Chem 276(36):33297–33300
Takahashi K, Noda Y, Ohsawa I, Shirasawa T, Takahashi M (2014) Extended lifespan, reduced body size and leg skeletal muscle mass, and decreased mitochondrial function in clk-1 transgenic mice. Exp Gerontol 58:146–153. https://doi.org/10.1016/j.exger.2014.08.003
Takahashi M, Asaumi S, Honda S, Suzuki Y, Nakai D, Kuroyanagi H, Shimizu T, Honda Y, Shirasawa T (2001) Mouse coq7/clk-1 orthologue rescued slowed rhythmic behavior and extended life span of clk-1 longevity mutant in Caenorhabditis elegans. Biochem Biophys Res Commun 286(3):534–540
Takahashi M, Ogawara M, Shimizu T, Shirasawa T (2012a) Restoration of the behavioral rates and lifespan in clk-1 mutant nematodes in response to exogenous coenzyme Q(10). Exp Gerontol 47(3):276–279. https://doi.org/10.1016/j.exger.2011.12.012
Takahashi M, Shimizu T, Moriizumi E, Shirasawa T (2008) Clk-1 deficiency induces apoptosis associated with mitochondrial dysfunction in mouse embryos. Mech Ageing Dev 129(5):291–298
Takahashi M, Shimizu T, Shirasawa T (2012b) Reversal of slow growth and heartbeat through the restoration of mitochondrial function in clk-1-deficient mouse embryos by exogenous administration of coenzyme Q10. Exp Gerontol 47(6):425–431. https://doi.org/10.1016/j.exger.2012.03.008
Turunen M, Olsson J, Dallner G (2004) Metabolism and function of coenzyme Q. Biochim Biophys Acta 1660(1–2):171–199
Vajo Z, King LM, Jonassen T, Wilkin DJ, Ho N, Munnich A, Clarke CF, Francomano CA (1999) Conservation of the Caenorhabditis elegans timing gene clk-1 from yeast to human: a gene required for ubiquinone biosynthesis with potential implications for aging. Mamm Genome 10(10):1000–1004
Wang Y, Oxer D, Hekimi S (2015) Mitochondrial function and lifespan of mice with controlled ubiquinone biosynthesis. Nat Commun 6:6393. https://doi.org/10.1038/ncomms7393
Williams GC, Pleitropy N (1957) Selection and the evolution of senescence. Evolution 11:398–411
Wong A, Boutis P, Hekimi S (1995) Mutations in the clk-1 gene of Caenorhabditis elegans affect developmental and behavioral timing. Genetics 139(3):1247–1259
Yang W, Hekimi S (2010) A mitochondrial superoxide signal triggers increased longevity in Caenorhabditis elegans. PLoS Biol 8(12):e1000556. https://doi.org/10.1371/journal.pbio.1000556
Yu EA, Weaver DR (2011) Disrupting the circadian clock: gene-specific effects on aging, cancer, and other phenotypes. Aging 3(5):479–493
Zheng H, Lapointe J, Hekimi S (2010) Lifelong protection from global cerebral ischemia and reperfusion in long-lived Mclk1(+/)(−) mutants. Exp Neurol 223(2):557–565. https://doi.org/10.1016/j.expneurol.2010.02.002

Chapter 9
Reduced Coenzyme Q_{10} Decelerates Senescence and Age-Related Hearing Loss in Senescence-Accelerated Mice by Activating Mitochondrial Functions.

Jinko Sawashita, Xu Zhe, and Keiichi Higuchi

Abstract Coenzyme Q (CoQ) is present in all cellular and organelle membranes, in organisms ranging from yeast to humans. CoQ is synthesized exclusively in the mitochondrial inner membrane from farnesyl pyrophosphate via the mevalonate pathway. Meanwhile, CoQ in foods or medicines is converted to the reduced form ($CoQH_2$: ubiquinol) in small intestine epithelia before absorption. Previous studies in humans and rodents suggest that coenzyme Q_{10} (CoQ_{10}) supplementation mitigates cardiomyopathies, age-related declines in myocardial and arterial function, and some neurodegenerative disorders. CoQ_{10} also has beneficial effects in the aging process and lessens age-related hearing loss in animal models. Using Senescence-Accelerated Mouse Prone 1 (SAMP1) mice, we demonstrated that the reduced form of CoQ_{10} ($CoQ_{10}H_2$: ubiquinol-10) has more potent anti-aging effects than the oxidized form of CoQ_{10} (CoQ_{10}: ubiquinone-10). SAMP1 mice receiving lifelong supplementation with either 0.2 or 0.5% $CoQ_{10}H_2$ had lower senescence grading scores than untreated control mice. Microarrays containing 45,100 probe sets identified several peroxisome proliferator-activated receptor-α (PPAR-α)-associated genes

J. Sawashita (✉)
Institute for Biomedical Sciences, Shinshu University, Matsumoto, Japan

Department of Aging Biology, Institute of Pathogenesis and Disease Prevention, Shinshu University Graduate School of Medicine, Matsumoto, Japan

Present Affiliation: Supplement Business Division, Pharma & Supplemental Nutrition Solutions Vehicle, Kaneka Corporation, Osaka, Japan
e-mail: Jinko.Sawashita@kaneka.co.jp

X. Zhe
Department of Aging Biology, Institute of Pathogenesis and Disease Prevention, Shinshu University Graduate School of Medicine, Matsumoto, Japan

K. Higuchi
Institute for Biomedical Sciences, Shinshu University, Matsumoto, Japan

Department of Aging Biology, Institute of Pathogenesis and Disease Prevention, Shinshu University Graduate School of Medicine, Matsumoto, Japan
e-mail: keiichih@shinshu-u.ac.jp

G. López Lluch (ed.), *Coenzyme Q in Aging*,
https://doi.org/10.1007/978-3-030-45642-9_9

that were upregulated in the livers of the SAMP1 mice given $CoQ_{10}H_2$. Our recent results show that $CoQ_{10}H_2$ may enhance mitochondrial activity by increasing levels of sirtuin 1 (SIRT1), peroxisome proliferator-activated receptor c coactivator 1α (PGC-1α), and SIRT3 that protect against the progression of aging and age-related diseases.

Keywords Anti-aging · $CoQ_{10}H_2$ · Mitochondrial function · PGC-1α · SAMP1 mice · Sirtuin

9.1 Introduction

In eukaryotes ranging from yeast to humans coenzyme Q (CoQ) can exist in a variety of oxidation states, but the main reduced and oxidized forms are ubiquinol and ubiquinone, respectively (Fig. 9.1). Reduced CoQ, $CoQ_{10}H_2$ (ubiquinol-10), is an important component of ATP-producing bioenergetic pathways in mitochondria in humans. The total content of oxidized and reduced CoQ decreased in several organs associated with aging in humans (Fig. 9.2) (Kalén et al. 1989). Furthermore, levels of plasma $CoQ_{10}H_2$, which is the essential form that acts in mitochondria, are reduced in older people compared to younger or middle-aged people (Wada et al. 2007), as are levels of NAD(P)H: quinone oxidoreductase 1 (NQO1), an enzyme involved in the reduction of ubiquinone to ubiquinol (López-Lluch et al. 2010; Sohal and Forster 2007). Lowered plasma levels of $CoQ_{10}H_2$ also occur in metabolism-associated diseases and/or in conditions of hyper-oxidative stress, such as type II diabetes, kidney failure requiring hemodialysis, hepatic cancer, amyotrophic lateral sclerosis (ALS), Parkinson's disease and Down syndrome (Yamamoto and Yamashita 1999; Sohmiya et al. 2004, 2005; Nakazawa et al. 2005; Lim et al. 2006, 2008; Miles et al. 2007). Interestingly, dietary supplementation with CoQ_{10} did not improve plasma levels of $CoQ_{10}H_2$ in type II diabetes or kidney failure patients (Nakazawa et al. 2005; Lim et al. 2008), even though the $CoQ_{10}H_2$

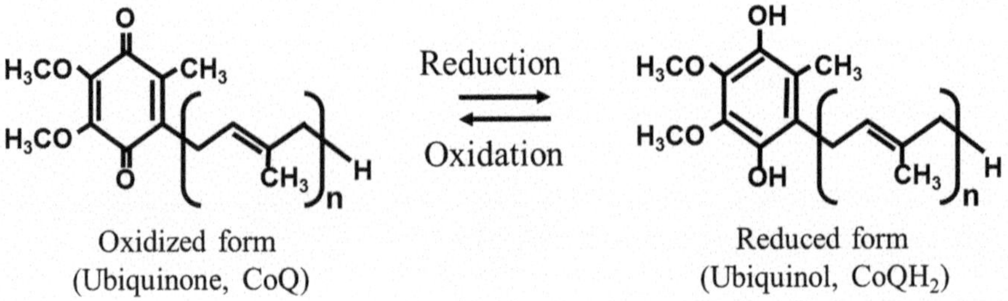

Fig. 9.1 Chemical structures of oxidized and reduced forms of coenzyme Q

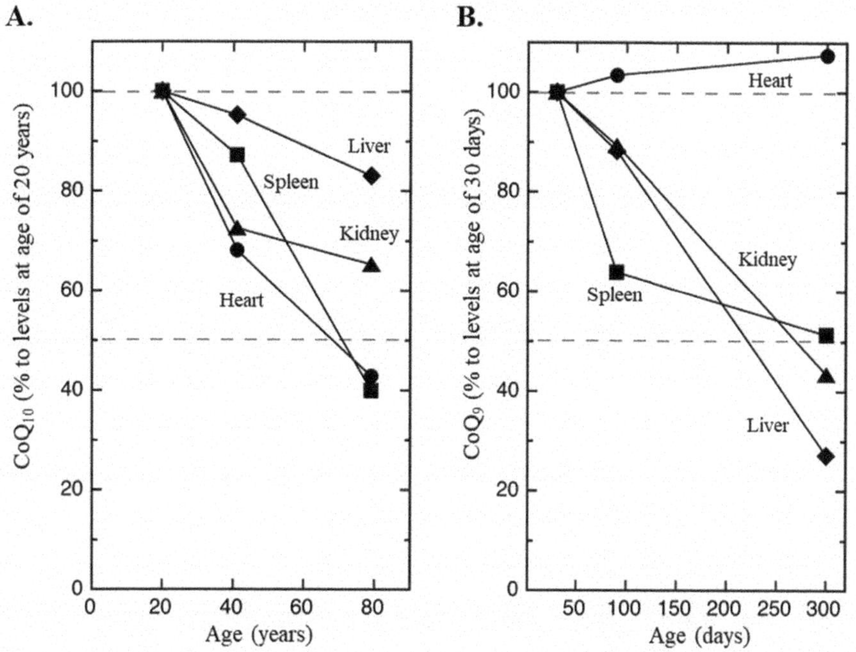

Fig. 9.2 Age-related decline in CoQ levels in human (**a**) and rat (**b**) organs. Reproduced from data published in Kalen et al. (1989)

supplementation improved remarkably in those levels and conditions of oxidative stress in Down syndrome patients (Miles et al. 2007). In patients that have conditions associated with hyperoxidative stress and aging, the rate of enzymatic conversion of CoQ_{10} to $CoQ_{10}H_2$ appears to be inadequate to cope with increased oxidative stress. In general, CoQ_{10} assimilated from the diet exists as an oxidized form of CoQ, so reduction of CoQ_{10} to $CoQ_{10}H_2$ is essential for ATP production. Recently, Bentinger et al. reported the pathway for $CoQ_{10}H_2$ biosynthesis from dietary lipids (Bentinger et al. 2010). But whether there are age-related changes in the activities of these enzymes is unknown. From these findings, we believe that dietary supplementation with $CoQ_{10}H_2$ rather than CoQ_{10}, may be a superior treatment for age-related conditions in humans and animals. In this chapter, we show our findings obtained in senescence-accelerated model mice (Takeda et al. 1997) and culture cell lines treated with $CoQ_{10}H_2$ supplementation.

9.2 Evaluation of Age-Related Parameters in Senescence-Accelerated Mouse

9.2.1 Senescence-Accelerated Mouse (SAM)

The Senescence-Accelerated Mouse Prone (SAMP) is a mouse model of accelerated aging established by Takeda et al. as recombinant-like inbred strains, which are composed of several SAMP (SAMP1, SAMP6, SAMP8 and SAMP10) strains and the Senescence-Accelerated Mouse Resistant 1 (SAMR1) strain (Takeda et al. 1997). Each SAMP strain shows characteristics of early occurrence and development of age-related diseases, such as amyloidosis, age-related hearing loss (AHL), senile osteoporosis, and learning and memory impairment (Takeda et al. 2013). Meanwhile, SAMR1 mice have physiological parameters and an average life span that is similar to general normal mouse strains, and can be used as a normal control for SAMP mice. The SAM series of mice are excellent models for studies on therapeutic and preventive treatments for age-related diseases.

9.2.2 Dietary Supplementation with CoQ_{10} and $CoQ_{10}H_2$ in SAMP1 Mice

In our studies, SAMP1 mice were used to study the anti-aging effects of CoQ_{10}. As mentioned above, SAMP1 mice have short life spans and pathological changes such as senile amyloidosis, renal atrophy, AHL, and emphysema (Fig. 9.3) (Takeda et al. 2013). Changes in CoQ_{10} and $CoQ_{10}H_2$ contents in mouse organs after dietary supplementation with 0.5% CoQ_{10} or $CoQ_{10}H_2$ were detectable (Fig. 9.4) (Schmelzer et al. 2010 and unpublished data). CoQ_{10} and $CoQ_{10}H_2$ were mainly concentrated in

SAMP
(Prone)

SAMR
(Resistant)

Fig. 9.3 SAMP1 (upper) and SAMR1 (lower) mouse at age of 12 months

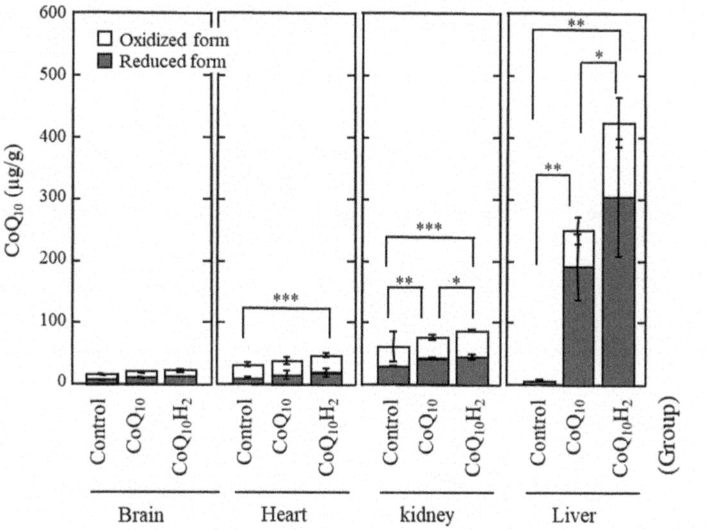

Fig. 9.4 CoQ_{10} levels in the brain, heart, kidneys, and liver of SAMP1 mice at age of 7 months with or without supplementation with 0.5% CoQ_{10} or $CoQ_{10}H_2$. Data are means ± SEM (N = 5). ∗: $p < 0.05$, ∗∗: $p < 0.01$, ∗∗∗: $p < 0.001$. Reproduced from Schmelzer et al. (2010) (the liver) and unpublished data (the others)

the liver, and the increases in total CoQ_{10} were significantly higher with $CoQ_{10}H_2$ supplementation relative to those seen for CoQ_{10}. This result indicated that the absorption efficiency of $CoQ_{10}H_2$ might be higher than that of CoQ_{10} in mice.

9.2.3 *Changes in Hepatic Gene Expression in SAMP1 Mice Supplemented with CoQ_{10} and $CoQ_{10}H_2$*

Gene expression levels of fatty acid synthesis- and inflammation-related factors in SAMP1 mouse liver were detected after 6 months of CoQ_{10} and $CoQ_{10}H_2$ dietary supplementation (Fig. 9.5) (Schmelzer et al. 2010). $CoQ_{10}H_2$ supplementation inhibited gene expression of fatty acid synthesis-related factors in SAMP1 mice, whereas expression of lipid metabolism-related and anti-inflammatory factors was increased relative to the control group. These changes were not observed for the CoQ_{10} group.

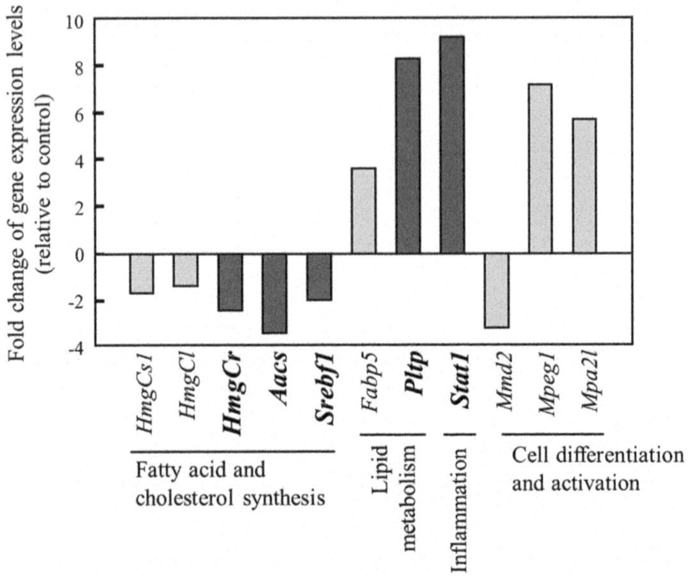

Fig. 9.5 Identification of $CoQ_{10}H_2$-sensitive genes and their functional connections in the livers of SAMP1 mouse. Genes presented in bold text and dark-colored bar indicate PPAR-α-related genes. Reproduced from Schmelzer et al. (2010)

9.3 Inhibitory Effect of Dietary CoQ_{10} Supplementation on Accelerated Aging in SAM Mice

9.3.1 Changes in Senescence Scores for SAMP1 Mice

The degree of senescence was evaluated by a grading system that includes 11 categories of general behavioral activity, as well as gross appearance of the skin, eyes, and spine, which are considered to be associated with the aging process (Hosokawa et al. 1984). This grading score system appears to be valid for evaluating the general degree of senescence in SAM mice. Grading was done at a fixed time (from 2 pm to 4 pm) by an observer who was blinded to the treatment group identities. In the 0.5% $CoQ_{10}H_2$ supplementation experiment, senescence grading scores for SAMP1 mice in all three groups increased in an age-dependent fashion beginning when the mice were 4 months old (Fig. 9.6) (Schmelzer et al. 2010). However, the accelerated aging of SAMP1 mice was inhibited by $CoQ_{10}H_2$ supplementation to a greater degree than that seen for the CoQ or control groups, which displayed no significant changes. Furthermore, senescence grading scores in SAMP1 mice were decelerated by the supplementation with $CoQ_{10}H_2$ of low-dosage (0.3%) (Yan et al. 2006; Tian et al. 2014).

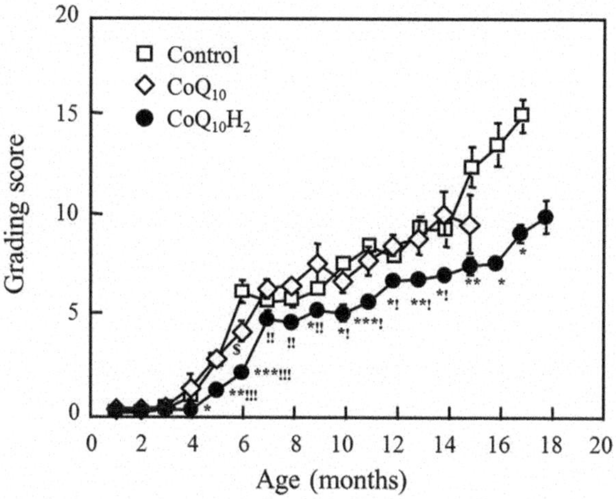

Fig. 9.6 $CoQ_{10}H_2$ slowed aging of SAMP1 mice. Age-related changes in senescence grading scores for SAMP1 mice supplemented with $CoQ_{10}H_2$ and CoQ_{10} diets are shown. ∗: $p < 0.05$, ∗∗: $p < 0.01$, ∗∗∗: $p < 0.001$ ($CoQ_{10}H_2$ vs. Control);!: $p < 0.05$,!!: $p < 0.01$,!!!: $p < 0.001$ ($CoQ_{10}H_2$ vs. CoQ_{10}). Reproduced from Schmelzer et al. (2010)

9.3.2 *Mitigation of AHL by $CoQ_{10}H_2$ Supplementation in SAMP1 Mice*

SAMP1 mice exhibit accelerated progression of many age-associated degenerative diseases, including AHL. We analysed the auditory brainstem response (ABR) in 2-, 7 , 13-, and 19-month-old SAMP1 mice to investigate whether $CoQ_{10}H_2$ supplementation could prevent AHL (Fig. 9.7a) (Tian et al. 2014). Control mice showed an age-associated increase in ABR hearing thresholds at high (32 kHz), middle (16 kHz), and low (8 kHz) frequencies (Fig. 9.7b, c, and d). Meanwhile, when $CoQ_{10}H_2$ supplementation was begun with one-month-old, significantly reduced hearing impairment was seen for all 3 frequencies. Beginning $CoQ_{10}H_2$ supplementation when the SAMP1 mice were 7 months old did not improve the high frequency AHL, but hearing impairments in the middle- and low frequency ranges were suppressed. Middle-frequency hearing impairment was improved to a certain extent even if mice were supplemented with $CoQ_{10}H_2$ beginning when the animals were 13 months old.

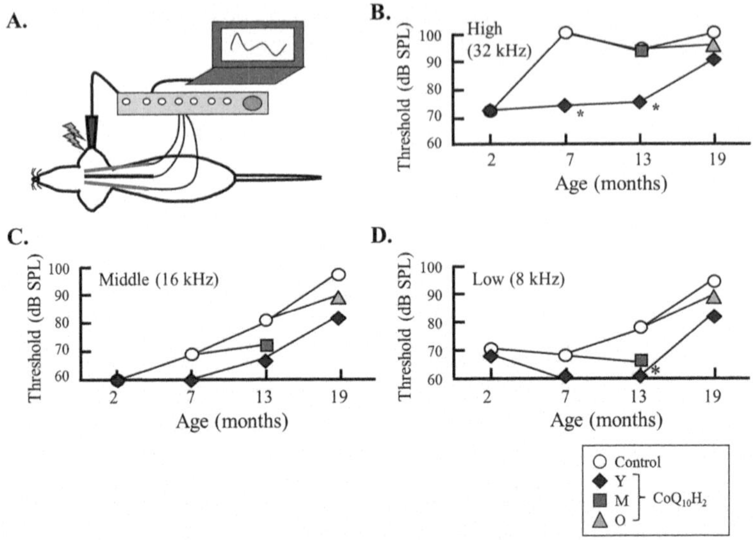

Fig. 9.7 $CoQ_{10}H_2$ delayed AHL progression in SAMP1 mice. ABR thresholds were determined using click stimuli at 60, 70, 80, 90 and 100 dB sound pressure levels (**a**). The threshold values were measured at high (32 kHz) (**b**), middle (16 kHz) (**c**), and low (8 kHz) (**d**) frequencies when SAMP1 mice were 2, 7, 13, and 19 months of age. The four treatment groups were: Control and $CoQ_{10}H_2$ supplementation begun at 1 (Y), 7 (M), and 13 (O) months (n = 2–5). $*p < 0.05$ (Control vs. $CoQ_{10}H_2$). Reproduced from Tian et al. (2014) with permission from the editorial

9.4 Anti-aging Mechanism of $CoQ_{10}H_2$

9.4.1 Antioxidant Effects of $CoQ_{10}H_2$

Liver malondialdehyde (MDA) levels were measured to assess lipid oxidative damage in the livers of control and $CoQ_{10}H_2$-supplemented SAMP1 mice (Fig. 9.8) (Tian et al. 2014). The results showed that oxidative damage of liver lipids increased with age in the control group, whereas this increased damage was suppressed in mice given $CoQ_{10}H_2$ supplementation (Fig. 9.8a). Similar changes were observed for oxidative damage of proteins and DNA (Fig. 9.8b, c). As the main antioxidant in cells, glutathione (GSH) plays a key physiological role in eliminating free radicals and protecting cells in the human body. Measurement of the conversion of GSH to oxidized glutathione (GSSG) can be used as a measure of oxidative stress. The ratios of GSH to GSSG in SAMP1 mice livers showed that the antioxidant ability of mice livers decreased dramatically with age. However, this decreasing trend could be effectively inhibited with $CoQ_{10}H_2$ supplementation (Fig. 9.8d).

Gene expression and protein activities of mitochondria-related genes in SAMP1 mice livers were next determined. The sirtuin family includes a series of important regulatory factors that are widely involved in physiological activities such as fatty acid oxidation, oxidative stress tolerance, and insulin secretion, which are closely

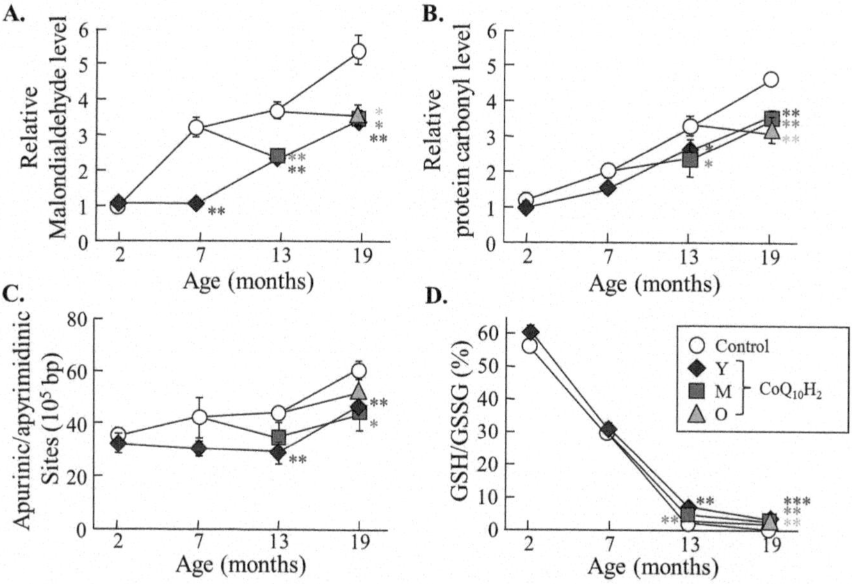

Fig. 9.8 $CoQ_{10}H_2$ decelerated the increase in age-related oxidative stress in SAMP1 mice. Western blot analysis of oxidative damage to lipids (MDA) (**a**) and proteins (protein carbonyl) (**b**) in SAMP1 mice livers. (**c**) Oxidative damage to DNA (apurinic/apyrimidinic sites) and (**d**) Liver GSH:GSSG ratios. The values are means ± SEM of mice in each age group (n = 3–5). The four treatment groups were: Control and $CoQ_{10}H_2$ supplementation begun at 1 (Y), 7 (M), and 13 (O) months (n = 3–5). *$p < 0.05$; **$p < 0.01$; ***$p < 0.001$ (Control vs. $CoQ_{10}H_2$). Reproduced from Tian et al. (2014) with permission from the editorial

related to metabolism-associated diseases, inflammation, and aging (Heranz et al. 2010). Some studies revealed that the promotion of enzyme activities of sirtuin family could increase lifespan (Howitz et al. 2003; Bordone and Guarente 2005; Finkel et al. 2009). In the livers of SAMP1 mouse, SIRT1 expression decreased with age and this decrease was reversed in the $CoQ_{10}H_2$-supplemented mice (Fig. 9.9a) (Tian et al. 2014). Peroxisome proliferator-activated receptor c coactivator 1α (PGC-1α) is a main regulatory factor of mitochondrial function. PGC-1α expression levels were decreased in the livers of 13- and 19-month-old control SAMP1 mice, but a comparison of the $CoQ_{10}H_2$-supplemented and control groups showed significant increases in PGC-1α expression for the $CoQ_{10}H_2$ group. Similar changes were seen for gene expression of other mitochondria-associated proteins, including estrogen related receptor alpha (ERRα), nuclear factor erythroid 2-related factor 2 (NRF2), and mitochondrial transcription factor A (TFAM). We next used Western blotting to determine the protein levels of the antioxidant enzymes SIRT3 and superoxide dismutase 2 (SOD2) in mitochondria. SIRT3 and SOD2 protein levels in the liver were enhanced when $CoQ_{10}H_2$ supplementation began when the mice were young or middle aged. In contrast, beginning $CoQ_{10}H_2$ supplementation when the mice reached old age did not affect the SIRT3 and SOD2 protein levels in the liver

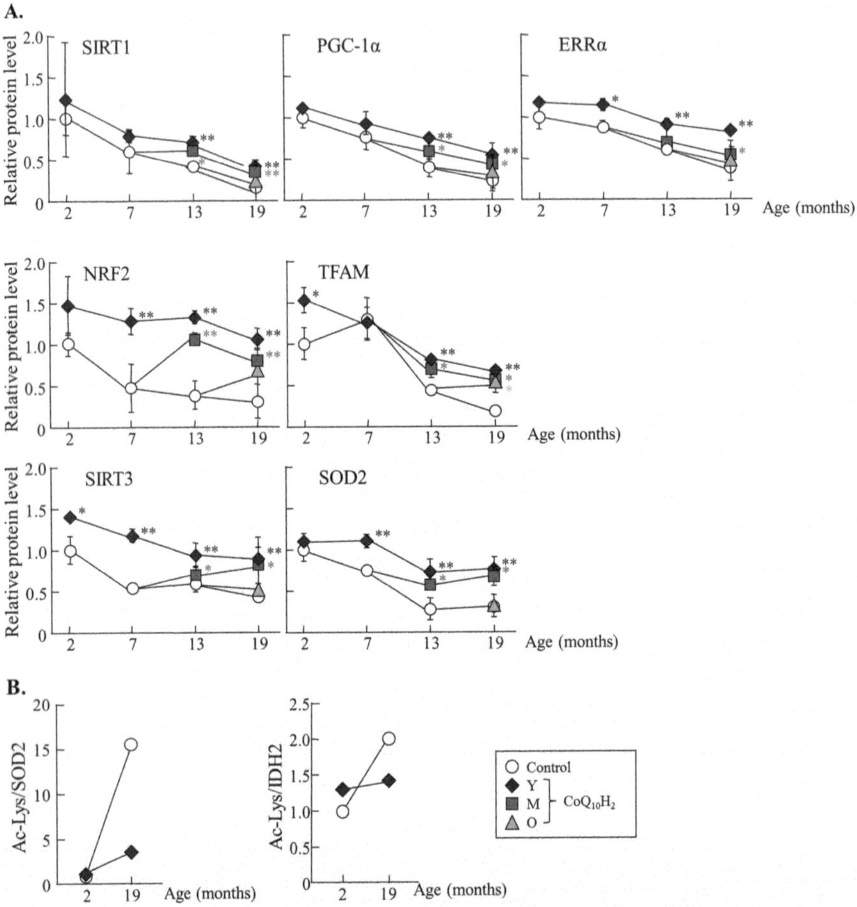

Fig. 9.9 $CoQ_{10}H_2$ supplementation decelerated age-related decline in the expression of proteins related to mitochondrial function. Western blot analyses for expression levels of SIRT1, PGC-1α, ERRα, NRF2, TFAM, SIRT3, and SOD2 protein in the livers of SAMP1 mice (**a**). Immunoblot analysis of SOD2 and IDH2 acetylation in the livers of SAMP1 mice (**b**). The values are means ± SEM of mice in each age group (n = 3–5). The four treatment groups were: Control and $CoQ_{10}H_2$ supplementation begun at 1 (Y), 7 (M), and 13 (O) months (n = 3–5). $*p < 0.05$; $**p < 0.01$ (Control vs. $CoQ_{10}H_2$). Reproduced from Tian et al. (2014) with permission from the editorial

relative to the control group. Since acetylation reduces the activity of SOD2 and isocitrate dehydrogenase 2 (IDH2), which both have antioxidant activities, we also determined the acetylation levels of these two proteins in the livers of SAMP1 mouse (Fig. 9.9b). $CoQ_{10}H_2$ supplementation could reduce acetylation levels of SOD2 and IDH2 and increase the antioxidant capacity of mitochondria compared with the control group.

To assess whether $CoQ_{10}H_2$ affected reactive oxygen species (ROS) production in liver cells, the changes in ROS in human hepatocellular carcinoma HepG2 cells supplemented with $CoQ_{10}H_2$ in the media were assessed (Fig. 9.10) (Tian et al. 2014). The addition of $CoQ_{10}H_2$ to HepG2 cell culture medium significantly reduced ROS production by the cells relative to that seen for untreated cells.

9.4.2 *Improvement of Mitochondrial Function by $CoQ_{10}H_2$*

The activities of mitochondrial respiratory chain complex 1 (Complex I) and VI (Complex VI) in SAMP1 mouse liver decreased gradually with age, but the downward trend improved with $CoQ_{10}H_2$ supplementation (Fig. 9.11a, b) (Tian et al. 2014). Meanwhile, ATP production fell as the mice aged, and this reduction was alleviated with $CoQ_{10}H_2$ supplementation (Fig. 9.11c).

Mitochondrial DNA (mtDNA) copy numbers can increase in response to increased ATP production demands. We found that mtDNA copy numbers were significantly increased in the $CoQ_{10}H_2$-supplemented SAMP1 mice group relative to the control group (Fig. 9.11d). Meanwhile, our studies confirmed that expression levels of the mitochondria generation-related genes such as NRF2 and TFAM increased with $CoQ_{10}H_2$ supplementation. Both results suggested that the number of mitochondria in the livers of SAMP1 mouse increased following supplementation with $CoQ_{10}H_2$. We showed that the number of mitochondria was indeed increased in our *in vivo* experiments. Moreover, Complex I activity and cellular oxygen consumption showed that $CoQ_{10}H_2$ supplementation can effectively improve mitochondrial function (Fig. 9.12).

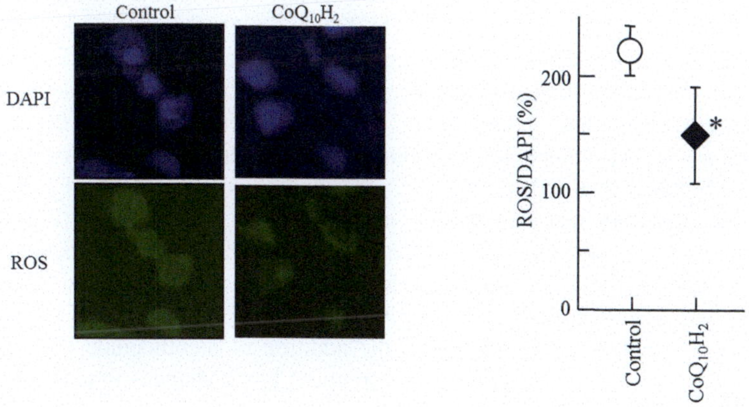

Fig. 9.10 ROS levels were decreased following $CoQ_{10}H_2$-treatment. ROS levels were measured in HepG2 cells with or without $CoQ_{10}H_2$ using laser fluorescence confocal microscopy (X200). DAPI was used to indicate nuclei. Quantification of ROS/DAPI is shown in the right panel. The values are means ± SEM (n = 5). $*p < 0.05$ (Control vs. $CoQ_{10}H_2$). Reproduced from Tian et al. (2014) with permission from the editorial

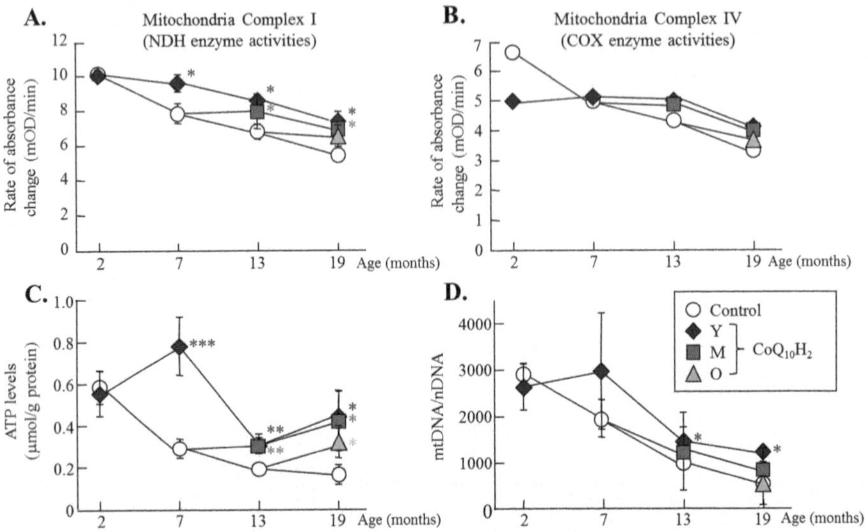

Fig. 9.11 $CoQ_{10}H_2$ decelerated age-related decline in SAMP1 mice mitochondrial function. Mitochondria complex I (NADH dehydrogenase; NDH) and IV (cytochrome c oxidase: COX) enzyme activities were measured in the livers of SAMP1 mice (**a**, **b**). ATP levels in the livers of SAMP1 mice fed control or $CoQ_{10}H_2$-supplemented diet (**c**). Analysis of mtDNA copy number per nuclear DNA in SAMP1 mice (**d**). The values are means ± SEM of mice in each age group (n = 3–5). The four treatment groups were: Control and $CoQ_{10}H_2$ supplementation begun at 1 (Y), 7 (M), and 13 (O) months (n = 3–5). $*p < 0.05$; $**p < 0.01$; $***p < 0.001$ (Control vs. $CoQ_{10}H_2$). Reproduced from Tian et al. (2014) with permission from the editorial

9.4.3 Regulation of the cAMP-SIRT Signaling Pathway by $CoQ_{10}H_2$

We confirmed that $CoQ_{10}H_2$ could increase SIRT1 gene expression and activity in the livers of SAMP1 mouse. This increase could in turn affect the activities of downstream factors that can improve the metabolic function and inhibit the accelerated aging of SAMP1 mice. We reproduced these results in SAMP1 mice cochleae (Fig. 9.13) (Tian et al. 2014). We also measured changes in levels of cAMP, an upstream regulatory factor of SIRT1, in the livers of SAMP1 mice (Fig. 9.14). cAMP production was increased in the livers of SAMP1 mice fed diets supplemented with $CoQ_{10}H_2$. The phosphorylation levels of the downstream factors of cAMP changed concurrently. We recapitulated these results in HepG2 cell culture experiments (Fig. 9.15). Furthermore, the addition of nicotinamide (NAM), the SIRT1 blocker, to the cell culture medium could block changes in active formed proteins induced by $CoQ_{10}H_2$ supplementation.

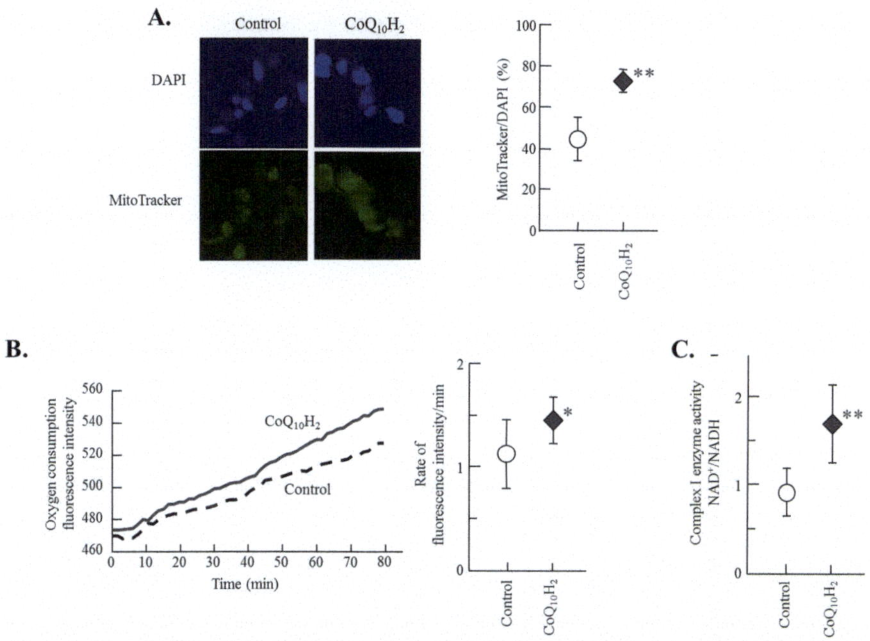

Fig. 9.12 $CoQ_{10}H_2$ increased mitochondrial metabolic activity in HepG2 cells. HepG2 cells with or without $CoQ_{10}H_2$ added to the media were incubated with MitoTracker Green to stain mitochondria, and then examined by laser fluorescence confocal microscopy (X200). DAPI was used to indicate nuclei. Quantification of MitoTracker/DAPI is shown at right (**a**). Oxygen consumption rates (rate of fluorescence intensity/min) were measured in HepG2 cells and are shown in panel (**b**). ELISA analysis to determine NAD and NADH levels in HepG2 cells (**c**). The values are means ± SEM (n = 5–10). *$p < 0.05$; **$p < 0.01$ (Control vs. $CoQ_{10}H_2$). Reproduced from Tian et al. (2014) with permission from the editorial

9.5 Discussion

Aging is a multifactorial process that depends on the individual and the environment, as well as lifestyle and other factors (Finkel et al. 2009). Although aging is a significant risk factor for the occurrence of many other human diseases, the basic mechanisms involved in the aging process remain unclear. Decreases in antioxidant substances with concurrent increases in pro-oxidant substances in the body may be important factors during the aging process (Bordone and Guarente 2005). Mitochondria are the main sources of ROS in cells, and ROS induced damage may lead to age-associated mitochondrial dysfunction that accumulates over time (Del Pozo-Cruz et al. 2014). Another age-related change is a decline in the ability of the mitochondria to synthesize ATP (Niklowitz et al. 2016).

CoQ is a lipid soluble micronutrient that is ubiquitous in the human body and a coenzyme of mitochondrial oxidative phosphorylation. In humans and animals, CoQ can improve immune function, strengthen anti-oxidation systems, and enhance vitality. With increasing age, CoQ_{10} contents in serum and tissues gradually decrease.

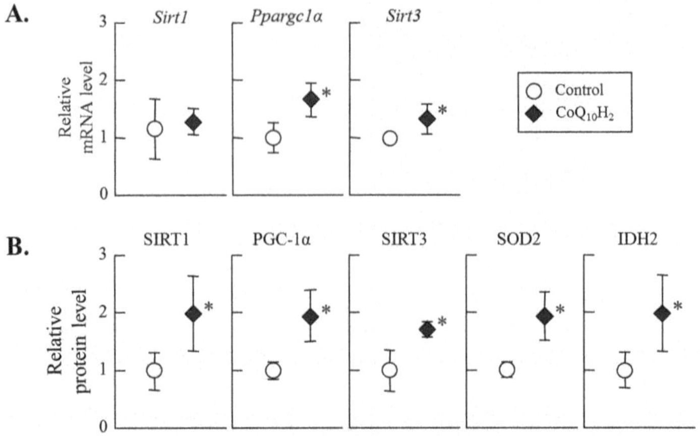

Fig. 9.13 $CoQ_{10}H_2$ regulates mitochondrial metabolic activity by increasing SIRT1, PGC-1α, and SIRT3 levels in the cochleae of SAMP1 mice. (**a**) Real-time PCR analysis of *Sirt1*, *Ppargc1α (Pgc-1α)*, and *Sirt3* mRNA expression levels in cochleae from 7-month-old SAMP1 mice fed a control or $CoQ_{10}H_2$-supplemented diet. (**b**) Western blot analysis of SIRT1, PGC-1α, SIRT3, SOD2, and IDH2 protein levels in cochleae from 10-month-old SAMP1 mice fed a control or $CoQ_{10}H_2$-supplemented diet. The values are means ± SEM of mice in each age group (n = 5). $*p < 0.05$ (Control vs. $CoQ_{10}H_2$). Reproduced from Tian et al. (2014) with permission from the editorial

Since tissue CoQ_{10} is mainly derived from *de novo* synthesis, the CoQ_{10} concentration in serum or plasma is often used as an important indicator to predict the content and function of CoQ_{10} in organs and the entire body in vivo. CoQ_{10} content in tissues is often closely correlated with the occurrence and development of various age-related diseases. For example, patients with heart failure have decreased CoQ_{10} content in cardiac tissues, and the degree of decrease correlated with an increased risk for heart failure related to left ventricle dysfunction (Harman 1957).

CoQ_{10} has a critical effect on mitochondrial oxidative phosphorylation and cellular ATP production (Sohal and Forster. 2007). The rate of oxidative phosphorylation in the respiratory chain depends on the CoQ_{10} concentration in the mitochondrial inner membrane, such that any subtle changes in CoQ_{10} levels may lead to significant changes in respiratory rates (Folkers et al. 1985). Meanwhile, the loss of CoQ_{10} may reduce ATP generation. A lack of CoQ_{10} can be manifested as pressure on the myocardial wall that may aggravate heart failure by increasing the energy demand, thus leading to an imbalance between energy supply and demand (Crane 2001). Heart failure patients also have a disrupted balance between oxidation and antioxidation that leads to a weakening of the antioxidant enzyme system and subsequent generation of large amounts of ROS (Littarru 1994). Increased ROS levels in patients with ischemic and non-ischemic heart failure can damage mitochondrial proteins, thus leading to lipid peroxidation that impairs mitochondrial respiratory chain function and accelerates destruction of CoQ_{10}, along with its antioxidant properties. This sequence of events can partly explain the low concentration of CoQ_{10} in

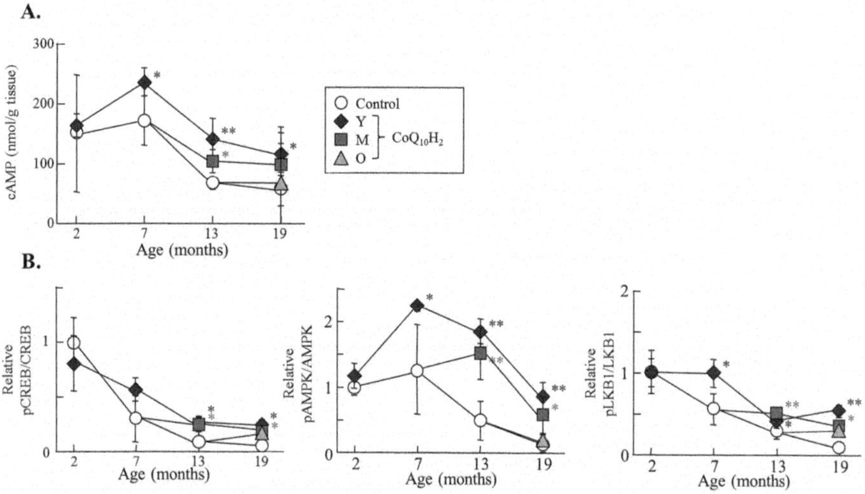

Fig. 9.14 $CoQ_{10}H_2$ supplementation suppressed age-related decreases in cAMP levels (**a**) and the expression of phosphorylated cAMP response elements binding protein (CREB), AMP-activated protein kinase (AMPK), and phosphorylated serine/threonine kinase B1 (LKB1) in the livers of SAMP1 mice (B). ELISA analysis of cAMP levels in the livers from 2-, 7-, 13-, and 19-month-old SAMP1 mice fed a control or $CoQ_{10}H_2$-supplemented diet (**a**). Western blot analysis of total and phosphorylated CREB, AMPK and LKB1 expression levels in livers from 2-, 7-, 13-, and 19-month-old SAMP1 mice fed a control or $CoQ_{10}H_2$ -supplemented diet (**b**). The four treatment groups were: Control and $CoQ_{10}H_2$ supplementation begun at 1 (Y), 7 (M), and 13 (O) months (n = 3–5). $*p < 0.05$; $**p < 0.01$ (Control vs. $CoQ_{10}H_2$). Reproduced from Tian et al. (2014) with permission from the editorial

cardiac tissue from heart failure patients, and also why CoQ$_{10}$ can reduce levels of oxygen free radicals (Harman 1957; Littarru 1994; van den Heuvel et al. 2000). Furthermore, CoQ$_{10}$ ensures the presence of nitrogen oxide (NO) in endothelial cells and prevents reaction of NO during lipid peroxidation that would increase vascular smooth muscle relaxation to avoid myocardial ischemia (Crane 2001; Ferrari et al. 2004; Hodgson and Watts 2003; Mugoni et al. 2013). Meanwhile, the antioxidant activity of CoQ$_{10}$ can contribute to the removal of ROS produced by exercise, improve SOD2 activity, delay muscle fatigue, and accelerate recovery from fatigue after exercise (Rosenfeldt et al. 2005). CoQ$_{10}$ supplementation can also improve the content and function of brain mitochondria that lessens the severity of learning disabilities in aged mice (Ochoa et al. 2011; Mortensen et al. 2014; Takahashi et al. 2016).

Our study revealed that $CoQ_{10}H_2$ increases cAMP levels and activates SIRT1 and PGC-1α expression by increasing levels of phosphorylated CREB, LKB1, and AMPK. Activated SIRT1 deacetylates and activates PGC-1α that in turn induces SIRT3 expression. SIRT3 then activates SOD2 and IDH2 via NRF2- and TFAM-mediated increases in mitochondrial gene expression. Activated SOD2, IDH2, and mitochondria genes increase the ratio of GSH:GSSG to decrease ROS levels and increase the activity of the mitochondrial electron transport chain complexes I and

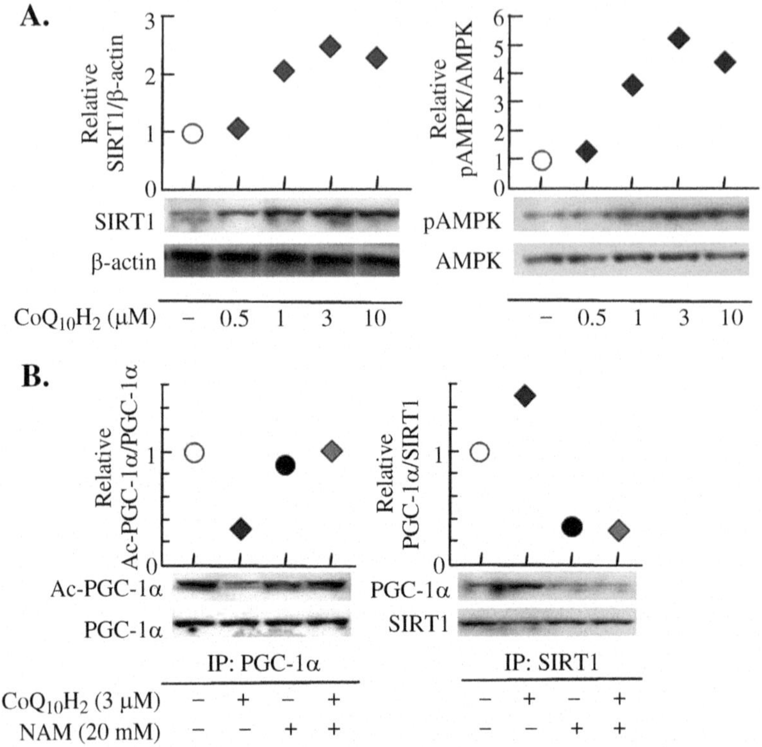

Fig. 9.15 $CoQ_{10}H_2$ promoted deacetylation of PGC-1α by SIRT1. HepG2 cells were treated with 0.5, 1, 3, or 10 μM $CoQ_{10}H_2$ for 48 h. Cell lysates (10 μg) were analyzed by Western blot (**a**). Deacetylated PGC-1α in HepG2 cells after treatment with or without 3 μM $CoQ_{10}H_2$ in the presence of the SIRT1 inhibitor nicotinamide (NAM, 20 mM) was determined by immunoblot analysis. Molecular interactions of SIRT1 and PGC-1α were determined after treatment with or without $CoQ_{10}H_2$ in the presence of nicotinamide (**b**). Reproduced from Tian et al. (2014) with permission from the editorial

IV. Together these effects could protect tissues from oxidative stress and prevent the accelerated senescence seen in SAMP1 mice (Fig. 9.16). We hypothesize that this mechanism involves $CoQ_{10}H_2$ effects on cAMP synthesis or decomposition. In addition, we showed that $CoQ_{10}H_2$ is more effectively absorbed in the intestine than is CoQ_{10}. These results suggest that $CoQ_{10}H_2$ could be used with preference over CoQ_{10} as a safe and efficient agent that has an important anti-aging effect in humans and animals.

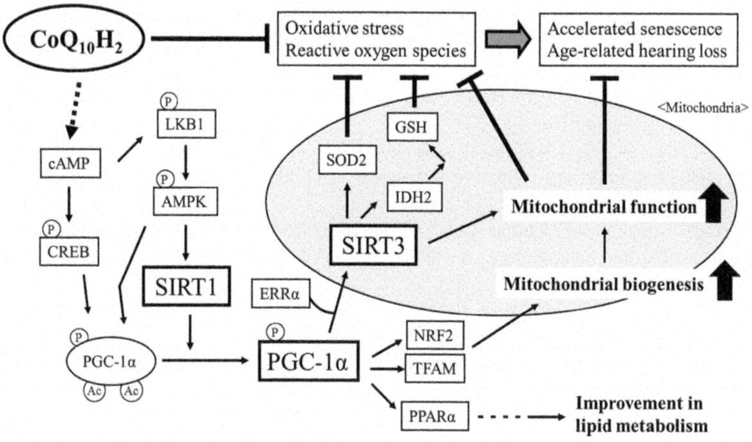

Fig. 9.16 Possible mechanism of the anti-aging effects by $CoQ_{10}H_2$ supplementation

References

Bentinger M, Tekle M, Dallner G (2010) Coenzyme Q–biosynthesis and functions. Biochem Biophys Res Commun 396:74–79

Bordone L, Guarente L (2005) Calorie restriction, SIRT1 and metabolism: understanding longevity. Nat Rev Mol Cell Biol 6:298–305

Crane FL (2001) Biochemical functions of coenzyme Q10. J Am Coll Nutri 20:591–598

Del Pozo-Cruz J, Rodríguez-Bies E, Ballesteros-Simarro M et al (2014) Physical activity affects plasma coenzyme Q10 levels differently in young and old humans. Biogerontology 15:199–211

Ferrari R, Guardigli G, Mele D et al (2004) Oxidative stress during myocardial ischaemia and heart failure. Curr Pharm Des 10:1699–1711

Finkel T, Deng CX, Mostoslavsky R (2009) Recent progress in the biology and physiology of sirtuins. Nature 460:587–591

Folkers K, Vadhanavikit S, Mortensen SA (1985) Biochemical rationale and myocardial tissue data on the effective therapy of cardiomyopathy with coenzyme Q10. Proc Natl Acad Sci USA 82:901–904

Harman D (1957) Aging: a theory based on free radical and radiation chemistry. J Gerontol 2:298–300

Heranz D, Munoz-Martin M, Canamero M et al (2010) Sirt1 improves healthy ageing and protects from metabolic syndrome-associated cancer. Nat Commun 12:1–3

Hodgson JM, Watts GF (2003) Can coenzyme Q10 improve vascular function and blood pressure? Potential for effective therapeutic reduction in vascular oxidative stress. Biofactors 18:129–136

Hosokawa M, Kasai R, Higuchi K et al (1984) Grading score system: a method for evaluation of the degree of senescence in senescence accelerated mouse (SAM). Mech Ageing Dev 26:91–102

Howitz KT, Botterman KJ, Cohen HY et al (2003) Small molecule activators of sirtuins extend Saccharomyces cerevisiae lifespan. Nature 425:191–196

Kalén A, Appelkvist EL, Dallner G (1989) Age-related changes in the lipid compositions of rat and human tissues. Lipids 24:579–584

Lim SC, Tan HH, Goh SK et al (2006) Oxidative burden in prediabetic and diabetic individuals: evidence from plasma coenzyme Q_{10}. Diabet Med 23:1344–1349

Lim SC, Lekshminarayanan R, Goh SK et al (2008) The effect of coenzyme Q 10 on microcirculatory endothelial function of subjects with type 2 diabetes mellitus. Atherosclerosis 196:966–969

Littarru GP (1994) Location and function of coenzyme Q in the respiratory chain. In: Energy and defense. Facts and perspectives on coenzyme Q10 in biology and medicine. Casa Editrice Scientifica Internazionale, Rome, pp 14–22

López-Lluch G, Rodríguez-Aguilera JC, Santos-Ocaña C et al (2010) Is coenzyme Q a key factor in aging? Mech Ageing Dev 131:225–235

Miles MV, Patterson BJ, Chalfonte-Evans ML et al (2007) Coenzyme Q10 (ubiquinol-10) supplementation improves oxidative imbalance in children with trisomy 21. Pediatr Neurol 37:398–403

Mortensen SA, Rosenfeldt F, Kumar A et al (2014) The effect of coenzyme Q10 on morbidity and mortality in chronic heart failure: results from Q-SYMBIO: a randomized double-blind trial. JACC Heart Fail 2:641–649

Mugoni V, Postel R, Catanzaro V et al (2013) Ubiad1 is an antioxidant enzyme that regulates eNOS activity by CoQ_{10} synthesis. Cell 152:504–518

Nakazawa R, Motohashi K, Azuma N (2005) Effect of coenzyme Q10 supplementation with vitamin E-bonded polysulfone dialyzer on the oxidative stress in long-term haemodialysis patients. In: Nephrology dialysis transplantation, vol 20. Oxford University Press, Oxford, pp V129–V129

Niklowitz P, Onur S, Fischer A et al (2016) Coenzyme Q10 serum concentration and redox status in European adults: influence of age, sex, and lipoprotein concentration. J Clin Bichem Nutri 58:240

Ochoa JJ, Pamplona R, Ramirez-Tortosa MC et al (2011) Age-related changes in brain mitochondrial DNA deletion and oxidative stress are differentially modulated by dietary fat type and coenzyme Q 10. Free Radic Biol Med 50:1053–1064

Rosenfeldt F, Marasco S, Lyon W et al (2005) Coenzyme Q 10 therapy before cardiac surgery improves mitochondrial function and in vitro contractility of myocardial tissue. J Thorac Cardiovasc Surg 129:25–32

Schmelzer C, Kubo H, Mori M et al (2010) Supplementation with the reduced form of Coenzyme Q10 decelerates phenotypic characteristics of senescence and induces a peroxisome proliferator-activated receptor-α gene expression signature in SAMP1 mice. Mol Nutr Food Res 54:805–815

Sohal RS, Forster MJ (2007) Coenzyme Q, oxidative stress and aging. Mitochondrion 7:S103–S111

Sohmiya M, Tanaka M, Tak NW et al (2004) Redox status of plasma coenzyme Q10 indicates elevated systemic oxidative stress in Parkinson's disease. J Neurol Sci 223:161–166

Sohmiya M, Tanaka M, Suzuki Y et al (2005) An increase of oxidized coenzyme Q-10 occurs in the plasma of sporadic ALS patients. J Neurol Sci 228:49–53

Takahashi K, Ohsawa I, Shirasawa T et al (2016) Early-onset motor impairment and increased accumulation of phosphorylated α-synuclein in the motor cortex of normal aging mice are ameliorated by coenzyme Q. Exp Gerontol 81:65–75

Takeda T, Hosokawa M, Higuchi K (1997) Senescence-accelerated mouse (SAM): a novel murine model of senescence. Exp Gerontol 32:105–109

Takeda T, Hosokawa M, Higuchi K (2013) Senescence-accelerated mouse (SAM). In: Takeda T, Akiguchi I, Higuchi K, et al. (eds) The Senescence-Accelerated Mouse (SAM): achievements and future directions. Elsevier, 580 p, pp 3–14

Tian G, Sawashita J, Kubo H et al (2014) Ubiquinol-10 supplementation activates mitochondria functions to decelerate senescence in senescence-accelerated mice. Antioxid Redox Signal 20:2606–2620

van den Heuvel AFM, van Veldhuisen DJ, van der Wall EE et al (2000) Regional myocardial blood flow reserve impairment and metabolic changes suggesting myocardial ischemia in patients with idiopathic dilated cardiomyopathy. J Am Coll Cardiol 35:19–28

Wada H, Goto H, Hagiwara SI et al (2007) Redox status of coenzyme Q_{10} is associated with chronological age. J Am Geriatr Soc 55:1141–1142

Yamamoto Y, Yamashita S (1999) Plasma ubiquinone to ubiquinol ratio in patients with hepatitis, cirrhosis, and hepatoma, and in patients treated with percutaneous transluminal coronary reperfusion. Biofactors 9:241–246

Yan J, Fujii K, Yao J et al (2006) Reduced coenzyme Q10 supplementation decelerates senescence in SAMP1 mice. Exp Gerontol 41:130–140

Part III
Age-Related Diseases and Coenzyme Q

Chapter 10
Coenzyme Q, mtDNA and Mitochondrial Dysfunction During Aging

José L. Quiles, Alfonso Varela-López, María D. Navarro-Hortal, and Maurizio Battino

Abstract The main sources of reactive oxygen species (ROS) in cells are mitochondria, whose components would be primary targets of ROS. Both facts are responsible for the key role of these organelles in aging according to the "mitochondrial theory of aging". Oxidative damage to mitochondrial DNA (mtDNA) is especially important since it would have the longest-term consequences impairing mitochondrial function. This would lead to a decrease in ATP production, but also to an increased ROS generation. In turn, CoQ, which acts as an electron carrier in mitochondria, is an essential factor for cell bioenergetics and an equilibrated CoQ pool is expected to perform a better electron flow adaptation. Moreover, it is a lipid-soluble antioxidant and efficiently prevents oxidation of DNA along with other macromolecules. Other interesting attributed roles include interaction with cell signaling cascades, anti-inflammatory activities and interference with programmed cell death. Due to this pleiotropic effect, most of interventions with CoQ have been focused on multiple processes related to mitochondria. In this sense, its effects have been investigated in mitochondrial diseases and pathological conditions related with aging whose patients have shown a higher frequency of mtDNA alterations. In addition, dietary CoQ also has been tested in combination with different diets rich in particular type of fatty acids due to the role of these in biological membranes and oxidative stress, as well as aging. This chapter aims to review the effect of CoQ on aging and mitochondrial dysfunction, with especial interest in their actions on mtDNA or the consequences of mtDNA alterations.

Keywords Diet · Mitochondrial diseases · mtDNA mutation · Oxidative stress · ROS · Ubiquinone

---

J. L. Quiles (✉) · A. Varela-López · M. D. Navarro-Hortal
Department of Physiology, Institute of Nutrition and Food Technology "Jose Mataix Verdú", University of Granada, Biomedical Research Center, Granada, Spain
e-mail: jlquiles@ugr.es

M. Battino
Dipartimento di Scienze Cliniche Specialistiche ed Odontostomatologiche (DISCO)-Sez. Biochimica, Facoltà di Medicina, Università Politecnica delle Marche, Ancona, Italy

G. López Lluch (ed.), *Coenzyme Q in Aging*,
https://doi.org/10.1007/978-3-030-45642-9_10

10.1 Introduction

There is a growing body of evidences indicating that mitochondria have a key role in aging phenomenon particularly in organs or tissues with an important dependence of aerobic metabolism. A major fact explaining this possible role is that mitochondria are the main source of reactive oxygen species (ROS) in most cells. An impairment of mitochondrial function led to lower production of ATP, but other important downstream consequences also should be considered. Among other, the increase in the generation of reactive oxygen species is usually considered as the most relevant.

In turn, oxidative stress also seems a major factor influencing mitochondria "health" since primary target of ROS would be mitochondrial components. Oxidative damage to mitochondrial DNA (mtDNA) would be especially important since it would have the longest-term consequences in mitochondrial function. For these reason, there are several theories of aging that have suggested a key role of mitochondria and in particular of mtDNA in this phenomenon.

This chapter aims to review the effect of coenzyme Q (CoQ) on aging and mitochondrial dysfunction, with especial interest in their actions on mtDNA or mtDNA mutations consequences. Additionally, to improve the understanding of implications in aging of oxidative damage to mtDNA, this chapter provides an approximation to mtDNA processes as well as an analysis of mechanisms that try to explaining their relationship with aging.

10.2 Mitochondrial DNA Features

Unlike the rest of organelles (except chloroplasts in vegetal cells), mitochondria had their own extrachromosomal DNA molecules. mtDNA presents structural and functional features very different respect than nuclear DNA (nDNA). For this reason, a preliminary approach to these unique features is needed to properly understand the relationship of mtDNA with mitochondrial functionality and aging.

MtDNA is a circular double-stranded molecule located in the mitochondrial matrix whose size ranged from 16,000 to 18,000 base pairs (bp) in vertebrates. Namely, human mtDNA size is 16569 bp and its sequence was determined in 1981 (Anderson et al. 1981). To differentiate the two mtDNA strands, it has been established the terms "heavy strand" (H-strand) and the "light strand" (L-strand) according to their GCT content. This is due to their behavior when strands are separated on denaturing cesium chloride gradients. MtDNA has no histones but rather packaged into nucleoids. These consist in stable protein-mtDNA macrocomplexes primarily associated to inner mitochondrial membrane (Wang and Bogenhagen 2006; Holt et al. 2007). Each nucleoid has an average diameter of 100 nm (Kukat et al. 2011) and they may be exchanged between mitochondria (Wang and Bogenhagen 2006; Holt et al. 2007).

Mitochondrial genome encodes 13 polypetides, two rRNAs and 22 tRNAs (Anderson et al. 1981). Encoded polypetides are subunits of mitochondrial electron transport chain (mtETC) complexes whereas rRNAs and tRNAs are required for the intramitochondrial translation of the protein-coding units. In contrast to nuclear, mitochondrial genome does not contain introns within mtDNA coding region and almost no noncoding nucleotides exist between genes. In vertebrates, an exception is a noncoding region closely associated to origin of H-strand DNA replication (O_H) that contains the transcription promotors (Clayton 1982, 1991). In most of cases (some species present two), each strand contain only a promoter. The transcription processes initiated from each of them yield two large polycistronic transcripts that are processed later to generate mature tRNAs, rRNAs, and mRNA by precise endonucleolytic cleavages. In most cases, such cleavages occur, immediately before and after a tRNA sequence. Depending on which strand acts as the template for transcription, promoters are designated as light-strand promoter (LSP) or heavy-strand promoter (HSP) (Ojala et al. 1981; Attardi and Schatz 1988; Clayton 1992). In addition to 13 mtDNA encoded subunits, there are other approximately 70 components of mitochondrial respiratory chain and other proteins that participate in mitochondrial metabolism and maintenance, which are encoded in the nuclear genome and require specialized import systems to be imported to mitochondria (Mokranjac and Neupert 2005).

Several molecules (usually from two to ten) of mtDNA coexist in a mitochondrion and there are mitochondria in every cell. Therefore, a cell possesses hundreds of copies of mtDNA, a condition termed as polyplasmy. When all mtDNA in a cell present the same sequence, the condition is known as homoplasmy. In contrast, heteroplasmy occurs when two or more different molecules of mtDNA can be also present in a cell or organism.

Because of zygote does not receive mitochondria from sperm in mammals or possible transmitted mitochondria are selected against during replication, all mitochondria are inherited from the mother. Then, mitochondria divide and proliferate during development, but also in adult life increasing mitochondrial mass. During this process known as mitochondrial biogenesis synthesis of new mitochondrial proteins, but also mtDNA, is required. It is known that biogenesis is under the control of nuclear factors, although the exact mechanism has not been fully unraveled yet.

10.2.1 Replication of Mitochondrial DNA

It is assumed that mtDNA replication is not necessary linked to the cell cycle (Clayton 1982) and mtDNA is continuously turned over. Thus, mitochondrial and nuclear genomes would be independently replicated (Bogenhagen and Clayton 1977). For this reason, mtDNA is also replicated in postmitotic cells, which has been also evidenced (Pohjoismäki et al. 2009). Notwithstanding, there is a increasing number of publications reporting some relationship between mitochondrial function and cell cycle (Arakaki et al. 2006; Owusu-Ansah et al. 2008; Mitra et al. 2009), which suggests a possible connection between mtDNA replication and the cell cycle.

The exact mechanism of human mtDNA replication is not completely known yet. The two most important replication models proposed until now are: the strand-asynchronous method that is the most traditional and the leading-lagging strand model (Holt et al. 2000; Fish et al. 2004). In both, DNA replication is initiating at O_H that is located downstream of the LSP in the D-loop region. In the first one, subsequent elongation of a nascent newly-synthesized H-strand leads to the parental H-strand displacement from the H-strand. Because of the origin of L-strand DNA replication ($O_{L)}$ is located approximately two thirds the genomic distance away from O_H on the mtDNA molecule, L-strand synthesis, that proceeds in the opposite direction, will not be initiated until almost two-thirds of new H-strand have been synthesized and O_L is exposed (Clayton 1992). Here, the adoption of a particular configuration by the H-strand allows a mitochondrial DNA primase initiate L-strand DNA synthesis (Wong and Clayton 1985a, b). In the second model, L-strand synthesis starts in a coordinately way shortly after replication in form of short Okazaki fragments that will be joined then (Yasukawa et al. 2006).

In any of the models, after DNA strand synthesis, the two daughter molecules will be separated, RNA primers removed, and remaining DNA gaps will be filled and ligated. An additional step introducing superhelical turns into the closed molecule also occurs (Shadel y Clayton 1997). The existence of a model does not necessarily exclude the other and their occurrence may depend on cell type. It has been suggested that cells requiring rapid mtDNA synthesis present a strand-displacement mechanism whereas the leading-lagging strand one would be more prevalent in cells which are in a steady-state (Holt et al. 2000; Fish et al. 2004; Jacobs et al. 2006; Tuppen et al. 2010).

Regardless of the model, various nuclear DNA (nDNA)-encoded proteins are needed to form the mitochondrial replisome and accomplish mtDNA replication. These include a 5′–3′ DNA helicase named Twinkle, some mitochondrial SSB proteins (mt-SBB) and the polymerase γ (Polγ) that contains two subunits, one catalytic with 5′-3′ exonuclease activity (PolγA) and other processivity (PolγB) (Korhonen et al. 2004).

10.2.2 Mutations and Mitochondrial DNA

Molecular defects in mtDNA have a significant role in human disease and aging and they have been found in each type of mitochondrial gene. MtDNA mutations range from single base changes in the genome (point mutations) up to large rearrangements (deletions and duplications). In turn, these changes in mtDNA sequence can be maternally inherited or somatic (i.e. created *in situ*). In general, deletions and duplications are most often sporadic or somatic whereas maternally inherited alterations are commonly point mutations (Leonard y Schapira 2000).

Alterations in mtDNA sequence have been strongly associated with deleterious effects on organisms. There are at least two reasons explaining this relationship. On the one hand mtDNA and has no introns, so that a random mutation will usually strike a coding DNA sequence. On the other hand, estimations indicate a 10–20 fold higher mutation frequency in human mtDNA than in nDNA (Brown et al. 1979). This higher rate could be due to the combination of two factors. On the one hand, maybe the number of systems of DNA repair is insufficient for all the damage that occurs, although there are a growing number of reports indicating that in mitochondria there are more enzyme activities for repair of damaged nucleotides of which it was believed at first. On the other hand, it seems that mtDNA has an increased susceptibility to mutation (Shadel and Clayton 1997). Several differential features are responsible for this susceptibility to mutation:

- Mitochondrial oxidative environment, mainly generated by free radicals generated at the electron transport chain, although there are other sources.
- The absence of protective histones, although mtDNA is packaged into protein-mtDNA aggregates termed nucleoids (Kukat et al. 2011) that are believed to protect mtDNA from chemical damage in some degree (Lagouge and Larsson 2013).
- Failure of proof-reading by mtDNA polymerases during mtDNA replication that usually also occurs although cell does not divide.
- Lack of recombination as consequence of maternal heritage that allows sequential accumulations of mutations through maternal lineages.

Mutations in mtDNA have received great interest because it is known that specific mtDNA mutations found in humans are likely causative in different diseases. Moreover, common neurodegenerative disorders and others disease often associated with aging also have been associated with mtDNA sequence alterations (Larsson and Clayton 1995). Because of cells present polyplasmy, normal and mutant mtDNA can coexist within the same cell. Actually, the existence of either completely normal or completely mutant mtDNA is rare. In this context, heteroplasmy is particularly important since it can allow an otherwise lethal mutation to persist. This is due to a certain minimal amount or threshold level is required to have deleterious effects in cell (Larsson 2010). In that sense, there are selection pressures at the molecular and cellular levels, as well as at the level of the organism itself. The proportion of mutant mitochondrial DNA required for the occurrence of a deleterious phenotype, known as the threshold effect, varies among persons, among organ systems, and within a given tissue.

After mtDNA sequence alterations are produced, several processes can modify their frequencies in the cell. In this sense, it has been suggested that changes in the frequencies of different mtDNA molecules follows principles of population genetics but rather Mendelian laws. Both, *de novo* and inherited mutations in mtDNA, if there are present in heteroplasmy, are subject to mitotic segregation. Consequently, frequency of different mtDNA molecules can shift in daughter cells since they are randomly segregated during mitosis. Thus, mutated mtDNA can increase with possible deleterious effects or decrease to disappear, particularly in fast-dividing tissues (Tuppen et al. 2010). Anyway, mechanisms for mitotic segregation need to be studied further. On the other hand, a mtDNA molecule may be replicated many times or not at all as a cell divide. If mtDNA molecules are selectively replicated, proportions of mutant and normal molecules in mother cells would be modified. In addition, it is important to note that replication of mtDNA also occurs in the absence of cell division. Thus, mtDNA is replicated also in postmitotic cells, so it can undergo similar types of segregation (Larsson 2010). It has been suggested that it is caused by random genetic drift, in conditions of relaxed mtDNA replication (Elson et al. 2001). Actually, expansion in postmitotic tissues, a preferential amplification of mtDNA mutations might occur termed clonal (Larsson et al. 1990; Weber et al. 1997).

In mammals, a rapid segregation in heteroplasmic mtDNA genotypes returning to homoplasmy in some descendants has been reported (Upholt and Dawid 1977; Olivo et al. 1983; Holt et al. 1989; Vilkki et al. 1990; Larsson et al. 1992; Blok et al. 1997; Brown 1997). The existence of a mtDNA bottleneck during development has been proposed to explain these observations (Tuppen et al. 2010). Different mechanisms by which this bottleneck is present have been hypothesized, but discussion about this topic remains. A relatively well-accepted hypothesis suggests that a marked reduction in mtDNA copy number would take place in the germ line leading to a genetic bottleneck during embryonic development (Jenuth et al. 1996; Cree et al. 2008). In contrast, other authors have suggested that, during oogenesis, there is a preferential replication of a particular mtDNA or a subgroup of them, but neither reduction of mtDNA copy number is produced in germ line (Cao et al. 2007). Other recently proposed explanation suggests that mtDNA subpopulation is selectively replicated during postnatal folliculogenesis, thus the mtDNA bottleneck would not occur during oogenesis. In single germ cells, mtDNA heteroplasmy and copy number vary throughout oogenesis (Wai et al. 2008), a finding that support that hypothesis. Anyway, more research is needed to clarify the mtDNA bottleneck exact nature (Tuppen et al. 2010).

10.2.3 Mitochondrial DNA Repair Systems

Although its importance is currently discussed (Richter et al. 1988; Hegler et al. 1993), oxidative damage occurs normally and can be elevated in cells and tissues (LeDoux et al. 1992; Mecocci et al. 1993, 1994; Driggers et al. 1993; Shigenaga et al. 1994). However, mitochondria have their own repair systems for damaged mtDNA that help to maintain mtDNA integrity, although their number seems to be more limited than in nucleus.

Base excision repair (BER) is one of the most studied mitochondrial mechanisms for mtDNA repair. In fact, intially it was though that short-patch BER was the unique pathway to repair mtDNA damage, especially oxidative damage (Stierum et al. 1999). This mechanism represents the main pathway for repairing oxidized modifications (Slupphaug et al. 2003), but it is also a primary pathway for alkylation and deamination-derived modifications repair (Dianov et al. 2001; Chan et al. 2006).

First step in BER is the cleaving the N-glycosidic bond leading to an abasic site. This reaction is catalyzed by different DNA glycosylases that are responsible to recognize modified bases and also present AP lyase activity to cleavage DNA backbone (Robertson et al. 2009). Among other, these include the uracil DNA glycosylase (UNG), the endonuclease III homolog (NTH1), and the 8-oxoguanine DNA glycosylase-1 (OGG1). OGG1 is particularly interesting since it is required for the recognition and cleavage of 8-oxoguanine (8-oxoG) from double-stranded DNA (Kuznetsov et al. 2005). In this step also participates the AP endonuclease (APE1) that cleaves on the immediate 5′ side of the apurinic/apyrimidinic (AP) site, leaving a 3′ hydroxyl and 5′-deoxyribose-5-phosphate (5′-dRP) residue (Masuda et al. 1998). Then, the resultant gap is filled with the correct nucleotide by the mitochondrial DNA polymerase -i.e. Polγ- (Ropp and Copeland 1996).

Here, it is possible to distinguish two BER pathways according to the number of nucleotides incorporated to the gap. When a single nucleotide is incorporated, the mechanism termed short-patch BER is relatively simple. In contrast, long-patch BER, which involves the incorporation of multiple nucleotides (commonly ranged from 2 to 7) is more complex and additional enzymatic activities are required (Robertson et al. 2009). Such enzymatic activities would deal with the exposure of the original DNA strand as a single-stranded overhang or a flap structure that is the main difficulty generated by the incorporation of several nucleotides (Xu et al. 2008). Finally, the nick generated is sealed by the mitochondrial DNA ligase, ligase III (Lakshmipathy and Campbell 1999a).

Other known nuclear DNA repair mechanisms has been proposed to exist in a mitochondrial version. It seem that the most clear additional mechanism is homologous recombination (LeDoux et al. 1992; Ling et al. 1995; Sage et al. 2010) that is the primary pathway to repair double-strand breaks. That plays a critical role in facilitating the progression of replication when advancing polymerase complex progress is blocked by the presence of a DNA lesion. There are also evidences for existence of mismatch repair (Mason et al. 2003) and non-homologous end-joining activities (Lakshmipathy and Campbell 1999b) at mitochondria but more research is needed to confirm them. Other hypothesized mechanisms especially useful to repair 8-oxoG have been nucleotide excision repair (Stevnsner et al. 2002) and translesion synthesis (Pinz et al. 1995; Graziewicz et al. 2004, 2007), although none enzyme activity related to them has been reported in mitochondria up to date.

10.3 Aging and Mitochondrial DNA

10.3.1 Aging and Mitochondrial DNA Mutations Relationship: A Conceptual Framework

Overall, different studies have indicated mutated mtDNA molecules accumulate with aging since elderly people has shown higher levels of somatic point mutations or deletions in mtDNA from different tissue types (Cortopassi and Arnheim 1990;

Simonetti et al. 1992; Laderman et al. 1996: Melov et al. 1999; Berneburg et al. 2004; Bender et al. 2006; Marín-García et al. 2006; Krishnan et al. 2008). This association between aging and mtDNA alterations also has been in found studies in rodents (Pikó et al. 1988; Quiles et al. 2006, 2010; Ochoa et al. 2011). However, discussion exists about what is the magnitude of such accumulation and its importance in aging. Still, rather low, the differences between young and old individuals in frequency of mtDNA mutations are statistically significant (Pikó et al. 1988). Additionally, an accumulation of multiple mtDNA deletions has also been reported in individuals with neurodegenerative diseases, such as Alzheimer and Parkinson's disease (Cortopassi et al. 1992; Coskun et al. 2004; Bender et al. 2006; Kraytsberg et al. 2006 Krishnan et al. 2008). Similarly, overall mtDNA heteroplasmy seem to increase with aging indicating that additional somatic mutations are continuously appearing during adult life (Pliss et al. 2011; Sondheimer et al. 2011; Diot et al. 2016). Most of these observations support the idea of mtDNA alterations and their subsequent accumulation during life are responsible or at least contribute to the senescent phenotype. Several proposed theories that related mitochondria and aging providing an explanation for this phenomenon.

As indicated above, mitochondrion is a major site of ROS production in the cell (especially at mtETC) which would makes mitochondria the prime targets for oxidative damage (Harman and others 1955; Miquel et al. 1980). This fact was taken into account by Harman and other authors (Harman and others 1955; Miquel et al. 1980) to considered mtDNA mutations to be the initiating, primary event in the aging process in their mitochondrial free radical theory of aging. According to that, a vicious cycle would be established whereby oxidative damage to mtDNA and other mitochondrial components leads to respiratory chain dysfunction, which in turn leads to increased generation of ROS, further facilitating respiratory chain components damage and thus creating a self-amplifying deterioration. The mitochondrial free radical theory of aging, thus, suggests the existence of a vicious cycle that results in an exponential increase in mtDNA mutations with time. Interestingly, Greaves et al. (2014), using next-generation sequencing, has reported that mtDNA mutation rate does not seem to increase with age (Greaves et al. 2014). This fact would be contradictory with the exponential increase in mtDNA mutations proposed by the mitochondrial free radical theory of aging.

More recently, new evidences have emerged that give more subtle roles beyond those as damaging agent to ROS. Actually, despite lifespan and ROS production is correlated, it has been shown that ROS are not directly responsible for aging (Sanz et al. 2010). However, it has been progressively appreciated that ROS also can function as signaling molecules, facilitating adaptation to stress in a wide variety of physiological situations (Sena et al. 2008). In this context, Hekimi and colleagues (2011) proposed the gradual ROS response hypothesis that suggests that "ROS generation is not a cause of aging, but rather represents a stress signal in response to age-dependent damage". Concerning mitochondria, when respiratory chain dysfunction coupled with moderate increases in ROS levels, these act as stress signal that activates protective quality control pathways improving mitochondria quality. These finding also have resulted in the mitohormesis hypothesis (Tapia 2006;

Ristow and Zarse 2010). Despite of the existence of protective quality control pathways, a continuous or strong dysfunction of the mitochondrial respiratory chain would lead to a substantial ROS accumulation. Consequently, protective and defense mechanisms against oxidative stress would be overwhelmed (Tapia 2006; Zelenka et al. 2015). Indeed, evidences suggest that mtDNA controls longevity (Sanz et al. 2010) which is consistent with this theory.

Lastly, possible link between mitochondria (and mtDNA) and other important event in aging process have been also proposed. In this sense, Ahmed et al. (2008) have suggested that telomerase protects mitochondria from mild oxidative stress. Other possible causes would include the repression of PGC-1 promoter by p53 activated as consequence of telomere dysfunction or TERT activity effect on mtDNA repair (Monickaraj et al 2012; Tyrka et al. 2015).

10.3.2 Generation and Accumulation of Mitochondrial DNA Mutations

Different mechanisms have been proposed to explain the accumulation of mtDNA mutations with aging. Oxidative damage to mtDNA is often assumed as main responsible for age-associated somatic mtDNA mutations generation, although there are other agents able to produce DNA lesions that also could accumulate that might be important under some conditions. If the amount of mtDNA (oxidative) damage overwhelm the mtDNA repair mechanism, a progressively accumulation of mtDNA alterations or mutations would occur with aging. The involvement of ROS in the creation of mtDNA mutations is central to the mitochondrial free radical theory of aging and it is supported by correlative data showing higher levels of somatic mtDNA mutations in older than in younger mammals including humans (Larsson 2010). It is known that DNA bases can suffer until 24 oxidative lesions different (Evans et al. 2004). There are also 13 additional major products of oxidative damage to the sugar moiety (Evans et al. 2004). In spite of this number, most of investigations has been focused on the guanine adduct 7,8-dihydro-8-oxo-deoxyguanosine (8-oxodG) (Evans et al. 2004) that is considered one of the most abundant oxidative lesions that accumulate in mtDNA over time. One consequence of 8-oxodG presence in DNA is the transversion with adenine during replication due to the mispairing of 8-oxoG, but the biological significance for the majority of the lesions remains unknow (Larsson 2010). It has been reported that accumulation of 8-oxoG in mtDNA occurs with age (Szczesny et al. 2003). Initially, it was thought that *in vivo* levels of 8-oxodG were very high (Richter et al. 1988), although then this finding was attributed to overestimation by methodological problems (Hamilton et al. 2001). More recently a sequencing study in mouse showed that mtDNA transversion mutations not increased with age, so oxidative damage may not be a major source of formation of mtDNA mutations (Ameur et al. 2011).

Although the exact steady state level of oxidative damage in mtDNA is variable among tissues and the importance is discussed in the literature (Richter et al. 1988; Hegler et al. 1993; Shadel and Clayton 1997), such damage occurs normally and can be elevated in cells and tissues under certain conditions. These include exposure to certain chemical agents (Driggers et al. 1993) and antiviral drugs (Lewis and Dalakas 1995), UV radiations (Berneburg et al. 2004; Krishnan et al. 2008; Birket y Birch-Machin 2007) or pathologies (Mecocci et al. 1994). In this context, dietary conditions could results especially interesting for mtDNA mutation implications in aging since certain nutritional conditions may be maintained over life. Recent experimental studies indicate that reduction in the degree of unsaturation of fatty acids in the diet induces less oxidative damage and alterations in mitochondrial DNA (mtDNA) in different tissues including liver (Quiles et al. 2006), brain (Ochoa et al. 2011) and heart (Quiles et al. 2010).

An alternative source of mtDNA mutations is pol γ that would produce somatic mtDNA mutations by slipped mispairing during mtDNA replication. Namely, these replication errors have suggested being an important mechanism for formation of mtDNA deletions (Madsen et al. 1993). In human mtDNA deletions that mainly occur between O_H and O_L, are typically flanked by short direct repeated sequences (Mita et al. 1990; Samuels et al. 2004; Bua et al. 2006), which supports this hypothesis. Moreover, an *in vitro* analysis of the mutations generated by wild-type Pol γ showed a good concordance with those observed *in vivo* in human, including a paucity of G:C to T:A transversions (Zheng et al. 2006). This hypothesis is also supported by mathematical modeling (Cortopassi and Arnheim 1990). Still, the fact of true turnover rate of mtDNA in mammalian tissues is largely unknown complicates studies in this area (Larsson 2010).

However, most important evidences in favor of this mechanism come from studies using a well-established knock-in murine model (Trifunovic et al. 2004; Kujoth et al. 2005). This has homozygous genotype for a mutated version of PolγA with increased proofreading activity, so it provides a critical test of the replication error hypothesis. Expression of the proof-reading deficient PolγA leads to a rapid accumulation of mtDNA point mutations and deletion during embryogenesis, which are clearly present in midgestation (Trifunovic et al. 2004). Moreover, in adult life, accumulation of mtDNA mutations goes on in a linear manner leading to the progressive and random accumulation of mtDNA point mutations during mitochondrial biogenesis (Trifunovic et al. 2004). Because of this amount of mtDNA mutations, it has been generally named as Polγ mutator mouse. Most of the mutations generated in mtDNA mutator mice are transitions (Trifunovic et al. 2004) and their pattern after germ line transmission resembles the mutation spectra found in natural populations of mice and humans (Stewart et al. 2008a, b). Because of mitochondrial free radical theory of aging predicts an exponential increase in the mutation burden throughout life; findings from these models are in contradiction with it. Interestingly, accumulation of mtDNA mutations has no major increase in oxidative damage in many different tissues in adults (Trifunovic et al. 2005).

Most researchers consider replication to be the most likely mechanism of deletion formation (Lloret et al. 2009; Lagouge and Larsson 2013), but Krishnan et al.

(2008) proposed that "mtDNA deletions arise during the repair of damaged mtDNA". Although they remain unclear, some mechanisms have been proposed to explain the impairment or restriction of repair machinery efficiency with aging. A possibility would be the age-associated decline in import capacity of the mitochondria. Accumulation into the mitochondrial intermembrane space and importation failure inside the mitochondrial matrix of an unprocessed form of DNA glycosylase OGG1 that is involved in BER of 8-oxoG has been proposed to occur with aging (Szczesny et al. 2003). This would explain why 8-oxoG is so abundant among oxidative lesions that accumulate in mtDNA. In fact, mice lacking this enzyme have increased levels of 8-oxodG in mtDNA (de Souza-Pinto et al. 2001). In turn, mispairing of 8-oxoG during replications would extend the mutation

Several processes cooperate to maintain mitochondrial quality, among which highlights mitophagy that is the only mechanism known to turn over whole mitochondrial genomes (Kim et al. 2012; Diot et al. 2016). As it is expected, to keep the pool of mitochondria healthy, replacement by biogenesis is needed that must be adequately coordinated with mitophagy. Although mtDNA turnover in differentiated tissues is not well defined, if this results affected by aging, accumulation of mutant mtDNA can occur. In this sense, it has been reported a decline of mitophagy with aging (Diot et al. 2015) thus disadvantages both the turnover of dysfunctional mitochondria and the production of fresh mitochondria. Interestingly, Greaves et al. (2014), using next-generation sequencing, have shown that mtDNA mutation rate could not increase with age, which enhances the importance of autophagy decline in mutations accumulation.

Along with the aforementioned mechanisms, changes in mitochondrial dynamics also are very important, as well as affecting fusion and fission of membranes, modulate mitochondrial turnover. Fission disrupted mitochondria segregation (Katajisto et al. 2015) whereas fusion would mix content of different mitochondria including mtDNA molecules (Chan 2012; Tam et al. 2013). When the frequency of fusion/fission cycles is reduced, mtDNA mutations tend to accumulate and there is less mtDNA mixing (Diot et al. 2016). In support of these mechanisms importance, a mathematical model by Tam et al. (2014) suggests that a combination of rapid mitochondrial fission, fusion and mitophagy can extend lifespan because mitochondrial function maintenance would be achieved.

10.3.3 The Impact of Mitochondrial DNA Alterations on Mitochondrial Function and Aging

Independently of the actual cause of a given mutation, it is possible to suppose at least some of the consequences of changes in mitochondrial DNA sequence. Most mtDNA sequence alterations are neutral polymorphisms (Ingman et al. 2000), but when this not occurs, their magnitude could be different depending on gene affected. Point mutations can affect to protein, tRNA, or rRNA genes within mtDNA. Phenotypical consequence of mutations in a protein-coding gene would be a functional alteration of a particular complex of mitochondrial respiratory chain to which the corresponding protein belongs (Tuppen et al. 2010). In turn, mutations in genes encoding for mt-tRNAs might impair overall translation of mtDNA by reducing functional mt-tRNAs availability (Tuppen et al. 2010). Regarding mtDNA rearrangements, it has been reported that most of them are large-scale deletions ranged from 1.3 to 8 kb that span several genes (Schon et al. 1989).

It is expected that accumulation of somatic mtDNA mutations would lead to mitochondria with respiratory chain deficiencies. Notwithstanding, an important feature of mtDNA further must be considered to understand the consequences of mtDNA alterations in cell and/or tissue. As mentioned, cells are polyplasmic for mtDNA, which implies that *de novo* somatic mutations in mtDNA would be in heteroplasmy, at least at beginning. Moreover, mutated mtDNA frequencies can vary dramatically between tissues (Shoffner et al. 1990; Goto et al. 1990). Actually, even in mitochondrial disorder patients, there is considerable clinical heterogeneity with mostly mtDNA mutations in heteroplasmy that are also considered highly recessive (Tuppen et al. 2010). This is particularly important for mutations causing lethal impairments that would be viable only in heteroplasmy. When heteroplasmy is present, there is a minimum critical frequency of mutated mtDNAs necessary to biochemical defects and tissue dysfunction become apparent. In humans, it has been reported that pathogenic mtDNA mutations only cause respiratory chain dysfunction when they are present above a certain threshold level, which is 60% for single large mtDNA deletions (Hayashi et al. 1991) and 90% for certain point mutations in tRNA genes (Chomyn et al. 1992). Therefore, threshold value varies for each mutation but it also differs amongst tissues according to the dependence on the oxidative metabolism presented by the tissue. It would be higher in tissues that need to obtain most energy from oxidative phosphorylation than in those that can rely on anaerobic glycolysis (Schultz and Harrington 2003). However, it is important to note that mutations in nuclear genes and in mitochondrial genes other than those in the respiratory chain also can lead to mitochondrial dysfunction. These are mainly nuclear genes encoding for respiratory chain subunits, as well as those controlling mtDNA structure and function (Leonard y Schapira 2000).

Overall, a moderate decline of respiratory chain function with age has been widely reported (Trounce et al. 1989). However, in many cases, respiratory chain deficient cells by accumulation of mtDNA alterations would represent only a part of the cells present in a tissue or organ. In addition, age-associated somatic mtDNA mutations tend to undergo clonal expansion and thereby cause focal respiratory chain deficiency. Focal respiratory chain deficiency is a ubiquitous phenomenon in human aging tissues (Müller-Höcker 1989, 1990; Trifunovic and Larsson 2008) supporting that mitochondrial dysfunction is important in human aging. Mosaic respiratory chain deficiency has been also found in many different types of aged tissues in humans including heart (Müller-Höcker 1989), skeletal muscle (Fayet et al. 2002; Bua et al. 2006; Park et al. 2009), hippocampal neurons (Cottrell et al. 2001), choroid plexus (Cottrell et al. 2001), midbrain dopaminergic neurons (Bender et al. 2006), and colon (Taylor et al. 2003). However, the responsibility of age-associated accumulation mtDNA alterations for oxidative phosphorylation impairment has been discussed since low levels of mutated mtDNA has been found in aged humans. Still, there are evidences in favor of mtDNA effects on mitochondria function. Initially, clonal accumulation of deleted mtDNA was associated with focal respiratory chain deficiency in skeletal muscle fiber segments (Fayet et al. 2002). Similarly, it has been found that mtDNA deletions are common (Bender et al. 2006) in respiratory chain-deficient dopaminergic neurons (Reeve et al. 2009). Studies in rats, rhesus monkeys, and humans have shown that accumulation of deleted mtDNA colocalizes with respiratory chain deficiency (Wanagat et al. 2001; Bua et al. 2006). A severe respiratory chain dysfunction has been also found in cardiomyocytes from Tfam homozygous knockout mice that present an associated mtDNA depletion (Wang et al. 2001).

In humans, most published cases until now show that clonally expanded mutations are single large mtDNA deletions. The deletions differ in various respiratory chain-deficient cells of the same tissue, which is in agreement with their somatic nature (Larsson 2010). Along with deletions, clonally expanded point mutations also have been found in other tissues from aging subjects. For instance, in colonic crypts from elderly humans which show focal respiratory chain deficiency in more than 15% of all colonic crypts (Taylor et al. 2003). It has been suggested that they would be originated in the crypts stem cells and they clonally expand with division. It has been suggested that accumulated mutation type by a particular tissues depend on its mitotic activity.

Various experimental models have improved the understanding of the functional consequences of mtDNA mutations and their molecular mechanisms. Biochemical effects of mtDNA mutations have been well described in all experimental systems and are invariably characterized by lower mitochondrial respiration, compromised mtETC complex activity, and reduced ATP synthesis. However, mitochondria also play important roles in different cellular pathways beyond ATP production. These include apoptosis and nucleotide synthesis, calcium regulation (Smeitink et al. 2006; Diot et al. 2016). Therefore alterations in mitochondrial function might affect to different processes that can modulate aging at distinct levels. Oxidative stress has been studied in various animal models that accumulate mtDNA mutations with associated mitochondrial dysfunction. In Tfam homozygous knockout mice there are an initial increased ROS production as consequence of mtDNA depletion and respiratory chain dysfunction (Wang et al. 2001). In contrast, mouse strains knockout for complex I Ndufs4 protein (Kruse et al. 2008) and apoptosis inducer factor (AIF) (Pospisilik et al. 2007) do not have substantially increased ROS production or oxidative damage. Evidence of oxidative stress was almost also absent in Polγ "mutator" (Kujoth et al. 2005; Trifunovic et al. 2005; Niu et al. 2007), although there are rare exceptions (Geromel et al. 2001). In contrast, Tfam heterozygous knockout mice, which also undergo mild mtDNA depletion exhibit increased oxidative mtDNA damage susceptibility (Woo et al. 2012). In addition, apoptotic cell loss can be a common feature in respiratory chain deficiency (Trifunovic and Larsson 2008). These mutator mice displayed a massive increase in apoptosis (Kujoth et al. 2005; Trifunovic et al. 2005; Niu et al. 2007) that has been also observed in mice conditionally knockout for Tfam that led to abolished mtDNA expression (Wang et al. 2001).

The accumulation of deficient mitochondria in a cell can lead to compensatory increase in mitochondria number. This would be mediated by the increase in mitochondrial biogenesis, which in turn leads to an increased number of mtDNA copy number in the cell. In skeletal muscle, it has been reported that ragged-red fibers, which are featured by an extraordinary accumulation of mitochondria, have very high proportions of mutated mtDNA in comparison with adjacent normal-appearing fibers (Moraes et al. 1992). It has been suggested that so massive amount of mitochondria is due to the activated mitochondrial biogenesis indeed, which is a futile response in this case. This is expected since new mtDNA molecules also harbor the same mutations, thus, many additional mitochondria remain dysfunctional. Moreover, this compensatory increase in copy number could lead to mutated mtDNA accumulation (Elson et al. 2001). Still, it has been reported that a higher number of mitochondria compensates for a decrease of oxidative phosphorylation capacity in associated to a mitochondrial myopathy in mouse skeletal muscle (Wredenberg et al. 2002; Wenz et al. 2008).

More recently, some investigations have been directed particularly to stem cells. Neural stem cells from mtDNA mutator mouse showed decreased renewal *in vitro* and quiescent pools of neural stem cells were decreased, whereas the haematopoietic stem cells showed a skewed lineage differentiation leading to anaemia and lymphopenia (Ahlqvist et al. 2012). In contrast, other model known as mtDNA "deletor" mice (Tyynismaa et al. 2005) that accumulate large-scale mtDNA deletions in postmitotic tissues and exhibited a similar late-onset respiratory chain deficiency not present any signs of premature aging as the mtDNA mutator mouse. This last probably is correlated with the fact that they have no similar somatic stem cell phenotypes (Ahlqvist et al. 2012). Therefore, consequences of mtDNA mutations in stem cells may explain at least in part the aging phenotypes.

Consequences of impaired mitochondrial respiratory chain function derived from accumulation of mtDNA mutations, combined or separately, would contribute to age-associated organ dysfunction and disease onset. Studies have suggested an association of deleted mtDNA with areas of fiber atrophy and splitting, thus, that mitochondrial dysfunction have a role in age-associated sarcopenia (Pak et al. 2003). Similarly, frequency of mtDNA mutation is higher in patients with parkinson´s disease that in age-matched controls (Bender et al. 2006). Concerning cancer, colon (Polyak et al. 1998) and prostate (Chinnery et al. 2002) cancer cases has been also associated to mtDNA mutations. Curiously, different evidences suggest that the most important factors in determining clinical symptoms are not the size and location of the deletions but tissue distribution (Zeviani et al. 1988; Moraes et al. 1995; Vielhaber et al. 2000).

The Polγ mutator mouse also results useful in this aspect since certainly shows that high levels of mtDNA mutations cause a phenotype that included shortened life-span, weight loss, osteoporosis, kyphosis, reduced subcutaneous fat, alopecia, reduced fertility, and cardiac hypertrophy (Trifunovic et al. 2004). This suggests a link between mtDNA mutations and aging phenotypes in mammals. However, the existence of a premature aging syndrome does not necessarily implies that mtDNA mutation levels found in normal aging are high enough to cause aging-related pathology. Additional experiments to test whether a decrease in somatic mtDNA mutations extends lifespan need to be done to confirm that. In contrast, the mtDNA mutator mice show no signs of premature aging (Tyynismaa et al. 2005) in spite of similarities in mutations accumulations and mitochondrial dysfunctions. Because of between featured stem cells mentioned above (Ahlqvist et al. 2012), it has been suggested that somatic stem cell dysfunction has a crucial role of in generating the progeroid phenotype seen in mtDNA mutator mice.

In some occasions, the consequences in mitochondria and cell physiology of mtDNA mutation accumulation led enhanced aging alterations or age-associated diseases progression. Actually, abnormalities of mtDNA have been described in several diseases and high levels of deletions and point mutations cause human mitochondrial disease or syndromes. It has been characterised up to 250 pathogenic mtDNA mutations (point mutations and rearrangements) (Schaefer et al. 2008) causing a wide variety of diseases with a heterogeneity of phenotypes and a variable age of onset (McFarland et al. 2007). Although many mutations are heteroplasmic, there are also an increasing number of pathogenic homoplasmic mutations, often affecting just a single tissue and characterized by incomplete penetrance (McFarland et al. 2002, 2004, 2007; Temperley et al. 2003; Taylor et al. 2003; Yang et al. 2009). In concordance with previous observation, mitochondrial disorders share common cellular consequences including a decreased ATP production, an increased reliance on alternative anaerobic energy sources, and an increased production of reactive oxygen species. Regardless alteration responsible for these, studies in patients with mitochondrial disease were thus able to establish a clear cause and-effect relationship between mtDNA mutations and respiratory chain dysfunction. In addition, an increased ROS production has been described in different models with mutations associated to any of these diseases (Wong et al. 2002; Baracca et al. 2007; Li et al. 2008).

10.4 Therapies Based on Coenzyme Q Against Diseases Associated with mtDNA Alterations

Despite the role of the alterations in mtDNA in aging have not clarified yet, different treatments have been tested to retard aging or to attenuate aging consequences, which, among other possible effects, can prevent mtDNA alterations. CoQ, usually CoQ_{10}, has been used with this aim in humans and in different experimental models with this aim. Traditionally, the interest in this molecule usually comes from two main roles or activities. On the one hand, CoQ is an essential factor for cell bioenergetics as consequence of its activity as electron carrier in mitochondria. Actually, it has been proposed that an equilibrated CoQ pool may perform a better electron flow adaptation than a higher or lower CoQ pool by keeping a better mitochondrial homeostasis control (López-Lluch et al. 2010). In addition, it also seems to affect protein complex activity and structure (López-Lluch et al. 2010). On the other hand, CoQ also is considered as an endogenously synthetized lipid-soluble antioxidant in biological membranes and it has been shown to efficiently prevent oxidation of DNA along with other macromolecules (Ernster and Forsmark-Andrée 1993). However, other roles that also result interesting for aging have been reported. These include interaction with cell signaling cascades, certain anti-inflammatory activities and even the prevention of events leading to programmed cell death. As consequence of the possible pleiotropic effect of CoQ on cell, many interventions did not aimed specifically to attenuate accumulation of mtDNA mutations. Instead, most of studies have focused on different processes related to mitochondria that have been associated with the generalized role of this organelle in aging. However, the present section of the chapter will be mainly devoted to those studies that evaluated the effect of CoQ on mtDNA.

The simplest approach has been the administration of CoQ to subjects with mitochondrial diseases or syndromes due to one or more specific mtDNA mutations (both, point mutations and deletions). The objective of this treatment was to stop the progression or to reduce some of the symptoms of these diseases. There are two features that make CoQ_{10} particularly popular in the management of patients with these disorders: its already mentioned rol as component of mtETC and its action as antioxidant, together with its well-documented safety, even at very high doses. Different trials have been carried out in this sense (Bresolin et al. 1990; Chan et al. 1998; Abe et al. 1999; Glover et al. 2010). In most of cases syndromes considered were mitochondrial myopathy, encephalopathy, lactic acidosis, and stroke-like episodes (MELAS), but also other rare myopathies and/or encephalopathies. In general, main parameters tested in related to mitochondrial function has been post-exercise serum (Bresolin et al. 1990; Abe et al. 1999; Glover et al. 2010) or platelets (Bresolin et al. 1990) lactate levels, which were reduced by CoQ_{10} supplementation (Bresolin et al. 1990; Abe et al. 1999; Glover et al. 2010), although not in all studies (Bresolin et al. 1990). Similarly, a study showed a reduction in lactate/pyruvate ratio that also proved to be the clinically most useful parameter in the evaluation and monitoring of mitochondrial function (Chan et al. 1998). Differences in CoQ dosage and treatment duration explain contradictory results in most cases, but in a study (Bresolin et al. 1990) observed also inter-individual differences for the same treatment that need to be clarified. It has been reported that responsiveness to treatment was apparently not related to CoQ_{10} levels in serum and platelets or to the presence or absence of mtDNA deletions (Bresolin et al. 1990). Furthermore, when they have been evaluated, it did not affect other clinically relevant variables such as strength or resting lactate (Glover et al. 2010). Therefore, these trials suggest that CoQ_{10} supplementation in relative high amount offers some improve of mitochondrial function, but the relevance for overall health is not confirmed. Likewise, most studies had a very small sample size and a rigorous placebo-controlled trial is still lacking. In addition to the above mentioned studies, there are also a set of studies using CoQ_{10} usually as a component of a "cocktail" that also includes L-carnitine, vitamin B complex, vitamin C and vitamin K_1 (Marriage et al. 2004).

To understand possible mechanisms under the effect of CoQ treatment on mitochondrial function more research is still needed. Nevertheless, an *in vitro* assay suggested that reduction of oxidative stress and inhibition of apoptosis signaling cascades could be implicated. With more detail, it has been reported that preincubation with CoQ_{10} reduced both, ROS production and activation of caspase 3, after induction by ultraviolet light in cybrids carrying mtDNA with a large-scale deletions associated to chronic progressive external ophthalmoplegia (i.e., 4366-bp and 4977-bp large-scale deletions) (Lee et al. 2005). A particular studied mitochondrial disorder is maternally inherited diabetes mellitus and deafness (MIDD) that is featured by progressive insulin secretory defect and neurosensory deafness. In a randomized-controlled trial, daily oral administration of 150 mg of CoQ_{10} for 3 years led to higher insulin secretory response than in the control group. Likewise, it improved lactate levels and prevented progressive hearing loss, although other diabetic complications and clinical symptoms remained unchanged (Suzuki et al. 1998).

In addition, dietary CoQ has shown to enhance electron transfer and ATP synthesis in some pathological situations related with aging such as cardiac failure (Rosenfeldt et al. 2005; Molyneux et al. 2009), Parkinson's disease (Beal 1999; Shults 2003; Young et al. 2007; Thomas and Beal 2010), Alzheimer's disease (Dumont et al. 2010; Yang et al. 2010; Dumont and Beal 2011) and Friedreich's ataxia (Hart et al. 2005). Although they are not specifically mitochondrial disorders, patients affected with most of these pathologies have shown a higher frequency of mtDNA alterations. In relation to these diseases, it has been reported that presence of CoQ_{10} restored the activity of impaired respiratory chain complexes I and IV in cultured fibroblasts from Parkinson´s patients. Some beneficial CoQ effects have been also observed in patients affected by HIV. This pathology is associated with alterations in the amount of mtDNA, as well as with presence of lipodystrophy and peripheral neuropathy with mitochondrial toxicity induced by reverse-transcriptase inhibitors. The administration of 100 mg of CoQ twice a day for 3 months improved the general condition and well-being in asymptomatic HIV-infected patients. However, the treatment aggravated pain in patients with peripheral neuropathy and it did not change mtDNA levels in fat and peripheral blood mononuclear cells (Rabing Christensen et al. 2004).

10.5 Coenzyme Q, Dietary Fat and Aging in Relation to mtDNA Alterations

Another approach to the study of CoQ in relation to mtDNA alterations has been the life-long dietary administration of low dosages of the molecule to rodents, in order to investigate some aspects of the interaction between nutrition and aging, mainly in relation to dietary fat. Dietary fat has been shown to be particularly interesting because of the importance of phospholipid acyl chain of mitochondrial membrane in their susceptibility to oxidative damage as well as in membrane function and structure. This is due to the fatty acids that form them present different chemical reactivity (Pamplona 2008). Unsaturated fatty acids are more susceptible to damage from ROS molecules owing to the high presence of unstable electrons near their double bonds, and also because its sensitivity to lipid peroxidation is greater as molecules have more double bond (Bielski et al. 1983; Holman 1954). Further, they also can participate in free radical chain reactions and lipid peroxidation product would produce covalent modifications of other macromolecules as proteins and DNA. Thus, a low degree of unsaturation in the fatty acids of biological membranes would decrease their sensitivity to lipid peroxidation, which, in turn, can protect damage other lipooxidation-derivative molecules (Mataix et al. 1998). In fact, some studies in mammals have shown that fatty acids unsaturation degree in biological membranes of various tissues is negatively correlated with longevity (Pamplona 2008; Pamplona et al. 2000).

There are enough evidences indicating that fatty acids present in the diet modify the lipid profile of biological membranes, including mitochondrial membranes (Huertas et al. 1991, 1999; Ochoa et al. 2001; Quiles et al. 1999). Thus, dietary fat affects the structure and mitochondrial function, as well as its susceptibility to oxidative stress. In this sense, if we could build "customized" biological membranes depending on the type of dietary fat, maybe we could positively change the way in which the organs age. This working hypothesis represented a new approach to the study of aging from the point of view of nutrition, and had important implications for aging phenomenon study (González-Alonso et al. 2015a). This was the basis for the work of our research group in a series of experiments performed on a rat model of aging for the last 20 years. In these studies, male Wistar rats were life-long maintained on different diets with different fat sources (virgin olive oil, sunflower oil or fish oil) which notably varying in their unsaturated fatty acids profiles to evaluate how this component of the diet affected to aging of different tissues and organs. Because of mitochondria role in aging and oxidative stress, evaluations were focused on mitochondrial aspects as ultrastructural alterations, mtDNA and/or respiratory chain functionality, as well as oxidative stress (including oxidative damage and antioxidant defense components) (González-Alonso et al. 2015a). As consequence of CoQ importance in mitochondria and oxidative stress, most of the experiments were carried out by using these dietary fats without or with a supplement of CoQ_{10}.

In early interventions, rats were fed diets based on AIN-93 (Reeves 1997; Reeves et al. 1993) criteria but with different dietary fat source virgin olive oil (rich in MUFA) or sunflower oil (rich in n-6 PUFA). Animals were sacrificed at different ages to study how dietary fat and CoQ modulated aging in different tissues (Quiles et al. 2002, 2004a, b, 2005, 2006; Ochoa et al. 2003, 2011). In this context, mitochondria isolated from three different tissues, liver, heart and skeletal muscle, were compared at 6, 12, 18, and 24 months of age. Lipid peroxidation markers used (i.e. hydroperoxides levels) indicated that, in general, postmitotic tissues (i.e. heart as skeletal muscle) were more prone to suffer oxidation, but n-6 PUFA-rich diets led to a higher degree of membrane polyunsaturation and peroxidation. In addition, the degree of polyunsaturation in mitochondria was found to correlate with those in diet, confirming a very good degree of membrane adaption to diet (Ochoa et al. 2003). Similar experiments also showed a worsening of aging effects by n-6 PUFA on different tissue and/or markers. These included total antioxidant capacity and DNA double-strand breaks which, respectively, decreased and increased in all animals as they age. In spite of all this effects of n-6 PUFA can affect to onset of some diseases, particularly those associated to aging, no changes in mean or maximal lifespan were observed (Quiles et al. 2004b).

In liver, ROS-mediated damage products (Hydroperoxides and TBARS) relative amounts were higher in 24 months old animals than in those aged 6 months but only in those receiving n-6 PUFA-rich diets, whereas rats fed virgin olive oil showed the lowest values at both ages. In most of case these levels correlates with activities of antioxidant enzymes (SOD, catalase and GPX) and concentrations of lipophilic antioxidant (α-tocopherol and CoQ). This suggests that this tissue as it ages triggers protection mechanism against oxidative stress probably as response to higher levels ROS or ROS-mediated damage products (Quiles et al. 2006). Interestingly, this study were even more focused on mitochondria an effects of diet and aging on mitochondrial ultrastructure and mtDNA were also evaluated. Namely, possible effects on mtDNA were evaluated using a particular deletion in the region encoded for mtETC complex I components (*Nd4* gene) since it has been suggested that is one of the complexes most affected by aging (Sanz et al. 2006). An age-related increase in mtDNA deletion frequency was observed in all animals but this was higher in rats fed sunflower oil. Likewise, old animals fed on n-6 PUFA rich diet displayed a lower crests number and higher circularity, factors that have been linked to a reduced functionality of mitochondria (Quiles et al. 2006). These findings, thus, revealed a relationship among ROS production and alterations of ultrastructure and mtDNA with aging at liver mitochondria in rats. But the most interesting was to see how these aspects (including accumulation of mtDNA deletion), which could be defining the appearance of aging phenotype, could be modulated through diet by choosing more or less unsaturated fat source and, which gives rise to the possibility modular aging through diet.

Based on negative consequences of n-6 PUFA intake found in above mentioned investigations, other studies were carried out where two experimental groups received similar sunflower oil-based diets, but with or without a supplementation on CoQ_{10} to reach a daily dosage of 0.7 g/kg (Ochoa et al. 2005; Quiles et al. 2004a, 2005). In heart, long-term supplementation with $CoQ_{10,}$ led to lower hydroperoxide levels, higher content of lipophilic antioxidants (α-tocopherol and coenzyme Q), and a higher catalase activity. Also, a slightly lower decrease in certain key activities for mitochondrial function when animals with age of 6, 12, or 24 months were compared (Ochoa et al. 2005). At the systemic level, an age-associated increase in nDNA strand breaks in peripheral blood lymphocytes was observed. This increase associated to age was lower in animals supplemented on CoQ_{10}. If it is assumed that main cause of such breaks is oxidative damage, this suggests that CoQ_{10} by means of both, reactive species scavenging and antioxidant recycling, protects DNA against oxidative damage. It could be possible that also can occur in mitochondria at least under certain conditions. However, this finding could also result indirectly from of CoQ_{10} effects on mitochondria or other organelles that would reduce ROS levels and consequent attacks to nDNA (Quiles et al. 2005). In liver, similar effects on DNA double-strand breaks, CoQ levels at mitochondrial membrane have been reported. Lastly, it has also been noted that CoQ_{10}-supplemented animals reached a significantly higher mean life span and a significantly higher maximum life span (Quiles et al. 2004a). This emphasized the importance of oxidative stress, DNA damage and mitochondria in aging since CoQ has shown effect on all them (Ochoa et al. 2005; Quiles et al. 2005).

According to previous finding, it seems that life-long supplementation with CoQ_{10} of n-6 PUFA-rich diet resulted interesting to attenuate aging consequences, but it was necessary to check if CoQ_{10} led even to better results than virgin olive oil. For this reason, additional experiments similar to the previous one but including also a group fed on a virgin olive oil-based diet were carried out (Ochoa et al. 2011; Quiles et al. 2010). Thus, three diets rich in MUFA, n-6 PUFA and CoQ_{10}-supplemented n-6 PUFA were compared. Because of previous results and their importance in ROS generation and aging, mitochondria received a greater attention and mtDNA and ultrastructure were also analyzed in most of cases along with ROS and antioxidants levels. Again, the frequency of a specific deletion in mtDNA corresponding to the mETC complex I of the was used as marker of mtDNA alterations, This experimental design was used to evaluate diet and aging interaction in two tissues, both postmitotic, brain and heart.

In heart, animals fed virgin olive oil showed a lower increase in the frequency of studied mtDNA deletion than those fed sunflower oil. However, the addition of CoQ_{10} to the n-6 PUFA-rich fat source (i.e. sunflower oil) reduced the difference between young and old animals although the lowest values were present by MUFA-fed animals (Quiles et al. 2010). Concerning mitochondrial ultrastructure, dietary fat used had similar effects (Quiles et al. 2010) to those achieved in liver tissue in absence of CoQ_{10} (Quiles et al. 2006), whereas CoQ_{10} treatment led to lower mitochondrial perimeter in this case (Quiles et al. 2010). CoQ_{10} also prevented the decrease in cytochrome C oxidase activity and mtETC complex I levels suggested for old subjects fed on the same dietary fat. Therefore, it would prevent mitochondrial respiratory chain dysfunction in some degree. Aged animals receiving CoQ also showed lower hydroperoxide levels than those fed on sunflower or virgin olive oil not supplemented (Quiles et al. 2010). This would suggest that CoQ contributes to decrease oxidative stress, although there are several possible mechanisms. In any case, the effect found for dietary CoQ_{10} on either ultrastructure, mtDNA and some respiratory chain components would alleviate ROS production associated to age.

Very similar aspects were also evaluated in brain, but in this case in the experiment performed an additional group consisted in a virgin olive oil-based diet supplemented with $CoQ_{10.}$ In this, mtDNA deletion was higher in old groups fed on n-6 PUFA-rich diets but no age-associated differences were found for animals fed virgin olive oil. However, in this case CoQ_{10} did not show effects on mtDNA deletions in animals fed on sunflower oil-based diet at 24 months. In relation to oxidative stress markers, CoQ led to lower values of lipid peroxidation (hydroperoxides) at 24 months, although the lowest values were found in the two virgin olive oil fed groups (Ochoa et al. 2011).

These organs (heart and brain) are clear affected by aging and their alteration lead to overall health impairment reducing longevity. Moreover mitochondrial alterations and oxidative stress are key aspects in aging of these organs as it has been previously reported. Altogether, these findings revealed that CoQ, at least under certain conditions, can modulate aging effects on different tissues affecting to mtDNA, but also to mitochondrial ultrastructure and ROS production. Again, a key finding is the possibility of modulating mtDNA mutations associated to age through diet.

In more recent experiments, a third diet type has started to be compared with diets similar to previously described studies. So, as a new fat source namely fish oil, very rich in n-3 PUFA, was used. In addition, fat content was the half of the amount used in previous experiments (4% *versus* 8% w/w) according to more recent actualization of AIN93 criteria (Reeves 1997) until that moment. As previously, some additional experiments were carried out to test the effects of these diets under CoQ_{10} supplementation. Moreover, new organ/tissues, not previously studied, like pancreas and periodontum, were included in the experiments. In pancreas, it has been reported that dietary fat affected to endocrine and exocrine pancreas in a different way (Roche et al. 2014). In 24-months-old animals, n-6 PUFA rich-diets consumption was associated with a greater number of β-cells that correlated with an increase in insulin content and hyperleptinemia (Roche et al. 2014), signs that have been described in obesity, glucose intolerance, insulin resistance, disruption of adipoinsular axis or prediabetes (Sattar et al. 2008). Concerning exocrine compartment, old rats fed with n-3 PUFA-rich diets (Roche et al. 2014) led to histological features resembling those observed in pancreatic fibrosis in elderly people (Klöppel et al. 2004). In other experiments, it was observed that dietary CoQ_{10} improved endocrine pancreas structure and in particular β-cell mass from rat fed on n-6 PUFA resembling positive effects of virgin olive oil (González-Alonso et al. 2015b). Because of importance of mitochondria in this organ, CoQ_{10} effect could be mediated by effect on mtDNA previously reported (Quiles et al. 2010). However, oxidative damage or alterations of mtDNA sequence have not been directly analyzed yet. In a study focused on the pancreas of 24 months old rat fed on these diet, the profile of serum fatty acids confirmed, that animals an adaptation to the diet at 6 months of age since they resembled lipid profile of the diets. The percentages of circulating MUFA were significantly higher in rats fed virgin olive oil; the highest levels of n-6 PUFA were achieved in rats fed with sunflower oil, and the highest levels of n-3 PUFA were found in those rats fed fish oil (Roche et al. 2014; González-Alonso et al. 2015b). Moreover, the effect of this fat sources and CoQ on some bone metabolism markers at serum and age-associated alveolar bone have been also studied (Bullon et al. 2013). Alveolar bone loss is a major clinical outcome of periodontitis (Page and Kornman 1997), a disease with high prevalence in elderly people that in the last years has been associated with systemic diseases such as atherosclerosis and metabolic syndrome that would have oxidative stress as potential link (Bullon et al. 2009, 2011). Again, feeding on an n-6 PUFA-rich diet led to worse consequences in health since it was associated to the highest age-associated alveolar bone loss (Bullon et al. 2013). Although mtDNA alterations were not directly measured, expression of genes LC3 and ATG5 that are implicated in autophagy and the biogenesis markers Tfam and PGC-1α suggests that both processes increase with aging in gingival tissue, but not in animals fed n-6 PUFA. The combination of both processes would reduce or prevent accumulation of damage in mtDNA and its possible consequences. Moreover, this effect also was associated affecting to some mtETC components and antioxidant enzymes expression. In other study, CoQ supplementation eliminated differences in age-associated alveolar bone loss among dietary groups (Varela-Lopez et al. 2015). CoQ_{10} had no effect on age-associated changes in expression of genes of autophagy markers in rats fed on n-6 PUFA-rich diet (Varela-Lopez et al. 2015). An increase in the expression of the biogenesis marker Tfam was observed in n-6 PUFA fed animals indicating that there was an increase in mitochondria and probably in mtDNA copies. Although this mechanism possibly does not reduce the accumulation of altered or mutated mtDNA molecules, it seems that it can compensate, at least in part, the associated loss of function, as it has been reported for skeletal muscle. Sumarizing all these experiments on aged rats, it could be concluded that the basis for a putative beneficial effect of CoQ on mtDNA disturbances could be the enhancement of the cellular antioxidant protection systems in cell membranes where CoQ preventing lipid peroxidation and consequently reduced oxidative stress and mtDNA damage by ROS.

Finally, in healthy humans, comparisons between CoQ_{10} supplementation to diet have been also established following a cross-over design, although only for a short period of time (4 weeks). In this regard, elderly subjects following a Mediterranean diet (rich in MUFA) supplemented and not with CoQ_{10} or a Western diet rich in SFA were studied (Yubero-Serrano et al. 2010, 2012; Gutierrez-Mariscal et al. 2011, 2014; González-Guardia et al. 2015). Some postprandial oxidative stress marker levels were reduced by CoQ_{10} addition to the MUFA-rich diet (Yubero-Serrano et al. 2010, 2012). Interestingly, dietary CoQ also improved DNA repair systems (Gutierrez-Mariscal et al. 2011; Yubero-Serrano et al. 2012). These results suggest that CoQ may also protect mtDNA against the accumulation of mutations by this mechanism in addition to the prevention of ROS-mediated damage discussed for the rat models.

Acknowledgments Dr. Alfonso Varela-López was recipient of a "Contrato Puente" from the University of Granada during 2017. At present is recipient of a research grant at universities or centers abroad from the Alfonso Martin Escudero Foundation. María Dolores Navarro Hortal has been awarded a Scholarship of Initiation to Research for undergraduate students of the Own Plan of the University of Granada during the academic year 2016–2017 and a Scholarship for Student Collaboration in University Departments for the academic year 2017–2018 in Research in University Departments of the Ministry of Education, Culture and Sports 2017/2018 call. Authors acknowledge to the University of Granada and the Autonomous Government of Andalusia for partial support of the research team.

References

Abe K, Matsuo Y, Kadekawa J, Inoue S, Yanagihara T (1999) Effect of coenzyme Q10 in patients with mitochondrial myopathy, encephalopathy, lactic acidosis, and stroke-like episodes (MELAS): evaluation by noninvasive tissue oximetry. J Neurol Sci 162:65–68

Ahlqvist KJ, Hämäläinen RH, Yatsuga S, Uutela M, Terzioglu M, Götz A, Forsström S, Salven P, Angers-Loustau A, Kopra OH, Tyynismaa H, Larsson N-G, Wartiovaara K, Prolla T, Trifunovic A, Suomalainen A (2012) Somatic progenitor cell vulnerability to mitochondrial DNA mutagenesis underlies progeroid phenotypes in Polg mutator mice. Cell Metab 15:100–109. https://doi.org/10.1016/j.cmet.2011.11.012

Ahmed S, Passos JF, Birket MJ, Beckmann T, Brings S, Peters H, Birch-Machin MA, von Zglinicki T, Saretzki G (2008) Telomerase does not counteract telomere shortening but protects mitochondrial function under oxidative stress. J Cell Sci 121 (7):1046–1053

Ameur A, Stewart JB, Freyer C, Hagström E, Ingman M, Larsson N-G, Gyllensten U (2011) Ultradeep sequencing of mouse mitochondrial DNA: mutational patterns and their origins. PLoS Genet 7:e1002028. https://doi.org/10.1371/journal.pgen.1002028

Anderson S, Bankier AT, Barrell BG, de Bruijn MH, Coulson AR, Drouin J, Eperon IC, Nierlich DP, Roe BA, Sanger F, Schreier PH, Smith AJ, Staden R, Young IG (1981) Sequence and organization of the human mitochondrial genome. Nature 290:457–465

Arakaki N, Nishihama T, Owaki H, Kuramoto Y, Suenaga M, Miyoshi E, Emoto Y, Shibata H, Shono M, Higuti T (2006) Dynamics of mitochondria during the cell cycle. Biol Pharm Bull 29:1962–1965

Attardi G, Schatz G (1988) Biogenesis of mitochondria. Annu Rev Cell Biol 4:289–333. https://doi.org/10.1146/annurev.cb.04.110188.001445

Baracca A, Sgarbi G, Mattiazzi M, Casalena G, Pagnotta E, Valentino ML, Moggio M, Lenaz G, Carelli V, Solaini G (2007) Biochemical phenotypes associated with the mitochondrial ATP6 gene mutations at nt8993. Biochim Biophys Acta BBA Bioenerg 1767:913–919. https://doi.org/10.1016/j.bbabio.2007.05.005

Beal MF (1999) Coenzyme Q10 administration and its potential for treatment of neurodegenerative diseases. Biofactors 9:261–266

Bender A, Krishnan KJ, Morris CM, Taylor GA, Reeve AK, Perry RH, Jaros E, Hersheson JS, Betts J, Klopstock T, Taylor RW, Turnbull DM (2006) High levels of mitochondrial DNA deletions in substantia nigra neurons in aging and Parkinson disease. Nat Genet 38:515–517. https://doi.org/10.1038/ng1769

Berneburg M, Plettenberg H, Medve-König K, Pfahlberg A, Gers-Barlag H, Gefeller O, Krutmann J (2004) Induction of the photoaging-associated mitochondrial common deletion in vivo in normal human skin. J Invest Dermatol 122:1277–1283. https://doi.org/10.1111/j.0022-202X.2004.22502.x

Bielski BH, Arudi RL, Sutherland MW (1983) A study of the reactivity of HO2/O2- with unsaturated fatty acids. J Biol Chem 258:4759–4761

Birket MJ, Birch-Machin MA, (2007) Ultraviolet radiation exposure accelerates the accumulation of the aging-dependent T414G mitochondrial DNA mutation in human skin. Aging Cell 6(4):557–564

Blok RB, Gook DA, Thorburn DR, Dahl H-HM (1997) Skewed segregation of the mtDNA nt 8993 (TrG) mutation in human oocytes. Am J Hum Genet 60:1495–1501. https://doi.org/10.1086/515453

Bogenhagen D, Clayton DA (1977) Mouse L cell mitochondrial DNA molecules are selected randomly for replication throughout the cell cycle. Cell 11:719–727. https://doi.org/10.1016/0092-8674(77)90286-0

Bresolin N, Doriguzzi C, Ponzetto C, Angelini C, Moroni I, Castelli E, Cossutta E, Binda A, Gallanti A, Gabellini S (1990) Ubidecarenone in the treatment of mitochondrial myopathies: a multi-center double-blind trial. J Neurol Sci 100:70–78

Brown GK (1997) Bottlenecks and beyond: mitochondrial DNA segregation in health and disease. J Inherit Metab Dis 20:2–8
Brown WM, George M, Wilson AC (1979) Rapid evolution of animal mitochondrial DNA. Proc Natl Acad Sci USA 76:1967–1971
Bua E, Johnson J, Herbst A, Delong B, McKenzie D, Salamat S, Aiken JM (2006) Mitochondrial DNA-deletion mutations accumulate intracellularly to detrimental levels in aged human skeletal muscle fibers. Am J Hum Genet 79:469–480. https://doi.org/10.1086/507132
Bullon P, Morillo JM, Ramírez-Tortosa MC, Quiles JL, Newman HN, Battino M (2009) Metabolic syndrome and periodontitis: is oxidative stress a common link? J Dent Res 88:503–518. https://doi.org/10.1177/0022034509337479
Bullon P, Cordero MD, Quiles JL, Morillo JM, Ramírez-Tortosa MC, Battino M (2011) Mitochondrial dysfunction promoted by Porphyromonas gingivalis lipopolysaccharide as a possible link between cardiovascular disease and periodontitis. Free Radic Biol Med 50:1336–1343. https://doi.org/10.1016/j.freeradbiomed.2011.02.018
Bullon P, Battino M, Varela-Lopez A, Perez-Lopez P, Granados-Principal S, Ramirez-Tortosa MC, Ochoa JJ, Cordero MD, Gonzalez-Alonso A, Ramirez-Tortosa CL, Rubini C, Zizzi A, Quiles JL (2013) Diets based on virgin olive oil or fish oil but not on sunflower oil prevent age-related alveolar bone resorption by mitochondrial-related mechanisms. PLoS One 8:e74234. https://doi.org/10.1371/journal.pone.0074234. eCollection 2013
Cao L, Shitara H, Horii T, Nagao Y, Imai H, Abe K, Hara T, Hayashi J-I, Yonekawa H (2007) The mitochondrial bottleneck occurs without reduction of mtDNA content in female mouse germ cells. Nat Genet 39:386–390. https://doi.org/10.1038/ng1970
Chan DC (2012) Fusion and fission: interlinked processes critical for mitochondrial health. Annu Rev Genet 46:265–287. https://doi.org/10.1146/annurev-genet-110410-132529
Chan A, Reichmann H, Kögel A, Beck A, Gold R (1998) Metabolic changes in patients with mitochondrial myopathies and effects of coenzyme Q10 therapy. J Neurol 245:681–685
Chan KKL, Zhang Q-M, Dianov GL (2006) Base excision repair fidelity in normal and cancer cells. Mutagenesis 21:173–178. https://doi.org/10.1093/mutage/gel020
Chinnery PF, Samuels DC, Elson J, Turnbull DM (2002) Accumulation of mitochondrial DNA mutations in ageing, cancer, and mitochondrial disease: is there a common mechanism? Lancet Lond Engl 360:1323–1325. https://doi.org/10.1016/S0140-6736(02)11310-9
Chomyn A, Martinuzzi A, Yoneda M, Daga A, Hurko O, Johns D, Lai ST, Nonaka I, Angelini C, Attardi G (1992) MELAS mutation in mtDNA binding site for transcription termination factor causes defects in protein synthesis and in respiration but no change in levels of upstream and downstream mature transcripts. Proc Natl Acad Sci USA 89:4221–4225
Clayton DA (1982) Replication of animal mitochondrial DNA. Cell 28:693–705
Clayton DA (1991) Replication and transcription of vertebrate mitochondrial DNA. Annu Rev Cell Biol 7:453–478. https://doi.org/10.1146/annurev.cb.07.110191.002321
Clayton DA (1992) Structure and function of the mitochondrial genome. J Inherit Metab Dis 15:439–447
Cortopassi GA, Arnheim N (1990) Detection of a specific mitochondrial DNA deletion in tissues of older humans. Nucleic Acids Res 18:6927–6933
Cortopassi GA, Shibata D, Soong NW, Arnheim N (1992) A pattern of accumulation of a somatic deletion of mitochondrial DNA in aging human tissues. Proc Natl Acad Sci U S A 89:7370–7374
Coskun PE, Beal MF, Wallace DC (2004) Alzheimer's brains harbor somatic mtDNA control-region mutations that suppress mitochondrial transcription and replication. Proc Natl Acad Sci U S A 101:10726–10731. https://doi.org/10.1073/pnas.0403649101
Cottrell DA, Blakely EL, Johnson MA, Ince PG, Turnbull DM (2001) Mitochondrial enzyme-deficient hippocampal neurons and choroidal cells in AD. Neurology 57:260–264
Cree LM, Samuels DC, de Sousa Lopes SC, Rajasimha HK, Wonnapinij P, Mann JR, Dahl H-HM, Chinnery PF (2008) A reduction of mitochondrial DNA molecules during embryogenesis explains the rapid segregation of genotypes. Nat Genet 40:249–254. https://doi.org/10.1038/ng.2007.63

de Souza-Pinto NC, Eide L, Hogue BA, Thybo T, Stevnsner T, Seeberg E, Klungland A, Bohr VA (2001) Repair of 8-oxodeoxyguanosine lesions in mitochondrial Dna depends on the Oxoguanine Dna Glycosylase (OGG1) gene and 8-oxoguanine accumulates in the mitochondrial Dna of OGG1-defective mice. Cancer Res 61:5378–5381

Dianov GL, Souza-Pinto N, Nyaga SG, Thybo T, Stevnsner T, Bohr VA (2001) Base excision repair in nuclear and mitochondrial DNA. Prog Nucleic Acid Res Mol Biol 68:285–297

Diot A, Hinks-Roberts A, Lodge T, Liao C, Dombi E, Morten K, Brady S, Fratter C, Carver J, Muir R, Davis R, Green CJ, Johnston I, Hilton-Jones D, Sue C, Mortiboys H, Poulton J (2015) A novel quantitative assay of mitophagy: combining high content fluorescence microscopy and mitochondrial DNA load to quantify mitophagy and identify novel pharmacological tools against pathogenic heteroplasmic mtDNA. Pharmacol Res 100:24–35. https://doi.org/10.1016/j.phrs.2015.07.014

Diot A, Morten K, Poulton J (2016) Mitophagy plays a central role in mitochondrial ageing. Mamm Genome 27(7–8):381–395

Driggers WJ, LeDoux SP, Wilson GL (1993) Repair of oxidative damage within the mitochondrial DNA of RINr 38 cells. J Biol Chem 268:22042–22045

Dumont M, Beal MF (2011) Neuroprotective strategies involving ROS in Alzheimer disease. Free Radic Biol Med 51:1014–1026. https://doi.org/10.1016/j.freeradbiomed.2010.11.026

Dumont M, Lin MT, Beal MF (2010) Mitochondria and antioxidant targeted therapeutic strategies for Alzheimer's disease. J Alzheimers Dis 20(Suppl 2):S633–S643. https://doi.org/10.3233/JAD-2010-100507

Elson JL, Samuels DC, Turnbull DM, Chinnery PF (2001) Random intracellular drift explains the clonal expansion of mitochondrial DNA mutations with age. Am J Hum Genet 68:802–806. https://doi.org/10.1086/318801

Ernster L, Forsmark-Andrée P (1993) Ubiquinol: an endogenous antioxidant in aerobic organisms. Clin Investig 71(S8)

Evans MD, Dizdaroglu M, Cooke MS (2004) Oxidative DNA damage and disease: induction, repair and significance. Mutat Res 567:1–61. https://doi.org/10.1016/j.mrrev.2003.11.001

Fayet G, Jansson M, Sternberg D, Moslemi AR, Blondy P, Lombès A, Fardeau M, Oldfors A (2002) Ageing muscle: clonal expansions of mitochondrial DNA point mutations and deletions cause focal impairment of mitochondrial function. Neuromuscul Disord NMD 12:484–493

Fish J, Raule N, Attardi G (2004) Discovery of a major D-loop replication origin reveals two modes of human mtDNA synthesis. Science 306:2098–2101. https://doi.org/10.1126/science.1102077

Geromel V, Kadhom N, Cebalos-Picot I, Ouari O, Polidori A, Munnich A, Rötig A, Rustin P (2001) Superoxide-induced massive apoptosis in cultured skin fibroblasts harboring the Neurogenic Ataxia Retinitis Pigmentosa (NARP) mutation in the ATPase-6 gene of the mitochondrial DNA. Hum Mol Genet 10:1221–1228

Glover EI, Martin J, Maher A, Thornhill RE, Moran GR, Tarnopolsky MA (2010) A randomized trial of coenzyme Q10 in mitochondrial disorders. Muscle Nerve 42:739–748. https://doi.org/10.1002/mus.21758

González-Alonso A, Pérez-López P, Varela-López A, Ramírez-Tortosa MC, Battino M, Quiles JL (2015a) Experimental evidence on the role of different types unsaturated fats in the diet on ageing. Rev Esp Geriatría Gerontol 50:285–288. https://doi.org/10.1016/j.regg.2015.05.003

González-Alonso A, Ramírez-Tortosa CL, Varela-López A, Roche E, Arribas MI, Ramírez-Tortosa MC, Giampieri F, Ochoa JJ, Quiles JL (2015b) Sunflower oil but not fish oil resembles positive effects of virgin olive oil on aged pancreas after life-long coenzyme Q addition. Int J Mol Sci 16:23425–23445. https://doi.org/10.3390/ijms161023425

González-Guardia L, Yubero-Serrano EM, Delgado-Lista J, Perez-Martinez P, Garcia-Rios A, Marin C, Camargo A, Delgado-Casado N, Roche HM, Perez-Jimenez F, Brennan L, López-Miranda J (2015) Effects of the Mediterranean diet supplemented with coenzyme Q_{10} on metabolomic profiles in elderly men and women. J Gerontol Biol Sci Med Sci 70:78–84. https://doi.org/10.1093/gerona/glu098

Goto YI, Nonaka I, Horai S (1990) A mutation in the tRNA Leu(UUR) gene associated with the MELAS subgroup of mitochondrial encephalomyopathies. Nature 348:651–653. https://doi.org/10.1038/348651a0

Graziewicz MA, Sayer JM, Jerina DM, Copeland WC (2004) Nucleotide incorporation by human DNA polymerase gamma opposite benzo[a]pyrene and benzo[c]phenanthrene diol epoxide adducts of deoxyguanosine and deoxyadenosine. Nucleic Acids Res 32:397–405. https://doi.org/10.1093/nar/gkh213

Graziewicz MA, Bienstock RJ, Copeland WC (2007) The DNA polymerase gamma Y955C disease variant associated with PEO and parkinsonism mediates the incorporation and translesion synthesis opposite 7,8-dihydro-8-oxo-2'-deoxyguanosine. Hum Mol Genet 16:2729–2739. https://doi.org/10.1093/hmg/ddm227

Greaves LC, Nooteboom M, Elson JL, Tuppen HAL, Taylor GA, Commane DM, Arasaradnam RP, Khrapko K, Taylor RW, Kirkwood TBL, Mathers JC, Turnbull DM (2014) Clonal expansion of early to mid-life mitochondrial DNA point mutations drives mitochondrial dysfunction during human ageing. PLoS Genet 10:e1004620. https://doi.org/10.1371/journal.pgen.1004620

Gutierrez-Mariscal FM, Perez-Martinez P, Delgado-Lista J, Yubero-Serrano EM, Camargo A, Delgado-Casado N, Cruz-Teno C, Santos-Gonzalez M, Rodriguez-Cantalejo F, Castaño JP, Villalba-Montoro JM, Fuentes F, Perez-Jimenez F, Lopez-Miranda J (2011) Mediterranean diet supplemented with coenzyme Q_{10} induces postprandial changes in p53 in response to oxidative DNA damage in elderly subjects. Age Drodr 34:389–403. https://doi.org/10.1007/s11357-011-9229-1

Gutierrez-Mariscal FM, Yubero-Serrano EM, Rangel-Zúñiga OA, Marín C, García-Rios A, Perez-Martinez P, Delgado-Lista J, Malagón MM, Tinahones FJ, Pérez-Jimenez F, López-Miranda J (2014) Postprandial activation of P53-dependent DNA repair is Modified by Mediterranean diet supplemented with coenzyme Q_{10} in elderly subjects. J Gerontol Biol Sci Med Sci 69:886–893. https://doi.org/10.1093/gerona/glt174

Hamilton ML, Guo Z, Fuller CD, Van Remmen H, Ward WF, Austad SN, Troyer DA, Thompson I, Richardson A (2001) A reliable assessment of 8-oxo-2-deoxyguanosine levels in nuclear and mitochondrial DNA using the sodium iodide method to isolate DNA. Nucleic Acids Res 29:2117–2126

Harman D, others (1955) Aging: a theory based on free radical and radiation chemistry

Hart PE, Lodi R, Rajagopalan B, Bradley JL, Crilley JG, Turner C, Blamire AM, Manners D, Styles P, Schapira AHV, Cooper JM (2005) Antioxidant treatment of patients with Friedreich ataxia: four-year follow-up. Arch Neurol 62:621–626. https://doi.org/10.1001/archneur.62.4.621

Hayashi J-I, Ohta S, Kikuchi A, Takemitsu M, Goto Y-I, Nonaka I (1991) Introduction of disease-related mitochondrial DNA deletions into HeLa cells lacking mitochondrial DNA results in mitochondrial dysfunction. Proc Natl Acad Sci USA 88:10614–10618

Hegler J, Bittner D, Boiteux S, Epe B (1993) Quantification of oxidative DNA modifications in mitochondria. Carcinogenesis 14:2309–2312

Hekimi S, Lapointe J, Wen Y (2011) Taking a "good" look at free radicals in the aging process. Trends Cell Biol 21:569–576. https://doi.org/10.1016/j.tcb.2011.06.008

Holman RT (1954) Autoxidation of fats and related substances. In: Holman RT, Lundberg WO, Malkin T (eds) Progress in chemistry of fats and other lipids. Pergamon Press, London, pp 51–98

Holt IJ, Miller DH, Harding AE (1989) Genetic heterogeneity and mitochondrial DNA heteroplasmy in Leber's hereditary optic neuropathy. J Med Genet 26:739–743. https://doi.org/10.1136/jmg.26.12.739

Holt IJ, Lorimer HE, Jacobs HT (2000) Coupled leading- and lagging-strand synthesis of mammalian mitochondrial DNA. Cell 100:515–524. https://doi.org/10.1016/S0092-8674(00)80688-1

Holt IJ, He J, Mao CC, Boyd-Kirkup JD, Martinsson P, Sembongi H, Reyes A, Spelbrink JN (2007) Mammalian mitochondrial nucleoids: organizing an independently minded genome. Mitochondrion 7:311–321. https://doi.org/10.1016/j.mito.2007.06.004

Huertas JR, Battino M, Lenaz G, Mataix FJ (1991) Changes in mitochondrial and microsomal rat liver coenzyme Q_9 and Q_{10} content induced by dietary fat and endogenous lipid peroxidation. FEBS Lett 287:89–92. https://doi.org/10.1016/0014-5793(91)80022-U

Huertas JR, Martínez-Velasco E, Ibáñez S, López-Frías M, Ochoa JJ, Quiles JL, Parenti-Castelli G, Mataix J, Lenaz G (1999) Virgin olive oil protect heart mitochondria from peroxidative damage during aging. BioFactors 9:337–343. https://doi.org/10.1002/biof.5520090233

Ingman M, Kaessmann H, Pääbo S, Gyllensten U (2000) Mitochondrial genome variation and the origin of modern humans. Nature 408:708–713. https://doi.org/10.1038/35047064

Jacobs LJAM, Wert G de, JPM G, Coo IFM de, HJM S (2006) The transmission of OXPHOS disease and methods to prevent this. Hum Reprod Update 12:119–136. https://doi.org/10.1093/humupd/dmi042

Jenuth JP, Peterson AC, Fu K, Shoubridge EA (1996) Random genetic drift in the female germline explains the rapid segregation of mammalian mitochondrial DNA. Nat Genet 14:146–151. https://doi.org/10.1038/ng1096-146

Katajisto P, Döhla J, Chaffer CL, Pentinmikko N, Marjanovic N, Iqbal S, Zoncu R, Chen W, Weinberg RA, Sabatini DM (2015) Stem cells. Asymmetric apportioning of aged mitochondria between daughter cells is required for stemness. Science 348:340–343. https://doi.org/10.1126/science.1260384

Kim TY, Wang D, Kim AK, Lau E, Lin AJ, Liem DA, Zhang J, Zong NC, Lam MP, Ping P (2012) Metabolic labeling reveals proteome dynamics of mouse mitochondria. Mol Cell Proteomics 11(12):1586–1594. https://doi.org/10.1074/mcp.M112.021162

Klöppel G, Detlefsen S, Feyerabend B (2004) Fibrosis of the pancreas: The initial tissue damage and the resulting pattern. Virchows Arch 445:1–8. https://doi.org/10.1007/s00428-004-1021-5

Korhonen JA, Pham XH, Pellegrini M, Falkenberg M (2004) Reconstitution of a minimal mtDNA replisome in vitro. EMBO J 23:2423–2429. https://doi.org/10.1038/sj.emboj.7600257

Kraytsberg Y, Kudryavtseva E, McKee AC, Geula C, Kowall NW, Khrapko K (2006) Mitochondrial DNA deletions are abundant and cause functional impairment in aged human substantia nigra neurons. Nat Genet 38:518–520. https://doi.org/10.1038/ng1778

Krishnan KJ, Reeve AK, Samuels DC, Chinnery PF, Blackwood JK, Taylor RW, Wanrooij S, Spelbrink JN, Lightowlers RN, Turnbull DM (2008) What causes mitochondrial DNA deletions in human cells? Nat Genet 40:275–279. https://doi.org/10.1038/ng.f.94

Kruse SE, Watt WC, Marcinek DJ, Kapur RP, Schenkman KA, Palmiter RD (2008) Mice with mitochondrial complex I deficiency develop a fatal encephalomyopathy. Cell Metab 7:312–320. https://doi.org/10.1016/j.cmet.2008.02.004

Kujoth GC, Hiona A, Pugh TD, Someya S, Panzer K, Wohlgemuth SE, Hofer T, Seo AY, Sullivan R, Jobling WA, Morrow JD, Van Remmen H, Sedivy JM, Yamasoba T, Tanokura M, Weindruch R, Leeuwenburgh C, Prolla TA (2005) Mitochondrial DNA mutations, oxidative stress, and apoptosis in mammalian aging. Science 309:481–484. https://doi.org/10.1126/science.1112125

Kukat C, Wurm CA, Spåhr H, Falkenberg M, Larsson N-G, Jakobs S (2011) Super-resolution microscopy reveals that mammalian mitochondrial nucleoids have a uniform size and frequently contain a single copy of mtDNA. Proc Natl Acad Sci U S A 108:13534–13539. https://doi.org/10.1073/pnas.1109263108

Kuznetsov NA, Koval VV, Zharkov DO, Nevinsky GA, Douglas KT, Fedorova OS (2005) Kinetics of substrate recognition and cleavage by human 8-oxoguanine-DNA glycosylase. Nucleic Acids Res 33:3919–3931. https://doi.org/10.1093/nar/gki694

Laderman KA, Penny JR, Mazzucchelli F, Bresolin N, Scarlato G, Attardi G (1996) Aging-dependent functional alterations of mitochondrial DNA (mtDNA) from human fibroblasts transferred into mtDNA-less cells. J Biol Chem 271:15891–15897

Lagouge M, Larsson N-G (2013) The role of mitochondrial DNA mutations and free radicals in disease and ageing. J Intern Med 273:529–543. https://doi.org/10.1111/joim.12055

Lakshmipathy U, Campbell C (1999a) The human DNA ligase III gene encodes nuclear and mitochondrial proteins. Mol Cell Biol 19:3869–3876

Lakshmipathy U, Campbell C (1999b) Double strand break rejoining by mammalian mitochondrial extracts. Nucleic Acids Res 27:1198–1204

Larsson N-G (2010) Somatic mitochondrial DNA mutations in mammalian aging. Annu Rev Biochem 79:683–706. https://doi.org/10.1146/annurev-biochem-060408-093701

Larsson NG, Clayton DA (1995) Molecular genetic aspects of human mitochondrial disorders. Annu Rev Genet 29:151–178. https://doi.org/10.1146/annurev.ge.29.120195.001055

Larsson N, Holme E, Kristiansson B, Oldfors A, Tulinius M (1990) Progressive increase of the mutated mitochondrial DNA fraction in Kearns-Sayre syndrome. Pediatr Res 28:131–136. https://doi.org/10.1203/00006450-199008000-00011

Larsson NG, Tulinius MH, Holme E, Oldfors A, Andersen O, Wahlström J, Aasly J (1992) Segregation and manifestations of the mtDNA tRNA(Lys) A-->G(8344) mutation of myoclonus epilepsy and ragged-red fibers (MERRF) syndrome. Am J Hum Genet 51:1201–1212

LeDoux SP, Wilson GL, Beecham EJ, Stevnsner T, Wassermann K, Bohr VA (1992) Repair of mitochondrial DNA after various types of DNA damage in Chinese hamster ovary cells. Carcinogenesis 13:1967–1973

Lee C-F, Liu C-Y, Chen S-M, Sikorska M, Lin C-Y, Chen T-L, Wei Y-H (2005) Attenuation of UV-Induced apoptosis by Coenzyme Q in human cells harboring large-scale deletion of mitochondrial DNA. Ann N Y Acad Sci 1042(1):429–438

Leonard JV, Schapira AHV (2000) Mitochondrial respiratory chain disorders I: mitochondrial DNA defects. Lancet 355(9200):299–304

Lewis W, Dalakas MC (1995) Mitochondrial toxicity of antiviral drugs. Nat Med 1:417–422

Li J, Zhou K, Meng X, Wu Q, Li S, Liu Y, Wang J (2008) Increased ROS generation and SOD activity in heteroplasmic tissues of transmitochondrial mice with A3243G mitochondrial DNA mutation. Genet Mol Res GMR 7:1054–1062

Ling F, Makishima F, Morishima N, Shibata T (1995) A nuclear mutation defective in mitochondrial recombination in yeast. EMBO J 14:4090–4101

Lloret A, Badía M-C, Mora NJ, Pallardó FV, Alonso M-D, Viña J (2009) Vitamin E paradox in Alzheimer's disease: it does not prevent loss of cognition and may even be detrimental. J Alzheimers Dis JAD 17:143–149. https://doi.org/10.3233/JAD-2009-1033

López-Lluch G, Rodríguez-Aguilera JC, Santos-Ocaña C, Navas P (2010) Is coenzyme Q a key factor in aging? Mech Ageing Dev 131:225–235. https://doi.org/10.1016/j.mad.2010.02.003

Madsen CS, Ghivizzani SC, Hauswirth WW (1993) In vivo and in vitro evidence for slipped mispairing in mammalian mitochondria. Proc Natl Acad Sci U S A 90:7671–7675

Marín-García J, Pi Y, Goldenthal MJ (2006) Mitochondrial-nuclear cross-talk in the aging and failing heart. Cardiovasc Drugs Ther Spons Int Soc Cardiovasc Pharmacother 20:477–491. https://doi.org/10.1007/s10557-006-0584-6

Marriage BJ, Clandinin MT, Macdonald IM, Glerum DM (2004) Cofactor treatment improves ATP synthetic capacity in patients with oxidative phosphorylation disorders. Mol Genet Metab 81:263–272. https://doi.org/10.1016/j.ymgme.2003.12.008

Mason PA, Matheson EC, Hall AG, Lightowlers RN (2003) Mismatch repair activity in mammalian mitochondria. Nucleic Acids Res 31:1052–1058

Masuda Y, Bennett RA, Demple B (1998) Dynamics of the interaction of human apurinic endonuclease (Ape1) with its substrate and product. J Biol Chem 273:30352–30359

Mataix J, Quiles JL, Huertas JR, Battino M, Mañas M (1998) Tissue specific interactions of exercise, dietary fatty acids, and vitamin E in lipid peroxidation. Free Radic Biol Med 24:511–521. https://doi.org/10.1016/S0891-5849(97)00288-8

McFarland R, Clark KM, Morris AAM, Taylor RW, Macphail S, Lightowlers RN, Turnbull DM (2002) Multiple neonatal deaths due to a homoplasmic mitochondrial DNA mutation. Nat Genet 30:145–146. https://doi.org/10.1038/ng819

McFarland R, Schaefer AM, Gardner JL, Lynn S, Hayes CM, Barron MJ, Walker M, Chinnery PF, Taylor RW, Turnbull DM (2004) Familial myopathy: new insights into the T14709C mitochondrial tRNA mutation. Ann Neurol 55:478–484. https://doi.org/10.1002/ana.20004

McFarland R, Chinnery PF, Blakely EL, Schaefer AM, Morris AAM, Foster SM, Tuppen HAL, Ramesh V, Dorman PJ, Turnbull DM, Taylor RW (2007) Homoplasmy, heteroplasmy, and mitochondrial dystonia. Neurology 69:911–916. https://doi.org/10.1212/01.wnl.0000267843.10977.4a

Mecocci P, MacGarvey U, Kaufman AE, Koontz D, Shoffner JM, Wallace DC, Beal MF (1993) Oxidative damage to mitochondrial DNA shows marked age-dependent increases in human brain. Ann Neurol 34:609–616. https://doi.org/10.1002/ana.410340416

Mecocci P, MacGarvey U, Beal MF (1994) Oxidative damage to mitochondrial DNA is increased in Alzheimer's disease. Ann Neurol 36:747–751. https://doi.org/10.1002/ana.410360510

Melov S, Schneider JA, Coskun PE, Bennett DA, Wallace DC (1999) Mitochondrial DNA rearrangements in aging human brain and in situ PCR of mtDNA. Neurobiol Aging 20:565–571

Miquel J, Economos AC, Fleming J, Johnson JE (1980) Mitochondrial role in cell aging. Exp Gerontol 15:575–591

Mita S, Rizzuto R, Moraes CT, Shanske S, Arnaudo E, Fabrizi GM, Koga Y, DiMauro S, Schon EA (1990) Recombination via flanking direct repeats is a major cause of large-scale deletions of human mitochondrial DNA. Nucleic Acids Res 18:561–567

Mitra K, Wunder C, Roysam B, Lin G, Lippincott-Schwartz J (2009) A hyperfused mitochondrial state achieved at G1–S regulates cyclin E buildup and entry into S phase. Proc Natl Acad Sci 106:11960–11965. https://doi.org/10.1073/pnas.0904875106

Mokranjac D, Neupert W (2005) Protein import into mitochondria. Biochem Soc Trans 33:1019–1023. https://doi.org/10.1042/BST20051019

Molyneux SL, Florkowski CM, Richards AM, Lever M, Young JM, George PM (2009) Coenzyme Q10; an adjunctive therapy for congestive heart failure? N Z Med J 122:74–79

Moraes CT, Ricci E, Petruzzella V, Shanske S, DiMauro S, Schon EA, Bonilla E (1992) Molecular analysis of the muscle pathology associated with mitochondrial DNA deletions. Nat Genet 1:359–367. https://doi.org/10.1038/ng0892-359

Moraes CT, Sciacco M, Ricci E, Tengan CH, Hao H, Bonilla E, Schon EA, DiMauro S (1995) Phenotype-genotype correlations in skeletal muscle of patients with mtDNA deletions. Muscle Nerve Suppl 3:S150–S153

Müller-Höcker J (1989) Cytochrome-c-oxidase deficient cardiomyocytes in the human heart—an age-related phenomenon. A histochemical ultracytochemical study. Am J Pathol 134:1167–1173

Müller-Höcker J (1990) Cytochrome c oxidase deficient fibres in the limb muscle and diaphragm of man without muscular disease: an age-related alteration. J Neurol Sci 100:14–21

Niu X, Trifunovic A, Larsson N-G, Canlon B (2007) Somatic mtDNA mutations cause progressive hearing loss in the mouse. Exp Cell Res 313:3924–3934. https://doi.org/10.1016/j.yexcr.2007.05.029

Ochoa JJ, Huertas JR, Quiles JL, Mataix J (2001) Dietary oils high in oleic acid, but with different non-glyceride contents, have different effects on lipid profiles and peroxidation in rabbit hepatic mitochondria. J Nutr Biochem 12:357–364. https://doi.org/10.1016/S0955-2863(01)00150-4

Ochoa JJ, Quiles JL, Ibáñez S, Martínez E, López-Frías M, Huertas JR, Mataix J (2003) Aging-related oxidative stress depends on dietary lipid source in rat postmitotic tissues. J Bioenerg Biomembr 35:267–275

Ochoa JJ, Quiles JL, Huertas JR, Mataix J (2005) Coenzyme Q10 protects from aging-related oxidative stress and improves mitochondrial function in heart of rats fed a polyunsaturated fatty acid (PUFA)-rich diet. J Gerontol Biol Sci Med Sci 60:970–975

Ochoa JJ, Pamplona R, Ramirez-Tortosa MC, Granados-Principal S, Perez-Lopez P, Naudí A, Portero-Otin M, López-Frías M, Battino M, Quiles JL (2011) Age-related changes in brain mitochondrial DNA deletion and oxidative stress are differentially modulated by dietary fat type and coenzyme Q_{10}. Free Radic Biol Med 50:1053–1064. https://doi.org/10.1016/j.freeradbiomed.2011.02.004

Ojala D, Montoya J, Attardi G (1981) tRNA punctuation model of RNA processing in human mitochondria. Nature 290:470–474

Olivo PD, Van de Walle MJ, Laipis PJ, Hauswirth WW (1983) Nucleotide sequence evidence for rapid genotypic shifts in the bovine mitochondrial DNA D-loop. Nature 306:400–402. https://doi.org/10.1038/306400a0

Owusu-Ansah E, Yavari A, Mandal S, Banerjee U (2008) Distinct mitochondrial retrograde signals control the G1-S cell cycle checkpoint. Nat Genet 40:356–361. https://doi.org/10.1038/ng.2007.50

Page RC, Kornman KS (1997) The pathogenesis of human periodontitis: an introduction. Periodontology 2000 14:9–11. https://doi.org/10.1111/j.1600-0757.1997.tb00189.x

Pak JW, Herbst A, Bua E, Gokey N, McKenzie D, Aiken JM (2003) Mitochondrial DNA mutations as a fundamental mechanism in physiological declines associated with aging. Aging Cell 2:1–7

Pamplona R (2008) Membrane phospholipids, lipoxidative damage and molecu-lar integrity: a causal role in aging and longevity. Biochim Biophys Acta 1777:1249–1262. https://doi.org/10.1016/j.bbabio.2008.07.003

Pamplona R, Portero-Otín M, Ruiz C, Gredilla R, Herrero A, Barja G (2000) Double bond content of phospholipids and lipid peroxidation negatively correlate with maximum longevity in the heart of mammals. Mech Ageing Dev 112:169–183. https://doi.org/10.1016/S0047-6374(99)00045-7

Park KS, Lim JW, Kim H (2009) Inhibitory mechanism of omega-3 fatty acids in pancreatic inflammation and apoptosis. Ann N Y Acad Sci 1171:421–427. https://doi.org/10.1111/j.1749-6632.2009.04887.x

Pikó L, Hougham AJ, Bulpitt KJ (1988) Studies of sequence heterogeneity of mitochondrial DNA from rat and mouse tissues: evidence for an increased frequency of deletions/additions with aging. Mech Ageing Dev 43:279–293

Pinz KG, Shibutani S, Bogenhagen DF (1995) Action of mitochondrial DNA polymerase gamma at sites of base loss or oxidative damage. J Biol Chem 270:9202–9206

Pliss L, Brakmanis A, Ranka R, Elferts D, Krumina A, Baumanis V (2011) The link between mitochondrial DNA hypervariable segment I heteroplasmy and ageing among genetically unrelated Latvians. Exp Gerontol 46:560–568. https://doi.org/10.1016/j.exger.2011.02.016

Pohjoismäki JLO, Goffart S, Tyynismaa H, Willcox S, Ide T, Kang D, Suomalainen A, Karhunen PJ, Griffith JD, Holt IJ, Jacobs HT (2009) Human heart mitochondrial DNA is organized in complex catenated networks containing abundant four-way junctions and replication forks. J Biol Chem 284:21446–21457. https://doi.org/10.1074/jbc.M109.016600

Polyak K, Li Y, Zhu H, Lengauer C, Willson JK, Markowitz SD, Trush MA, Kinzler KW, Vogelstein B (1998) Somatic mutations of the mitochondrial genome in human colorectal tumours. Nat Genet 20:291–293. https://doi.org/10.1038/3108

Pospisilik JA, Knauf C, Joza N, Benit P, Orthofer M, Cani PD, Ebersberger I, Nakashima T, Sarao R, Neely G, Esterbauer H, Kozlov A, Kahn CR, Kroemer G, Rustin P, Burcelin R, Penninger JM (2007) Targeted deletion of AIF decreases mitochondrial oxidative phosphorylation and protects from obesity and diabetes. Cell 131:476–491. https://doi.org/10.1016/j.cell.2007.08.047

Quiles JL, Huertas JR, Mañas M, Battino M, Mataix J (1999) Physical exercise affects the lipid profile of mitochondrial membranes in rats fed with virgin olive oil or sunflower oil. Br J Nutr 81:21–24

Quiles JL, Martínez E, Ibáñez S, Ochoa JJ, Martín Y, López-Frías M, Huertas JR, Mataix J (2002) Ageing-related tissue-specific alterations in mitocondrial composition and function are modulated by dietary fat type in the rat. J Bioenerg Biomembr 34:517–524

Quiles JL, Ochoa JJ, Huertas JR, Mataix J (2004a) Coenzyme Q supplementation protects from age-related DNA double-strand breaks and increases lifespan in rats fed on a PUFA-rich diet. Exp Gerontol 39:189–194. https://doi.org/10.1016/j.exger.2003.10.002

Quiles JL, Ochoa JJ, Ramirez-Tortosa C, Battino M, Huertas JR, Martín Y, Mataix J (2004b) Dietary fat type (virgin olive vs. sunflower oils) affects age-related changes in DNA double-strand-breaks, antioxidant capacity and blood lipids in rats. Exp Gerontol 39:1189–1198. https://doi.org/10.1016/j.exger.2004.05.002

Quiles JL, Ochoa JJ, Battino M, Gutierrez-Rios P, Nepomuceno EA, Frías ML, Huertas JR, Mataix J (2005) Life-long supplementation with a low dosage of coenzyme Q_{10} in the rat: effects on antioxidant status and DNA damage. Biofactors 25:73–86

Quiles JL, Ochoa JJ, Ramírez-Tortosa MC, Huertas JR, Mataix J (2006) Age-related mitocondrial DNA deletion in rat liver depends on dietary fat unsaturation. J Gerontol A Biol Sci Med Sci 61:107–114

Quiles JL, Pamplona R, Ramirez-Tortosa MC, Naudí A, Portero-Otin M, Araujo-Nepomuceno E, López-Frías M, Battino M, Ochoa JJ (2010) Coenzyme Q addition to an n-6 PUFA-rich diet resembles benefits on age-related mitochondrial DNA deletion and oxidative stress of a MUFA-rich diet in rat heart. Mech Ageing Dev 131:38–47. https://doi.org/10.1016/j.mad.2009.11.004

Rabing Christensen E, Stegger M, Jensen-Fangel S, Laursen AL, Ostergaard L (2004) Mitochondrial DNA levels in fat and blood cells from patients with lipodystrophy or peripheral neuropathy and the effect of 90 days of high-dose coenzyme Q treatment: a randomized, double-blind, placebo-controlled pilot study. Clin Infect Dis 39:1371–1379. https://doi.org/10.1086/424881

Reeves PG (1997) Components of the AIN-93 diets as improvements in the AIN-76A diet. J Nut 127:838S–841S

Reeves PG, Nielsen FH, Fahey GC (1993) AIN-93 purified diets for laboratory rodents: final report of the American Institute of Nutrition ad hoc writing committee on the reformulation of the AIN-76A rodent diet. J Nutr 123:1939–1951

Reeve AK, Krishnan KJ, Taylor G, Elson JL, Bender A, Taylor RW, Morris CM, Turnbull DM (2009) The low abundance of clonally expanded mitochondrial DNA point mutations in aged substantia nigra neurons. Aging Cell 8(4):496–498

Richter C, Park JW, Ames BN (1988) Normal oxidative damage to mitochondrial and nuclear DNA is extensive. Proc Natl Acad Sci U S A 85:6465–6467

Ristow M, Zarse K (2010) How increased oxidative stress promotes longevity and metabolic health: the concept of mitochondrial hormesis (mitohormesis). Exp Gerontol 45:410–418. https://doi.org/10.1016/j.exger.2010.03.014

Robertson AB, Klungland A, Rognes T, Leiros I (2009) DNA repair in mammalian cells: Base excision repair: the long and short of it. Cell Mol Life Sci CMLS 66:981–993. https://doi.org/10.1007/s00018-009-8736-z

Roche E, Ramírez-Tortosa CL, Arribas MI, Ochoa JJ, Sirvent-Belando JE, Battino M, Ramírez-Tortosa MC, González-Alonso A, Pérez-López MP, Quiles JL (2014) Comparative analysis of pancreatic changes in aged rats fed life long with sunflower, fish, or olive oils. J Gerontol A Biol Sci Med Sci 69:934–944. https://doi.org/10.1093/gerona/glt157

Ropp PA, Copeland WC (1996) Cloning and characterization of the human mitochondrial DNA polymerase, DNA polymerase gamma. Genomics 36:449–458. https://doi.org/10.1006/geno.1996.0490

Rosenfeldt F, Marasco S, Lyon W, Wowk M, Sheeran F, Bailey M, Esmore D, Davis B, Pick A, Rabinov M, Smith J, Nagley P, Pepe S (2005) Coenzyme Q_{10} therapy before cardiac surgery improves mitochondrial function and in vitro contractility of myocardial tissue. J Thorac Cardiovasc Surg 129:25–32. https://doi.org/10.1016/j.jtcvs.2004.03.034

Sage JM, Gildemeister OS, Knight KL (2010) Discovery of a novel function for human Rad51: maintenance of the mitochondrial genome. J Biol Chem 285:18984–18990. https://doi.org/10.1074/jbc.M109.099846

Samuels DC, Schon EA, Chinnery PF (2004) Two direct repeats cause most human mtDNA deletions. Trends Genet TIG 20:393–398. https://doi.org/10.1016/j.tig.2004.07.003

Sanz A, Caro P, Sanchez JG, Barja G (2006) Effect of lipid restriction on mitochondrial free radical production and oxidative DNA damage. Ann N Acad Sci 1067:200–209. https://doi.org/10.1196/annals.1354.024

Sanz A, Fernández-Ayala DJM, Stefanatos RK, Jacobs HT (2010) Mitochondrial ROS production correlates with, but does not directly regulate lifespan in Drosophila. Aging 2:200–223. https://doi.org/10.18632/aging.100137

Sattar N, Wannamethee SG, Forouhi NG (2008) Novel biochemical risk factors for type 2 diabetes: pathogenic insights or prediction possibilities? Diabetologia 51:926–940. https://doi.org/10.1007/s00125-008-0954-7

Schaefer AM, McFarland R, Blakely EL, He L, Whittaker RG, Taylor RW, Chinnery PF, Turnbull DM (2008) Prevalence of mitochondrial DNA disease in adults. Ann Neurol 63:35–39. https://doi.org/10.1002/ana.21217

Schon EA, Rizzuto R, Moraes CT, Nakase H, Zeviani M, DiMauro S (1989) A direct repeat is a hotspot for large-scale deletion of human mitochondrial DNA. Science 244:346–349

Schultz DR, Harrington WJ Jr (2003) Apoptosis: programmed cell death at a molecular level. Semin Arthritis Rheum 32:345–369. https://doi.org/10.1053/sarh.2003.50005

Sena CM, Nunes E, Gomes A, Santos MS, Proença T, Martins MI, Seiça RM (2008) Supplementation of coenzyme Q_{10} and alpha-tocopherol lowers glycated hemoglobin level and lipid peroxidation in pancreas of diabetic rats. Nutr Res 28:113–121. https://doi.org/10.1016/j.nutres.2007.12.005

Shadel GS, Clayton DA (1997) Mitochondrial DNA maintenance in vertebrates. Annu Rev Biochem 66:409–435. https://doi.org/10.1146/annurev.biochem.66.1.409

Shigenaga MK, Hagen TM, Ames BN (1994) Oxidative damage and mitochondrial decay in aging. Proc Natl Acad Sci U S A 91:10771–10778

Shoffner JM, Lott MT, Lezza AMS, Seibel P, Ballinger SW, Wallace DC (1990) Myoclonic epilepsy and ragged-red fiber disease (MERRF) is associated with a mitochondrial DNA tRNALys mutation. Cell 61:931–937. https://doi.org/10.1002/(SICI)1097-4598(199703)20:3<271::AID-MUS2>3.0.CO;2-8

Shults CW (2003) Coenzyme Q_{10} in neurodegenerative diseases. Curr Med Chem 10:1917–1921

Simonetti S, Chen X, DiMauro S, Schon EA (1992) Accumulation of deletions in human mitochondrial DNA during normal aging: analysis by quantitative PCR. Biochim Biophys Acta 1180:113–122

Slupphaug G, Kavli B, Krokan HE (2003) The interacting pathways for prevention and repair of oxidative DNA damage. Mutat Res Mol Mech Mutagen 531:231–251. https://doi.org/10.1016/j.mrfmmm.2003.06.002

Smeitink JA, Zeviani M, Turnbull DM, Jacobs HT (2006) Mitochondrial medicine: a metabolic perspective on the pathology of oxidative phosphorylation disorders. Cell Metab 3:9–13. https://doi.org/10.1016/j.cmet.2005.12.001

Sondheimer N, Glatz CE, Tirone JE, Deardorff MA, Krieger AM, Hakonarson H (2011) Neutral mitochondrial heteroplasmy and the influence of aging. Hum Mol Genet 20:1653–1659. https://doi.org/10.1093/hmg/ddr043

Stevnsner T, Thorslund T, de Souza-Pinto NC, Bohr VA (2002) Mitochondrial repair of 8-oxoguanine and changes with aging. Exp Gerontol 37:1189–1196. https://doi.org/10.1016/S0531-5565(02)00142-0

Stewart JB, Freyer C, Elson JL, Larsson N-G (2008a) Purifying selection of mtDNA and its implications for understanding evolution and mitochondrial disease. Nat Rev Genet 9:657–662. https://doi.org/10.1038/nrg2396

Stewart JB, Freyer C, Elson JL, Wredenberg A, Cansu Z, Trifunovic A, Larsson N-G (2008b) Strong purifying selection in transmission of mammalian mitochondrial DNA. PLoS Biol 6:e10. https://doi.org/10.1371/journal.pbio.0060010

Stierum RH, Dianov GL, Bohr VA (1999) Single-nucleotide patch base excision repair of uracil in DNA by mitochondrial protein extracts. Nucleic Acids Res 27:3712–3719

Suzuki S, Hinokio Y, Ohtomo M, Hirai M, Hirai A, Chiba M, Kasuga S, Satoh Y, Akai H, Toyota T (1998) The effects of coenzyme Q10 treatment on maternally inherited diabetes mellitus and deafness, and mitochondrial DNA 3243 (A to G) mutation. Diabetologia 41:584–588. https://doi.org/10.1007/s001250050950

Szczesny B, Hazra TK, Papaconstantinou J, Mitra S, Boldogh I (2003) Age-dependent deficiency in import of mitochondrial DNA glycosylases required for repair of oxidatively damaged bases. Proc Natl Acad Sci U S A 100:10670–10675. https://doi.org/10.1073/pnas.1932854100

Tam ZY, Gruber J, Halliwell B, Gunawan R (2013) Mathematical modeling of the role of mitochondrial fusion and fission in mitochondrial DNA maintenance. PloS One 8:e76230. https://doi.org/10.1371/journal.pone.0076230

Tam ZY, Gruber J, Ng LF, Halliwell B, Gunawan R (2014) Effects of lithium on age-related decline in mitochondrial turnover and function in Caenorhabditis elegans. J Gerontol A Biol Sci Med Sci 69:810–820. https://doi.org/10.1093/gerona/glt210

Tapia PC (2006) Sublethal mitochondrial stress with an attendant stoichiometric augmentation of reactive oxygen species may precipitate many of the beneficial alterations in cellular physiology produced by caloric restriction, intermittent fasting, exercise and dietary phytonutrients: “Mitohormesis” for health and vitality. Med Hypotheses 66:832–843. https://doi.org/10.1016/j.mehy.2005.09.009

Taylor RW, Barron MJ, Borthwick GM, Gospel A, Chinnery PF, Samuels DC, Taylor GA, Plusa SM, Needham SJ, Greaves LC, Kirkwood TBL, Turnbull DM (2003) Mitochondrial DNA mutations in human colonic crypt stem cells. J Clin Invest 112:1351–1360. https://doi.org/10.1172/JCI19435

Temperley RJ, Seneca SH, Tonska K, Bartnik E, Bindoff LA, Lightowlers RN, Chrzanowska-Lightowlers ZMA (2003) Investigation of a pathogenic mtDNA microdeletion reveals a translation-dependent deadenylation decay pathway in human mitochondria. Hum Mol Genet 12:2341–2348. https://doi.org/10.1093/hmg/ddg238

Thomas B, Beal MF (2010) Mitochondrial therapies for Parkinson’s disease. Mov Disord 25(Suppl 1):S155–S160. https://doi.org/10.1002/mds.22781

Trifunovic A, Larsson N-G (2008) Mitochondrial dysfunction as a cause of ageing. J Intern Med 263:167–178. https://doi.org/10.1111/j.1365-2796.2007.01905.x

Trifunovic A, Wredenberg A, Falkenberg M, Spelbrink JN, Rovio AT, Bruder CE, Bohlooly-Y M, Gidlöf S, Oldfors A, Wibom R, Törnell J, Jacobs HT, Larsson N-G (2004) Premature ageing in mice expressing defective mitochondrial DNA polymerase. Nature 429:417–423. https://doi.org/10.1038/nature02517

Trifunovic A, Hansson A, Wredenberg A, Rovio AT, Dufour E, Khvorostov I, Spelbrink JN, Wibom R, Jacobs HT, Larsson N-G (2005) Somatic mtDNA mutations cause aging phenotypes without affecting reactive oxygen species production. Proc Natl Acad Sci U S A 102:17993–17998. https://doi.org/10.1073/pnas.0508886102

Trounce I, Byrne E, Marzuki S (1989) Decline in skeletal muscle mitochondrial respiratory chain function: possible factor in ageing. Lancet Lond Engl 1:637–639

Tuppen HAL, Blakely EL, Turnbull DM, Taylor RW (2010) Mitochondrial DNA mutations and human disease. Biochim Biophys Acta 1797:113–128. https://doi.org/10.1016/j.bbabio.2009.09.005

Tyrka AR, Carpenter LL, Kao H-T, Porton B, Philip NS, Ridout SJ, Ridout KK, Price LH (2015) Association of telomere length and mitochondrial DNA copy number in a community sample of healthy adults. Exp Gerontol 66:17–20

Tyynismaa H, Mjosund KP, Wanrooij S, Lappalainen I, Ylikallio E, Jalanko A, Spelbrink JN, Paetau A, Suomalainen A (2005) Mutant mitochondrial helicase Twinkle causes multiple mtDNA deletions and a late-onset mitochondrial disease in mice. Proc Natl Acad Sci U S A 102:17687–17692. https://doi.org/10.1073/pnas.0505551102

Upholt WB, Dawid IB (1977) Mapping of mitochondrial DNA of individual sheep and goats: rapid evolution in the D loop region. Cell 11:571–583. https://doi.org/10.1016/0092-8674(77)90075-7

Varela-Lopez A, Bullon P, Battino M, Ramirez-Tortosa MC, Ochoa JJ, Cordero MD, Ramirez-Tortosa CL, Rubini C, Zizzi A, Quiles JL (2015) Coenzyme Q protects against age-related alveolar bone loss associated to n-6 PUFA rich-diets by modulating mitochondrial mechanisms. J Gerontol A Biol Sci Med Sci 71:593–600

Vielhaber S, Kunz D, Winkler K, Wiedemann FR, Kirches E, Feistner H, Heinze HJ, Elger CE, Schubert W, Kunz WS (2000) Mitochondrial DNA abnormalities in skeletal muscle of patients with sporadic amyotrophic lateral sclerosis. Brain J Neurol 123(Pt 7):1339–1348

Vilkki J, Savontaus ML, Nikoskelainen EK (1990) Segregation of mitochondrial genomes in a heteroplasmic lineage with Leber hereditary optic neuroretinopathy. Am J Hum Genet 47:95–100

Wai T, Teoli D, Shoubridge EA (2008) The mitochondrial DNA genetic bottleneck results from replication of a subpopulation of genomes. Nat Genet 40:1484–1488. https://doi.org/10.1038/ng.258

Wanagat J, Cao Z, Pathare P, Aiken JM (2001) Mitochondrial DNA deletion mutations colocalize with segmental electron transport system abnormalities, muscle fiber atrophy, fiber splitting, and oxidative damage in sarcopenia. FASEB J Off Publ Fed Am Soc Exp Biol 15:322–332. https://doi.org/10.1096/fj.00-0320com

Wang Y, Bogenhagen DF (2006) Human mitochondrial DNA nucleoids are linked to protein folding machinery and metabolic enzymes at the mitochondrial inner membrane. J Biol Chem 281:25791–25802. https://doi.org/10.1074/jbc.M604501200

Wang J, Silva JP, Gustafsson CM, Rustin P, Larsson NG (2001) Increased in vivo apoptosis in cells lacking mitochondrial DNA gene expression. Proc Natl Acad Sci U S A 98:4038–4043. https://doi.org/10.1073/pnas.061038798

Weber K, Wilson JN, Taylor L, Brierley E, Johnson MA, Turnbull DM, Bindoff LA (1997) A new mtDNA mutation showing accumulation with time and restriction to skeletal muscle. Am J Hum Genet 60:373–380

Wenz T, Diaz F, Speigelman BM, Moraes CT (2008) Activation of the PPAR/PGC-1alpha pathway pre- vents a bioenergetic deficit and effectively improves a mitochondrial myopathy phenotype. Cell Metab 8:249–256. https://doi.org/10.1016/j.cmet.2008.07.006

Wong TW, Clayton DA (1985a) Isolation and characterization of a DNA primase from human mitochondria. J Biol Chem 260:11530–11535

Wong TW, Clayton DA (1985b) In vitro replication of human mitochondrial DNA: accurate initiation at the origin of light-strand synthesis. Cell 42:951–958

Wong A, Cavelier L, Collins-Schramm HE, Seldin MF, McGrogan M, Savontaus M-L, Cortopassi GA (2002) Differentiation-specific effects of LHON mutations introduced into neuronal NT2 cells. Hum Mol Genet 11:431–438. https://doi.org/10.1093/hmg/11.4.431

Woo DK, Green PD, Santos JH, D'Souza AD, Walther Z, Martin WD, Christian BE, Chandel NS, Shadel GS (2012) Mitochondrial genome instability and ROS enhance intestinal tumorigenesis in APC(Min/+) mice. Am J Pathol 180:24–31. https://doi.org/10.1016/j.ajpath.2011.10.003

Wredenberg A, Wibom R, Wilhelmsson H, Graff C, Wiener HH, Burden SJ, Oldfors A, Westerblad H, Larsson NG (2002) Increased mitochondrial mass in mitochondrial myopathy mice. Proc Natl Acad Sci USA 99:15066–15071. https://doi.org/10.1073/pnas.232591499

Xu G, Herzig M, Rotrekl V, Walter CA (2008) Base excision repair, aging and health span. Mech Ageing Dev 129:366–382. https://doi.org/10.1016/j.mad.2008.03.001

Yang J, Zhu Y, Tong Y, Chen L, Liu L, Zhang Z, Wang X, Huang D, Qiu W, Zhuang S, Ma X (2009) Confirmation of the mitochondrial ND1 gene mutation G3635A as a primary LHON mutation. Biochem Biophys Res Commun 386:50–54. https://doi.org/10.1016/j.bbrc.2009.05.127

Yang X, Dai G, Li G, Yang ES (2010) Coenzyme Q_{10} reduces beta-amyloid plaque in an APP/PS1 transgenic mouse model of Alzheimer's disease. J Mol Neurosci 41:110–113. https://doi.org/10.1007/s12031-009-9297-1

Yasukawa T, Reyes A, Cluett TJ, Yang M-Y, Bowmaker M, Jacobs HT, Holt IJ (2006) Replication of vertebrate mitochondrial DNA entails transient ribonucleotide incorporation throughout the lagging strand. EMBO J 25:5358–5371. https://doi.org/10.1038/sj.emboj.7601392

Young AJ, Johnson S, Steffens DC, Doraiswamy PM (2007) Coenzyme Q_{10}: a review of its promise as a neuroprotectant. CNS Spectr 12:62–68

Yubero-Serrano EM, Delgado-Casado N, Delgado-Lista J, Perez-Martinez P, Tasset-Cuevas I, Santos-Gonzalez M, Caballero J, Garcia-Rios A, Marin C, Gutierrez-Mariscal FM, Fuentes F, Villalba JM, Tunez I, Perez-Jimenez F, Lopez-Miranda J (2010) Postprandial antioxidant effect of the Mediterranean diet supplemented with coenzyme Q_{10} in elderly men and women. Age Drodr 33:579–590. https://doi.org/10.1007/s11357-010-9199-8

Yubero-Serrano EM, Gonzalez-Guardia L, Rangel-Zuñiga O, Delgado-Lista J, Gutierrez-Mariscal FM, Perez-Martinez P, Delgado-Casado N, Cruz-Teno C, Tinahones FJ, Villalba JM, Perez-Jimenez F, Lopez-Miranda J (2012) Mediterranean diet supplemented with coenzyme Q_{10} modifies the expression of proinflammatory and endoplasmic reticulum stress–related genes in elderly men and women. J Gerontol Biol Sci Med Sci 67A:3–10. https://doi.org/10.1093/gerona/glr167

Zelenka J, Dvořák A, Alán L (2015) L-lactate protects skin fibroblasts against aging-associated mitochondrial dysfunction via mitohormesis. Oxid Med Cell Longev 2015:351698. https://doi.org/10.1155/2015/351698

Zeviani M, Moraes CT, DiMauro S, Nakase H, Bonilla E, Schon EA, Rowland LP (1988) Deletions of mitochondrial DNA in Kearns-Sayre syndrome. Neurology 38:1339–1346

Zheng W, Khrapko K, Coller HA, Thilly WG, Copeland WC (2006) Origins of human mitochondrial point mutations as DNA polymerase gamma-mediated errors. Mutat Res 599:11–20. https://doi.org/10.1016/j.mrfmmm.2005.12.012

Chapter 11
Coenzyme Q_{10} and Metabolic Syndrome

Juan Diego Hernández-Camacho

Abstract Metabolic syndrome (MS) has become a global health issue due to affect a high percentage of people in most of the countries. MS can be defined as the presence of three of the following factors: obesity, high triglyceride and cholesterol levels, low HDL cholesterol, high blood pressure or high fasting plasma glucose. All these factors increase the risk of cardiovascular disease, diabetes type II, some kind of cancers, sleep abnormalities or physical incapacity among other. Several factors have been identified in the aetiology of MS such as dietary patterns, sedentary lifestyle, genetic background, microbiota, socioeconomic status or age. Different treatments have been proposed for the treatment of MS, but, until today, there is no efficient solution. CoQ_{10} has emerged as a potential way in MS treatment endorsed by several clinical trials have shown improvements in lipid profile, glucose control, insulin homeostasis and hypertension control in MS patients. The molecular mechanism that could explain these improvements would be the antioxidant capacity of CoQ_{10} inhibiting oxidative stress that it is present in MS. Additionally, the proportion of $CoQ_{10}H_2$ could be also a crucial role in the protection again MS components. Furthermore, CoQ_{10} administration could be also helpful in the management of mitochondrial dysfunction associated to MS.

Keywords Mitochondria · Coenzyme Q_{10} · Oxidative stress · Metabolic syndrome · Insulin resistance · Aging

11.1 Introduction to Metabolic Syndrome

Metabolic syndrome (MS) is considered a recent concept because it was only thirty years ago when the MS components were grouped as a common pathology. The World Health Organization proposed as MS concept the presence of hypertension,

J. D. Hernández-Camacho (✉)
Universidad Pablo de Olavide, Centro Andaluz de Biología del Desarrollo, CABD-CSIC, CIBERER Instituto de Salud Carlos III, Sevilla, Spain
e-mail: jdhercam@upo.es

G. López Lluch (ed.), *Coenzyme Q in Aging*,
https://doi.org/10.1007/978-3-030-45642-9_11

glucose intolerance, obesity and dyslipemia in 1999. The National Cholesterol Education Program Adult Treatment Panel III (NEPATP III) defined the MS as the presence of three of the next risk factors: obesity, high triglyceride levels, low HDL cholesterol, high blood pressure and high fasting plasma glucose (Expert Panel on Detection and Treatment of High Blood Cholesterol in 2001; National Cholesterol Education Program Expert Panel on Detection and Treatment of High Blood Cholesterol in 2002). Additionally, the Japan Society for the study of Obesity (JASSO) and the International Diabetes Federation (IDF) contemplate obesity as a crucial factor to diagnose MS. The American Heart Association (AHA), the National Heart, Lung and Blood Institute (NHLBI) and the IDF describe MS when a patient suffers three of these factors, being a high waist circumference a key required factor.

MS is present in almost all countries independently of the developmental status. The prevalence of MS in Europe and USA is the 25% of the population (Pais et al. 2009), specifically in Spain MS is the 31% (Fernandez-Berges et al. 2012). In developing countries, MS is also a public health problem, in Brazil one in three adults suffer MS (Turi et al. 2016). Oriental countries which has been characterized traditionally for having a low prevalence of MS, have recently reported a 27.4% of population suffering MS as in China (Song et al. 2015).

A huge cluster of different factors has been associated with MS etiology such as familiar relationships (Lee et al. 2011). Children present 3.31 higher risk to present MS when both parents suffer MS. Being member of a specific ethnic group can also affect MS occurrence (Philco et al. 2012). Age and sex are also related with MS (Kuk and Ardern 2010; Sheu et al. 2006), and also genetic background (Hotta et al. 2011). Even microbiota has been related with MS (Mazidi et al. 2016). However, the main factors that have been traditionally related with MS are an inadequate dietary pattern and sedentarism (Pan and Pratt 2008).

MS increases comorbidities of other chronic pathologies such as some types of cancers (Pais et al. 2009), irritable bowel syndrome (Guo et al. 2014), chronic kidney disease (Watanabe et al. 2010), polycystic ovary syndrome (De Leo et al. 2009), cardiovascular disease (Fernandez-Berges et al. 2012), sleep apnea (Sasanabe et al. 2006) and mental disorders (Babic et al. 2010). Furthermore, MS also affects physical capacity (Senechal et al. 2012) conditioning the quality of life of patients permanently.

Prevalence of metabolic syndrome increases with age, especially in sedentary people showing high BMI and weight gain (Roos et al. 2017). In aged people, MS increases the risk of cardiovascular disease as indicated a recent work with Portuguese population (Ribeiro et al. 2018). In elderly people, MS has been associated with frailty increasing the risk of the functional decline and dependence (Buchmann et al. 2019) probably by affecting loss of muscle (Dominguez and Barbagallo 2016) and bone mineral density (Eckstein et al. 2016). Furthermore, MS has been also associated with the evolution of cognitive decline associated with aging although in this case the relationship needs to be confirmed by further research (Gomez et al. 2018; Philippou et al. 2018).

MS includes a group different clinical conditions that has a global impact on health. Different factors are involved in the development of MS and its associated chronic diseases. Nowadays, many different approaches have been proposed for MS management including drugs, nutritional treatments, exercise and surgical procedures, but MS is still a crucial issue for the health systems.

11.2 Coenzyme Q_{10} in Metabolic Syndrome

Patients suffering MS show metabolic dysregulation leading to chronic inflammation and oxidative stress characterized by mitochondrial dysfunction. Taken into consideration the essential role of CoQ_{10} in mitochondrial physiology and as membrane and plasma lipoproteins antioxidant its use as part of the treatment of MS can be considered (Lopez-Lluch 2019; Lopez-Lluch et al. 2018; Lopez-Lluch et al. 2010). In fact, as CoQ_{10} is an essential compound of human body and a key component of cell bioenergetics and antioxidant systems, its supplementation in cardiometabolic disorders has been recently considered (Zozina et al. 2018).

In spite of the role in bioenergetics and antioxidant status, the studies about the effect of CoQ_{10} in MS are contradictory. In patients with diabetes, one of the components of MS, treatment with CoQ_{10} decreases fasting blood glucose indicating a presumed positive effect on insulin-resistance (Moradi et al. 2016). In contrast, another paper did not find any effect of CoQ_{10} supplementation on glycaemic control neither lipid profile or blood pressure in humans (Suksomboon et al. 2015). However, this study did reported a significant reduction in triglycerides levels. Interestingly, a gender-depending effect can be also considered since in women with type 2 diabetes, CoQ_{10} ingestion reduced total cholesterol, LDL-cholesterol, HOMA insulin resistance, ferritin and glucose levels while HDL-cholesterol level was increased (Gholami et al. 2019). Further, it has been also reported that supplementation with CoQ_{10} improves endothelial dysfunction in type 2 diabetic patients by affecting local vascular oxidative stress (Hamilton et al. 2009). Reduction in plasma lipoprotein(a) levels, associated with cardiovascular risk, was also found in subjects supplemented with CoQ_{10} (Sahebkar et al. 2016) indicating a protective effect in the release of markers of damage associated with MS.

Regarding lipid profile, Pirro and colleagues analysed the effect of CoQ_{10} intake in combination with other nutraceutical compounds in humans obtaining a decrease on total cholesterol, LDL-cholesterol, triglycerides and glucose while HDL-cholesterol level increased (Pirro et al. 2016). Another recent study examined the effect of CoQ_{10} in combination with red yeast rice administration for 2 months in patients suffering MS. This study also found a decrease in blood pressure, total cholesterol, triglycerides, LDL-cholesterol and glucose levels (Mazza et al. 2018). However, these studies did not assess the effect of CoQ_{10} alone, without other compounds, and therefore, a synergic effect of these compounds cannot be discarded.

In general, most of the studies demonstrate that supplementation with CoQ_{10} produces positive effects in many of the factors involved in MS. In fact, Zhang and collaborators examined randomized controlled trials including 765 patients and reported that CoQ_{10} intake decreased HbA1c, fasting glucose, triglycerides and improved HDL-cholesterol, but they did not reported a clear effect in fasting insulin (Zhang et al. 2018). Another study revised randomized controlled trials studied CoQ_{10} intake in overweight and obese patients with diabetes type 2, including fourteen trials with 693 subjects. The authors reported a decrease in fasting plasma glucose, HbA1c and triglycerides without reporting adverse reactions associated to CoQ_{10} intake (Huang et al. 2018). On the other hand, another meta-analysis examine the effect of CoQ_{10} administration on another MS component, obesity (Saboori et al. 2019). The authors analysed clinical trials that studied CoQ_{10} on body weight and body mass, reporting no beneficial effect of coenzyme Q_{10} on the parameters analysed. Taken together, these studies indicate that CoQ_{10} is a safe treatment in MS affecting several of the biochemical markers of this disease but without showing effects on body weight.

The use of CoQ_{10} has been also studied for the treatment of hypertension, another important factor in MS. Hypertension has been associated with an increase in oxidative stress through the production of superoxide radical's levels. Superoxide radical reacts with nitric oxide leading to the production of peroxynitrite. The concomitant reduction in the levels of nitric oxide would affect to the capacity of endothelium to relax the smooth muscle resulting in vasoconstriction and consequently in increase in blood pressure (Grunfeld et al. 1995). A randomized, double-blind, placebo-controlled trial analysed the effect of CoQ_{10} administration (60 mg day) in a population of men and women with isolated systolic hypertension (Burke et al. 2001). The effect was a reduction of systolic blood pressure in subject who received CoQ_{10}. The same result was found in another clinical trial in which CoQ_{10} was supplemented with monacolin K, a component of red rice (Mazza et al. 2018). Apart of the reducing effect of LDL-cholesterol, triglycerides and glucose, treated MS patients also showed a reduction of both systolic and diastolic blood pressures. Further, a recent meta-analysis of randomized controlled clinical trials including seventeen trials with 684 participants confirmed the reduction of systolic blood pressure after treatment with CoQ_{10} (Tabrizi et al. 2018).

In summary, these studies demonstrate that CoQ_{10} is a promising dietary supplement in the treatment of MS. However, the data available are still limited and there is a lack of well-designed and well-powered randomized controlled trials analysing short and long-term effects.

11.3 Molecular Mechanism of Coenzyme Q_{10} in Metabolic Syndrome

In general, the mechanistic effects of CoQ_{10} in MS can be obtained from the studies using animal models, mainly rodents. It is clear that one of the principal mechanisms is based on the antioxidant capacity of CoQ_{10} in plasma and endothelium. In fact, in rats supplemented with CoQ_{10} (100 mg/kg of body weight) the oxidative stress serum profile improved including the levels of malondialdehyde and thiol groups (Chis et al. 2019). Another study examined the effect of CoQ_{10} administration in senescence-accelerated mice resulting in the inhibition of oxidative stress and the slow down the process of aging, delaying the progression of type 2 diabetes and, in this case, also obesity (Xu et al. 2017). In contrast with human studies, the effect of CoQ_{10} in obesity in these rats can be associated with high doses and the animal model used. Further, in rats, HFD induces oxidation of haemoglobin and oxidation of CoQ_{10}, supplementation with CoQ_{10} reduced both, the oxidation of haemoglobin and CoQ_{10} indicating that maintenance of high levels of CoQ_{10} can prevent oxidative damage in HFD (Orlando et al. 2014) (Table 11.1).

In this section, we highlight the molecular effects of CoQ_{10} involved in the development of MS.

11.3.1 Antioxidant Function in Plasma

High cholesterol levels are associated ~~in~~ with MS. Its oxidation is prevented by $CoQ_{10}H_2$ as its main activity in plasma, this function prevents oxidative damage in cardiovascular system reducing cardiovascular risk (Thomas et al. 1997). Interestingly, we found that sedentary lifestyle and high BMI was inversely associated with the levels of CoQ_{10} and directly with the levels of oxidative damage in plasma (Del Pozo-Cruz et al. 2014). In fact, the supplementation with CoQ_{10} has been considered as the treatment in many metabolic diseases that affect oxidative damage affecting plasma, cells and tissues (Zozina et al. 2018; Lopez-Lluch et al. 2018; Hernandez-Camacho et al. 2018).

MS is accompanied by the increase of oxidative and nitrative damage and inflammation (Kunitomo et al. 2008). One of the main effects of CoQ_{10} in the progression of MS can be attributed to its antioxidant activity in cell membranes and plasma lipoproteins (Lopez-Lluch et al. 2010). The inverse relationship between CoQ_{10} levels in plasma and oxidative damage of low density lipoprotein (LDL) particles can contribute to the prevention of the atherosclerotic damage and the delay in the progression of cardiovascular disease associated with MS (Thomas et al. 1997). In a rat model of MS, treatment with CoQ_{10} attenuated the increase of LDL oxidation, levels of 3-nitrotyrosine as marker of nitrative stress, 3-chlorotyrosine as marker of myeloperoxidase activity and C-reactive protein as marker of inflammation in plasma (Kunitomo et al. 2008). In general, the antioxidant capacity of CoQ_{10} was proposed to reduce the cardiovascular risk in MS.

It has been found that plasma levels of reduced CoQ_{10} decrease in diabetic patients suffering haemodialysis or ambulatory peritoneal dialysis but not in patients showing hyperlipidaemia (McDonnell and Archbold 1996). However, in a further study, a decrease in the levels of reduced CoQ_{10} were associated higher oxidative damage in plasma of patients showing hyperlipidaemia (Kontush et al. 1997). In agreement with this study, in plasma isolated from patients with primary hypercholesterolemia the lag-phase of plasma LDL oxidation was minimal in comparison with normal patients (Lankin et al. 2003). In these samples reduction of CoQ_{10}

Table 11.1 Study of CoQ_{10} effect on MS in animal models and humans

References	MS component	CoQ_{10} dose	Outcome	Side effects
Hamilton et al. (2009)	Endothelial dysfunction in type 2 diabetes humans.	200 mg/day CoQ_{10}	Increase on brachial artery flow-mediated dilatation.	No reported.
Mazza et al. (2018)	Blood pressure and lipid and glucose profile in MS human patients.	30 mg/day of CoQ_{10} and red yeast rice 10 mg/day	Reduction in blood pressure, Total cholesterol, triglycerides, LDL cholesterol and glucose levels.	No side effects.
Gholami et al. (2019)	Women with type 2 diabetes mellitus.	100 mg/day CoQ_{10}	Decrease levels of fasting blood sugar, Homeostatic model assessment Insulin resistance, ferritin, total cholesterol, LDL cholesterol. Increase levels of HDL cholesterol.	No reported.
Burke et al. (2001)	Men and women with isolate systolic hypertension.	60 mg/ twice per day CoQ_{10}	Reduction in systolic blood pressure.	No reported.
Zarei et al. (2018)	Women with type 2 diabetes mellitus.	100 mg/day CoQ_{10}	Increase on total antioxidant capacity, catalase activity and quantitative insulin sensitivity check index. Reduction in fasting blood sugar.	No side effects.
Raygan et al. (2016)	Human patients with MS (overweight-obese, type 2 diabetes mellitus patients with coronary heart disease, 40–85 years old).	100 mg/day CoQ_{10}	Reduction on insulin levels in plasma and homeostatic model assessment Insulin resistance. Improvement in plasma antioxidant capacity.	No side effects.
Young et al. (2012)	Human patients with MS and an inadequate blood pressure control.	100 mg/ twice per day CoQ_{10}	No effect on in systolic or diastolic pressure or heart rate.	No side effects.
Kunitomo et al. (2008)	SHR/NDmcr-cp (SHR/cp) rats, animal model of MS.	0.07%–0.7% CoQ_{10} of their diet.	Decrease in oxidative stress markers (ox-LDL, 8-OHdG and 3 nitrotyrosine) and inflammatory markers (hsCRP).	No reported.
Feillet-Coudray et al. (2014)	Young male Spargue-Dawley rats.	375 μmoles of MitoQ per kg of diet	Decrease of body weight gain and improvement on glucose intolerance.	No reported.

(continued)

Table 11.1 (continued)

References	MS component	CoQ_{10} dose	Outcome	Side effects
Orlando et al. (2014)	Wistar Ottawa Karlsburg W (WOKW) rats.	3 mg/100 g b.w. of CoQ_{10}	Prevented methaemoglobin formation and endogenous CoQ_{10} oxidation.	No reported.
Prangthip et al. (2016)	Diabetic rats.	Ubiquinol-10 or Ubiquinone-10 5 mg/kg/day	Decreased oxidative stress, blood glucose and blood pressure.	No side effects.
Chis et al. (2019)	Wistar rat inducing MS.	100 mg/kg/day CoQ_{10}	Improvement in oxidative stress response.	No reported.

levels increases the oxidation of LDL indicating the importance of $CoQ_{10}H_2$ in the protection of these lipoproteins against oxidation (Lankin et al. 2003).

Further, patients showing hypertension had lower proportion of $CoQ_{10}H_2$ than those without indicating a putative role of oxidative damage or plasma CoQ_{10} levels with kidney function (Kontush et al. 1997).

In relationship with this, obesity in children is associated with lower plasma levels of lipophilic antioxidants such as α-tocopherol or β-carotene (Strauss 1999). However, although oxidized LDL in plasma increase in obese children no changes in CoQ_{10} levels were found (Menke et al. 2004). This study probably demonstrate an age-related difference in the progression of the levels of CoQ_{10} in metabolic syndrome.

11.3.2 Antioxidant Function in Endothelial Cells

The presence of CoQ_{10} in adequate concentrations in tissues has been considered an important factor in limiting oxidative and nitrosative damage in vivo (Hodgson and Watts 2003). In the arterial wall, oxidative and nitrosidative stress contribute to the increase of blood pressure and vascular dysfunction, two main factors involved in MS. As it has been indicated before, the positive effect of CoQ_{10} on hypertension can be due to the capacity to scavenge superoxide radicals. Reduction in the levels of superoxide can increase the levels of nitric oxide by blocking the reaction to produce peroxinitrite. Higher levels of peroxynitrite would maintain the capacity of endothelium to relax the smooth muscle preventing hypertension (Grunfeld et al. 1995). A recent study on cardiovascular regulation of hypertension indicates that CoQ_{10} reduces the production of superoxide levels by abolishing NADPH-oxidase activation in endothelial cells of rats fed with fructose as model of MS (Chen et al. 2019). However, we cannot discard the role of CoQ_{10} as superoxide scavenger through the activity of NQO1, a plasma-membrane associated CoQ_{10}-dependent enzyme (Ross and Siegel 2017).

It is clear that prevention of oxidative damage of endothelial cells can protect against cardiovascular the disease progression associated with metabolic syndrome. In a rat model of ischemia/reperfusion damage, demonstrated that treatment with CoQ_{10} prevented the oxidative damage of endothelial cells after reperfusion indicating a key role of CoQ_{10} in oxidative damage protection in these cells (Yokoyama et al. 1996). This protective role can be associated with the improvement of endothelial function found in MS patients (Hamilton et al. 2009). In a double-bling crossover study, 23 statin-treated type 2 diabetic patients were treated with CoQ_{10} or placebo. Treatment with CoQ_{10} improved endothelial dysfunction possibly by reducing local vascular oxidative stress (Hamilton et al. 2009).

Further, mitochondrial targeted derivatives of CoQ_{10} such as MitoQ have been proposed to rescue endothelial cells against mitochondrial damage associated with aging through an antioxidant mechanism. Incubation with MitoQ prevented the susceptibility of endothelial cells against acute mitochondrial damage (Gioscia-Ryan et al. 2014).

11.3.3 Prevention of Mitochondrial Damage

Mitochondrial dysfunction has been also associated with MS. MS produces the accumulation of fatty acids in muscle and liver, and this is accompanied by defects in mitochondrial fatty acid oxidation (Parish and Petersen 2005). Without being clear if mitochondrial dysfunction is a consequence or part of the mechanism of type 2 diabetes, mitochondrial dysfunction seems to be a common denominator in insulin resistance found in this disease, in obesity and MS (Abdul-Ghani and DeFronzo 2008). Mitochondrial abnormalities are also associated with cardiac contractile dysfunction associated with obesity, type 2 diabetes and insulin resistance (Bugger and Abel 2008). In this relationship mitochondrial stress and unfolded protein response (mtUPR) can play an important role (Hu and Liu 2011). The use of therapies or mechanisms to improve the capacity of mitochondria to reduce stress and reduce mtUPR can be considered an effective approach to improve mitochondrial dysfunction associated with many of the processes associated with MS (Hu and Liu 2011).

The importance of CoQ_{10} in the bioenergetics of mitochondria makes it a good candidate to reduce mitochondrial dysfunction associated with MS. Taken into consideration the positive effect of CoQ_{10} supplementation in chronic statin-treated patients, some authors have considered that supplementation with CoQ_{10} can be useful in the treatment of obesity, oxidative stress and inflammation associated with MS (Alam and Rahman 2014; Salminen et al. 2012). Supplementation with CoQ_{10} would improve the metabolic capacity of mitochondria by restoring the capacity of the electron transport chain to maintain the ATP-generating proton motive force (Alam and Rahman 2014). Supplementation with CoQ_{10} can also modulate muscle lipid profile and improves mitochondrial respiration in obesogenic diet-fed rats (Coudray et al. 2016). All these results indicate that the role of CoQ_{10} in mitochondria is important in the progression of MS and can reduce the damaging factors triggered by this disease.

MitoQ can exert antioxidant activity in mitochondria. Mitochondrial targeted CoQ_{10}-derived compounds such as MitoQ can also improve the progression and features of MS in rat models (Feillet-Coudray et al. 2014). MitoQ activity can break the vicious cycle stablished between the increase in ROS production from mitochondrial dysfunction and the impairment of many of the factors involved in MS (Mitchell and Darley-Usmar 2012). In a mouse model of MS, 14 weeks of MitoQ treatment prevented the increase of adiposity, levels of cholesterol and triglycerides associated with MS indicating the importance of mitochondrial function and the prevention of its oxidative damage in the progression of this disease (Mercer et al. 2012). In a Zucker obese fatty (ZOF) rats, the structural abnormalities in mitochondria and oxidative stress was reduced by treatment with mitochondrial targeted antioxidants such as MitoQ and also MitoTempol indicating the importance of oxidative stress in mitochondrial dysfunction during MS (Pung et al. 2012). MitoQ treatment also reduced the ROS production in leukocytes from type 2 diabetic patients improving the capacity of these cells to interact with endothelium. The antioxidant capacity of MitoQ reduced proinflammatory signals in leukocytes such as NF-κB and TNF-α pathways (Escribano-Lopez et al. 2016). All these evidence have indicated that the reduction of generation of ROS by mitochondria and the derived signalling can reduce many of the factors involved in MS and then, the evolution of diabetes, cardiovascular disease and hepatosteatosis (Mitchell and Darley-Usmar 2012).

11.4 Conclusions

The importance of CoQ_{10} in the progression of MS and in its treatment is based on its antioxidant and bioenergetic role in the cells. It is clear that MS is accompanied by several dysfunctions that affects many important activities in the cell, tissues and organism. As a key component in bioenergetics in cells, the role of CoQ_{10} in mitochondrial activity can be crucial in the regulation of carbohydrate and lipid metabolism associated with many bioenergetics problems in MS.

References

Abdul-Ghani MA, DeFronzo RA (2008) Mitochondrial dysfunction, insulin resistance, and type 2 diabetes mellitus. Curr Diab Rep 8(3):173–178. https://doi.org/10.1007/s11892-008-0030-1

Alam MA, Rahman MM (2014) Mitochondrial dysfunction in obesity: potential benefit and mechanism of Co-enzyme Q10 supplementation in metabolic syndrome. J Diabetes Metab Disord 13:60. https://doi.org/10.1186/2251-6581-13-60

Babic D, Maslov B, Martinac M, Nikolic K, Uzun S, Kozumplik O (2010) Bipolar disorder and metabolic syndrome: comorbidity or side effects of treatment of bipolar disorder. Psychiatr Danub 22(1):75–78

Buchmann N, Spira D, Konig M, Demuth I, Steinhagen-Thiessen E (2019) Frailty and the metabolic syndrome – results of the Berlin Aging Study II (BASE-II). J Frailty Aging 8(4):169–175. https://doi.org/10.14283/jfa.2019.15

Bugger H, Abel ED (2008) Molecular mechanisms for myocardial mitochondrial dysfunction in the metabolic syndrome. Clin Sci (Lond) 114(3):195–210. https://doi.org/10.1042/CS20070166

Burke BE, Neuenschwander R, Olson RD (2001) Randomized, double-blind, placebo-controlled trial of coenzyme Q10 in isolated systolic hypertension. South Med J 94(11):1112–1117. https://doi.org/10.1097/00007611-200111000-00015

Chen HH, Yeh TC, Cheng PW, Ho WY, Ho CY, Lai CC, Sun GC, Tseng CJ (2019) Antihypertensive potential of coenzyme Q10 via free radical scavenging and enhanced Akt-nNOS signaling in the nucleus Tractus Solitarii in rats. Mol Nutr Food Res 63(6):e1801042. https://doi.org/10.1002/mnfr.201801042

Chis BA, Chis AF, Muresan A, Fodor D (2019) Q10 coenzyme supplementation can improve oxidative stress response to exercise in metabolic syndrome in rats. Int J Vitam Nutr Res 90:1–9. https://doi.org/10.1024/0300-9831/a000301

Coudray C, Fouret G, Lambert K, Ferreri C, Rieusset J, Blachnio-Zabielska A, Lecomte J, Ebabe Elle R, Badia E, Murphy MP, Feillet-Coudray C (2016) A mitochondrial-targeted ubiquinone modulates muscle lipid profile and improves mitochondrial respiration in obesogenic diet-fed rats. Br J Nutr 115(7):1155–1166. https://doi.org/10.1017/S0007114515005528

De Leo V, Musacchio MC, Palermo V, Di Sabatino A, Morgante G, Petraglia F (2009) Polycystic ovary syndrome and metabolic comorbidities: therapeutic options. Drugs Today (Barc) 45(10):763–775. https://doi.org/10.1358/dot.2009.45.10.1429463

Del Pozo-Cruz J, Rodriguez-Bies E, Navas-Enamorado I, Del Pozo-Cruz B, Navas P, Lopez-Lluch G (2014) Relationship between functional capacity and body mass index with plasma coenzyme Q10 and oxidative damage in community-dwelling elderly-people. Exp Gerontol 52:46–54. https://doi.org/10.1016/j.exger.2014.01.026

Dominguez LJ, Barbagallo M (2016) The biology of the metabolic syndrome and aging. Curr Opin Clin Nutr Metab Care 19(1):5–11. https://doi.org/10.1097/MCO.0000000000000243

Eckstein N, Buchmann N, Demuth I, Steinhagen-Thiessen E, Nikolov J, Spira D, Eckardt R, Norman K (2016) Association between metabolic syndrome and bone mineral density--data from the Berlin aging study II (BASE-II). Gerontology 62(3):337–344. https://doi.org/10.1159/000434678

Escribano-Lopez I, Diaz-Morales N, Rovira-Llopis S, de Maranon AM, Orden S, Alvarez A, Banuls C, Rocha M, Murphy MP, Hernandez-Mijares A, Victor VM (2016) The mitochondria-targeted antioxidant MitoQ modulates oxidative stress, inflammation and leukocyte-endothelium interactions in leukocytes isolated from type 2 diabetic patients. Redox Biol 10:200–205. https://doi.org/10.1016/j.redox.2016.10.017

Expert Panel on Detection E, Treatment of High Blood Cholesterol in A (2001) Executive summary of the third report of the National Cholesterol Education Program (NCEP) expert panel on detection, evaluation, and treatment of high blood cholesterol in adults (adult treatment panel III). JAMA 285(19):2486–2497. https://doi.org/10.1001/jama.285.19.2486

Feillet-Coudray C, Fouret G, Ebabe Elle R, Rieusset J, Bonafos B, Chabi B, Crouzier D, Zarkovic K, Zarkovic N, Ramos J, Badia E, Murphy MP, Cristol JP, Coudray C (2014) The mitochondrial-targeted antioxidant MitoQ ameliorates metabolic syndrome features in obesogenic diet-fed rats better than Apocynin or Allopurinol. Free Radic Res 48(10):1232–1246. https://doi.org/10.3109/10715762.2014.945079

Fernandez-Berges D, Cabrera de Leon A, Sanz H, Elosua R, Guembe MJ, Alzamora M, Vega-Alonso T, Felix-Redondo FJ, Ortiz-Marron H, Rigo F, Lama C, Gavrila D, Segura-Fragoso A, Lozano L, Marrugat J (2012) Metabolic syndrome in Spain: prevalence and coronary risk associated with harmonized definition and WHO proposal. DARIOS study. Rev Esp Cardiol (Engl Ed) 65(3):241–248. https://doi.org/10.1016/j.recesp.2011.10.015

Gholami M, Rezvanfar MR, Delavar M, Abdollahi M, Khosrowbeygi A (2019) Effects of coenzyme Q10 supplementation on serum values of gamma-glutamyl transferase, Pseudocholinesterase, bilirubin, ferritin, and high-sensitivity C-reactive protein in women with type 2 diabetes. Exp Clin Endocrinol Diabetes 127(5):311–319. https://doi.org/10.1055/s-0043-124183

Gioscia-Ryan RA, LaRocca TJ, Sindler AL, Zigler MC, Murphy MP, Seals DR (2014) Mitochondria-targeted antioxidant (MitoQ) ameliorates age-related arterial endothelial dysfunction in mice. J Physiol 592(12):2549–2561. https://doi.org/10.1113/jphysiol.2013.268680

Gomez G, Beason-Held LL, Bilgel M, An Y, Wong DF, Studenski S, Ferrucci L, Resnick SM (2018) Metabolic syndrome and amyloid accumulation in the aging brain. J Alzheimers Dis 65(2):629–639. https://doi.org/10.3233/JAD-180297

Grunfeld S, Hamilton CA, Mesaros S, McClain SW, Dominiczak AF, Bohr DF, Malinski T (1995) Role of superoxide in the depressed nitric oxide production by the endothelium of genetically hypertensive rats. Hypertension 26(6 Pt 1):854–857. https://doi.org/10.1161/01.hyp.26.6.854

Guo Y, Niu K, Momma H, Kobayashi Y, Chujo M, Otomo A, Fukudo S, Nagatomi R (2014) Irritable bowel syndrome is positively related to metabolic syndrome: a population-based cross-sectional study. PLoS One 9(11):e112289. https://doi.org/10.1371/journal.pone.0112289

Hamilton SJ, Chew GT, Watts GF (2009) Coenzyme Q10 improves endothelial dysfunction in statin-treated type 2 diabetic patients. Diabetes Care 32(5):810–812. https://doi.org/10.2337/dc08-1736

Hernandez-Camacho JD, Bernier M, Lopez-Lluch G, Navas P (2018) Coenzyme Q10 supplementation in aging and disease. Front Physiol 9:44. https://doi.org/10.3389/fphys.2018.00044

Hodgson JM, Watts GF (2003) Can coenzyme Q10 improve vascular function and blood pressure? Potential for effective therapeutic reduction in vascular oxidative stress. Biofactors 18(1–4):129–136. https://doi.org/10.1002/biof.5520180215

Hotta K, Kitamoto T, Kitamoto A, Mizusawa S, Matsuo T, Nakata Y, Kamohara S, Miyatake N, Kotani K, Komatsu R, Itoh N, Mineo I, Wada J, Yoneda M, Nakajima A, Funahashi T, Miyazaki S, Tokunaga K, Masuzaki H, Ueno T, Hamaguchi K, Tanaka K, Yamada K, Hanafusa T, Oikawa S, Yoshimatsu H, Sakata T, Matsuzawa Y, Nakao K, Sekine A (2011) Association of variations in the FTO, SCG3 and MTMR9 genes with metabolic syndrome in a Japanese population. J Hum Genet 56(9):647–651. https://doi.org/10.1038/jhg.2011.74

Hu F, Liu F (2011) Mitochondrial stress: a bridge between mitochondrial dysfunction and metabolic diseases? Cell Signal 23(10):1528–1533. https://doi.org/10.1016/j.cellsig.2011.05.008

Huang H, Chi H, Liao D, Zou Y (2018) Effects of coenzyme Q10 on cardiovascular and metabolic biomarkers in overweight and obese patients with type 2 diabetes mellitus: a pooled analysis. Diabetes Metab Syndr Obes 11:875–886. https://doi.org/10.2147/DMSO.S184301

Kontush A, Reich A, Baum K, Spranger T, Finckh B, Kohlschutter A, Beisiegel U (1997) Plasma ubiquinol-10 is decreased in patients with hyperlipidaemia. Atherosclerosis 129(1):119–126. https://doi.org/10.1016/s0021-9150(96)06021-2

Kuk JL, Ardern CI (2010) Age and sex differences in the clustering of metabolic syndrome factors: association with mortality risk. Diabetes Care 33(11):2457–2461. https://doi.org/10.2337/dc10-0942

Kunitomo M, Yamaguchi Y, Kagota S, Otsubo K (2008) Beneficial effect of coenzyme Q10 on increased oxidative and nitrative stress and inflammation and individual metabolic components developing in a rat model of metabolic syndrome. J Pharmacol Sci 107(2):128–137. https://doi.org/10.1254/jphs.fp0072365

Lankin VZ, Konovalova GG, Tikhaze AK, Nezhdanova IB, Olfer'ev AM, Kukharchuk VV (2003) Oxidation of plasma low-density lipoproteins from coronary patients with various forms of hypercholesterolemia. Bull Exp Biol Med 136(1):42–45. https://doi.org/10.1023/a:1026032728328

Lee MH, Kim HC, Thomas GN, Ahn SV, Hur NW, Choi DP, Suh I (2011) Familial concordance of metabolic syndrome in Korean population--Korean National Health and Nutrition Examination Survey 2005. Diabetes Res Clin Pract 93(3):430–436. https://doi.org/10.1016/j.diabres.2011.06.002

Lopez-Lluch G (2019) The important role of CoQ_{10} in aging. Antioxidants (Basel) 8(12). https://doi.org/10.3390/antiox8120570

Lopez-Lluch G, Hernandez-Camacho JD, Fernandez-Ayala DJM, Navas P (2018) Mitochondrial dysfunction in metabolism and ageing: shared mechanisms and outcomes? Biogerontology 19(6):461–480. https://doi.org/10.1007/s10522-018-9768-2

Lopez-Lluch G, Rodriguez-Aguilera JC, Santos-Ocana C, Navas P (2010) Is coenzyme Q a key factor in aging? Mech Ageing Dev 131(4):225–235. https://doi.org/10.1016/j.mad.2010.02.003

Mazidi M, Rezaie P, Kengne AP, Mobarhan MG, Ferns GA (2016) Gut microbiome and metabolic syndrome. Diabetes Metab Syndr 10(2 Suppl 1):S150–S157. https://doi.org/10.1016/j.dsx.2016.01.024

Mazza A, Lenti S, Schiavon L, Di Giacomo E, Tomasi M, Manunta R, Torin G, Townsend DM, Rubello D (2018) Effect of Monacolin K and CoQ_{10} supplementation in hypertensive and hypercholesterolemic subjects with metabolic syndrome. Biomed Pharmacother 105:992–996. https://doi.org/10.1016/j.biopha.2018.06.076

McDonnell MG, Archbold GP (1996) Plasma ubiquinol/cholesterol ratios in patients with hyperlipidaemia, those with diabetes mellitus and in patients requiring dialysis. Clin Chim Acta 253(1–2):117–126. https://doi.org/10.1016/0009-8981(96)06357-7

Menke T, Niklowitz P, de Sousa G, Reinehr T, Andler W (2004) Comparison of coenzyme Q10 plasma levels in obese and normal weight children. Clin Chim Acta 349(1–2):121–127. https://doi.org/10.1016/j.cccn.2004.06.015

Mercer JR, Yu E, Figg N, Cheng KK, Prime TA, Griffin JL, Masoodi M, Vidal-Puig A, Murphy MP, Bennett MR (2012) The mitochondria-targeted antioxidant MitoQ decreases features of the metabolic syndrome in ATM+/−/ApoE−/− mice. Free Radic Biol Med 52(5):841–849. https://doi.org/10.1016/j.freeradbiomed.2011.11.026

Mitchell T, Darley-Usmar V (2012) Metabolic syndrome and mitochondrial dysfunction: insights from preclinical studies with a mitochondrially targeted antioxidant. Free Radic Biol Med 52(5):838–840. https://doi.org/10.1016/j.freeradbiomed.2011.12.014

Moradi M, Haghighatdoost F, Feizi A, Larijani B, Azadbakht L (2016) Effect of Coenzyme Q10 supplementation on diabetes biomarkers: a systematic review and meta-analysis of randomized controlled clinical trials. Arch Iran Med 19 (8):588–596. 0161908/AIM.0012

National Cholesterol Education Program Expert Panel on Detection E, Treatment of High Blood Cholesterol in A (2002) Third report of the National Cholesterol Education Program (NCEP) expert panel on detection, evaluation, and treatment of high blood cholesterol in adults (adult treatment panel III) final report. Circulation 106(25):3143–3421

Orlando P, Silvestri S, Bruge F, Tiano L, Kloting I, Falcioni G, Polidori C (2014) High-fat diet-induced met-hemoglobin formation in rats prone (WOKW) or resistant (DA) to the metabolic syndrome: effect of CoQ_{10} supplementation. Biofactors 40(6):603–609. https://doi.org/10.1002/biof.1190

Pais R, Silaghi H, Silaghi AC, Rusu ML, Dumitrascu DL (2009) Metabolic syndrome and risk of subsequent colorectal cancer. World J Gastroenterol 15(41):5141–5148. https://doi.org/10.3748/wjg.15.5141

Pan Y, Pratt CA (2008) Metabolic syndrome and its association with diet and physical activity in US adolescents. J Am Diet Assoc 108(2):276–286;. discussion 286. https://doi.org/10.1016/j.jada.2007.10.049

Parish R, Petersen KF (2005) Mitochondrial dysfunction and type 2 diabetes. Curr Diab Rep 5(3):177–183. https://doi.org/10.1007/s11892-005-0006-3

Philco LP, Seron SP, Munoz NS, Navia BP, Lanas ZF (2012) Risk factors for metabolic syndrome in a case control study in Temuco, Chile. Rev Med Chil 140(3):334–339. https://doi.org/10.4067/S0034-98872012000300008

Philippou E, Michaelides MP, Constantinidou F (2018) The role of metabolic syndrome factors on cognition using latent variable modeling: the neurocognitive study on aging. J Clin Exp Neuropsychol 40(10):1030–1043. https://doi.org/10.1080/13803395.2018.1483487

Pirro M, Mannarino MR, Bianconi V, Simental-Mendia LE, Bagaglia F, Mannarino E, Sahebkar A (2016) The effects of a nutraceutical combination on plasma lipids and glucose: a systematic review and meta-analysis of randomized controlled trials. Pharmacol Res 110:76–88. https://doi.org/10.1016/j.phrs.2016.04.021

Prangthip P, Kettawan A, Posuwan J, Okuno M, Okamoto T (2016) An improvement of oxidative stress in diabetic rats by Ubiquinone-10 and Ubiquinol-10 and bioavailability after short- and long-term coenzyme Q10 supplementation. J Diet Suppl 13(6):647–659. https://doi.org/10.3109/19390211.2016.1164788

Pung YF, Rocic P, Murphy MP, Smith RA, Hafemeister J, Ohanyan V, Guarini G, Yin L, Chilian WM (2012) Resolution of mitochondrial oxidative stress rescues coronary collateral growth in Zucker obese fatty rats. Arterioscler Thromb Vasc Biol 32(2):325–334. https://doi.org/10.1161/ATVBAHA.111.241802

Raygan F, Rezavandi Z, Dadkhah Tehrani S, Farrokhian A, Asemi Z (2016) The effects of coenzyme Q10 administration on glucose homeostasis parameters, lipid profiles, biomarkers of inflammation and oxidative stress in patients with metabolic syndrome. Eur J Nutr 55(8):2357–2364. https://doi.org/10.1007/s00394-015-1042-7

Ribeiro AS, Seixas R, Galvez JM, Climent V (2018) Cardiovascular risk factors: is the metabolic syndrome related to aging? Epidemiology in a Portuguese population. Diabetes Metab Syndr 12(6):885–891. https://doi.org/10.1016/j.dsx.2018.05.010

Roos V, Elmstahl S, Ingelsson E, Sundstrom J, Arnlov J, Lind L (2017) Metabolic syndrome development during aging with special reference to obesity without the metabolic syndrome. Metab Syndr Relat Disord 15(1):36–43. https://doi.org/10.1089/met.2016.0082

Ross D, Siegel D (2017) Functions of NQO1 in cellular protection and CoQ_{10} metabolism and its potential role as a redox sensitive molecular switch. Front Physiol 8:595. https://doi.org/10.3389/fphys.2017.00595

Saboori S, Rad EY, Mardani M, Khosroshahi MZ, Nouri Y, Falahi E (2019) Effect of Q10 supplementation on body weight and body mass index: a systematic review and meta-analysis of randomized controlled clinical trials. Diabetes Metab Syndr 13(2):1179–1185. https://doi.org/10.1016/j.dsx.2019.01.047

Sahebkar A, Simental-Mendia LE, Stefanutti C, Pirro M (2016) Supplementation with coenzyme Q10 reduces plasma lipoprotein(a) concentrations but not other lipid indices: a systematic review and meta-analysis. Pharmacol Res 105:198–209. https://doi.org/10.1016/j.phrs.2016.01.030

Salminen A, Ojala J, Kaarniranta K, Kauppinen A (2012) Mitochondrial dysfunction and oxidative stress activate inflammasomes: impact on the aging process and age-related diseases. Cell Mol Life Sci 69(18):2999–3013. https://doi.org/10.1007/s00018-012-0962-0

Sasanabe R, Banno K, Otake K, Hasegawa R, Usui K, Morita M, Shiomi T (2006) Metabolic syndrome in Japanese patients with obstructive sleep apnea syndrome. Hypertens Res 29(5):315–322. https://doi.org/10.1291/hypres.29.315

Senechal M, Bouchard DR, Dionne IJ, Brochu M (2012) Lifestyle habits and physical capacity in patients with moderate or severe metabolic syndrome. Metab Syndr Relat Disord 10(3):232–240. https://doi.org/10.1089/met.2011.0136

Sheu WH, Chuang SY, Lee WJ, Tsai ST, Chou P, Chen CH (2006) Predictors of incident diabetes, metabolic syndrome in middle-aged adults: a 10-year follow-up study from Kinmen, Taiwan. Diabetes Res Clin Pract 74(2):162–168. https://doi.org/10.1016/j.diabres.2006.03.011

Song QB, Zhao Y, Liu YQ, Zhang J, Xin SJ, Dong GH (2015) Sex difference in the prevalence of metabolic syndrome and cardiovascular-related risk factors in urban adults from 33 communities of China: the CHPSNE study. Diab Vasc Dis Res 12(3):189–198. https://doi.org/10.1177/1479164114562410

Strauss RS (1999) Comparison of serum concentrations of alpha-tocopherol and beta-carotene in a cross-sectional sample of obese and nonobese children (NHANES III). National Health and Nutrition Examination Survey. J Pediatr 134(2):160–165. https://doi.org/10.1016/s0022-3476(99)70409-9

Suksomboon N, Poolsup N, Juanak N (2015) Effects of coenzyme Q10 supplementation on metabolic profile in diabetes: a systematic review and meta-analysis. J Clin Pharm Ther 40(4):413–418. https://doi.org/10.1111/jcpt.12280

Tabrizi R, Akbari M, Sharifi N, Lankarani KB, Moosazadeh M, Kolahdooz F, Taghizadeh M, Asemi Z (2018) The effects of coenzyme Q10 supplementation on blood pressures among patients with metabolic diseases: a systematic review and meta-analysis of randomized controlled trials. High Blood Press Cardiovasc Prev 25(1):41–50. https://doi.org/10.1007/s40292-018-0247-2

Thomas SR, Neuzil J, Stocker R (1997) Inhibition of LDL oxidation by ubiquinol-10. A protective mechanism for coenzyme Q in atherogenesis? Mol Aspects Med 18 Suppl:S85–S103

Turi BC, Codogno JS, Fernandes RA, Monteiro HL (2016) Low levels of physical activity and metabolic syndrome: cross-sectional study in the Brazilian public health system. Cien Saude Colet 21(4):1043–1050. https://doi.org/10.1590/1413-81232015214.23042015

Watanabe H, Obata H, Watanabe T, Sasaki S, Nagai K, Aizawa Y (2010) Metabolic syndrome and risk of development of chronic kidney disease: the Niigata preventive medicine study. Diabetes Metab Res Rev 26(1):26–32. https://doi.org/10.1002/dmrr.1058

Xu Z, Huo J, Ding X, Yang M, Li L, Dai J, Hosoe K, Kubo H, Mori M, Higuchi K, Sawashita J (2017) Coenzyme Q10 improves lipid metabolism and ameliorates obesity by regulating CaMKII-mediated PDE4 inhibition. Sci Rep 7(1):8253. https://doi.org/10.1038/s41598-017-08899-7

Yokoyama H, Lingle DM, Crestanello JA, Kamelgard J, Kott BR, Momeni R, Millili J, Mortensen SA, Whitman GJ (1996) Coenzyme Q10 protects coronary endothelial function from ischemia reperfusion injury via an antioxidant effect. Surgery 120(2):189–196. https://doi.org/10.1016/s0039-6060(96)80287-x

Young JM, Florkowski CM, Molyneux SL, McEwan RG, Frampton CM, Nicholls MG, Scott RS, George PM (2012) A randomized, double-blind, placebo-controlled crossover study of coenzyme Q10 therapy in hypertensive patients with the metabolic syndrome. Am J Hypertens 25(2):261–270. https://doi.org/10.1038/ajh.2011.209

Zarei P, Rezvanfar MR, Ansarihadipour H, Delavar M, Abdollahi M, Khosrowbeygi A (2018) Effects of coenzyme Q10 supplementation on the serum levels of amylase, adenosine deaminase, catalase, and total antioxidant capacity in women with type 2 diabetes mellitus: a randomized, double-blind placebo-controlled trial. J Res Med Sci 23:91. https://doi.org/10.4103/jrms.JRMS_970_17

Zhang SY, Yang KL, Zeng LT, Wu XH, Huang HY (2018) Effectiveness of coenzyme Q10 supplementation for type 2 diabetes mellitus: a systematic review and meta-analysis. Int J Endocrinol 2018:6484839–6484811. https://doi.org/10.1155/2018/6484839

Zozina VI, Covantev S, Goroshko OA, Krasnykh LM, Kukes VG (2018) Coenzyme Q10 in cardiovascular and metabolic diseases: current state of the problem. Curr Cardiol Rev 14(3):164–174. https://doi.org/10.2174/1573403X14666180416115428

Chapter 12
Coenzyme Q and Age-Related Neurodegenerative Disorders: Parkinson and Alzheimer Diseases

Francisco J. Alcaín, Javier Domínguez, Mario Durán-Prado, and Julia Vaamonde

Abstract Parkinson's Disease (PD) and Alzheimer's Disease (AD) are the two most common neurodegenerative diseases in the elderly. Both are proteinopathies that interact with mitochondria, which generate reactive oxygen species (ROS) and leads to mitochondria-activated programmed neuronal death. Several lines of evidence suggest that oxidative stress and mitochondrial dysfunction play central roles in the onset and progression of both diseases. Redox status of coenzyme Q_{10} (CoQ) in the plasma or cerebrospinal fluid are altered in PD and AD patients; as such, neuroprotective strategies targeting mitochondria such as the use of a supplement containing CoQ have been proposed as treatment. Preclinal data in cellular and animal models have yielded promising results, including the protection of mitochondria from biochemical insults and inhibition of α-synuclein aggregation in dopaminergic neurons in PD models and a reduction of Aβ burden in the cortex and hippocampus in an AD model. However, CoQ failed to elicit therapeutic effects in humans, likely because patients received treatment at late stages, indicating that the complexities of human disease cannot be fully recapitulated by animal models.

Keywords Coenzyme Q · Parkinson's disease · Alzheimer's disease · Neurodegeneration · Redox · Reactive oxygen species

F. J. Alcaín (✉) · M. Durán-Prado
Departamento de Ciencias Médicas, Facultad de Medicina, Universidad de Castilla-La Mancha, Ciudad Real, Spain

Grupo de Estrés Oxidativo y Neurodegeneración, CRIB, Universidad de Castilla-La Mancha, Ciudad Real, Spain
e-mail: FranciscoJ.Alcain@uclm.es

J. Domínguez · J. Vaamonde
Departamento de Ciencias Médicas, Facultad de Medicina, Universidad de Castilla-La Mancha, Ciudad Real, Spain

Servicio de Neurología, Hospital General Universitario de Ciudad Real, Ciudad Real, Spain

G. López Lluch (ed.), *Coenzyme Q in Aging*,
https://doi.org/10.1007/978-3-030-45642-9_12

12.1 Introduction

Parkinson's disease (PD) and Alzheimer's disease (AD) are the two most common neurodegenerative diseases in the elderly. Both affect about 1% of adults over the age of 60 years (Samii et al. 2004), but the latter represents 50%–75% of all cases of dementia and the disease risk doubles every 5 years after the age of 65 (Kawas 2003). The two diseases have distinct physiopathology: PD is mainly a movement disorder, whereas AD is characterized by deterioration of cognitive function and memory. However, they also share many similarities. For instance, age is the main risk factor for both diseases (Kawas 2003; Samii et al. 2004), which progress according to a defined spatiotemporal pattern, affecting first the dorsal motor nucleus of the glossopharyngeal and vagal nerves and anterior olfactory nucleus in PD (Braak et al. 2003) and starting in the entorhinal region and progressing to limbic areas in AD (Braak and Braak 1991). Both are proteinopathies that result from aberrant protein folding, processing, or degradation—in PD, this is manifested as the formation of Lewy neurites and Lewy bodies by α-synuclein (Braak et al. 2003) and in AD, amyloid peptide (Aβ) and tau form senile plaques and neurofibrillary tangles (Braak and Braak 1991). α-Synuclein and Aβ oligomers are more toxic than their respective fibrillary forms (Gosavi et al. 2002; Walsh et al. 2002) and generate reactive oxygen species (ROS) that further induce the formation of oligomers (Abou-Sleiman et al. 2003; Di Carlo 2010). Both proteins interact with and disrupt the function of mitochondria, thereby altering their morphology and dynamics and perturbing ATP production and calcium homeostasis, which induces mitochondria-activated programmed cell death (Duran-Prado et al. 2014; Franco-Iborra et al. 2016; Reddy and Beal 2008).

Oxidative stress and mitochondrial dysfunction play a central role in the onset and progression of PD and AD. Various therapeutic strategies involving antioxidants have been proposed to treat both diseases. For example, coenzyme Q_{10} (CoQ)—which acts as an electron acceptor in mitochondrial complex (C)-I and C-II and as an antioxidant—has shown positive effects in cellular and animal models of PD and AD. In fact, the redox status of CoQ in the plasma and cerebrospinal fluid (CSF) is altered in PD and AD patients (Gotz et al. 2000; Isobe et al. 2009). In this chapter, we review the rationale for the use of CoQ in the treatment of age-related neurodegenerative diseases and the results of clinical trials.

12.2 Parkinson Disease

12.2.1 Clinical Features and Causes of PD

The classic symptoms of PD include bradykinesia, rigidity, and rest tremor caused from progressive loss of dopaminergic (DAergic) neurons in the substantia nigra pars compacta (SNpc) region of the midbrain. Moreover, some non-motor symptoms as constipations, olfactory dysfunction, sleep disorders, hallucinations and dementia are present in 60% to 97% of PD patients (Martinez-Fernandez et al. 2016). Symptomatic therapy can slow disease progression, but to date there is no cure for PD. Strategies to delay onset or slow progression of PD is an important consideration of overall treatment. A histopathological hallmark is the presence of α-synuclein aggregates that form Lewy bodies and neurites in DAergic neurons (Braak et al. 2003). Most cases of PD are sporadic; strong evidence indicates that exposure to environmental factors plays a significant role in disease etiology (Dick et al. 2007; Tanner et al. 2011). Several genes have been linked to the familial form of PD, which accounts for 5% of all cases. Five genes—including *SNCA* encoding α-synuclein—are known to cause dominant monogenic PD, whereas mutations in five other genes underlie an autosomal recessive form of the disease. Among these genes is phosphatase and tensin homolog-induced putative kinase (PINK)1, which encodes a mitochondrial protein that protects cells against oxidative stress, regulates mitochondrial bioenergetics by modulating C-I activity, and promotes mitophagy of depolarized mitochondria (Ferreira and Massano 2016; Voigt et al. 2016). In addition, mitochondrial (mt)DNA deletion mutation rates have been found to be higher in DAergic neurons of PD patients as compared to those of healthy individuals (Chaturvedi and Flint Beal 2013).

12.2.2 Mechanisms of Neurodegeneration in PD

The first evidence of mitochondrial involvement in PD pathogenesis came from an illicit drug containing 1 methyl-4-phenyl-1,2,5,6-tetrahydropyridine (MPTP) that induced acute and irreversible parkinsonian syndrome in users (Langston et al. 1983). A metabolite of this compound, the 1-methyl-4-phenyl-pyridine was (MPP+), directly inhibits C-I. Monoamine oxidase (MAO) B present in glia surrounding DAergic neurons generates MPP+ from MPTP (Speciale 2002), which is released into the extracellular space and is taken up by DAergic neurons, and then mitochondria rapidly concentrates MPP+ to very high concentrations (millimolar) from the micromolar external concentrations, thus where it inhibits mitochondrial nicotinamide adenine dinucleotide (NADH) dehydrogenase (Ramsay and Singer 1986). MPP+ and rotenone, another C-I inhibitor, was shown to induce apoptosis and necrosis at low and high concentrations, respectively, thereby depleting ATP while increasing superoxide and ROS production via NADH activation (Hartley et al. 1994; Zawada et al. 2011). The role of mitochondrial dysfunction in the pathogenesis of PD was supported by the finding that C-I activity was decreased by 50% in platelet mitochondria purified from 10 patients with idiopathic PD (Parker et al. 1989). C-I activity was also found to be reduced in homogenates of the SNpc of PD patients relative to controls, whereas C-II and -III activities were normal (Gatt et al. 2016), suggesting a link between MPTP and sporadic PD (Schapira et al. 1989). Another study reported that C-II to -V levels in mitochondria were similar between PD patients and controls; however, C-I showed reduced expression and contained more oxidized catalytic subunits in PD, which was correlated with complex misassembly and dysfunction (Keeney et al. 2006). Furthermore, some C-I subunits were moderately downregulated in the striatum in 4 out of 5 patients with PD (Mizuno et al. 1989) and mitochondrial activity in the frontal cortex was reduced in patients (Parker et al. 2008). However, it was recently reported that C-I activity and mtDNA levels were reduced in the frontal cortex in PD with dementia but not in cases without dementia, suggesting C-I as a potential therapeutic target for treating or delaying dementia onset in PD patients (Gatt et al. 2016). Low activity in C-I, -II, and -IV have also been reported in skeletal muscle mitochondria of untreated and L-3,4-dihydroxyphenylalanine (L-DOPA)-treated PD patients (Bindoff et al. 1991; Blin et al. 1994).

MPTP was found to induce the DAergic degeneration in the same nuclei in a Parkinsonism mouse model that occurs in humans with PD (German et al. 1996), whereas rotenone—a highly lipophilic mitochondrial C-I inhibitor that passes through the blood–brain barrier—showed selective toxicity in DAergic neurons recapitulating the parkinsonism symptoms of PD in animal models, including bradykinesia and rigidity (Greenamyre et al. 2001); even a sublethal dose enhanced the toxicity of L-DOPA in these cells (Nakao et al. 1997; Pardo et al. 1995). Moreover, neurons in the affected brain region developed inclusions containing α-synuclein and ubiquitin; DAergic neuronal death increased the levels of oxidized proteins and caused glutathione depletion in the SNpc (Greenamyre et al. 2001). Rotenone and paraquat, which are used as pesticides or herbicides, can cross the blood–brain barrier and inhibit C-I, inducing parkinsonism in animal models; it has also been linked to PD in humans, even when exposures were truncated at 5, 10, or 15 years before the diagnosis. This effect was independent of race/ethnicity, cigarette use, state, or duration of disease (Tanner et al. 2011).

12.2.3 Consequences of C-I Dysfunction

Mitochondrial dysfunction increases oxidative stress, which is defined as a net increase in the levels of ROS, and the main source of ROS is the electron transport chain. Most O_2^- in mammalian mitochondria is produced by C-I. O_2^- dismutation by hydrogen peroxide yields H_2O_2, which passes through mitochondrial membranes into the cytosol. MAO located in the outer mitochondrial membrane is another important source of oxidative stress. Increases in ROS levels lead to oxidative modification of lipids, DNA, and proteins (Orrenius et al. 2007). Lipid peroxidation, oxidative DNA damage, and oxidation of C-I catalytic subunits in the SNpc have been reported in the PD brain (Alam et al. 1997; Dexter et al. 1986; Keeney et al. 2006), suggesting that dysregulation of bioenergetics due to oxidative stress leads to the death of SNpc neurons in PD (Jenner et al. 1992).

The functional consequences of the C-I defect in PD was studied using cybrid cells. Platelets islolated from PD patients with mild to moderately advanced symptoms receiving L-DOPA therapy or from controls were fused with SH-SY5Y human neuroblastoma cells lacking mtDNA (i.e., ρ^{0} cells) to obtain cybrids cells. Those that were generated from PD patient platelets produced more oxygen radicals and showed greater sensitivity to the toxicity of MPP+ than control cybrids (Swerdlow et al. 1996), and had higher levels of ROS in the cytosol, which was associated with increased α-synuclein oligomerization showing an increase in carbonyl groups in PD cybrids compared to control, linking C-I mitochondrial dysfunction to α-synuclein oligomerization (Esteves et al. 2009). Furthermore, the expression of the transcription factor peroxisome proliferator-activated rgeceptor co-activator (PGC)-1α—a master regulator of mitochondrial biogenesis, function, and dynamics that controls mitochondrial size and number—was downregulated while that of nuclear factor (NF)-κB was upregulated in PD cybrids (Esteves et al. 2010). NF-κB is activated by ROS and is localized in the nuclei of DAergic neurons in PD patients (Hunot et al. 1997). PGC-1α overexpression mitigated the loss of DAergic neurons induced by α-synuclein and rotenone in the A53T and MTPT PD mouse models (Mudo et al. 2012; Zheng et al. 2010) and prevented α-synuclein oligomerization (Eschbach et al. 2015).

Genetically or pharmacologically induced C-I dysfunction decreases mitochondrial membrane potential ($\Delta\psi m$), which drives ATP production (Fontaine et al. 1998; Morais et al. 2014). A reduction in $\Delta\psi m$ in PINK1$^{-/-}$ cells is associated with increased opening of the mitochondrial permeability transition pore in primary mouse embryonic fibroblasts (Gautier et al. 2012), as well as mitochondrial swelling and rupture of the outer mitochondrial membrane, leading to the release of pro-apoptotic molecules such as cytochrome c into the cytosol (Rao et al. 2014). Pharmacological inhibition of C-I by MPTP has also been shown to stimulate cytochrome c release induced by B cell lymphoma (Bcl)-2-associated X protein (Bax) (Perier et al. 2005).

12.2.4 Role of CoQ Supplementation in Cellular Models of PD

Antioxidants may protect DAergic neurons in the SNpc against cell death caused by free radicals originating from dysfunctional C-I. DAergic neurons in the SNpc are thought to be more sensitive to oxidative stress than other neurons; treatment of rat embryonic mesencephalic neuron cultures with a low concentration of rotenone induced mitochondrial membrane depolarization and death in DAergic but not other types of neuron from the ventral mesencephalon, an effect that was abolished by pretreatment with CoQ (Moon et al. 2005). CoQ also preserved mitochondrial C-I and tyrosine hydroxylase activity in primary mesencephalic cultures treated with MPTP and L-DOPA, thereby improving neuronal survival (Gille et al. 2004). An increase in ROS levels can lead to opening of the mitochondrial permeability transition pore and the inhibition of opening halts the apoptosis execution-related events, for that it has been hypothesized that the pretreatment with CoQ could prevent apoptosis independently of its free radical scavenging ability but related to direct inhibition of mitochondrial permeability transition pore opening because other free radical scavenger such as N-acetylcysteine glutathione and vitamin C did not (Moon et al. 2005; Papucci et al. 2003). Ubisol-CoQ_{10}, a water-soluble form of CoQ, was also found to block mitochondrial permeability transition pore opening (Kumari et al. 2016). CoQ prevented mitochondrial depolarization and inhibited cytochrome c release into the cytosol by directly suppressing Bax-induced mitochondrial dysfunction in isolated mitochondria from two human cell lines (Naderi et al. 2006). CoQ-treated primary mesencephalic cultures from mouse embryos reversed the decrease in hexokinase activity induced by MPTP, thereby increasing the levels of metabolites required for the Krebs cycle and ATP production (Gille et al. 2004), overcoming the bioenergetics deficit found in DA neurons.

In C-I, electrons are donated to CoQ for transport to ubiquinol:cytochrome-c oxidoreductase; however, CoQ is itself and antioxidant that can scavenge ROS in mitochondria. CoQ administration increased electron transport chain activity both *in vitro* and *in vivo* (Ohhara et al. 1981; Schneider et al. 1982). Thus, combined treatment with CoQ and nicotinamide partly blocked MPTP-induced mild and moderate depletion of DA in a mouse model of PD, but not when resulted in severe dopamine depletions of approximately 80% (Schulz et al. 1995). CoQ level in platelet mitochondria isolated from PD patients was lower than that in age−/sex-matched control subjects and was correlated with C-I activity (Shults et al. 1997). Additionally, abnormal mitochondrial function was also observed in fibroblasts from PD patients along with a mild deficiency in both C-I and -IV activities; the latter was restored by treatment with CoQ (Winkler-Stuck et al. 2004).

CoQ has been shown to exert neuroprotective effects in cellular models of PD. In human neuroblastoma SH-SY5Y cells, paraquat induced oxidative stress and neuronal death; however, pretreatment with water-soluble CoQ protected cells against oxidative stress as well as decreases in $\Delta\psi m$ and ATP production and paraquat-induced apoptosis (McCarthy et al. 2004). Moreover, oxidative stress induced by H_2O_2 resulted in a reduction of $\Delta\psi m$ leading to mitochondrial dysfunction, which was abrogated by CoQ (Somayajulu et al. 2005). Similar to the effects in PD cybrids (Esteves et al. 2009), oxidative stress resulting from treatment with rotenone or from L-buthionine sulfoxamine-induced glutathione depletion in SH-SY5Y cells was correlated with an increase in α-synuclein expression, which was abolished by co-treatment with CoQ and glutathione (Esteves et al. 2009; Shavali et al. 2004).

Total iron content in the SNpc is increased in PD patients, which causes oxidative stress-induced cell death and α-synuclein deposition (Kooncumchoo et al. 2006). In cultured human dopaminergic neurons SK-N-SH, iron-induced mitochondrial damage resulted in CoQ depletion, ROS production, NF-κB activation, and Bcl-2 downregulation, effects that were abolished by pretreatment with CoQ and deferoxamine, a potent iron chelator (Kooncumchoo et al. 2006). Chronic exposure of these cells to rotenone also decreased CoQ levels, C-I activity, and $\Delta\psi m$, effects that were reversed by CoQ treatment (Sharma et al. 2006).

12.2.5 Role of CoQ Supplementation in Animal Models of PD

Animal models of PD are generated by neurotoxic chemicals—including 6-hydroxydopamine, MPTP, and agricultural herbicides such as rotenone and paraquat—that selectively affect DAergic neurons in the SNpc system. These models recapitulate the classic features of PD such as C-I inhibition, oxidative stress, iron accumulation in SNpc, degeneration of and Lewy body formation in DAergic neurons, and α-synuclein phosphorylation, although the neurodegeneration progress rapidly and not always presented α-synuclein aggregates (Antony et al. 2011; Blesa et al. 2012; Le et al. 2014). CoQ levels decline with age and are downregulated in neurodegenerative diseases (Lopez-Lluch et al. 2010). Oral administration of CoQ at 200 mg/kg/day in rats increased mitochondrial concentrations of CoQ in the cerebral cortex and exerted neuroprotective effects in both, a pharmacological rat model and a transgenic mouse model of familial ALS (Matthews et al. 1998). CoQ (200 mg/kg/day) administration in a mouse model of MPTP-induced PD increased the striatal level of DA and the density of tyrosine hydroxylase-positive fibers in the caudal striatum in aged mice (Beal et al. 1998); oral CoQ supplementation was more effective in aged as compared to young animals (Schulz et al. 1995). CoQ also showed additive effects with creatine in attenuating neuronal degeneration and reducing lipid peroxidation and α-synuclein aggregation in SNpc neurons of MPTP-treated mice (Yang et al. 2009).

In an acute or semichronic MPTP model of Parkinsonism, oral administration of CoQ either as ubiquinone or ubiquinol (oxidized and reduced forms, respectively) mitigated the loss of DAergic neurons and formation of cytosolic α-synuclein aggregates in SNpc. Reduced CoQ was found to exert more potent protective effects than the oxidized form (Cleren et al. 2008). Chronic exposure to the organophosphate pesticide dichlorvos caused nigrostriatal DA degeneration and α-synuclein aggregation and reduced mitochondrial C-I and -IV activities in rats; CoQ supplementation protected against these effects and reversed dichlorvos-induced catalepsy and motor dysfunction (Binukumar et al. 2011).

CoQ is insoluble in water and is therefore poorly absorbed, and the maximum concentration increases nonlinearly with dose; as such, various formulations have been developed to increase CoQ solubility and absorption (Villalba et al. 2010). The water-soluble form ubisol-Q_{10} has greater bioavailability and contains two potent antioxidants (CoQ and a derivatized form of α-tocopherol, i.e., vitamin E) and can cross the brain–blood barrier, resulting in an increase in CoQ level by 30%–50% 3 h after administration (Muthukumaran et al. 2014; Sikorska et al. 2014). This CoQ formulation has been shown to be neuroprotective when used prophylactically and also had therapeutic effects in two pharmacological models of PD (MPTP in mice and paraquat in rats). When the formulation was administrated in the drinking water after the DA neurodegeneration started by MPTP or paraquat injection was able to protect the remains DA neurons form cell death, which otherwise would be dyed. Ubisol-CoQ_{10} also improved motor performance in these animals. However, continuous supplementation was required to prevent the resumption of neuronal death (Muthukumaran et al. 2014; Sikorska et al. 2014).

In the last decade, several genetic models of PD have been generated in mice, including those overexpressing dominant mutations in α-synuclein, carrying recessive genes known to cause PD in humans, and knockout lines. In the two most widely used models (A53T and A30P), α-synuclein transgenes with mutations associated with familial PD recapitulate the early motor and autonomic dysfunction and olfactory impairment that characterize the disease. However, neither of these models exhibit loss of DAergic neurons in the SNpc (Antony et al. 2011; Blesa et al. 2012; Le et al. 2014) and, in our knowledge, they have not been used to examine the effects of CoQ supplementation on PD progression.

12.2.6 CoQ as Biomarker for PD

Reductions in C-I and -II/III activities in PD patients have been correlated with lower levels of CoQ (Shults et al. 1997). Moreover, levels of the reduced form were lower in platelet mitochondria from PD patients; total CoQ level was also lower and the ratio of oxidized to reduced CoQ (%CoQ) was higher in PD (Gorgone et al. 2012; Gotz et al. 2000; Sohmiya et al. 2004), suggesting a systemic increase in oxidative stress since reduced CoQ is a more efficient antioxidant than the oxidized form (Crane 2001). However, another study reported that CoQ levels did not differ between PD patients and controls, but was correlated with disease duration and motor score (Jimenez-Jimenez et al. 2000). A screening for oxidative stress markers in PD revealed lower levels of CoQ and α-tocopherol and higher levels of oxidative stress markers in the plasma and CSF of PD patients (Buhmann et al. 2004), while CoQ deficiency was observed at a higher frequency in PD patients, underscoring its utility as a peripheral biomarker (Mischley et al. 2012). However, no differences were found between patients and controls in terms of the levels of α-tocopherol, which is reduced from α-tocopheroxyl radical by CoQ (Lopez-Lluch et al. 2010). The % CoQ in the CSF was also found to be higher in PD as compared to control subjects, and was positively correlated with disease duration and level of oxidative stress (Isobe et al. 2010a).

Human tissue contains measurable amounts of CoQ, but the levels in the brain are low, and about 80% of the total content is in the oxidized form (Aberg et al. 1992). CoQ concentration also varies according to brain region, with low levels in the SNpc and cerebellum and higher levels in the cortex and striatum (Hargreaves et al. 2008). CoQ levels were found to be reduced in the cortex of PD patients relative to that of control subjects (Hargreaves et al. 2008).

12.2.7 CoQ Treatment in PD: Clinical Trials

Neuroprotection preserves nervous system structure and function (Schapira and Olanow 2004), and agents that can protect the PD patient brain against disease progression remains a major focus of research. A significant limitation in studies of PD patients has been the lack of a generally accepted surrogate endpoint that neuronal loss; validated endpoints need to be developed. Accurate early diagnosis and improved knowledge of disease progression will facilitate clinical trials of potential neuroprotective agents, since at the time of clinical diagnosis this has already progressed considerably. Therefore, methods for identifying presymptomatic patients for clinical trials are needed, and trials that include long-term follow-up must be implemented to obtain conclusive evidence of the neuroprotective effects of a drug.

Mitochondrial dysfunction is involved in the pathogenesis of PD and is a potential therapeutic target for halting the progressive degeneration of DAergic neurons in PD (Abou-Sleiman et al. 2003; Perier and Vila 2012). Many studies have reported low COQ or high oxidized COQ levels in PD patients relative to controls (Buhmann et al. 2004; Gorgone et al. 2012; Gotz et al. 2000; Isobe et al. 2009; Mischley et al. 2012; Shults et al. 1997; Sohmiya et al. 2004). Moreover, CoQ administration was shown to preserve mitochondrial function and to reduce the loss of DAergic neurons in preclinical PD models (Schulz et al. 1995), indicating that it can be effective for treating PD patients.

A multicenter, randomized, parallel-group, placebo-controlled, doubled-blind clinical trial of 80 patients with early PD who did not require treatment for their disability was carried out to determine whether a range of CoQ dosages (300/ 600/ 1200 mg/day) was safe and well tolerated and could slow functional decline (Shults et al. 2002). The primary response variable was a change in total Unified PD Rating Scale (UPDRS) score. Patients who received CoQ_{10} showed less disability than placebo-treated controls; although the difference was not statistically significant, the results suggested that CoQ slowed functional decline in PD and was safe at dosages of up to 1200 mg/day. In CoQ-treated patients, plasma levels of CoQ were increased, reaching about 4.6 μM in patients receiving the 1200 mg/day dosage, along with the combined activities in mitochondrial platelets of C-I and C-III, which are dependent on endogenous CoQ. In 2006, the Quality Standards Subcommittee of the American Academy of Neurology defined key issues in the management of PD relating to neuroprotective strategies and alternative treatments and made evidence-based treatment recommendations (Suchowersky et al. 2006); it was also concluded that the study by Shults et al. (Shults et al. 2002) was designed to determine safety and tolerability in the dose range of 300–1200 mg/day, and did not demonstrate a clear neuroprotective benefit. On the other hand, in a review of four randomized, double-blind, placebo-controlled trials with a total of 452 patients that compared CoQ to placebo for patients who suffered early and midstage were reported improvements in activities of daily living, UPDRS, and Schwab and England scores in patients treated with CoQ at 1200 mg/day for 16 months as compared to placebo groups (Liu et al. 2011). More recently, the Parkinson Study Group

QE3 examined again whether administration CoQ in the early stages of PD (i.e., who had been diagnosed within 5 years and were not expected to require DA therapy for at least 3 months) could slow disease progression. Participants (600) were randomly to receive placebo, 1200 mg/d of CoQ_{10}, or 2400 mg/d of CoQ_{10}; all participants received 1200 UI/D of vitamin E. Participants were observed for 16 months or until a disability requiring dopaminergic treatment. Change in total UPDRS score from baseline to final visit was evaluated. Conclusions were found that CoQ was safe and well tolerated but did not have specific clinical benefits (Beal et al. 2014).

An open-label, dose-escalation study that examined the effects of CoQ (400–1400 mg/day) on oxidative stress and clinical outcomes in 16 patients with early PD. Each dose (400, 800, 1200 and 1400 mg/day) was consumed daily for 2 weeks. They found that CoQ was well tolerated, but only higher doses (1200 and 1400 mg/day) increased the ratio of ubiquinol to ubiquinone and was associated with improved total UPDRS scores. Subjects showing improvement in symptoms following treatment had lower baseline levels of plasma ubiquinol and decreased F2-isoprostane per unit arachidonate, suggesting that therapeutic response to CoQ depends on baseline levels of ubiquinol and whether oxidative damage is alleviated by the administered dosage of CoQ (Seet et al. 2014). The plasma concentrations of other antioxidants were similar between PD patients and controls, and there were no deficiencies in vitamin E, selenium, lipoic acid, or glutathione levels (Mischley et al. 2012), indicating that antioxidant status in PD can be a tool to select patients that respond more favorably to antioxidant therapy.

The effectiveness of CoQ is determined to a greater extent by its redox state than by the total CoQ amount. One function of ubiquinol is to recycle α-tocopheroxyl radical to its reduced form (α-tocopherol) along with other antioxidants in plasma and membranes.(Lopez-Lluch et al. 2010). Intestinal absorption also depends on CoQ redox status. Ubiquinone, the oxidized form of CoQ, is reduced to ubiquinol as it is absorbed; maximum concentration increased nonlinearly with dose and reached a steady state 2 weeks after the start of treatment (2400 mg/day) (Bhagavan and Chopra 2007; Bhagavan et al. 2007). Ubiquinol, the reduced form of CoQ, showed superior absorption. When healthy human received a supplementation of 200 m/day of ubiquinone or ubiquinol, the baseline increased from about 1 μm/L to 3 μm/L using ubiquinone, up to 5.1 μm/L using ubiquinol and the ratio ubiquinol total CoQ only increased significantly in ubiquinol supplementation.(Langsjoen and Langsjoen 2014). CoQ concentrations that are higher than normal plasma levels are necessary for it to reach peripheral tissues and cross the blood–brain barrier. The safety and clinical effects of orally administered ubiquinol (300 mg/day) were evaluated in a randomized, double-blind, placebo-controlled pilot trial of 31 Japanese patients with PD. Ubiquinol was found to be safe and well tolerated, and total UPDRS score improved relative to controls (4.2 ± 8.2 for the ubiquinol group and 2.9 ± 8.9 for thr placebo) and also were significant differences in subtotal scores for finger and hand movements during "off" phase. There were no significant differences in other motor symptoms such as tremor, rigidity, foot tap or arising from a chair (Yoritaka et al. 2015). Importantly, the improvement disappeared 8 weeks post-treatment, likely due to the return of plasma CoQ to basal levels (Hosoe et al. 2007).

Autopsies and animal studies have shown that many factors contribute to neuronal death in PD, including oxidative stress, mitochondrial dysfunction, inflammation, and excitotoxicity and apoptosis. Therapies that are developed to slow disease progression must target these pathogenic mechanisms; in addition, a major goal in PD research is to determine the precise cause of neuronal death in PD and identify specific molecular targets. The development of drugs that might modify disease progression is one of the most critical goals in PD research.

12.3 Alzheimer's Disease

12.3.1 Clinical Features and Causes of AD

Although in rare cases AD can occur before the age of 60, it is generally a disease associated with old age.The incidence and prevalence of AD increase exponentially with age, essentially doubling in prevalence every 5 years after the age of 65 years (Knopman et al. 2001). The most essential and often earliest clinical manifestation of AD is selective memory impairment, although there are exceptions. While some treatments can ameliorate symptoms of AD, there is presently no cure, and the disease inevitably progresses in all patients. In addition to memory, AD patients exhibit deficits in other cognitive domains such as executive dysfunction and visuospatial impairment, which are often observed at relatively early stages, as well as language and behavioral symptoms that are manifested later in the disease course (Dubois et al. 2007).

The neuropathological features of AD consist of a gradual loss of neurons and their synapses along with accumulation of senile plaques and tangles, formed by Aβ peptide and hyperphosphorylated tau protein deposits, respectively (Swerdlow 2007). Like PD, about 1% of AD cases have a familial component whereas 99% are sporadic and have a late onset. Many factors contribute to the disease, which can develop silently for decades before clinical symptoms become apparent. The amyloid cascade hypothesis synthesizes histopathological and genetic information and is the hypothesis central proposed to explain the onset and progression of the disease (Karran et al. 2011). Autosomal dominant mutations that cause early-onset familial AD affect three genes involved in the proteolytic processing of amyloid precursor protein (APP), including APP and proteases presenilin (PSEN)1 and PSEN2 forming the catalytic core of the γ-secretase complex. Alterations in the processing of APP by PSEN1 and PSEN2 results in the production of pathological APP aggregates (Goate et al. 1991; Levy-Lahad et al. 1995; Rogaev et al. 1995; Sherrington et al. 1995).

Under normal conditions, APP processing starts with the generation of a carboxy terminal fragment (CTF83) by α-secretase and later by the γ-secretase forming a non-amylogenic peptide. APP can also be cleaved first by the β-secretase complex (BACE1)—which generates the CTF99 fragment—and later by γ-secretase, which produces the Aβ peptide (Chow et al. 2010). Mutations in APP can increase the rate of proteolysis by BACE1; additionally, mutations in PSEN1 and PSEN2 that cause the formation of longer Aβ peptides can lead to accumulation of toxic Aβ species (Karran et al. 2011).

In the healthy brain, Aβ is cleared by CSF and peripheral blood vessels. A major risk factor for sporadic AD is homozygosity for APOE4 alleles, given that this protein interferes with Aβ clearance, lowering the age of the onset of AD (Meyer et al. 1998). The accumulation of Aβ in the brain leads to the formation of senile plaques and consequently cognitive dysfunction, although this progression is non-linear with time whereas neurofibrillary tangles exhibited a regular distribution (Braak and Braak 1991). Neuronal loss is correlated with the development of neurofibrillary tangles whereas amyloid burden were not related to neuronal loss (Gomez-Isla et al. 1997), but amyloid aggregation precedes in the time an can trigger tau pathology (Karran et al. 2011). Neurotoxicity is mediated to a greater extent by soluble forms of Aβ as compared to aggregates (Karran et al. 2011; Scheltens et al. 2016). In the familial AD, the accumulation of Aβ is caused by autosomal dominant mutations affecting mainly to presenilin genes that originate an overproduction of the amyloid that cannot been cleared form brain. The causes of sporadic AD are unknown but oxidative stress associated with aging induces a conformational change in PSEN1 to a pathogenic conformation in the mouse model of AD Tg2576 that alters APP processing (Wahlster et al. 2013). Oxidative damage is higher during early stages of AD and declines with disease progression (Nunomura et al. 2001), and has been shown to induce BACE1 expression and activity in vitro (Tamagno et al. 2002).

12.3.2 Mechanisms of Neurodegeneration in AD

Aβ generates oxidative stress in the presence of iron and copper ions, leading to neuronal membrane lipid peroxidation that affects ceramide and cholesterol metabolism, and to the formation of 4 hydroxynonenal (4HNE), a neurotoxic hydroperoxide that oxidizes neprilysin, the main peptidase that degrades Aβ (Cutler et al. 2004; Jana and Pahan 2004). Neprilysin levels were found to be downregulated in the mid-temporal gyrus of patients exhibiting mild cognitive impairment relative to healthy control subjects (Huang et al. 2012). Aβ is taken up by cells and targeted to mitochondria where it affects outer and inner membrane functions and generates oxidative stress and induces mitochondrial permeability transition pore opening, leading to apoptosis (Cha et al. 2012; Duran-Prado et al. 2014; Hansson Petersen et al. 2008). Moreover, oligomeric Aβ induces glycogen synthase 3 kinase expression in cultured neurons and the brains of APP/tau-overexpressing mice and AD patients; this results in tau hyperphosphorylation and microtubule disruption, impairment of memory and long-term potentiation (LTP), and neuronal apoptosis (DaRocha-Souto et al. 2012; Hernandez et al. 2002).

When the neuronal Aβ production exceeds the safe limit and toxic Aβ oligomers accumulate above a threshold level, they are taken up by the astrocytes enveloping neuronal processes via Aβ oligomer-binding acetylcholine receptors (Dal Pra et al. 2015), leading to the release of glutamate that is internalized by neuronal synapses (Talantova et al. 2013). The discharged glutamate activates extrasynaptic N-methyl-D-aspartate receptors (NMDA) and the resultant Ca^{2+} influx generates toxic levels of nitric oxide that induce synaptic depression in hippocampal slices and lead to the loss of synaptic spines (Talantova et al. 2013).

12.3.3 AD and Endothelial Dysfunction

Aβ oligomers are cleared from the brain via several pathways, including proteolytic degradation by neprilysin and insulin-degrading enzyme, uptake by astrocytes and microglia, passive flow into the CSF, and sequestration in vascular compartments by low-density lipoprotein receptor-related protein 1 (Braak and Del Tredici 2011; Yasojima et al. 2001). Aβ is toxic to neurons but also to endothelial cells; recent evidence suggests that cerebrovascular dysregulation also contributes to the onset and progression of neurodegenerative diseases including AD because major cardiovascular risk are also mayor risk for AD (Grammas 2011; Iadecola 2004; Karran et al. 2011). Furthermore, Aβ deposited in cerebral blood vessels can cause cerebral amyloid angiopathy.

Oxidative stress in cerebral blood vessels precedes the oxidative stress in brain and the formation of amyloid plaques in animal models of AD (Park et al. 2004). Aβ exerts its deleterious effects in endothelial cells via the activation of nicotinamide adenine dinucleotide phosphate (NADPH) oxidase (Park et al. 2005) reduces cerebral flow producing hypoperfusion into the brain parenchyma and increases the expression of hypoxia inducible factor 1-α, which in turn leads to upregulation of BACE1 (Guglielmotto et al. 2009; Zhang et al. 2007). Reduced cerebral blood flow induces BACE1 expression in neurons, leading to the production of Aβ and increased oxidative stress, which causes endothelial cell damage (Chami and Checler 2012). Cerebrovascular volume was found to be reduced in the 3xTg-AD mouse model of AD (Bourasset et al. 2009) and in humans, a decrease in cerebral blood flow has been observed prior to the onset of dementia (Iadecola 2004; Ruitenberg et al. 2005).

Aβ can cause damage to endothelial cells in the cerebral microvasculature via mechanisms that involve increases in oxidative stress and calcium entry that induce necrosis and apoptosis (Duran-Prado et al. 2014). Aβ along with other vascular risk factors such as high glucose, hypertension, and low-density lipoprotein (LDL) oxidation exerts deleterious effects in endothelial cells via NADPH oxidase activation; CoQ can protect against cell damage by blocking Aβ translocation to mitochondria and inhibiting NADH activation (Duran-Prado et al. 2014; Park et al. 2005; Tsai et al. 2011; Tsai et al. 2012; Tsuneki et al. 2007; Tsuneki et al. 2013). CoQ can also prevent LDL oxidation in vitro (Ahmadvand et al. 2013); this may reduce the risk of atherosclerosis, which has been linked to AD. Dietary intake of CoQ and vitamin E has been shown to be anti-atherogenic in apolipoprotein E knockout mice (Thomas et al. 2001). Different isoforms of NADPH oxidase exist in neurons, astrocytes, and microglia that are activated by the same stimuli that act on endothelial NADPH oxidase and Aβ can cause NADPH oxidase-dependent ROS production in neurons (Hernandes and Britto 2012; Sorce and Krause 2009). However, whether CoQ inhibits NADPH oxidase activation in neurons have to be tested. A recent review of published literature dealing on CoQ treatment of cardiovascular disorders of ageing concluded that CoQ is an effective antihypertensive agent that acts by suppressing oxidized LDL-induced endothelial dysfunction via reduction of plasma inflammatory marker levels in patients with diabetes mellitus (Yang et al. 2015).

12.3.4 Role of CoQ Supplementation in Cellular and Animal Models of AD

The amyloid cascade hypothesis postulates that Aβ slowly accumulates in the brain parenchyma over many years until a threshold is reached and before manifestation of clinical features of AD; thus, even a modest suppression in Aβ production can delay AD onset (Karran et al. 2011). Given that AD is a neurodegenerative disease in which oxidative stress occurs early on and causes Aβ accumulation in mitochondria, neuroprotective strategies that target mitochondria such as CoQ supplementation have been proposed as treatment. Indeed, CoQ_{10} protected against Aβ-induced dysfunction of mitochondria isolated from the brains of aging diabetic rats, which inhibited Aβ-induced increases in hydrogen peroxide levels (Moreira et al. 2005). BACE1 activation in NT2 neurons is mediated by oxidative stress (Tamagno et al. 2002), and Aβ accumulation in mitochondria of neurons and endothelial cells induced an increase in ROS levels and mitochondrial dysfunction including mitochondrial permeability transition pore opening and cytochrome c release, leading to cell death (Cha et al. 2012; Duran-Prado et al. 2014; Hansson Petersen et al. 2008). Aβ-mediated induction of oxidative stress is partly due to hydrogen peroxide formation (Behl et al. 1994); this leads to neuronal apoptosis involving downregulation of Bcl-2 and upregulation of Bax protein levels, as well as release of cytochrome c into the cytosol (Tamagno et al. 2003). CoQ pretreatment for a week prior to ischemia was shown to protect hippocampal CA1 neurons from cell death by promoting Bcl-2 and suppressing Bax expression (Zamani et al. 2012). Bcl-2 is located in mitochondria and can prevent oxidative stress locally induced by Aβ (Bruce-Keller et al. 1998). CoQ was shown to protect neural stem cells against hypoxia-reperfusion by increasing the levels of survival-related factors including Bcl-2 and reducing the expression of cell death-related proteins such as caspase-3 (Park et al. 2012). The oxidative stress-induced activation of the apoptotic pathway associated with Aβ has been linked to accumulation of ceramide and cholesterol in the plasma membrane, which involves activation of Mg^{2+}-dependent sphingomyelinase (Cutler et al. 2004); this was found to be inhibited in CoQ-enriched membranes (Bello et al. 2005). In isolated brain mitochondria, Aβ-stimulated hydrogen peroxide production decreased respiratory control ratio and ATP levels, effects that were abrogated by CoQ treatment (Moreira et al. 2005). The decrease in mitochondrial membrane potential in neurons caused by ROS was also prevented by CoQ treatment (Somayajulu et al. 2005), while suppressing ROS formation in addition to CoQ treatment activated phosphoinositide 3-kinase-regulated survival proteins and prevented apoptosis of Aβ-injured neurons (Choi et al. 2012; Choi et al. 2013).

CoQ has been shown to decrease amyloid deposition in animal models of AD. In L235P PSEN-1 mutant mice, increased oxidative stress in the cortex was mitigated by treatment with CoQ_{10} (1200 mg/kg/day) for 60 days, while Aβ deposition was decreased by about 23% (Yang et al. 2008). CoQ also reduced Aβ burden in the cortex and hippocampus of APP/PS1 double transgenic mice (Yang et al. 2010) and prevented atrophy in hemispheres and hippocampus in aging transgenic mice (Li et al. 2008). Importantly, the reduced oxidative stress and Aβ burden were correlated with behavioral improvements in the Tg19959 mouse model (Dumont et al. 2011). Furthemore, both the hypoxia and plaque burden, which colocalize, were abolished by CoQred plus ascorbate supplementation since prodromal stages of disease in the 3xTg-AD mice model. Also, the thickening in the cerebrovascular basement membrane was avoided (Frontiñan-Rubio et al. 2018).

Aβ injury is partially mediated by the release of excessive amounts of glutamate from astrocytes that activate NMDA receptors involved in neuronal plasticity (Talantova et al. 2013). The activation of NMDA receptors induced a rapid increase in superoxide production via NADPH oxidase complex activation (Brennan et al. 2009). The 4-aminopyridine-induced release by glutamate from rat cerebrocortical nerve terminals was shown to be inhibited by CoQ in a concentration-dependent manner (Chang et al. 2012).

Aβ can also directly activate NADPH oxidase in endothelial cells, an effect that is abrogated by CoQ supplementation (Tsai et al. 2011; Tsai et al. 2012; Tsuneki et al. 2007; Tsuneki et al. 2013). However, NMDA receptor activation is also implicated in the loss of dendritic spines (Talantova et al. 2013); this along with cognitive decline was prevented by MitoQ, a mitochondria-targeted antioxidant, in 3xTg-AD mouse model of AD (McManus et al. 2011). MitoQ had similar effects on the impairment of LTP in the hippocampus of Tg2576 mice, another mouse model of AD (Ma et al. 2011).

12.3.5 CoQ as Biomarker for AD

Markers of oxidative stress can be useful for diagnosing AD in the earliest stages of the disease. Some studies have shown that oxidative stress damage occurs in mild cognitive impairment and AD patients who exhibit diminished antioxidant defenses and higher levels of oxidative damage markers (Padurariu et al. 2010; Puertas et al.

2012). However, the published data about serum CoQ concentration obtained from AD patient are conflicting. Bustos et al. reported in 2000 that CoQ serum levels in AD or vascular dementia patients were not related to the risk of dementia (de Bustos et al. 2000). These results were later corroborated by Battino et al., who neither found significantly differences for CoQ and vitamin E contents in plasma from patients and control subjects (Battino et al. 2003). In another study, CSF concentrations of oxidized and reduced CoQ were similar between treated AD patients and controls, but % CoQ was higher in the former. It was also found that %CoQ was not correlated with AD severity but was inversely correlated with disease duration (Isobe et al. 2010b), and two recent studies made in Japan others have reported could be useful for predicting the development of dementia in Japanese general population and is inversely associated with risk of disabling dementia (Momiyama 2014; Yamagishi et al. 2014).

12.3.6 CoQ Treatment in AD: Clinical Trials

Mitochondrial dysfunction and oxidative damage have been linked to AD pathogenesis. Based on these rationale, several open clinical trials have investigated the usefulness and efficacy of CoQ or its analog idebenone in the treatment of AD, with mixed findings.

Idebenone (2,3-dimethoxy-5-methyl-6-[10-hydroxydecyl]-1,4-benzoquinone) is a synthetic CoQ analog with low hydrophobicity that functions, similarly to CoQ, as an antioxidant and electron carrier in the respiratory chain. A double blind, placebo-controlled trial of 45 mg idebenone twice daily for 4 months in 102 elderly AD patients demonstrated clinical benefits in terms of memory, attention and behavior and disease progression (Senin et al. 1992). Bergamasco and colleagues conducted a multicenter, randomized, placebo-controlled, double blind trial of idebenone treatment for 90 days in 92 patients, followed by optional long-term treatment, and found idebenone to be effective on memory, attention, orientation, and to slow the disease progression (Bergamasco et al. 1994).

Another study that randomized 300 patients to receive idebenone 30 mg, 90 mg daily, or placebo for 6 months, showed that idebenone improved global response, and the AD Assessment Scale cognitive (ADAS-Cog) and noncognitive (ADAS-Noncog) scores (Weyer et al. 1997). Others have reported in a two-year trial that patients who initiated idebenone treatment at an earlier time point showed greater improvement than those who started after 1 year of receiving a placebo, suggesting disease-modifying rather than symptomatic effects (Gutzmann and Hadler 1998). However, a 1-year, multicenter, double-blind, placebo-controlled randomized trial of idebenone (120, 240, or 360 mg daily) in 536 AD patients in which ADAS-Cog and Clinical Global Impression of Change scores were used as endpoints found no significant differences in primary or secondary outcomes between groups (Thal et al. 2003).

The Alzheimer's Disease Cooperative Study initiated in 2006 is a phase I investigation evaluating the safety, tolerability and impact on biomarkers of antioxidant treatment of mild to moderate AD. Seventy-five patients with AD underwent treatment with either coenzyme Q 400 mg or vitamin E 800 IU plus vitamin C 200 mg plus α-lipoic acid 600 mg or placebo three times a day. The primary endpoint was changes in CSF biomarkers after 4 months' therapy and secondary outcome measures were the CSF levels of β-amyloid 40 and β-amyloid 42. Although the study found that CoQ was safe and well tolerated in patients with AD, the absence of a biomarker signal in CSF suggests that CoQ, at the dose tested, did not improve indices of oxidative stress or neurodegeneration (Galasko et al. 2012).

12.4 Conclusions

Oxidative stress damage occurs in PD and AD patients exhibiting decreased levels of the main antioxidant defenses and upregulation of oxidative damage markers. As such, the use of antioxidants such as CoQ is an attractive strategy for treatment of these neurodegenerative disorders. However, although CoQ has shown promising results in cellular and animal models, it has failed to show consistent therapeutic effects in humans. One reason for this is that animal models cannot recapitulate all the pathological aspects of diseases. For example, mouse AD models do not present the extensive neuronal loss observed in AD patients nor follow a comparable time course duration. In murine models the time-course is ranged between weeks to months, but in human is ranged between years to decades. Moreover, like other antioxidants, CoQ may have protective rather than curative effects, and by the time patients are examined for the first time by a neurologist, the disease has usually progressed over many years from a long asymptomatic phase to impaired neuronal function and connectivity. To this end, identifying early and reliable biomarkers will allow antioxidant treatment to be initiated before irreversible brain injury occurs. Despite their limitations, animal models provide a useful system for evaluating the efficacy of drugs that could prevent or delay progression of PD or AD to dementia.

Acknowledgments This work was supported by grants from the Instituto de Salud Carlos III (no. PI13/00907 to JV), Junta de Comunidades de Castilla La-Mancha (no. PPII-2014-010-P to FJA), and University of Castilla La-Mancha (no. GI20163390 to FJA).

References

Aberg F, Appelkvist EL, Dallner G, Ernster L (1992) Distribution and redox state of ubiquinones in rat and human tissues. Arch Biochem Biophys 295(2):230–234

Abou-Sleiman PM, Healy DG, Quinn N, Lees AJ, Wood NW (2003) The role of pathogenic DJ-1 mutations in Parkinson's disease. Ann Neurol 54:283–286

Ahmadvand H, Mabuchi H, Nohara A, Kobayahi J, Kawashiri MA (2013) Effects of coenzyme Q(10) on LDL oxidation in vitro. Acta Med Iran 51(1):12–18

Alam ZI, Jenner A, Daniel SE, Lees AJ, Cairns N, Marsden CD, Jenner P, Halliwell B (1997) Oxidative DNA damage in the parkinsonian brain: an apparent selective increase in 8-hydroxyguanine levels in substantia nigra. J Neurochem 69(3):1196–1203

Antony PM, Diederich NJ, Balling R (2011) Parkinson's disease mouse models in translational research. Mamm Genome 22(7–8):401–419. https://doi.org/10.1007/s00335-011-9330-x

Battino M, Bompadre S, Leone L, Devecchi E, Degiuli A, D'Agostino F, Cambie G, D'Agostino M, Faggi L, Colturani G, Gorini A, Villa RF (2003) Coenzyme Q, vitamin E and Apo-E alleles in Alzheimer disease. Biofactors 18(1–4):277–281

Beal MF, Matthews RT, Tieleman A, Shults CW (1998) Coenzyme Q10 attenuates the 1-methyl-4-phenyl-1,2,3,tetrahydropyridine (MPTP) induced loss of striatal dopamine and dopaminergic axons in aged mice. Brain Res 783 (1):109–114

Beal MF, Oakes D, Shoulson I, Henchcliffe C, Galpern WR, Haas R, Juncos JL, Nutt JG, Voss TS, Ravina B, Shults CM, Helles K, Snively V, Lew MF, Griebner B, Watts A, Gao S, Pourcher E, Bond L, Kompoliti K, Agarwal P, Sia C, Jog M, Cole L, Sultana M, Kurlan R, Richard I, Deeley C, Waters CH, Figueroa A, Arkun A, Brodsky M, Ondo WG, Hunter CB, Jimenez-Shahed J, Palao A, Miyasaki JM, So J, Tetrud J, Reys L, Smith K, Singer C, Blenke A, Russell DS, Cotto C, Friedman JH, Lannon M, Zhang L, Drasby E, Kumar R, Subramanian T, Ford DS, Grimes DA, Cote D, Conway J, Siderowf AD, Evatt ML, Sommerfeld B, Lieberman AN, Okun MS, Rodriguez RL, Merritt S, Swartz CL, Martin WR, King P, Stover N, Guthrie S, Watts RL, Ahmed A, Fernandez HH, Winters A, Mari Z, Dawson TM, Dunlop B, Feigin AS, Shannon B, Nirenberg MJ, Ogg M, Ellias SA, Thomas CA, Frei K, Bodis-Wollner I, Glazman S, Mayer T, Hauser RA, Pahwa R, Langhammer A, Ranawaya R, Derwent L, Sethi KD, Farrow B, Prakash R, Litvan I, Robinson A, Sahay A, Gartner M, Hinson VK, Markind S, Pelikan M, Perlmutter JS, Hartlein J, Molho E, Evans S, Adler CH, Duffy A, Lind M, Elmer L, Davis K, Spears J, Wilson S, Leehey MA, Hermanowicz N, Niswonger S, Shill HA, Obradov S, Rajput A, Cowper M, Lessig S, Song D, Fontaine D, Zadikoff C, Williams K, Blindauer KA, Bergholte J, Propsom CS, Stacy MA, Field J, Mihaila D, Chilton M, Uc EY, Sieren J, Simon DK, Kraics L, Silver A, Boyd JT, Hamill RW, Ingvoldstad C, Young J, Thomas K, Kostyk SK, Wojcieszek J, Pfeiffer RF, Panisset M, Beland M, Reich SG, Cines M, Zappala N, Rivest J, Zweig R, Lumina LP, Hilliard CL, Grill S, Kellermann M, Tuite P, Rolandelli S, Kang UJ, Rao J, Cook MM, Severt L, Boyar K (2014) A randomized clinical trial of high-dosage coenzyme Q10 in early Parkinson disease: no evidence of benefit. JAMA Neurol 71(5):543–552. https://doi.org/10.1001/jamaneurol.2014.131

Behl C, Davis JB, Lesley R, Schubert D (1994) Hydrogen peroxide mediates amyloid beta protein toxicity. Cell 77(6):817–827

Bello RI, Gomez-Diaz C, Buron MI, Alcain FJ, Navas P, Villalba JM (2005) Enhanced anti-oxidant protection of liver membranes in long-lived rats fed on a coenzyme Q10-supplemented diet. Exp Gerontol 40(8–9):694–706. https://doi.org/10.1016/j.exger.2005.07.003

Bergamasco B, Scarzella L, La Commare P (1994) Idebenone, a new drug in the treatment of cognitive impairment in patients with dementia of the Alzheimer type. Funct Neurol 9(3):161–168

Bhagavan HN, Chopra RK (2007) Plasma coenzyme Q10 response to oral ingestion of coenzyme Q10 formulations. Mitochondrion 7(Suppl):S78–S88. https://doi.org/10.1016/j.mito.2007.03.003

Bhagavan HN, Chopra RK, Craft NE, Chitchumroonchokchai C, Failla ML (2007) Assessment of coenzyme Q10 absorption using an in vitro digestion-Caco-2 cell model. Int J Pharm 333(1–2):112–117. https://doi.org/10.1016/j.ijpharm.2006.10.007

Bindoff LA, Birch-Machin MA, Cartlidge NE, Parker WD Jr, Turnbull DM (1991) Respiratory chain abnormalities in skeletal muscle from patients with Parkinson's disease. J Neurol Sci 104(2):203–208

Binukumar BK, Gupta N, Bal A, Gill KD (2011) Protection of dichlorvos induced oxidative stress and nigrostriatal neuronal death by chronic coenzyme Q10 pretreatment. Toxicol Appl Pharmacol 256(1):73–82. https://doi.org/10.1016/j.taap.2011.07.015

Blesa J, Phani S, Jackson-Lewis V, Przedborski S (2012) Classic and new animal models of Parkinson's disease. J Biomed Biotechnol 2012:845618. https://doi.org/10.1155/2012/845618

Blin O, Desnuelle C, Rascol O, Borg M, Peyro Saint Paul H, Azulay JP, Bille F, Figarella D, Coulom F, Pellissier JF et al (1994) Mitochondrial respiratory failure in skeletal muscle from patients with Parkinson's disease and multiple system atrophy. J Neurol Sci 125(1):95–101

Bourasset F, Ouellet M, Tremblay C, Julien C, Do TM, Oddo S, LaFerla F, Calon F (2009) Reduction of the cerebrovascular volume in a transgenic mouse model of Alzheimer's disease. Neuropharmacology 56(4):808–813

Braak H, Braak E (1991) Neuropathological stageing of Alzheimer-related changes. Acta Neuropathol 82(4):239–259

Braak H, Del Tredici K (2011) Alzheimer's pathogenesis: is there neuron-to-neuron propagation? Acta Neuropathol 121(5):589–595. https://doi.org/10.1007/s00401-011-0825-z

Braak H, Del Tredici K, Rub U, de Vos RA, Jansen Steur EN, Braak E (2003) Staging of brain pathology related to sporadic Parkinson's disease. Neurobiol Aging 24(2):197–211

Brennan AM, Suh SW, Won SJ, Narasimhan P, Kauppinen TM, Lee H, Edling Y, Chan PH, Swanson RA (2009) NADPH oxidase is the primary source of superoxide induced by NMDA receptor activation. Nat Neurosci 12(7):857–863. https://doi.org/10.1038/nn.2334

Bruce-Keller AJ, Begley JG, Fu W, Butterfield DA, Bredesen DE, Hutchins JB, Hensley K, Mattson MP (1998) Bcl-2 protects isolated plasma and mitochondrial membranes against lipid peroxidation induced by hydrogen peroxide and amyloid beta-peptide. J Neurochem 70(1):31–39

Buhmann C, Arlt S, Kontush A, Moller-Bertram T, Sperber S, Oechsner M, Stuerenburg HJ, Beisiegel U (2004) Plasma and CSF markers of oxidative stress are increased in Parkinson's disease and influenced by antiparkinsonian medication. Neurobiol Dis 15(1):160–170

Cha MY, Han SH, Son SM, Hong HS, Choi YJ, Byun J, Mook-Jung I (2012) Mitochondria-specific accumulation of amyloid beta induces mitochondrial dysfunction leading to apoptotic cell death. PLoS One 7(4):e34929. https://doi.org/10.1371/journal.pone.0034929

Chami L, Checler F (2012) BACE1 is at the crossroad of a toxic vicious cycle involving cellular stress and beta-amyloid production in Alzheimer's disease. Mol Neurodegener 7:52. https://doi.org/10.1186/1750-1326-7-52

Chang Y, Huang SK, Wang SJ (2012) Coenzyme Q10 inhibits the release of glutamate in rat cerebrocortical nerve terminals by suppression of voltage-dependent calcium influx and mitogen-activated protein kinase signaling pathway. J Agric Food Chem 60(48):11909–11918. https://doi.org/10.1021/jf302875k

Chaturvedi RK, Flint Beal M (2013) Mitochondrial diseases of the brain. Free Radic Biol Med 63:1–29. https://doi.org/10.1016/j.freeradbiomed.2013.03.018

Choi H, Park HH, Koh SH, Choi NY, Yu HJ, Park J, Lee YJ, Lee KY (2012) Coenzyme Q10 protects against amyloid beta-induced neuronal cell death by inhibiting oxidative stress and activating the P13K pathway. Neurotoxicology 33(1):85–90. https://doi.org/10.1016/j.neuro.2011.12.005

Choi H, Park HH, Lee KY, Choi NY, Yu HJ, Lee YJ, Park J, Huh YM, Lee SH, Koh SH (2013) Coenzyme Q10 restores amyloid beta-inhibited proliferation of neural stem cells by activating the PI3K pathway. Stem Cells Dev 22(15):2112–2120. https://doi.org/10.1089/scd.2012.0604

Chow VW, Mattson MP, Wong PC, Gleichmann M (2010) An overview of APP processing enzymes and products. NeuroMolecular Med 12(1):1–12. https://doi.org/10.1007/s12017-009-8104-z

Cleren C, Yang L, Lorenzo B, Calingasan NY, Schomer A, Sireci A, Wille EJ, Beal MF (2008) Therapeutic effects of coenzyme Q10 (CoQ_{10}) and reduced CoQ_{10} in the MPTP model of parkinsonism. J Neurochem 104(6):1613–1621. https://doi.org/10.1111/j.1471-4159.2007.05097.x

Crane FL (2001) Biochemical functions of coenzyme Q10. J Am Coll Nutr 20(6):591–598

Cutler RG, Kelly J, Storie K, Pedersen WA, Tammara A, Hatanpaa K, Troncoso JC, Mattson MP (2004) Involvement of oxidative stress-induced abnormalities in ceramide and

cholesterol metabolism in brain aging and Alzheimer's disease. Proc Natl Acad Sci U S A 101(7):2070–2075. https://doi.org/10.1073/pnas.0305799101
Dal Pra I, Chiarini A, Gui L, Chakravarthy B, Pacchiana R, Gardenal E, Whitfield JF, Armato U (2015) Do astrocytes collaborate with neurons in spreading the "infectious" abeta and Tau drivers of Alzheimer's disease? Neuroscientist 21(1):9–29. https://doi.org/10.1177/1073858414529828
DaRocha-Souto B, Coma M, Perez-Nievas BG, Scotton TC, Siao M, Sanchez-Ferrer P, Hashimoto T, Fan Z, Hudry E, Barroeta I, Sereno L, Rodriguez M, Sanchez MB, Hyman BT, Gomez-Isla T (2012) Activation of glycogen synthase kinase-3 beta mediates beta-amyloid induced neuritic damage in Alzheimer's disease. Neurobiol Dis 45(1):425–437. https://doi.org/10.1016/j.nbd.2011.09.002
de Bustos F, Molina JA, Jimenez-Jimenez FJ, Garcia-Redondo A, Gomez-Escalonilla C, Porta-Etessam J, Berbel A, Zurdo M, Barcenilla B, Parrilla G, Enriquez-de-Salamanca R, Arenas J (2000) Serum levels of coenzyme Q10 in patients with Alzheimer's disease. J Neural Transm (Vienna) 107(2):233–239. https://doi.org/10.1007/s007020050019
Dexter D, Carter C, Agid F, Agid Y, Lees AJ, Jenner P, Marsden CD (1986) Lipid peroxidation as cause of nigral cell death in Parkinson's disease. Lancet 2(8507):639–640
Di Carlo M (2010) Beta amyloid peptide: from different aggregation forms to the activation of different biochemical pathways. Eur Biophys J 39(6):877–888. https://doi.org/10.1007/s00249-009-0439-8
Dick FD, De Palma G, Ahmadi A, Scott NW, Prescott GJ, Bennett J, Semple S, Dick S, Counsell C, Mozzoni P, Haites N, Wettinger SB, Mutti A, Otelea M, Seaton A, Soderkvist P, Felice A (2007) Environmental risk factors for Parkinson's disease and parkinsonism: the Geoparkinson study. Occup Environ Med 64(10):666–672. https://doi.org/10.1136/oem.2006.027003
Dubois B, Feldman HH, Jacova C, Dekosky ST, Barberger-Gateau P, Cummings J, Delacourte A, Galasko D, Gauthier S, Jicha G, Meguro K, O'Brien J, Pasquier F, Robert P, Rossor M, Salloway S, Stern Y, Visser PJ, Scheltens P (2007) Research criteria for the diagnosis of Alzheimer's disease: revising the NINCDS-ADRDA criteria. Lancet Neurol 6(8):734–746. https://doi.org/10.1016/S1474-4422(07)70178-3
Dumont M, Kipiani K, Yu F, Wille E, Katz M, Calingasan NY, Gouras GK, Lin MT, Beal MF (2011) Coenzyme Q10 decreases amyloid pathology and improves behavior in a transgenic mouse model of Alzheimer's disease. J Alzheimer's Dis 27(1):211–223. https://doi.org/10.3233/JAD-2011-110209
Duran-Prado M, Frontinan J, Santiago-Mora R, Peinado JR, Parrado-Fernandez C, Gomez-Almagro MV, Moreno M, Lopez-Dominguez JA, Villalba JM, Alcain FJ (2014) Coenzyme Q10 protects human endothelial cells from beta-amyloid uptake and oxidative stress-induced injury. PLoS One 9(10):e109223. https://doi.org/10.1371/journal.pone.0109223
Eschbach J, von Einem B, Muller K, Bayer H, Scheffold A, Morrison BE, Rudolph KL, Thal DR, Witting A, Weydt P, Otto M, Fauler M, Liss B, McLean PJ, Spada AR, Ludolph AC, Weishaupt JH, Danzer KM (2015) Mutual exacerbation of peroxisome proliferator-activated receptor gamma coactivator 1alpha deregulation and alpha-synuclein oligomerization. Ann Neurol 77(1):15–32. https://doi.org/10.1002/ana.24294
Esteves AR, Arduino DM, Swerdlow RH, Oliveira CR, Cardoso SM (2009) Oxidative stress involvement in alpha-synuclein oligomerization in Parkinson's disease cybrids. Antioxid Redox Signal 11(3):439–448. https://doi.org/10.1089/ARS.2008.2247
Esteves AR, Lu J, Rodova M, Onyango I, Lezi E, Dubinsky R, Lyons KE, Pahwa R, Burns JM, Cardoso SM, Swerdlow RH (2010) Mitochondrial respiration and respiration-associated proteins in cell lines created through Parkinson's subject mitochondrial transfer. J Neurochem 113(3):674–682. https://doi.org/10.1111/j.1471-4159.2010.06631.x
Ferreira M, Massano J (2016) An updated review of Parkinson's disease genetics and clinicopathological correlations. Acta Neurol Scand. https://doi.org/10.1111/ane.12616
Fontaine E, Eriksson O, Ichas F, Bernardi P (1998) Regulation of the permeability transition pore in skeletal muscle mitochondria. Modulation by electron flow through the respiratory chain complex i. J Biol Chem 273(20):12662–12668

Franco-Iborra S, Vila M, Perier C (2016) The Parkinson disease mitochondrial hypothesis: where are we at? Neuroscientist 22(3):266–277. https://doi.org/10.1177/1073858415574600
Frontiñan-Rubio J, Sancho-Bielsa FJ, Peinado JR, LaFerla FM, Gimenez-Llort L, Duran-Prado M, Alcain FJ (2018) Sex-dependent co-occurrence of hypoxia and beta-amyloid plaques in hippocampus and entorhinal cortex is reversed by long-term treatment with ubiquinol and ascorbic acid in the 3xTg-AD mouse model of Alzheimer's disease. Mol Cell Neurosci 92:67–81. https://doi.org/10.1016/j.mcn.2018.06.005
Galasko DR, Peskind E, Clark CM, Quinn JF, Ringman JM, Jicha GA, Cotman C, Cottrell B, Montine TJ, Thomas RG, Aisen P (2012) Antioxidants for Alzheimer disease: a randomized clinical trial with cerebrospinal fluid biomarker measures. Arch Neurol 69(7):836–841. https://doi.org/10.1001/archneurol.2012.85
Gatt AP, Duncan OF, Attems J, Francis PT, Ballard CG, Bateman JM (2016) Dementia in Parkinson's disease is associated with enhanced mitochondrial complex I deficiency. Mov Disord 31(3):352–359. https://doi.org/10.1002/mds.26513
Gautier CA, Giaime E, Caballero E, Nunez L, Song Z, Chan D, Villalobos C, Shen J (2012) Regulation of mitochondrial permeability transition pore by PINK1. Mol Neurodegener 7:22. https://doi.org/10.1186/1750-1326-7-22
German DC, Nelson EL, Liang CL, Speciale SG, Sinton CM, Sonsalla PK (1996) The neurotoxin MPTP causes degeneration of specific nucleus A8, A9 and A10 dopaminergic neurons in the mouse. Neurodegeneration 5(4):299–312
Gille G, Hung ST, Reichmann H, Rausch WD (2004) Oxidative stress to dopaminergic neurons as models of Parkinson's disease. Ann N Y Acad Sci 1018:533–540. https://doi.org/10.1196/annals.1296.066
Goate A, Chartier-Harlin MC, Mullan M, Brown J, Crawford F, Fidani L, Giuffra L, Haynes A, Irving N, James L et al (1991) Segregation of a missense mutation in the amyloid precursor protein gene with familial Alzheimer's disease. Nature 349(6311):704–706. https://doi.org/10.1038/349704a0
Gomez-Isla T, Hollister R, West H, Mui S, Growdon JH, Petersen RC, Parisi JE, Hyman BT (1997) Neuronal loss correlates with but exceeds neurofibrillary tangles in Alzheimer's disease. Ann Neurol 41(1):17–24. https://doi.org/10.1002/ana.410410106
Gorgone G, Curro M, Ferlazzo N, Parisi G, Parnetti L, Belcastro V, Tambasco N, Rossi A, Pisani F, Calabresi P, Ientile R, Caccamo D (2012) Coenzyme Q10, hyperhomocysteinemia and MTHFR C677T polymorphism in levodopa-treated Parkinson's disease patients. NeuroMolecular Med 14(1):84–90. https://doi.org/10.1007/s12017-012-8174-1
Gosavi N, Lee HJ, Lee JS, Patel S, Lee SJ (2002) Golgi fragmentation occurs in the cells with prefibrillar alpha-synuclein aggregates and precedes the formation of fibrillar inclusion. J Biol Chem 277(50):48984–48992. https://doi.org/10.1074/jbc.M208194200
Gotz ME, Gerstner A, Harth R, Dirr A, Janetzky B, Kuhn W, Riederer P, Gerlach M (2000) Altered redox state of platelet coenzyme Q10 in Parkinson's disease. J Neural Transm (Vienna) 107(1):41–48. https://doi.org/10.1007/s007020050003
Grammas P (2011) Neurovascular dysfunction, inflammation and endothelial activation: implications for the pathogenesis of Alzheimer's disease. J Neuroinflammation 8:26. https://doi.org/10.1186/1742-2094-8-26
Greenamyre JT, Sherer TB, Betarbet R, Panov AV (2001) Complex I and Parkinson's disease. IUBMB Life 52(3–5):135–141. https://doi.org/10.1080/15216540152845939
Guglielmotto M, Aragno M, Autelli R, Giliberto L, Novo E, Colombatto S, Danni O, Parola M, Smith MA, Perry G, Tamagno E, Tabaton M (2009) The up-regulation of BACE1 mediated by hypoxia and ischemic injury: role of oxidative stress and HIF1alpha. J Neurochem 108(4):1045–1056. https://doi.org/10.1111/j.1471-4159.2008.05858.x
Gutzmann H, Hadler D (1998) Sustained efficacy and safety of idebenone in the treatment of Alzheimer's disease: update on a 2-year double-blind multicentre study. J Neural Transm Suppl 54:301–310

Hansson Petersen CA, Alikhani N, Behbahani H, Wiehager B, Pavlov PF, Alafuzoff I, Leinonen V, Ito A, Winblad B, Glaser E, Ankarcrona M (2008) The amyloid beta-peptide is imported into mitochondria via the TOM import machinery and localized to mitochondrial cristae. Proc Natl Acad Sci U S A 105(35):13145–13150. https://doi.org/10.1073/pnas.0806192105

Hargreaves IP, Lane A, Sleiman PM (2008) The coenzyme Q10 status of the brain regions of Parkinson's disease patients. Neurosci Lett 447(1):17–19. https://doi.org/10.1016/j.neulet.2008.09.069

Hartley A, Stone JM, Heron C, Cooper JM, Schapira AH (1994) Complex I inhibitors induce dose-dependent apoptosis in PC12 cells: relevance to Parkinson's disease. J Neurochem 63(5):1987–1990

Hernandes MS, Britto LR (2012) NADPH oxidase and neurodegeneration. Curr Neuropharmacol 10(4):321–327. https://doi.org/10.2174/157015912804143540

Hernandez F, Borrell J, Guaza C, Avila J, Lucas JJ (2002) Spatial learning deficit in transgenic mice that conditionally over-express GSK-3beta in the brain but do not form tau filaments. J Neurochem 83(6):1529–1533

Hosoe K, Kitano M, Kishida H, Kubo H, Fujii K, Kitahara M (2007) Study on safety and bioavailability of ubiquinol (Kaneka QH) after single and 4-week multiple oral administration to healthy volunteers. Regul Toxicol Pharmacol 47(1):19–28. https://doi.org/10.1016/j.yrtph.2006.07.001

Huang JY, Hafez DM, James BD, Bennett DA, Marr RA (2012) Altered NEP2 expression and activity in mild cognitive impairment and Alzheimer's disease. J Alzheimer's Dis 28(2):433–441. https://doi.org/10.3233/JAD-2011-111307

Hunot S, Brugg B, Ricard D, Michel PP, Muriel MP, Ruberg M, Faucheux BA, Agid Y, Hirsch EC (1997) Nuclear translocation of NF-kappaB is increased in dopaminergic neurons of patients with parkinson disease. Proc Natl Acad Sci U S A 94(14):7531–7536

Iadecola C (2004) Neurovascular regulation in the normal brain and in Alzheimer's disease. Nat Rev Neurosci 5(5):347–360. https://doi.org/10.1038/nrn1387

Isobe C, Abe T, Terayama Y (2009) Increase in the oxidized/total coenzyme Q-10 ratio in the cerebrospinal fluid of Alzheimer's disease patients. Dement Geriatr Cogn Disord 28(5):449–454. https://doi.org/10.1159/000256209

Isobe C, Abe T, Terayama Y (2010a) Levels of reduced and oxidized coenzyme Q-10 and 8-hydroxy-2′-deoxyguanosine in the cerebrospinal fluid of patients with living Parkinson's disease demonstrate that mitochondrial oxidative damage and/or oxidative DNA damage contributes to the neurodegenerative process. Neurosci Lett 469(1):159–163. https://doi.org/10.1016/j.neulet.2009.11.065

Isobe C, Abe T, Terayama Y (2010b) Levels of reduced and oxidized coenzyme Q-10 and 8-hydroxy-2′-deoxyguanosine in the CSF of patients with Alzheimer's disease demonstrate that mitochondrial oxidative damage and/or oxidative DNA damage contributes to the neurodegenerative process. J Neurol 257(3):399–404. https://doi.org/10.1007/s00415-009-5333-x

Jana A, Pahan K (2004) Fibrillar amyloid-beta peptides kill human primary neurons via NADPH oxidase-mediated activation of neutral sphingomyelinase. Implications for Alzheimer's disease. J Biol Chem 279(49):51451–51459. https://doi.org/10.1074/jbc.M404635200

Jenner P, Dexter DT, Sian J, Schapira AH, Marsden CD (1992) Oxidative stress as a cause of nigral cell death in Parkinson's disease and incidental Lewy body disease. The Royal Kings and Queens Parkinson's Disease Research Group. Ann Neurol 32(Suppl):S82–S87

Jimenez-Jimenez FJ, Molina JA, de Bustos F, Garcia-Redondo A, Gomez-Escalonilla C, Martinez-Salio A, Berbel A, Camacho A, Zurdo M, Barcenilla B, Enriquez de Salamanca R, Arenas J (2000) Serum levels of coenzyme Q10 in patients with Parkinson's disease. J Neural Transm (Vienna) 107(2):177–181. https://doi.org/10.1007/s007020050015

Karran E, Mercken M, De Strooper B (2011) The amyloid cascade hypothesis for Alzheimer's disease: an appraisal for the development of therapeutics. Nat Rev Drug Discov 10(9):698–712. https://doi.org/10.1038/nrd3505

Kawas CH (2003) Clinical practice. Early Alzheimer's disease. N Engl J Med 349(11):1056–1063. https://doi.org/10.1056/NEJMcp022295

Keeney PM, Xie J, Capaldi RA, Bennett JP Jr (2006) Parkinson's disease brain mitochondrial complex I has oxidatively damaged subunits and is functionally impaired and misassembled. J Neurosci 26(19):5256–5264. https://doi.org/10.1523/JNEUROSCI.0984-06.2006

Knopman DS, DeKosky ST, Cummings JL, Chui H, Corey-Bloom J, Relkin N, Small GW, Miller B, Stevens JC (2001) Practice parameter: diagnosis of dementia (an evidence-based review). Report of the quality standards Subcommittee of the American Academy of neurology. Neurology 56 (9):1143–1153

Kooncumchoo P, Sharma S, Porter J, Govitrapong P, Ebadi M (2006) Coenzyme Q(10) provides neuroprotection in iron-induced apoptosis in dopaminergic neurons. J Mol Neurosci 28(2):125–141. https://doi.org/10.1385/JMN:28:2:125

Kumari S, Mehta SL, Milledge GZ, Huang X, Li H, Li PA (2016) Ubisol-Q10 prevents glutamate-induced cell death by blocking mitochondrial fragmentation and permeability transition pore opening. Int J Biol Sci 12(6):688–700. https://doi.org/10.7150/ijbs.13589

Langsjoen PH, Langsjoen AM (2014) Comparison study of plasma coenzyme Q10 levels in healthy subjects supplemented with ubiquinol versus ubiquinone. Clin Pharmacol Drug Dev 3(1):13–17. https://doi.org/10.1002/cpdd.73

Langston JW, Ballard P, Tetrud JW, Irwin I (1983) Chronic parkinsonism in humans due to a product of meperidine-analog synthesis. Science 219(4587):979–980

Le W, Sayana P, Jankovic J (2014) Animal models of Parkinson's disease: a gateway to therapeutics? Neurotherapeutics 11(1):92–110. https://doi.org/10.1007/s13311-013-0234-1

Levy-Lahad E, Wasco W, Poorkaj P, Romano DM, Oshima J, Pettingell WH, Yu CE, Jondro PD, Schmidt SD, Wang K et al (1995) Candidate gene for the chromosome 1 familial Alzheimer's disease locus. Science 269(5226):973–977

Li G, Jack CR, Yang XF, Yang ES (2008) Diet supplement CoQ_{10} delays brain atrophy in aged transgenic mice with mutations in the amyloid precursor protein: an in vivo volume MRI study. Biofactors 32(1–4):169–178

Liu J, Wang L, Zhan SY, Xia Y (2011) Coenzyme Q10 for Parkinson's disease. Cochrane Database Syst Rev 12:CD008150. https://doi.org/10.1002/14651858.CD008150.pub2

Lopez-Lluch G, Rodriguez-Aguilera JC, Santos-Ocana C, Navas P (2010) Is coenzyme Q a key factor in aging? Mech Ageing Dev 131(4):225–235. https://doi.org/10.1016/j.mad.2010.02.003

Ma T, Hoeffer CA, Wong H, Massaad CA, Zhou P, Iadecola C, Murphy MP, Pautler RG, Klann E (2011) Amyloid beta-induced impairments in hippocampal synaptic plasticity are rescued by decreasing mitochondrial superoxide. J Neurosci 31(15):5589–5595. https://doi.org/10.1523/JNEUROSCI.6566-10.2011

Martinez-Fernandez R, Schmitt E, Martinez-Martin P, Krack P (2016) The hidden sister of motor fluctuations in Parkinson's disease: a review on nonmotor fluctuations. Mov Disord 31(8):1080–1094. https://doi.org/10.1002/mds.26731

Matthews RT, Yang L, Browne S, Baik M, Beal MF (1998) Coenzyme Q10 administration increases brain mitochondrial concentrations and exerts neuroprotective effects. Proc Natl Acad Sci U S A 95(15):8892–8897

McCarthy S, Somayajulu M, Sikorska M, Borowy-Borowski H, Pandey S (2004) Paraquat induces oxidative stress and neuronal cell death; neuroprotection by water-soluble coenzyme Q10. Toxicol Appl Pharmacol 201(1):21–31. https://doi.org/10.1016/j.taap.2004.04.019

McManus MJ, Murphy MP, Franklin JL (2011) The mitochondria-targeted antioxidant MitoQ prevents loss of spatial memory retention and early neuropathology in a transgenic mouse model of Alzheimer's disease. J Neurosci 31(44):15703–15715. https://doi.org/10.1523/JNEUROSCI.0552-11.2011

Meyer MR, Tschanz JT, Norton MC, Welsh-Bohmer KA, Steffens DC, Wyse BW, Breitner JC (1998) APOE genotype predicts when--not whether--one is predisposed to develop Alzheimer disease. Nat Genet 19(4):321–322. https://doi.org/10.1038/1206

Mischley LK, Allen J, Bradley R (2012) Coenzyme Q10 deficiency in patients with Parkinson's disease. J Neurol Sci 318(1–2):72–75. https://doi.org/10.1016/j.jns.2012.03.023
Mizuno Y, Ohta S, Tanaka M, Takamiya S, Suzuki K, Sato T, Oya H, Ozawa T, Kagawa Y (1989) Deficiencies in complex I subunits of the respiratory chain in Parkinson's disease. Biochem Biophys Res Commun 163(3):1450–1455
Momiyama Y (2014) Serum coenzyme Q10 levels as a predictor of dementia in a Japanese general population. Atherosclerosis 237(2):433–434. https://doi.org/10.1016/j.atherosclerosis.2014.08.056
Moon Y, Lee KH, Park JH, Geum D, Kim K (2005) Mitochondrial membrane depolarization and the selective death of dopaminergic neurons by rotenone: protective effect of coenzyme Q10. J Neurochem 93(5):1199–1208. https://doi.org/10.1111/j.1471-4159.2005.03112.x
Morais VA, Haddad D, Craessaerts K, De Bock PJ, Swerts J, Vilain S, Aerts L, Overbergh L, Grunewald A, Seibler P, Klein C, Gevaert K, Verstreken P, De Strooper B (2014) PINK1 loss-of-function mutations affect mitochondrial complex I activity via NdufA10 ubiquinone uncoupling. Science 344(6180):203–207. https://doi.org/10.1126/science.1249161
Moreira PI, Santos MS, Sena C, Nunes E, Seica R, Oliveira CR (2005) CoQ_{10} therapy attenuates amyloid beta-peptide toxicity in brain mitochondria isolated from aged diabetic rats. Exp Neurol 196(1):112–119. https://doi.org/10.1016/j.expneurol.2005.07.012
Mudo G, Makela J, Di Liberto V, Tselykh TV, Olivieri M, Piepponen P, Eriksson O, Malkia A, Bonomo A, Kairisalo M, Aguirre JA, Korhonen L, Belluardo N, Lindholm D (2012) Transgenic expression and activation of PGC-1alpha protect dopaminergic neurons in the MPTP mouse model of Parkinson's disease. Cell Mol Life Sci 69(7):1153–1165. https://doi.org/10.1007/s00018-011-0850-z
Muthukumaran K, Leahy S, Harrison K, Sikorska M, Sandhu JK, Cohen J, Keshan C, Lopatin D, Miller H, Borowy-Borowski H, Lanthier P, Weinstock S, Pandey S (2014) Orally delivered water soluble Coenzyme Q10 (Ubisol-Q10) blocks on-going neurodegeneration in rats exposed to paraquat: potential for therapeutic application in Parkinson's disease. BMC Neurosci 15:21. https://doi.org/10.1186/1471-2202-15-21
Naderi J, Somayajulu-Nitu M, Mukerji A, Sharda P, Sikorska M, Borowy-Borowski H, Antonsson B, Pandey S (2006) Water-soluble formulation of Coenzyme Q10 inhibits Bax-induced destabilization of mitochondria in mammalian cells. Apoptosis 11(8):1359–1369. https://doi.org/10.1007/s10495-006-8417-4
Nakao N, Nakai K, Itakura T (1997) Metabolic inhibition enhances selective toxicity of L-DOPA toward mesencephalic dopamine neurons in vitro. Brain Res 777(1–2):202–209
Nunomura A, Perry G, Aliev G, Hirai K, Takeda A, Balraj EK, Jones PK, Ghanbari H, Wataya T, Shimohama S, Chiba S, Atwood CS, Petersen RB, Smith MA (2001) Oxidative damage is the earliest event in Alzheimer disease. J Neuropathol Exp Neurol 60(8):759–767
Ohhara H, Kanaide H, Nakamura M (1981) A protective effect of coenzyme Q10 on the adriamycin-induced cardiotoxicity in the isolated perfused rat heart. J Mol Cell Cardiol 13(8):741–752
Orrenius S, Gogvadze V, Zhivotovsky B (2007) Mitochondrial oxidative stress: implications for cell death. Annu Rev Pharmacol Toxicol 47:143–183. https://doi.org/10.1146/annurev.pharmtox.47.120505.105122
Paduraru M, Ciobica A, Hritcu L, Stoica B, Bild W, Stefanescu C (2010) Changes of some oxidative stress markers in the serum of patients with mild cognitive impairment and Alzheimer's disease. Neurosci Lett 469(1):6–10. https://doi.org/10.1016/j.neulet.2009.11.033
Papucci L, Schiavone N, Witort E, Donnini M, Lapucci A, Tempestini A, Formigli L, Zecchi-Orlandini S, Orlandini G, Carella G, Brancato R, Capaccioli S (2003) Coenzyme q10 prevents apoptosis by inhibiting mitochondrial depolarization independently of its free radical scavenging property. J Biol Chem 278(30):28220–28228. https://doi.org/10.1074/jbc.M302297200
Pardo B, Mena MA, Casarejos MJ, Paino CL, De Yebenes JG (1995) Toxic effects of L-DOPA on mesencephalic cell cultures: protection with antioxidants. Brain Res 682(1–2):133–143

Park J, Park HH, Choi H, Kim YS, Yu HJ, Lee KY, Lee YJ, Kim SH, Koh SH (2012) Coenzyme Q10 protects neural stem cells against hypoxia by enhancing survival signals. Brain Res 1478:64–73. https://doi.org/10.1016/j.brainres.2012.08.025

Park L, Anrather J, Forster C, Kazama K, Carlson GA, Iadecola C (2004) Abeta-induced vascular oxidative stress and attenuation of functional hyperemia in mouse somatosensory cortex. J Cereb Blood Flow Metab 24(3):334–342. https://doi.org/10.1097/01.WCB.0000105800.49957.1E

Park L, Anrather J, Zhou P, Frys K, Pitstick R, Younkin S, Carlson GA, Iadecola C (2005) NADPH-oxidase-derived reactive oxygen species mediate the cerebrovascular dysfunction induced by the amyloid beta peptide. J Neurosci 25(7):1769–1777. https://doi.org/10.1523/JNEUROSCI.5207-04.2005

Parker WD Jr, Boyson SJ, Parks JK (1989) Abnormalities of the electron transport chain in idiopathic Parkinson's disease. Ann Neurol 26(6):719–723. https://doi.org/10.1002/ana.410260606

Parker WD Jr, Parks JK, Swerdlow RH (2008) Complex I deficiency in Parkinson's disease frontal cortex. Brain Res 1189:215–218. https://doi.org/10.1016/j.brainres.2007.10.061

Perier C, Tieu K, Guegan C, Caspersen C, Jackson-Lewis V, Carelli V, Martinuzzi A, Hirano M, Przedborski S, Vila M (2005) Complex I deficiency primes Bax-dependent neuronal apoptosis through mitochondrial oxidative damage. Proc Natl Acad Sci U S A 102(52):19126–19131. https://doi.org/10.1073/pnas.0508215102

Perier C, Vila M (2012) Mitochondrial biology and Parkinson's disease. Cold Spring Harb Perspect Med 2(2):a009332. https://doi.org/10.1101/cshperspect.a009332

Puertas MC, Martinez-Martos JM, Cobo MP, Carrera MP, Mayas MD, Ramirez-Exposito MJ (2012) Plasma oxidative stress parameters in men and women with early stage Alzheimer type dementia. Exp Gerontol 47(8):625–630. https://doi.org/10.1016/j.exger.2012.05.019

Ramsay RR, Singer TP (1986) Energy-dependent uptake of N-methyl-4-phenylpyridinium, the neurotoxic metabolite of 1-methyl-4-phenyl-1,2,3,6-tetrahydropyridine, by mitochondria. J Biol Chem 261(17):7585–7587

Rao VK, Carlson EA, Yan SS (2014) Mitochondrial permeability transition pore is a potential drug target for neurodegeneration. Biochim Biophys Acta 1842(8):1267–1272. https://doi.org/10.1016/j.bbadis.2013.09.003

Reddy PH, Beal MF (2008) Amyloid beta, mitochondrial dysfunction and synaptic damage: implications for cognitive decline in aging and Alzheimer's disease. Trends Mol Med 14(2):45–53. https://doi.org/10.1016/j.molmed.2007.12.002

Rogaev EI, Sherrington R, Rogaeva EA, Levesque G, Ikeda M, Liang Y, Chi H, Lin C, Holman K, Tsuda T et al (1995) Familial Alzheimer's disease in kindreds with missense mutations in a gene on chromosome 1 related to the Alzheimer's disease type 3 gene. Nature 376(6543):775–778. https://doi.org/10.1038/376775a0

Ruitenberg A, den Heijer T, Bakker SL, van Swieten JC, Koudstaal PJ, Hofman A, Breteler MM (2005) Cerebral hypoperfusion and clinical onset of dementia: the Rotterdam study. Ann Neurol 57(6):789–794. https://doi.org/10.1002/ana.20493

Samii A, Nutt JG, Ransom BR (2004) Parkinson's disease. Lancet 363(9423):1783–1793. https://doi.org/10.1016/S0140-6736(04)16305-8

Schapira AH, Cooper JM, Dexter D, Jenner P, Clark JB, Marsden CD (1989) Mitochondrial complex I deficiency in Parkinson's disease. Lancet 1(8649):1269

Schapira AH, Olanow CW (2004) Neuroprotection in Parkinson disease: mysteries, myths, and misconceptions. JAMA 291(3):358–364. https://doi.org/10.1001/jama.291.3.358

Scheltens P, Blennow K, Breteler MM, de Strooper B, Frisoni GB, Salloway S, Van der Flier WM (2016) Alzheimer's disease. Lancet 388(10043):505–517. https://doi.org/10.1016/S0140-6736(15)01124-1

Schneider H, Lemasters JJ, Hackenbrock CR (1982) Lateral diffusion of ubiquinone during electron transfer in phospholipid- and ubiquinone-enriched mitochondrial membranes. J Biol Chem 257(18):10789–10793

Schulz JB, Henshaw DR, Matthews RT, Beal MF (1995) Coenzyme Q10 and nicotinamide and a free radical spin trap protect against MPTP neurotoxicity. Exp Neurol 132(2):279–283

Seet RC, Lim EC, Tan JJ, Quek AM, Chow AW, Chong WL, Ng MP, Ong CN, Halliwell B (2014) Does high-dose coenzyme Q10 improve oxidative damage and clinical outcomes in Parkinson's disease? Antioxid Redox Signal 21(2):211–217. https://doi.org/10.1089/ars.2013.5801

Senin U, Parnetti L, Barbagallo-Sangiorgi G, Bartorelli L, Bocola V, Capurso A, Cuzzupoli M, Denaro M, Marigliano V, Tammaro AE, Fioravanti M (1992) Idebenone in senile dementia of Alzheimer type: a multicentre study. Arch Gerontol Geriatr 15(3):249–260

Sharma SK, El Refaey H, Ebadi M (2006) Complex-1 activity and 18F-DOPA uptake in genetically engineered mouse model of Parkinson's disease and the neuroprotective role of coenzyme Q10. Brain Res Bull 70(1):22–32. https://doi.org/10.1016/j.brainresbull.2005.11.019

Shavali S, Carlson EC, Swinscoe JC, Ebadi M (2004) 1-Benzyl-1,2,3,4-tetrahydroisoquinoline, a parkinsonism-inducing endogenous toxin, increases alpha-synuclein expression and causes nuclear damage in human dopaminergic cells. J Neurosci Res 76(4):563–571. https://doi.org/10.1002/jnr.20082

Sherrington R, Rogaev EI, Liang Y, Rogaeva EA, Levesque G, Ikeda M, Chi H, Lin C, Li G, Holman K, Tsuda T, Mar L, Foncin JF, Bruni AC, Montesi MP, Sorbi S, Rainero I, Pinessi L, Nee L, Chumakov I, Pollen D, Brookes A, Sanseau P, Polinsky RJ, Wasco W, Da Silva HA, Haines JL, Perkicak-Vance MA, Tanzi RE, Roses AD, Fraser PE, Rommens JM, St George-Hyslop PH (1995) Cloning of a gene bearing missense mutations in early-onset familial Alzheimer's disease. Nature 375(6534):754–760. https://doi.org/10.1038/375754a0

Shults CW, Haas RH, Passov D, Beal MF (1997) Coenzyme Q10 levels correlate with the activities of complexes I and II/III in mitochondria from parkinsonian and nonparkinsonian subjects. Ann Neurol 42(2):261–264. https://doi.org/10.1002/ana.410420221

Shults CW, Oakes D, Kieburtz K, Beal MF, Haas R, Plumb S, Juncos JL, Nutt J, Shoulson I, Carter J, Kompoliti K, Perlmutter JS, Reich S, Stern M, Watts RL, Kurlan R, Molho E, Harrison M, Lew M (2002) Effects of coenzyme Q10 in early Parkinson disease: evidence of slowing of the functional decline. Arch Neurol 59(10):1541–1550

Sikorska M, Lanthier P, Miller H, Beyers M, Sodja C, Zurakowski B, Gangaraju S, Pandey S, Sandhu JK (2014) Nanomicellar formulation of coenzyme Q10 (Ubisol-Q10) effectively blocks ongoing neurodegeneration in the mouse 1-methyl-4-phenyl-1,2,3,6-tetrahydropyridine model: potential use as an adjuvant treatment in Parkinson's disease. Neurobiol Aging 35(10):2329–2346. https://doi.org/10.1016/j.neurobiolaging.2014.03.032

Sohmiya M, Tanaka M, Tak NW, Yanagisawa M, Tanino Y, Suzuki Y, Okamoto K, Yamamoto Y (2004) Redox status of plasma coenzyme Q10 indicates elevated systemic oxidative stress in Parkinson's disease. J Neurol Sci 223(2):161–166. https://doi.org/10.1016/j.jns.2004.05.007

Somayajulu M, McCarthy S, Hung M, Sikorska M, Borowy-Borowski H, Pandey S (2005) Role of mitochondria in neuronal cell death induced by oxidative stress; neuroprotection by coenzyme Q10. Neurobiol Dis 18(3):618–627. https://doi.org/10.1016/j.nbd.2004.10.021

Sorce S, Krause KH (2009) NOX enzymes in the central nervous system: from signaling to disease. Antioxid Redox Signal 11(10):2481–2504. https://doi.org/10.1089/ARS.2009.2578

Speciale SG (2002) MPTP: insights into parkinsonian neurodegeneration. Neurotoxicol Teratol 24(5):607–620

Suchowersky O, Gronseth G, Perlmutter J, Reich S, Zesiewicz T, Weiner WJ (2006) Practice parameter: neuroprotective strategies and alternative therapies for Parkinson disease (an evidence-based review): report of the quality standards Subcommittee of the American Academy of neurology. Neurology 66 (7):976–982. doi:https://doi.org/10.1212/01.wnl.0000206363.57955.1b

Swerdlow RH (2007) Pathogenesis of Alzheimer's disease. Clin Interv Aging 2(3):347–359

Swerdlow RH, Parks JK, Miller SW, Tuttle JB, Trimmer PA, Sheehan JP, Bennett JP Jr, Davis RE, Parker WD Jr (1996) Origin and functional consequences of the complex I defect in Parkinson's disease. Ann Neurol 40(4):663–671. https://doi.org/10.1002/ana.410400417

Talantova M, Sanz-Blasco S, Zhang X, Xia P, Akhtar MW, Okamoto S, Dziewczapolski G, Nakamura T, Cao G, Pratt AE, Kang YJ, Tu S, Molokanova E, McKercher SR, Hires SA, Sason H, Stouffer DG, Buczynski MW, Solomon JP, Michael S, Powers ET, Kelly JW, Roberts

A, Tong G, Fang-Newmeyer T, Parker J, Holland EA, Zhang D, Nakanishi N, Chen HS, Wolosker H, Wang Y, Parsons LH, Ambasudhan R, Masliah E, Heinemann SF, Pina-Crespo JC, Lipton SA (2013) Abeta induces astrocytic glutamate release, extrasynaptic NMDA receptor activation, and synaptic loss. Proc Natl Acad Sci U S A 110(27):E2518–E2527. https://doi.org/10.1073/pnas.1306832110

Tamagno E, Bardini P, Obbili A, Vitali A, Borghi R, Zaccheo D, Pronzato MA, Danni O, Smith MA, Perry G, Tabaton M (2002) Oxidative stress increases expression and activity of BACE in NT2 neurons. Neurobiol Dis 10(3):279–288

Tamagno E, Parola M, Guglielmotto M, Santoro G, Bardini P, Marra L, Tabaton M, Danni O (2003) Multiple signaling events in amyloid beta-induced, oxidative stress-dependent neuronal apoptosis. Free Radic Biol Med 35(1):45–58

Tanner CM, Kamel F, Ross GW, Hoppin JA, Goldman SM, Korell M, Marras C, Bhudhikanok GS, Kasten M, Chade AR, Comyns K, Richards MB, Meng C, Priestley B, Fernandez HH, Cambi F, Umbach DM, Blair A, Sandler DP, Langston JW (2011) Rotenone, paraquat, and Parkinson's disease. Environ Health Perspect 119(6):866–872. https://doi.org/10.1289/ehp.1002839

Thal LJ, Grundman M, Berg J, Ernstrom K, Margolin R, Pfeiffer E, Weiner MF, Zamrini E, Thomas RG (2003) Idebenone treatment fails to slow cognitive decline in Alzheimer's disease. Neurology 61(11):1498–1502

Thomas SR, Leichtweis SB, Pettersson K, Croft KD, Mori TA, Brown AJ, Stocker R (2001) Dietary cosupplementation with vitamin E and coenzyme Q(10) inhibits atherosclerosis in apolipoprotein E gene knockout mice. Arterioscler Thromb Vasc Biol 21(4):585–593

Tsai KL, Chen LH, Chiou SH, Chiou GY, Chen YC, Chou HY, Chen LK, Chen HY, Chiu TH, Tsai CS, Ou HC, Kao CL (2011) Coenzyme Q10 suppresses oxLDL-induced endothelial oxidative injuries by the modulation of LOX-1-mediated ROS generation via the AMPK/PKC/NADPH oxidase signaling pathway. Mol Nutr Food Res 55(Suppl 2):S227–S240. https://doi.org/10.1002/mnfr.201100147

Tsai KL, Huang YH, Kao CL, Yang DM, Lee HC, Chou HY, Chen YC, Chiou GY, Chen LH, Yang YP, Chiu TH, Tsai CS, Ou HC, Chiou SH (2012) A novel mechanism of coenzyme Q10 protects against human endothelial cells from oxidative stress-induced injury by modulating NO-related pathways. J Nutr Biochem 23(5):458–468. https://doi.org/10.1016/j.jnutbio.2011.01.011

Tsuneki H, Sekizaki N, Suzuki T, Kobayashi S, Wada T, Okamoto T, Kimura I, Sasaoka T (2007) Coenzyme Q10 prevents high glucose-induced oxidative stress in human umbilical vein endothelial cells. Eur J Pharmacol 566(1–3):1–10. https://doi.org/10.1016/j.ejphar.2007.03.006

Tsuneki H, Tokai E, Suzuki T, Seki T, Okubo K, Wada T, Okamoto T, Koya S, Kimura I, Sasaoka T (2013) Protective effects of coenzyme Q10 against angiotensin II-induced oxidative stress in human umbilical vein endothelial cells. Eur J Pharmacol 701(1–3):218–227. https://doi.org/10.1016/j.ejphar.2012.12.027

Villalba JM, Parrado C, Santos-Gonzalez M, Alcain FJ (2010) Therapeutic use of coenzyme Q10 and coenzyme Q10-related compounds and formulations. Expert Opin Investig Drugs 19(4):535–554. https://doi.org/10.1517/13543781003727495

Voigt A, Berlemann LA, Winklhofer KF (2016) The mitochondrial kinase PINK1: functions beyond mitophagy. J Neurochem. https://doi.org/10.1111/jnc.13655

Wahlster L, Arimon M, Nasser-Ghodsi N, Post KL, Serrano-Pozo A, Uemura K, Berezovska O (2013) Presenilin-1 adopts pathogenic conformation in normal aging and in sporadic Alzheimer's disease. Acta Neuropathol 125(2):187–199. https://doi.org/10.1007/s00401-012-1065-6

Walsh DM, Klyubin I, Fadeeva JV, Cullen WK, Anwyl R, Wolfe MS, Rowan MJ, Selkoe DJ (2002) Naturally secreted oligomers of amyloid beta protein potently inhibit hippocampal long-term potentiation in vivo. Nature 416(6880):535–539. https://doi.org/10.1038/416535a

Weyer G, Babej-Dolle RM, Hadler D, Hofmann S, Herrmann WM (1997) A controlled study of 2 doses of idebenone in the treatment of Alzheimer's disease. Neuropsychobiology 36(2):73–82

Winkler-Stuck K, Wiedemann FR, Wallesch CW, Kunz WS (2004) Effect of coenzyme Q10 on the mitochondrial function of skin fibroblasts from Parkinson patients. J Neurol Sci 220(1–2):41–48. https://doi.org/10.1016/j.jns.2004.02.003

Yamagishi K, Ikeda A, Moriyama Y, Chei CL, Noda H, Umesawa M, Cui R, Nagao M, Kitamura A, Yamamoto Y, Asada T, Iso H (2014) Serum coenzyme Q10 and risk of disabling dementia: the circulatory risk in communities study (CIRCS). Atherosclerosis 237(2):400–403. https://doi.org/10.1016/j.atherosclerosis.2014.09.017

Yang L, Calingasan NY, Wille EJ, Cormier K, Smith K, Ferrante RJ, Beal MF (2009) Combination therapy with coenzyme Q10 and creatine produces additive neuroprotective effects in models of Parkinson's and Huntington's diseases. J Neurochem 109(5):1427–1439. https://doi.org/10.1111/j.1471-4159.2009.06074.x

Yang X, Dai G, Li G, Yang ES (2010) Coenzyme Q10 reduces beta-amyloid plaque in an APP/PS1 transgenic mouse model of Alzheimer's disease. J Mol Neurosci 41(1):110–113. https://doi.org/10.1007/s12031-009-9297-1

Yang X, Yang Y, Li G, Wang J, Yang ES (2008) Coenzyme Q10 attenuates beta-amyloid pathology in the aged transgenic mice with Alzheimer presenilin 1 mutation. J Mol Neurosci 34(2):165–171. https://doi.org/10.1007/s12031-007-9033-7

Yang YK, Wang LP, Chen L, Yao XP, Yang KQ, Gao LG, Zhou XL (2015) Coenzyme Q10 treatment of cardiovascular disorders of ageing including heart failure, hypertension and endothelial dysfunction. Clinica chimica acta; international journal of clinical chemistry 450:83-89. https://doi.org/10.1016/j.cca.2015.08.002

Yasojima K, McGeer EG, McGeer PL (2001) Relationship between beta amyloid peptide generating molecules and neprilysin in Alzheimer disease and normal brain. Brain Res 919(1):115–121

Yoritaka A, Kawajiri S, Yamamoto Y, Nakahara T, Ando M, Hashimoto K, Nagase M, Saito Y, Hattori N (2015) Randomized, double-blind, placebo-controlled pilot trial of reduced coenzyme Q10 for Parkinson's disease. Parkinsonism Relat Disord 21(8):911–916. https://doi.org/10.1016/j.parkreldis.2015.05.022

Zamani M, Katebi M, Mehdizadeh M, Mohamadzadeh F, Soleimani M (2012) Coenzyme Q10 protects hippocampal neurons against ischemia/reperfusion injury via modulation of BAX/Bcl-2 expression. Basic Clin Neurosci 3:5–10

Zawada WM, Banninger GP, Thornton J, Marriott B, Cantu D, Rachubinski AL, Das M, Griffin WS, Jones SM (2011) Generation of reactive oxygen species in 1-methyl-4-phenylpyridinium (MPP+) treated dopaminergic neurons occurs as an NADPH oxidase-dependent two-wave cascade. J Neuroinflammation 8:129. https://doi.org/10.1186/1742-2094-8-129

Zhang X, Zhou K, Wang R, Cui J, Lipton SA, Liao FF, Xu H, Zhang YW (2007) Hypoxia-inducible factor 1alpha (HIF-1alpha)-mediated hypoxia increases BACE1 expression and beta-amyloid generation. J Biol Chem 282(15):10873–10880. https://doi.org/10.1074/jbc.M608856200

Zheng B, Liao Z, Locascio JJ, Lesniak KA, Roderick SS, Watt ML, Eklund AC, Zhang-James Y, Kim PD, Hauser MA, Grunblatt E, Moran LB, Mandel SA, Riederer P, Miller RM, Federoff HJ, Wullner U, Papapetropoulos S, Youdim MB, Cantuti-Castelvetri I, Young AB, Vance JM, Davis RL, Hedreen JC, Adler CH, Beach TG, Graeber MB, Middleton FA, Rochet JC, Scherzer CR (2010) PGC-1alpha, a potential therapeutic target for early intervention in Parkinson's disease. Science translational medicine 2 (52):52ra73. https://doi.org/10.1126/scitranslmed.3001059

Chapter 13
Immunosenescence and CoQ_{10}

Juan Diego Hernández-Camacho, Catherine Meza-Torres, and Guillermo López-Lluch

Abstract The decline of the immune system associated to aging is becoming a global issue nowadays due to the increase of lifespan. Immunosenescence is the deteriorated adaptative capacity of the immune system leading to increase in morbidity and mortality in elderly population. Aging has been related to a progressive affectation of mitochondria such as increase of mtDNA mutation, deregulation of electron transport chain, increase in ROS production and reduction in ATP production. Immunosenescence has been associated with aging associated mitochondrial dysfunction linked to chronic inflammation that induce higher oxidative stress levels and mitochondrial damage deteriorating the already compromised immune system. CoQ_{10} administration could be considered a beneficial strategy in order to reduce the progression of age-related diseases. Highlighting the important role of CoQ_{10} on mitochondrial function it has been proposed that this unique antioxidant soluble lipid could be helpful in the management of immunosenescence. High levels of CoQ_{10} have been related with a delay of immunosenescence attributable to a reduction in DNA damage. Other prolongevity factors, such as caloric restriction and exercise, have been considered for their potential effect in delaying aging effect and immunosenescence mainly affecting mitochondrial physiology and biogenesis and antioxidant activity.

Keywords Mitochondria · Coenzyme Q_{10} · Oxidative stress · Immunosenescence · Caloric restriction · Exercise

---

J. D. Hernández-Camacho · C. Meza-Torres · G. López-Lluch (✉)
Centro Andaluz de Biología del Desarrollo, CABD-CSIC, CIBERER, Instituto de Salud Carlos III, Universidad Pablo de Olavide, Sevilla, Spain
e-mail: glopllu@upo.es

G. López Lluch (ed.), *Coenzyme Q in Aging*,
https://doi.org/10.1007/978-3-030-45642-9_13

13.1 Introduction to the Concept of Immunosenescence

One of the most important physiological alterations happening during aging is the deterioration of the immune system, a process known as immunosenescence (Currie 1992). Immunosenescence was negligible in the last centuries due to the lower life expectancy since it is associated with elderly people. Therefore, the importance of immunosenescence has emerged as a consequence of the extension of the lifespan of individuals (De Martinis et al. 2007). In the past, the immune system of the majority of humans was active during 30–50 years; however, nowadays the immune system of most of the individuals must be active for 80 or more years. This longer activity of a very complex system probably had been not contemplated from an evolutionary point of view that did not considered the chronic effect of antigens as stressors (Franceschi et al. 1999). For this reason, immunosenescence been also defined as the result of a remodeling of the immune system as consequence of continuous adaptations to deteriorative changes over the time (De Martinis et al. 2007).

Immunosenescence has been considered as responsible of increases in morbidity and mortality observed in elderly people (Currie 1992). It has been demonstrated that it is associated with an increased risk of suffering infections with decreased probability of surviving acute episodes (Hazeldine and Lord 2015). During aging the innate and adaptive immune responses changes, decreasing the ability to respond to bacterial and viral pathogens (Panda et al. 2009). Lifelong antigenic load has been considered one of the main factors in immunosenescence since it decreases the numbers of virgin T-cells and at the same time accumulates expanded clones of memory and effector cells in the immunological space. In fact, the pattern of some specific cells (Helper T-cells ($CD4^+$) and Cytotoxic T-cells ($CD8^+$)) in elderly people is comparable with individuals suffering persistent infections or autoimmunity, indicating the importance of the antigenic load (Panda et al. 2009). To this pattern, we have to add that elderly people also present a progressive decrease in the production of T-cells in thymus, affecting both $CD4^+$ and $CD8^+$, due to thymus degeneration (Currie 1992).

Immunosenescence can be modulated by prolongevity factors. Exercise improves immunological and anti-inflammatory response decreasing morbidity and mortality from specific diseases (Turner 2016). In the case of caloric restriction (CR), another known prolongevity factor, studies in humans are controversial since long-term caloric restriction procedures in humans are very complex. In aged long-lived non-human primates caloric restriction delays T-cell senescence indicating a putative role in the progression of immunosenescence (Messaoudi et al. 2006). In the case of bioactive compounds such as polyphenols, also considered prolongevity effectors, the results are conflicting although research indicate the improvement of the immune function and reduction of proinflammatory profiles (Petersen et al. 2018; Zhang et al. 2012; Baeza et al. 2010).

13.2 Mitochondria and Immunosenescence

Human mitochondrial DNA (mtDNA) is a 16.6-kb circular DNA that contains only 37 genes. Twenty-two genes codify transfer RNAs and two other specify ribosomal RNAs. Only 13 genes encode polypeptides belonging to the respiratory chain (OXPHOS) system (Howell et al. 1992). The respiratory complexes also contain approximately 70 nuclear DNA-encoded structural subunits that are synthesized in the cytosol and are imported into the organelle, where they are co-assembled with the mtDNA-encoded subunits into the respective holoenzymes (van den Heuvel and Smeitink 2001).

Mitochondria follow the rules of population genetics. To understand the etiology and pathology of mitochondrial-related disorders six aspects are considered as essential (Taylor and Turnbull 2005). First, they are maternally inherited. Second, in cells, hundreds of mitochondria contains thousands of mitochondrial genomes. Third, a single cell can contains two or more mtDNA genotypes because the mutation of mtDNA population; this means that in mitochondria we can found high levels of heteroplasmy. Fourth, if the mutation causes a pathogenic phenotype, the proportion of mutated mtDNA indicates the severity of the pathology but not necessarily in a lineal fashion. Fifth, mtDNA replication and inheritance in lineages of somatic cells is stochastic, this causes changes in mutational loads along the life of the individual affecting different cells and tissues (mitotic segregation). Sixth, the level of energy requirements of different cells indicate that the level of heteroplasmy affects differently cells and organs (Taylor and Turnbull 2005).

Mitochondria are the major intracellular sources of oxygen free radicals, referred as reactive oxygen species (ROS). ROS cause cumulative damage to cellular constituents (DNA, RNA, proteins and lipids) that is postulated to result in aging and eventual death as indicated in the mitochondrial free radical theory of aging (Cadenas and Davies 2000). As mitochondrial organelles play in the central role in apoptosis and in ATP production, age-associated defects in mitochondrial function are postulated to produce consequences in the cell function. Many of these age-associated defects are associated with the increase of mtDNA mutations and the dysfunction of the electron transport chain causing more electron leaking, more release of ROS with less capacity to produce ATP (Lopez-Lluch et al. 2008; Lopez-Lluch et al. 2010). In fact, it is accepted that during human aging, decline in mitochondrial bioenergetics function can be in part attributed to the accumulation of genetic abnormalities within the mtDNA (Ross et al. 2002).

Regarding immune system, the information about the effect of mitochondrial dysfunction in its activity during aging is relatively scarce. Ross et al. (2002) did not find an age-related mutation of mtDNA in *in vitro* T cell clones or in *ex vivo* lymphocyte DNA samples (Ross et al. 2002). The authors indicate that their results do not support the hypothesis of a damage of mtDNA with immunosenescence. However, a recent revision hypothesize that immunosenescence can be partially related to the progression of mitochondrial dysfunction during aging (McGuire 2019). This mitochondrial dysfunction can be specially affecting T-helper $CD4^+$

lymphocytes and be caused by defective mitochondrial turnover (Bektas et al. 2019). Further, response of NK cells to IL-2 signaling is defective in old individuals affecting mitochondrial function and producing higher oxidative stress than NK from young individuals (Miranda et al. 2018). It seems clear that maintenance of mitochondrial function during aging can affect the activity of the immune system and that mitochondrial dysfunction is a factor to be taken into consideration in the evolution of immunosenescence.

A very recent study demonstrate the importance of mtDNA damage in the activity of the immune system. Dysregulation of mtDNA metabolism can be important in the promotion of inflammation (Zhong et al. 2019). The proinflammatory profile found during aging has been associated with the progression of many age-associated diseases. ROS could directly or indirectly activate mechanisms that increase the release of proinflammatory mediators (Vitale et al. 2013). ROS-induced DNA damage, particularly could be involved in production of IL-1β, IL-18, TNFα and interferon (Zhong et al. 2019). Through the activation of inflammasomes, ROS and the release of mtDNA, process proIL-1β and IL-18 to their respective final forms that are released from cells inducing inflammatory responses. Therefore, a high ROS production mainly from mitochondrial dysfunction for a long time, may connect oxidative stress and inflammation aging theories (Vitale et al. 2013). Disruption of mitochondrial dynamics may lead in higher ROS production and accumulation of damaged mitochondria that can aggravate the induction of proinflammatory processes impairing the progression of age-related diseases (Caruso et al. 2009).

Inflammaging theory speculates that aging is associated with a chronic inflammation status regulated by genes. Additionally, this idea is related to the deterioration of the immune system. Elderly people present less capacity for defense against new antigens and higher incidence of chronic inflammatory reactions. Inflammaging can be defined as a low grade chronic inflammation during aging. This status has been related with increases of morbidity and mortality in elderly people. Although it can be prevented and cured, nutritional and exercise interventions are proposed to lower morbidity and chronic inflammation (Franceschi and Campisi 2014).

A low grade inflammation status, determined by increased levels of inflammatory biomarkers (C-reactive protein or interleukin-6), is related with aging (Franceschi and Campisi 2014). Damaged macromolecules or cells accumulated with age by both higher production and/or by inadequate elimination. Damaged organelle components or free radicals could be recognized as danger signals (DAMPs) and consequently induce immune actions for physiological repair mainly producing inflammation. Theses immune responses could become chronic and maladaptive (Franceschi and Campisi 2014).

Mitochondria is a key point in inflammation through the activation of NLRP3 inflammasome. This is a multiprotein complex that could activate pro-caspase-1 as a response to cellular danger signals resulting in the secretion of proinflammatory cytokines (IL-1β and IL-18) (Franceschi and Campisi 2014). As it has been mentioned before, mitochondrial dysfunction generates cardiolipin, found only in bacterial and mitochondrial membrane, can activate the proinflammatory pathway of NLRP3 inflammasome.

It has been demonstrated that spleen lymphocytes from old mice showed lower respiration rates and mitochondrial membrane potential than lymphocytes from young mice (Rottenberg and Wu 1997). Lymphocytes from old animals presented a higher susceptibility of mitochondrial permeability transition and higher induction of apoptosis by mitochondrial inhibitor, cyclosporine, facilitating immunosenescence status (Rottenberg and Wu 1997).

All these evidence indicate that mitochondrial activity is essential for the balanced activity of the immune system. Mitochondrial dysfunction associated with aging impairs adaptive immune response at the same time that increases inflammatory response. Mitochondrial dysfunction and its bioenergetics effects can be responsible of immunosenescence.

13.3 CoQ_{10} and Immunosenescence

CoQ_{10} (2,3-dimethoxy-5-methy-6-decaprenyl-1,4-benzoquinone) plays a key role in mitochondrial bioenergetics and as main antioxidant in cell membranes including mitochondria (Lopez-Lluch et al. 2010). Deficiency of CoQ_{10} can be associated with the progression of many age-related diseases and supplementation could help to delay their progression (Hernandez-Camacho et al. 2018). As has been indicated previously, mitochondrial dysfunction is associated with the evolution of immunosenescence. Taken into consideration the essential role of CoQ_{10} in mitochondrial function (Rotig et al. 2000) and the decrease of CoQ levels during aging specially in thymus (Bliznakov et al. 1978), maintenance of its levels during aging could be an important factor to prevent immunosenescence. The antioxidant function of CoQ_{10} and the prevention of oxidative damage in mitochondria is another aspect to be taken into consideration in immunosenescence. In fact, supplementation of diet with CoQ_{10} prevents oxidative DNA damage (Quiles et al. 2005) and specially in human lymphocytes (Tomasetti et al. 1999). Further, supplementation also enhances the recovery of the lymphocytes from oxidative DNA damage (Tomasetti et al. 2001) and increase mitochondrial function in immune cells (Jhun et al. 2016).

Long time ago, supplementation with CoQ_{10} in mice demonstrated the capacity to reverse the suppression of the humoral primary response in old mice indicating the capacity to reduce immunosenescence (Bliznakov 1978). Interestingly, supplementation with CoQ_{10} also enhanced the capacity of immune cells to attack mastocytoma cancer cells in an effect associated with immunorestoration of the capacity of the immune system to fight against cancer (Kawase et al. 1978). This effect was attributed to the capacity of CoQ_{10} to restore mitochondrial function in depressed immune cells (Niitani et al. 1979). These and other studied concluded that mitochondrial activity of CoQ_{10} is essential for the optimal function of the immune system (Folkers and Wolaniuk 1985).

Immune deficiency and mitochondrial disease could be related with internal CoQ_{10} deficiency associated with higher recurrence of severe infections (Farough et al. 2014). This clinical situation could be improved with CoQ_{10} oral administration resulting in less frequent severe infections. CoQ_{10} supplementation for 12 weeks affected innate immune components related with the pathogenesis of diabetes and long term complications. On the other hand, supplementation with CoQ_{10} decreased serum hBD2 (human beta defensins 2) levels and changed activation markers and distribution in peripheral natural killer cells in diabetes mellitus type 1 patients, both factor involved in cardiovascular disease development (Brauner et al. 2014). Treatment with CoQ_{10} up-regulated the activation of receptor NKG2D and increased the proportion of $CD56^{bright}$ NK cells in T1DM patients. While the proinflammatory marker, hBD2, decreases, indicating a reduction in inflammatory status associate to diabetes. In fact, it has been proposed that CoQ_{10} could represent a novel way in order to prevent long term inflammatory immune complications associated to this disease (Brauner et al. 2014). Furthermore, natural killer cell function can be affected by vitamin E and ubiquinone reductions during aging (Gorczynski and Terzioglu 2008).

An interesting relationship between CoQ_{10} levels and the activity of the immune system has been obtained from studies with *D. melanogaster* (Cheng et al. 2011). Insects defective in gene *COQ2*, a CoQ synthesis factor, present a higher sensitivity to bacterial and fungal infections and CoQ_{10} supplementation partially rescued the immune function. Surprisingly, wild type drosophila presented more susceptibility to viral infections when they were supplemented with CoQ_{10}. These works indicate that alterations in CoQ genes could have repercussions in the susceptibility to infections although improving of the immune system can occurs maintaining CoQ levels in a normal range.

Supplementation with CoQ has presented radioprotection after chronic whole body irradiation (Novoselova et al. 1998). This irradiation caused changes in splenic T lymphocytes proliferation and chronic irradiation reduced immune response of T lymphocytes and produced changes in DNA of these cells (Novoselova et al. 1995). Supplementation with CoQ_9 restored the defects produced by chronic radiation indicating a protective effect of CoQ. Another research determined the effects of CoQ_{10} and vitamin B_6 (pyridoxine) ingestion together and separately (Folkers et al. 1993). Authors observed that blood levels of CoQ_{10} and immunoglobulin G were increased when CoQ_{10} were ingested alone or with pyridoxine. Blood levels of T4-lymphocytes and the ratio of T4/T8 lymphocytes increased when CoQ_{10} and vitamin B_6 were supplemented together or separately. The authors believed that these applications could be beneficial on acquired immune deficiency syndrome, infectious diseases or cancer, since this same effect was found in humans (Folkers et al. 1991).

13.4 CoQ Prevents DNA Damage in Immunosenescence

Regarding the effect of oxidative damage of mitochondrial and nuclear DNA in the activity of lymphocytes, it was demonstrated long time ago that high levels of CoQ_{10} in these cells prevent oxidative damage both in vitro (Tomasetti et al. 1999) and in vivo (Tomasetti et al. 2001). Then, if DNA damage is associated with immunosenescence (Ross et al. 2002), high levels of CoQ_{10} can delay immunosenescence progression by reducing DNA damage and probably the activation of proinflammatory processes by decreasing DAMPs. In studies performed with patients suffering mitochondrial diseases, supplementation with CoQ_{10} in leukocytes reduced the level of nuclear damage although without affecting the level of DNA damage (Migliore et al. 2004). However, in these cases, mitochondrial pathology of the individuals can produce higher DNA damage that cannot reflect the normal evolution of damage during aging. Interestingly, in studies performed in rats, chronic supplementation with low amounts of CoQ_{10} in the diet produced protection against oxidative damage in DNA in peripheral lymphocytes at the same time that increases life-span (Quiles et al. 2004; Quiles et al. 2005).

One important issue in immune system function is that the activation of the T-cell receptor induces signals that activate the mitochondrial biogenesis and changes mitochondrial activity including amplification of mtDNA indicating the importance of mitochondria in the activation of these cells (D'Souza et al. 2007). If aging accumulates mtDNA damage in these cells, activation of mitochondria from high damaged mtDNA-harboring cells would end in dysfunctional lymphocytes.

It seems clear that CoQ_{10} as other natural occurring antioxidants such as α-tocopherol can reduce inflammatory response in mice induced by LPS in a model of acute inflammation by modulating the activation of the NF-κB signaling pathway indicating the importance of maintaining the levels of lipid antioxidants in the regulation of inflammation (Novoselova et al. 2009).

On the other hand, CoQ_{10} has recently shown immunomodulatory properties in colitis in a mice model. Supplementation with CoQ_{10} reduced colitis effect at the histological level and reduced the levels of the proinflammatory cytokine IL-17 and the levels of anti-inflammatory cytokines and the phosphorylation of STAT-3, indicating modulation of activating pathways (Lee et al. 2017). The effect was associated with the activation of the AMPK and FOXP3 regulatory pathways involved in mitochondrial turnover.

13.5 Prolongevity Effectors Affect the Immune System

Prolongevity interventions can modulate the evolution of immunosenescence in organismos. The most powerful interventions to delay aging are caloric restriction (CR), physical activity and some dietary bioactive compounds. In the case of CR, the studies in humans show controversial results. Long-term CR was recently not associated with delay in immunologic aging but this study only considered the length of telomeres in T-cells as marker of immunosenescence (Tomiyama et al. 2017). However, studies performed in aged long-lived nonhuman primates have demonstrated that CR delays T-cell senescence, indicating a putative role in the progression of immunosenescence (Messaoudi et al. 2006). CR has been shown to affect $CD4^+$ and $CD8^+$ T-lymphocytes in the spleen, mesenteric lymph nodes, peripheral blood, thymus and salivary glands (Jolly 2004). Previous studies performed in mice demonstrate that CR prevented the thymic deterioration and improved the reponse of T-cells to IL-2 (Yang et al. 2009). The effect on these animals was suggested to be associated with the improvement in signal transduction that activates cells (Pahlavani 2004). Taken into consideration the importance of bioenergetics in the activity of T-cells (Choi et al. 2017) and the effect of CR on mitochondrial activity and turnover (Lopez-Lluch et al. 2006) and in CoQ_{10}-dependent antioxidant activities (De Cabo et al. 2004; Lopez-Lluch et al. 2005), the maintenance of mitochondrial activity in the immune system can be one of the main effect by which CR delays immunosenescence.

Exercise, a well-known anti-aging factor, also shows anti-immunosenescence activity by improving immunological and anti-inflammatory response decreasing morbidity and mortality from specific diseases showing an immunological and inflammatory etiology (Turner 2016). Although the type of exercise more appropriate to regulate immune system in the elderly is under controversy (Abd El-Kader and Al-Shreef 2018; Cao Dinh et al. 2019), it seems clear that physical activity can improve immunological response in old mice (Lee et al. 2019) and in humans (Davison et al. 2016; Sellami et al. 2018). Consequently, we can consider that the practice of physical activity or exercise in elderly people is an important strategy to prevent immunosenescence. In association with the immune modulatory effect of exercise, we can consider that at least part of the positive effect can be due to a decrease in dysfunctional mitochondrial by incrementing mitochondrial biogenesis (Lumini et al. 2008). Further, we cannot discard the association of the immunomodulatory effect of exercise with the stimulation of the production of CoQ levels as we demonstrated in mice muscle (Rodriguez-Bies et al. 2015; Rodriguez-Bies et al. 2010) or in active old people showing higher CoQ_{10} levels in plasma (Del Pozo-Cruz et al. 2014). Physical activity would then improve the activity of mitochondria in lymphocytes and also maintain high CoQ_{10} levels for bioenergetics homeostasis and antioxidant capacity.

Bioactive compounds such as polyphenols are able to affect mitochondrial dysfunction and to regulate CoQ_{10} synthesis and CoQ_{10}-dependent antioxidant activities (Baur et al. 2006; Tung et al. 2014). Further, resveratrol, a known polyphenol, also increases mitochondrial mass and regulates mitochondrial activity (Baur et al. 2006; Rodriguez-Bies et al. 2016). Interestingly, CoQ_{10} and resveratrol exert synergistic activity in the prevention of inflammatory insult in a model of muscle damage in mice, indicating immunomodulatory effects (Potgieter et al. 2011). In fact, bioactive compounds have shown positive effects of the immune function in many different models of immunological dysfunction such as ovariectomized mice (Baeza et al. 2010), normal aging (Yuan et al. 2012) or in senescence accelerated mice (Zhang et al. 2012). In humans, extracts rich in bioactive compounds are able to activate immune cells (Tumova et al. 2017) although with other extracts no improvement of the immune system activity was found (Hunter et al. 2012).

In general, it seems clear that prolongevity effectors can improve immune system in elderly people although the mechanism of action remains elusive. Taken into consideration the importance of bioenergetics, mitochondrial activity and antioxidant protection in the immune response, we can speculate that CoQ_{10} homeostasis can play an important role in this response and in the effect of these factors.

13.6 Concluding Remarks

Taken all these studies in consideration, it seems that CoQ_{10} can be an important factor in the maintenance of the activity of the immune system during aging. We can speculate that changes in the levels of CoQ_{10} during aging could be related to changes observed in immunosenescence and proinflammatory profile. Levels of CoQ decreases with age (Onur et al. 2014). Further, ubiquinone levels in plasma increases with age probably in response to a higher oxidative stress or a decrease in reducing capacity in elderly (Takahashi et al. 2019). These alterations could affect to the susceptibility of elderly people to infectious diseases and to the response to their immune systems. The evidence accumulated to date indicate that supplementation with CoQ_{10} can improve the activity of the immune system during aging affecting mitochondrial dysfunction, and antioxidant capacity (Fig. 13.1). These effects can ameliorate the activity of the acquired immune response at the same time that reduce chronic inflammation.

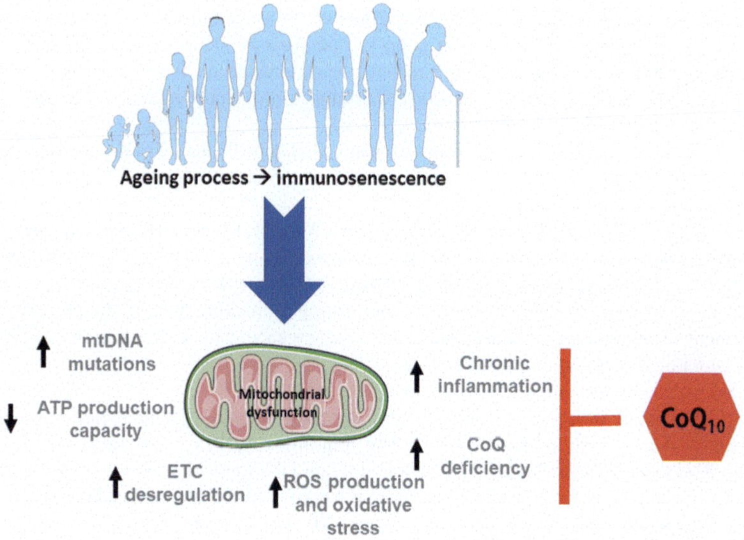

Fig. 13.1 Mitochondrial dysfunction is associated with immunosenescence. Mitochondrial dysfunction affects many aspects of the immune system and probably the evolution of immunosenescence. Supplementation with CoQ_{10} reduces mtDNA mutations, improves ATP production and activity of the ETC, reduces ROS production and oxidative stress and, in general, reduces chronic inflammation

References

Abd El-Kader SM, Al-Shreef FM (2018) Inflammatory cytokines and immune system modulation by aerobic versus resisted exercise training for elderly. Afr Health Sci 18(1):120–131. https://doi.org/10.4314/ahs.v18i1.16

Baeza I, De Castro NM, Arranz L, De la Fuente M (2010) Soybean and green tea polyphenols improve immune function and redox status in very old ovariectomized mice. Rejuvenation Res 13(6):665–674. https://doi.org/10.1089/rej.2010.1049

Baur JA, Pearson KJ, Price NL, Jamieson HA, Lerin C, Kalra A, Prabhu VV, Allard JS, Lopez-Lluch G, Lewis K, Pistell PJ, Poosala S, Becker KG, Boss O, Gwinn D, Wang M, Ramaswamy S, Fishbein KW, Spencer RG, Lakatta EG, Le Couteur D, Shaw RJ, Navas P, Puigserver P, Ingram DK, de Cabo R, Sinclair DA (2006) Resveratrol improves health and survival of mice on a high-calorie diet. Nature 444(7117):337–342. https://doi.org/10.1038/nature05354

Bektas A, Schurman SH, Gonzalez-Freire M, Dunn CA, Singh AK, Macian F, Cuervo AM, Sen R, Ferrucci L (2019) Age-associated changes in human CD4(+) T cells point to mitochondrial dysfunction consequent to impaired autophagy. Aging (Albany NY) 11. https://doi.org/10.18632/aging.102438

Bliznakov EG (1978) Immunological senescence in mice and its reversal by coenzyme Q10. Mech Ageing Dev 7(3):189–197. https://doi.org/10.1016/0047-6374(78)90065-9

Bliznakov EG, Watanabe T, Saji S, Folkers K (1978) Coenzyme Q deficiency in aged mice. J Med 9(4):337–346

Brauner H, Luthje P, Grunler J, Ekberg NR, Dallner G, Brismar K, Brauner A (2014) Markers of innate immune activity in patients with type 1 and type 2 diabetes mellitus and the effect of the anti-oxidant coenzyme Q10 on inflammatory activity. Clin Exp Immunol 177(2):478–482. https://doi.org/10.1111/cei.12316

Cadenas E, Davies KJ (2000) Mitochondrial free radical generation, oxidative stress, and aging. Free Radic Biol Med 29(3–4):222–230. https://doi.org/10.1016/s0891-5849(00)00317-8

Cao Dinh H, Njemini R, Onyema OO, Beyer I, Liberman K, De Dobbeleer L, Renmans W, Vander Meeren S, Jochmans K, Delaere A, Knoop V, Bautmans I (2019) Strength endurance training but not intensive strength training reduces senescence-prone T cells in peripheral blood in community-dwelling elderly women. J Gerontol A Biol Sci Med Sci 74(12):1870–1878. https://doi.org/10.1093/gerona/gly229

Caruso C, Buffa S, Candore G, Colonna-Romano G, Dunn-Walters D, Kipling D, Pawelec G (2009) Mechanisms of immunosenescence. Immun Ageing 6:10. https://doi.org/10.1186/1742-4933-6-10

Cheng W, Song C, Anjum KM, Chen M, Li D, Zhou H, Wang W, Chen J (2011) Coenzyme Q plays opposing roles on bacteria/fungi and viruses in Drosophila innate immunity. Int J Immunogenet 38(4):331–337. https://doi.org/10.1111/j.1744-313X.2011.01012.x

Choi IY, Lee C, Longo VD (2017) Nutrition and fasting mimicking diets in the prevention and treatment of autoimmune diseases and immunosenescence. Mol Cell Endocrinol 455:4–12. https://doi.org/10.1016/j.mce.2017.01.042

Currie MS (1992) Immunosenescence. Compr Ther 18(11):26–34

Davison G, Kehaya C, Wyn Jones A (2016) Nutritional and physical activity interventions to improve immunity. Am J Lifestyle Med 10(3):152–169. https://doi.org/10.1177/1559827614557773

De Cabo R, Cabello R, Rios M, Lopez-Lluch G, Ingram DK, Lane MA, Navas P (2004) Calorie restriction attenuates age-related alterations in the plasma membrane antioxidant system in rat liver. Exp Gerontol 39(3):297–304. https://doi.org/10.1016/j.exger.2003.12.003

De Martinis M, Franceschi C, Monti D, Ginaldi L (2007) Apoptosis remodeling in immunosenescence: implications for strategies to delay ageing. Curr Med Chem 14(13):1389–1397. https://doi.org/10.2174/092986707780831122

Del Pozo-Cruz J, Rodriguez-Bies E, Ballesteros-Simarro M, Navas-Enamorado I, Tung BT, Navas P, Lopez-Lluch G (2014) Physical activity affects plasma coenzyme Q10 levels differently in young and old humans. Biogerontology 15(2):199–211. https://doi.org/10.1007/s10522-013-9491-y

D'Souza AD, Parikh N, Kaech SM, Shadel GS (2007) Convergence of multiple signaling pathways is required to coordinately up-regulate mtDNA and mitochondrial biogenesis during T cell activation. Mitochondrion 7(6):374–385. https://doi.org/10.1016/j.mito.2007.08.001

Farough S, Karaa A, Walker MA, Slate N, Dasu T, Verbsky J, Fusunyan R, Canapari C, Kinane TB, Van Cleave J, Sweetser DA, Sims KB, Walter JE (2014) Coenzyme Q10 and immunity: a case report and new implications for treatment of recurrent infections in metabolic diseases. Clin Immunol 155(2):209–212. https://doi.org/10.1016/j.clim.2014.09.010

Folkers K, Wolaniuk A (1985) Research on coenzyme Q10 in clinical medicine and in immunomodulation. Drugs Exp Clin Res 11(8):539–545

Folkers K, Hanioka T, Xia LJ, McRee JT Jr, Langsjoen P (1991) Coenzyme Q10 increases T4/T8 ratios of lymphocytes in ordinary subjects and relevance to patients having the AIDS related complex. Biochem Biophys Res Commun 176(2):786–791. https://doi.org/10.1016/s0006-291x(05)80254-2

Folkers K, Morita M, McRee J Jr (1993) The activities of coenzyme Q10 and vitamin B6 for immune responses. Biochem Biophys Res Commun 193(1):88–92. https://doi.org/10.1006/bbrc.1993.1593

Franceschi C, Campisi J (2014) Chronic inflammation (inflammaging) and its potential contribution to age-associated diseases. J Gerontol A Biol Sci Med Sci 69(Suppl 1):S4–S9. https://doi.org/10.1093/gerona/glu057

Franceschi C, Valensin S, Fagnoni F, Barbi C, Bonafe M (1999) Biomarkers of immunosenescence within an evolutionary perspective: the challenge of heterogeneity and the role of antigenic load. Exp Gerontol 34(8):911–921. https://doi.org/10.1016/s0531-5565(99)00068-6

Gorczynski RM, Terzioglu E (2008) Aging and the immune system. Int Urol Nephrol 40(4):1117–1125. https://doi.org/10.1007/s11255-008-9412-1

Hazeldine J, Lord JM (2015) Innate immunesenescence: underlying mechanisms and clinical relevance. Biogerontology 16(2):187–201. https://doi.org/10.1007/s10522-014-9514-3

Hernandez-Camacho JD, Bernier M, Lopez-Lluch G, Navas P (2018) Coenzyme Q10 supplementation in aging and disease. Front Physiol 9:44. https://doi.org/10.3389/fphys.2018.00044

Howell N, McCullough DA, Kubacka I, Halvorson S, Mackey D (1992) The sequence of human mtDNA: the question of errors versus polymorphisms. Am J Hum Genet 50(6):1333–1340

Hunter DC, Skinner MA, Wolber FM, Booth CL, Loh JM, Wohlers M, Stevenson LM, Kruger MC (2012) Consumption of gold kiwifruit reduces severity and duration of selected upper respiratory tract infection symptoms and increases plasma vitamin C concentration in healthy older adults. Br J Nutr 108(7):1235–1245. https://doi.org/10.1017/S0007114511006659

Jhun J, Lee S, Kim SY, Na HS, Kim EK, Kim JK, Jeong JH, Park SH, Cho ML (2016) Combination therapy with metformin and coenzyme Q10 in murine experimental autoimmune arthritis. Immunopharmacol Immunotoxicol 38(2):103–112. https://doi.org/10.3109/08923973.2015.1122619

Jolly CA (2004) Dietary restriction and immune function. J Nutr 134(8):1853–1856. https://doi.org/10.1093/jn/134.8.1853

Kawase I, Niitani H, Saijo N, Sasaki H, Morita T (1978) Enhancing effect of coenzyme, Q10 on immunorestoration with Mycobacterium bovis BCG in tumor-bearing mice. Gan 69(4):493–497

Lee SY, Lee SH, Yang EJ, Kim JK, Kim EK, Jung K, Jung H, Lee K, Lee HH, Lee BI, Park SH, Shin DY, Cho ML (2017) Coenzyme Q10 inhibits Th17 and STAT3 signaling pathways to ameliorate colitis in mice. J Med Food 20(9):821–829. https://doi.org/10.1089/jmf.2016.3859

Lee JY, Paik IY, Kim JY (2019) Voluntary exercise reverses immune aging induced by oxidative stress in aging mice. Exp Gerontol 115:148–154. https://doi.org/10.1016/j.exger.2018.08.009

Lopez-Lluch G, Rios M, Lane MA, Navas P, de Cabo R (2005) Mouse liver plasma membrane redox system activity is altered by aging and modulated by calorie restriction. Age (Dordr) 27(2):153–160. https://doi.org/10.1007/s11357-005-2726-3

Lopez-Lluch G, Hunt N, Jones B, Zhu M, Jamieson H, Hilmer S, Cascajo MV, Allard J, Ingram DK, Navas P, de Cabo R (2006) Calorie restriction induces mitochondrial biogenesis and bioenergetic efficiency. Proc Natl Acad Sci U S A 103(6):1768–1773. https://doi.org/10.1073/pnas.0510452103

Lopez-Lluch G, Irusta PM, Navas P, de Cabo R (2008) Mitochondrial biogenesis and healthy aging. Exp Gerontol 43(9):813–819. https://doi.org/10.1016/j.exger.2008.06.014

Lopez-Lluch G, Rodriguez-Aguilera JC, Santos-Ocana C, Navas P (2010) Is coenzyme Q a key factor in aging? Mech Ageing Dev 131(4):225–235. https://doi.org/10.1016/j.mad.2010.02.003

Lumini JA, Magalhaes J, Oliveira PJ, Ascensao A (2008) Beneficial effects of exercise on muscle mitochondrial function in diabetes mellitus. Sports Med 38(9):735–750. https://doi.org/10.2165/00007256-200838090-00003

McGuire PJ (2019) Mitochondrial dysfunction and the aging immune system. Biology (Basel) 8(2). https://doi.org/10.3390/biology8020026

Messaoudi I, Warner J, Fischer M, Park B, Hill B, Mattison J, Lane MA, Roth GS, Ingram DK, Picker LJ, Douek DC, Mori M, Nikolich-Zugich J (2006) Delay of T cell senescence by caloric restriction in aged long-lived nonhuman primates. Proc Natl Acad Sci U S A 103(51):19448–19453. https://doi.org/10.1073/pnas.0606661103

Migliore L, Molinu S, Naccarati A, Mancuso M, Rocchi A, Siciliano G (2004) Evaluation of cytogenetic and DNA damage in mitochondrial disease patients: effects of coenzyme Q10 therapy. Mutagenesis 19(1):43–49. https://doi.org/10.1093/mutage/geg036

Miranda D, Jara C, Mejias S, Ahumada V, Cortez-San Martin M, Ibanez J, Hirsch S, Montoya M (2018) Deficient mitochondrial biogenesis in IL-2 activated NK cells correlates with impaired PGC1-alpha upregulation in elderly humans. Exp Gerontol 110:73–78. https://doi.org/10.1016/j.exger.2018.05.014

Niitani H, Kawase I, Taniguchi T, Saijo N, Irimajiri N (1979) Combined effect of BCG and coenzyme Q10 on ATP-ase activity and coenzyme Q content in spleen lymphocytes of tumor-bearing rats. Gan 70(3):315–322

Novoselova EG, Safonova MV, Gordon R, Semiletova NV (1995) Immune functions of spleen lymphocytes of rats subjected to chronic irradiation and antioxidant (ubiquinone Q-9) diet. Int J Radiat Biol 67(4):469–476. https://doi.org/10.1080/09553009514550541

Novoselova EG, Semiletova NV, Makar VR, Gordon R (1998) Early and long-term effects of chronic low-dose radiation and coenzyme Q diet on the proliferation of rat spleen T cells. Indian J Biochem Biophys 35(2):103–107

Novoselova EG, Lunin SM, Novoselova TV, Khrenov MO, Glushkova OV, Avkhacheva NV, Safronova VG, Fesenko EE (2009) Naturally occurring antioxidant nutrients reduce inflammatory response in mice. Eur J Pharmacol 615(1–3):234–240. https://doi.org/10.1016/j.ejphar.2009.05.004

Onur S, Niklowitz P, Fischer A, Metges CC, Grune T, Menke T, Rimbach G, Doring F (2014) A comparative study into alterations of coenzyme Q redox status in ageing pigs, mice, and worms. Biofactors 40(3):346–354. https://doi.org/10.1002/biof.1160

Pahlavani MA (2004) Influence of caloric restriction on aging immune system. J Nutr Health Aging 8(1):38–47

Panda A, Arjona A, Sapey E, Bai F, Fikrig E, Montgomery RR, Lord JM, Shaw AC (2009) Human innate immunosenescence: causes and consequences for immunity in old age. Trends Immunol 30(7):325–333. https://doi.org/10.1016/j.it.2009.05.004

Petersen KS, Marnewick JL, Smith C (2018) Grape polyphenols corrects ageing-related detriments in neutrophil functionality via modulation of specific molecular targets. Inflammopharmacology 26(5):1349–1358. https://doi.org/10.1007/s10787-018-0511-z

Potgieter M, Pretorius E, Van der Merwe CF, Beukes M, Vieira WA, Auer RE, Auer M, Meyer S (2011) Histological assessment of SJL/J mice treated with the antioxidants coenzyme Q10 and resveratrol. Micron 42(3):275–282. https://doi.org/10.1016/j.micron.2010.10.001

Quiles JL, Ochoa JJ, Huertas JR, Mataix J (2004) Coenzyme Q supplementation protects from age-related DNA double-strand breaks and increases lifespan in rats fed on a PUFA-rich diet. Exp Gerontol 39(2):189–194. https://doi.org/10.1016/j.exger.2003.10.002

Quiles JL, Ochoa JJ, Battino M, Gutierrez-Rios P, Nepomuceno EA, Frias ML, Huertas JR, Mataix J (2005) Life-long supplementation with a low dosage of coenzyme Q10 in the rat: effects on antioxidant status and DNA damage. Biofactors 25(1–4):73–86. https://doi.org/10.1002/biof.5520250109

Rodriguez-Bies E, Santa-Cruz Calvo S, Fontan-Lozano A, Pena Amaro J, Berral de la Rosa FJ, Carrion AM, Navas P, Lopez-Lluch G (2010) Muscle physiology changes induced by every other day feeding and endurance exercise in mice: effects on physical performance. PLoS One 5(11):e13900. https://doi.org/10.1371/journal.pone.0013900

Rodriguez-Bies E, Navas P, Lopez-Lluch G (2015) Age-dependent effect of every-other-day feeding and aerobic exercise in ubiquinone levels and related antioxidant activities in mice muscle. J Gerontol A Biol Sci Med Sci 70(1):33–43. https://doi.org/10.1093/gerona/glu002

Rodriguez-Bies E, Tung BT, Navas P, Lopez-Lluch G (2016) Resveratrol primes the effects of physical activity in old mice. Br J Nutr 116(6):979–988. https://doi.org/10.1017/S0007114516002920

Ross OA, Hyland P, Curran MD, McIlhatton BP, Wikby A, Johansson B, Tompa A, Pawelec G, Barnett CR, Middleton D, Barnett YA (2002) Mitochondrial DNA damage in lymphocytes: a role in immunosenescence? Exp Gerontol 37(2–3):329–340. https://doi.org/10.1016/s0531-5565(01)00200-5

Rotig A, Appelkvist EL, Geromel V, Chretien D, Kadhom N, Edery P, Lebideau M, Dallner G, Munnich A, Ernster L, Rustin P (2000) Quinone-responsive multiple respiratory-chain dysfunction due to widespread coenzyme Q10 deficiency. Lancet 356(9227):391–395. https://doi.org/10.1016/S0140-6736(00)02531-9

Rottenberg H, Wu S (1997) Mitochondrial dysfunction in lymphocytes from old mice: enhanced activation of the permeability transition. Biochem Biophys Res Commun 240(1):68–74. https://doi.org/10.1006/bbrc.1997.7605

Sellami M, Gasmi M, Denham J, Hayes LD, Stratton D, Padulo J, Bragazzi N (2018) Effects of acute and chronic exercise on immunological parameters in the elderly aged: can physical activity counteract the effects of aging? Front Immunol 9:2187. https://doi.org/10.3389/fimmu.2018.02187

Takahashi T, Mine Y, Okamoto T (2019) Extracellular coenzyme Q10 (CoQ_{10}) is reduced to ubiquinol-10 by intact Hep G2 cells independent of intracellular CoQ_{10} reduction. Arch Biochem Biophys 672:108067. https://doi.org/10.1016/j.abb.2019.108067

Taylor RW, Turnbull DM (2005) Mitochondrial DNA mutations in human disease. Nat Rev Genet 6(5):389–402. https://doi.org/10.1038/nrg1606

Tomasetti M, Littarru GP, Stocker R, Alleva R (1999) Coenzyme Q10 enrichment decreases oxidative DNA damage in human lymphocytes. Free Radic Biol Med 27(9–10):1027–1032. https://doi.org/10.1016/s0891-5849(99)00132-x

Tomasetti M, Alleva R, Borghi B, Collins AR (2001) In vivo supplementation with coenzyme Q10 enhances the recovery of human lymphocytes from oxidative DNA damage. FASEB J 15(8):1425–1427. https://doi.org/10.1096/fj.00-0694fje

Tomiyama AJ, Milush JM, Lin J, Flynn JM, Kapahi P, Verdin E, Sinclair E, Melov S, Epel ES (2017) Long-term calorie restriction in humans is not associated with indices of delayed immunologic aging: a descriptive study. Nutr Healthy Aging 4(2):147–156. https://doi.org/10.3233/NHA-160017

Tumova L, Ducaiova Z, Cheel J, Vokral I, Sepulveda B, Vokurkova D (2017) Azorella compacta infusion activates human immune cells and scavenges free radicals in vitro. Pharmacogn Mag 13(50):260–264. https://doi.org/10.4103/0973-1296.204558

Tung BT, Rodriguez-Bies E, Ballesteros-Simarro M, Motilva V, Navas P, Lopez-Lluch G (2014) Modulation of endogenous antioxidant activity by resveratrol and exercise in mouse liver is age dependent. J Gerontol A Biol Sci Med Sci 69(4):398–409. https://doi.org/10.1093/gerona/glt102

Turner JE (2016) Is immunosenescence influenced by our lifetime "dose" of exercise? Biogerontology 17(3):581–602. https://doi.org/10.1007/s10522-016-9642-z

van den Heuvel L, Smeitink J (2001) The oxidative phosphorylation (OXPHOS) system: nuclear genes and human genetic diseases. BioEssays 23(6):518–525. https://doi.org/10.1002/bies.1071

Vitale G, Salvioli S, Franceschi C (2013) Oxidative stress and the ageing endocrine system. Nat Rev Endocrinol 9(4):228–240. https://doi.org/10.1038/nrendo.2013.29

Yang H, Youm YH, Dixit VD (2009) Inhibition of thymic adipogenesis by caloric restriction is coupled with reduction in age-related thymic involution. J Immunol 183(5):3040–3052. https://doi.org/10.4049/jimmunol.0900562

Yuan J, Lu L, Zhang Z, Zhang S (2012) Dietary intake of resveratrol enhances the adaptive immunity of aged rats. Rejuvenation Res 15(5):507–515. https://doi.org/10.1089/rej.2012.1321

Zhang L, Shao WF, Yuan LF, Tu PF, Ma ZZ (2012) Decreasing pro-inflammatory cytokine and reversing the immunosenescence with extracts of Pu-erh tea in senescence accelerated mouse (SAM). Food Chem 135(4):2222–2228. https://doi.org/10.1016/j.foodchem.2012.07.033

Zhong F, Liang S, Zhong Z (2019) Emerging role of mitochondrial DNA as a major driver of inflammation and disease progression. Trends Immunol 40:1120–1133. https://doi.org/10.1016/j.it.2019.10.008

Chapter 14
Coenzyme Q_{10} in Fertility and Reproduction

Diana Vaamonde, Anthony C. Hackney, Carolina Algar-Santacruz, María José Garcia-Moreno, and Juan Manuel García-Manso

Abstract Coenzyme Q_{10} (CoQ_{10}) intake and supplementation has been directly and indirectly associated with physiological function relative to exercise, aging and reproduction. This chapter describes several significant aspects regarding biochemical properties and mechanism of action of CoQ_{10} in male and female fertility and reproduction. This effect is mainly through its action as an antioxidant, protecting against oxidative stress by controlling the levels of reactive oxygen species (ROS) associated with reproductive pathologies. Although some studies support the evidence of use of CoQ_{10} to improve fertility, the available literature is contradictory and conflicting due to lack of standardization regarding type, dosage and time frame of treatment with CoQ_{10} as well as the bio-specimen, the exercise protocol employed and the assays used to analyze these specimens. However, CoQ_{10} supplementation seems to be able to improve both male and female gamete physiology, conception and embryo development and pregnancy success, something that may be related to the protecting effect against ROS-related fertility issues. It seems it may, as well,

D. Vaamonde (✉)
Morphological Sciences Department. School of Medicine and Nursing, Universidad de Córdoba, Cordoba, Spain

International Network on Physical Exercise and Fertility, Cordoba, Spain
e-mail: fivresearch@yahoo.com

A. C. Hackney
Department of Exercise and Sport Science; Department of Nutrition, University of North Carolina, Chapel Hill, NC, USA

International Network on Physical Exercise and Fertility, Cordoba, Spain

C. Algar-Santacruz · M. J. Garcia-Moreno
Morphological Sciences Department. School of Medicine and Nursing, Universidad de Córdoba, Cordoba, Spain

J. M. García-Manso
Department of Physical Education, Universidad de Las Palmas de Gran Canaria, Santa María de Guía, Canary Islands, Spain

International Network on Physical Exercise and Fertility, Cordoba, Spain

G. López Lluch (ed.), *Coenzyme Q in Aging*,
https://doi.org/10.1007/978-3-030-45642-9_14

attenuate somewhat the negative impact of age on fertility, though discontinuation of treatment will result in cessation or diminution of such effect.

Keywords Fertility · Sperm · Oocytes · Reproduction

14.1 Introduction

Coenzyme Q_{10} (CoQ_{10}) is a chemical compound considered a pseudo-vitamin and is essential for the human body as it works both as an electron carrier/proton translocator in the respiratory chain (energy-promoting agent leading to ATP synthesis in the mitochondrial membrane) (Groneberg et al. 2005; Hernandez-Camacho et al. 2018) and as an antioxidant (Lopez-Lluch et al. 2010). Its function as an antioxidant is both direct by scavenging radicals (Smith et al. 1996) or indirect by regenerating vitamin E (Lewin and Lavon 1997). Energy production in the cell greatly depends on CoQ_{10} availability (Lewin and Lavon 1997). CoQ_{10} has also been shown to affect expression of genes involved in human cell signaling, metabolism, and transport affecting human fertility (Balercia et al. 2009b; Groneberg et al. 2005).

In this chapter, we show the importance of CoQ_{10} in human fertility and the effects of supplementation in the improvement of gamete production and fertilization.

14.2 Coenzyme Q_{10} in Fertility and Reproduction

Infertility affects approximately 1 out of 10 couples, male factors account for about 50% of these cases (Auger et al. 2001; Eskenazi et al. 2005; Lafuente et al. 2013) and female factors for the other 50% with many cases having both a female and a male component (De Kretser and Baker 1999; Safarinejad 2008, 2009). Though currently most causes behind infertility are known, still a significant proportion of cases are considered idiopathic as the etiology and pathogenesis are not well understood (Safarinejad 2009). Infertility cases are due to hormonal, environmental and nutritional factors that affect reproductive and gamete physiology (sperm and oocyte) and embryo development. Such factors ought to be evaluated, and in such context occupational risk factors (including exposure to heat, pesticides, chemicals, and heavy metals), lifestyle and dietary choices, and environmental factors (pesticides, xeno-estrogens) need to be examined as they may negatively impact fertility status (Auger et al. 2001; Eskenazi et al. 2005; Lafuente et al. 2013). Also, it is important to remember that age greatly affects fertility and increased age has been linked to chromosomal defects in the gametes of both men and women (Humm and Sakkas 2013).

14.3 Animal Models for the Effect of CoQ_{10} in Male Fertility

Different studies have been performed trying to assess the potential beneficial effect of CoQ supplementation in animal models and animal species for reproductive purposes.

Metabolic disease and age-associated accumulation of fat is associated with aging progression. To study the effect of metabolic disease in the quality of semen, rats were submitted to cholesterol and oxidized cholesterol rich food models to develop high levels of blood LDL and oxidized LDL. In these animals, the administration of CoQ_{10} and L-carnitine as a pretreatment before the diet resulted in improved levels of both sex hormones and semen parameters (Ghanbarzadeh et al. 2014a). The same authors show also similar results in rats with impaired spermatogenesis as a result of isoproterenol injury revealing that pretreatment with this combination decreased oxidative damage, increased LH and testosterone and also improved sperm parameters (Ghanbarzadeh et al. 2014b).

Another age-associated disease is type II diabetes. It has been documented that diabetes mellitus greatly affects reproductive function via increased ROS production and oxidative stress, which could cause an imbalance in bioavailability of energy for the spermatogenic process and thus impair testicular function (Amaral et al. 2008). In a rat model with Goto-Kakizaki rats (Amaral et al. 2006), it has been observed that these rats are less susceptible to lipid peroxidation due to higher levels of GSH and CoQ_9, suggesting that these rats have higher antioxidant capacity, maybe as a mitochondrial adaptive response (Palmeira et al. 2001).

In pigeons levels of CoQ correlated with the sperm concentration and the mean linear velocity whereas an inverse correlation was found with the motility percentage indicating a possible influence of CoQ in the quality of the sperm (Ducci et al. 2002). This effect could be due to the antioxidant effect of CoQ that tries to counteract oxidative damage in sperm and suggest the possible role of CoQ levels as fertility marker (Ducci et al. 2002).

In frozen/thawed samples of bovine sperm, the administration of a mixture of compounds including CoQ_{10} improves kinetic performance and intracellular pH; (Boni et al. 2017). In cooled equine semen, CoQ_{10} alone or in combination with α-tocopherol exerted protective effect during sperm cooling. Experiments performed only with supplemented CoQ_{10} exhibited the highest percentage of total motility whereas the group supplemented both with CoQ_{10} and α-tocopherol or only with α-tocopherol showed the lowest degree of lipid peroxidation (Nogueira et al. 2015). In bull's sperm, CoQ_{10} administration along with zinc and d-aspartate maintains sperm motility and prevents sperm DNA fragmentation. Oocytes fertilized by the treated sperm resulted in higher blastocyst rate and the formed blastocysts showed fewer apoptotic cells (Gualtieri et al. 2014).

As a means of improving the quality and functionality of gametes taken from animals postmortem, a combination of CoQ_{10} with glycerol and soybean lecithin was tested in caprine cauda epididymis sperm preservation. These experiments evidence the protection of sperm with good maintenance of motility, viability and membrane integrity (Datta et al. 2009).

In general, all these experiments performed in model animals demonstrate that CoQ_{10} can preserve sperm quality and improve male fertility.

14.4 CoQ_{10} and Male Fertility

Many of the studies about the effect of CoQ_{10} in fertility are not free of bias, even if minor, and that there is great heterogeneity in sample population (including cause of infertility, exclusion criteria, etc.), dosage and duration of administration. Furthermore, some of the studies have had a low number of patients which reduces their relevance. However, the results shown here are similar to those from other studies assessing other antioxidant compounds in fertility and sperm activity. Following we show the results of several studies about the effect of CoQ_{10} on sperm quality and male fertility.

In regards to male infertility/subfertility, the damaging effects of oxidative stress may be responsible for up to 30–80% of all cases. In cases of male idiopathic infertility, high seminal reactive oxygen species (ROS) levels and low antioxidant potential have been observed (Lafuente et al. 2013). Approximately, 25–40% of semen samples show high levels of ROS (Comhaire et al. 2000, 2005). Both infiltrating leukocytes and sperm produce ROS (Geva et al. 1998; Parinaud et al. 1997; Zalata et al. 1995). A relationship exists between ROS and leukocyte concentration (Lobascio et al. 2015) and between ROS and functional damage via lipid peroxidation (Balercia et al. 2009b; Griveau et al. 1995).

The loss of sperm cytoplasm during the maturational process, as a means of facilitating motility, results in loss of antioxidant defenses (Meldrum et al. 2016). Therefore, spermatozoa are susceptible to ROS (including hydrogen peroxide, superoxide anions and hydroxyl radicals) (Aitken and Krausz 2001; Balercia et al. 2009a). These compounds are counteracted by the antioxidant systems in sperm cells and seminal plasma (catalase, superoxide dismutase, glutathione peroxidase, and vitamin E) (Balercia et al. 2009b; Sheweita et al. 2005; Smith et al. 1996; Walczak-Jedrzejowska et al. 2013); that is, these are chain breaking compounds that appease the effects of free radicals. When these antioxidant systems are overwhelmed by ROS, oxidative damage may occur (Aitken and Krausz 2001; Balercia et al. 2009b) leading to negative effects on seminal characteristics and the sperm cells (oligospermia and asthenozoospermia) and increasing the risk of sperm dysfunction and male infertility, even in normozoospermic infertile men with unexplained infertility (Agarwal et al. 2003; Balercia et al. 2009a, b; Eskenazi et al. 2005; Lafuente et al. 2013; Sharma and Agarwal 1996; Ross et al. 2010).

ROS production, however, is a normal physiological process of cell metabolism for those cells which perform mitochondrial respiration (Lafuente et al. 2013), a process that is necessary in low levels for appropriate sperm function. They contribute to intracellular signaling, genetic regulation, sperm capacitation and sperm-oocyte fusion (Aitken 1997b; Balercia et al. 2009b). As such, superoxide anion radicals play an important role in the capacitation and acrosome reaction processes.

Because of inherent functions, sperm plasma membrane is highly specific in its lipidic structure, containing high levels of polyunsaturated fatty acids (Balercia et al. 2009b). This characteristic provides the sperm membrane with marked flexibility and fluidity, necessary for sperm-oocyte fusion (Aitken 1997a; Balercia et al. 2009b). Yet, at the same time, the membranes become highly susceptible to damage from excessive ROS levels, common in inflammatory and infectious processes as well as in a result of increased numbers of immature spermatogenic cells, etc.; since they are precisely the main target for lipid peroxidation (Balercia et al. 2009b; Eskenazi et al. 2005; Lafuente et al. 2013; Lenzi et al. 1996; Sharma and Agarwal 1996; Sheweita et al. 2005).

Sperm continuously undergo morphological and biochemical changes at several time points of development (testicular production and first maturational processes, epididymal storage and maturation, female tract capacitation). The aforementioned changes are intimately related to changes in plasma membrane biochemical composition affecting lipids, carbohydrates and proteins (Lafuente et al. 2013; Tulsiani 2000). During epididymal storage and maturation, unsaturation increases as sperm travel from the caput to the cauda, process which reveals active lipid metabolism (Balercia et al. 2009b; Lenzi et al. 1998). Noteworthy is the fact that the epididymis region is also a site of greater leukocyte invasion; and, it has been demonstrated that the time of permanence in the epididymis is greater in oligospermic patients and therefore the exposition to ROS is longer (Ford and Whittington 1998; Lafuente et al. 2013; Tapia et al. 2012). Oxidative stress mainly takes place at the epididymis and the transit through the different duct system, and the longer time between ejaculations, the longer time sperm (already have lost their cytoplasm and antioxidant defenses) are exposed to oxidative stress (Meldrum et al. 2016).

Membrane integrity has been reported to be maintained or improved by CoQ_{10} administration in caprine sperm as revealed by results in swelling tests (Datta et al. 2009). The same results have been reported and associated to decreased ROS formation such as hydrogen peroxide (Alleva et al. 1997) and thus avoiding membrane lipid peroxidation (Datta et al. 2009; Lafuente et al. 2013; Lewin and Lavon 1997). When lipid peroxidation occurs, membrane function is impaired, which in turn decreases membrane fluidity and thus sperm motility and sperm-oocyte binding ability; as a result, diminished fertilization may be expected (Agarwal et al. 2003; Talevi et al. 2013). The fact that sperm are deficient in intracellular antioxidant enzymes and possess limited DNA repair ability contributes to sperm being highly susceptible to oxidative stress related damage, with possible alteration in DNA, decreased mitochondrial membrane potential, and even possible cell death (Balercia et al. 2009b; Geva et al. 1998; Talevi et al. 2013; Wathes et al. 2007). Sperm are especially fragile due to features such as: DNA strand breaks (Irvine et al. 2000), mitochondrial DNA fragmentation (Donnelly et al. 2000) and poorly compacted chromatin (Balercia et al. 2009b; Sakkas et al. 1999). As a result of such sperm DNA damage, embryo development may be also impaired due to altered paternal genomes (Aitken et al. 2010; Talevi et al. 2013). As mentioned before, the greater exposure to oxidative stress mainly takes place at the epididymis and the transit through the different duct system once the sperm has lost most of its antioxidants, and also while at the female reproductive tract where different environmental conditions, such as a different pH, and functional modifications exist.

Impaired antioxidant capacity has been suggested to be related to male infertility (Macchia et al. 2010; Talevi et al. 2013) in contrast, men with high intake of antioxidants present with better semen quality and lower levels of sperm aneuploidy (Geva et al. 1996; Talevi et al. 2013). In this regard, numerous studies have recently addressed antioxidant supplementation in infertile men as a possible means of improving seminal parameters (Lafuente et al. 2013; Showell et al. 2011). The aging process seems to be a key mechanism behind alterations in sperm as age has marked influence on sperm DNA fragmentation and altered spermatogenesis (Humm and Sakkas 2013; Lafuente et al. 2013), and this process could be mediated by greater oxidative stress which has been seen to increase DNA fragmentation and alter spermatogenesis (Lafuente et al. 2013; Mahfouz et al. 2010; Meldrum et al. 2016). In addition, as a result of aging, it has been observed that the levels of CoQ_{10} decrease in all tissues (Nadjarzadeh et al. 2011), so, the observed higher cellular oxidation cannot be as easily counteracted. Although oocytes from younger women may have some repair capacity, even for the chromosomal defects in sperm, antioxidant intake is suspected to be beneficial for men over 40 (Meldrum et al. 2016). CoQ_{10} has been tested as one such possible supplementation to improve infertility due to its dual role as energy metabolism agent and liposoluble antioxidant (Lafuente et al. 2013; Showell et al. 2011).

In sperm cells, CoQ_{10} is found in the numerous mitochondria present in the midpiece, due to its clear role in energy production-related processes and sperm tail movement requiring high energy expenditure (Iwanier and Zachara 1995; Lewin and Lavon 1997). Therefore, CoQ_{10} bioavailability is essential for ATP synthesis and energy production and its deficiency may explain reduced sperm motility in some cases (Lewin and Lavon 1997). Its reduced form, ubiquinol, acts as an antioxidant preventing the lipid peroxidation mentioned in above paragraphs and preserving sperm integrity (Lewin and Lavon 1997).

14.4.1 *Coenzyme Q_{10} and Sperm*

As previously mentioned, ROS is the main cause of various alterations in sperm. Defective sperm generate larger amounts of ROS, and this correlates highly with impaired sperm motility (Lewin and Lavon 1997). Strong correlations have been found among seminal fluid content of ubiquinol and parameters reflecting sperm quality and function such as sperm count and motility (Gvozdjakova et al. 2013). Alterations in CoQ_{10} levels have been also observed in pathological conditions such as oligozoospermic or azoospermic, asthenozoospermia and varicocele (Balercia et al. 2002; Gvozdjakova et al. 2013; Mancini and Balercia 2011; Mancini et al. 1998). These alterations can in turn adversely affect fertilization.

Serum CoQ_{10} levels did not reveal any association with regards sperm parameters (Eroglu et al. 2014). On the other hand, different levels of seminal plasma CoQ_{10} have been observed between infertile and fertile populations (Li et al. 2006). It has been found that sperm from idiopathic asthenozoospermic samples have lower ubiquinol ($CoQ_{10}H_2$)/ubiquinone(CoQ_{10}) ratio (Balercia et al. 2002). The same finding was observed in seminal plasma in idiopathic asthenozoospermia and varicocele samples (Balercia et al. 2002; Mancini et al. 2005b). Such findings suggest that $CoQ_{10}H_2/CoQ_{10}$ ratio may be indicative of oxidative status and semen quality may be negatively affected by a reduction in this ratio (Balercia et al. 2002). Furthermore, normal morphology has been observed to correlate with the $CoQ_{10}H_2/CoQ_{10}$ ratio (Alleva et al. 1997). In infertile men, patients with better sperm motility also showed higher sperm concentration as well as CoQ_{10} and α-tocopherol concentrations and less sperm anomalies. TBARS, however, were increased with independence of motility dysfunction (Gvozdjakova et al. 2013). In men with idiopathic infertility, seminal CoQ_{10} levels correlated with sperm morphology while no correlation was found with regards sperm concentration or motility. On the other hand, some other studies did reveal a correlation between CoQ_{10} levels in seminal plasma and sperm motility (Balercia et al. 2009a, b; Mancini and Balercia 2011; Mancini et al. 2005a).

It can be hypothesized that, in certain circumstances, high oxidative stress in sperm cells can somehow over-consume CoQ_{10} in detriment of its bioenergetic role (Mancini and Balercia 2011). In fact, hydrogen peroxide levels have been observed to show inverse correlation with the $CoQ_{10}H_2/CoQ_{10}$ ratio (Balercia et al. 2009b; Mancini and Balercia 2011). It seems plausible that CoQ_{10} may inhibit the formation of hydrogen peroxide, as it has also been observed for plasma lipoproteins (Balercia et al. 2009b).

Though paradoxical, an inverse correlation between CoQ_{10} sperm content and sperm motility has also been reported (Angelitti et al. 1995) when comparing infertile to either fertile or potentially fertile patients, this strange paradox has been also observed in patients with varicocele (Mancini et al. 1998, 2003). This could be due to an increased demand for antioxidant effect of CoQ_{10} and less availability of this molecule for the ATP production process that sperm motility is dependent on. This finding however has not been supported, in patients other than varicocele patients, or in seminal plasma CoQ_{10} content as other authors have not been able to find such association (Eroglu et al. 2014; Gvozdjakova et al. 2013). In these cases, the found association was either direct between CoQ_{10} seminal plasma levels and motility (Gvozdjakova et al. 2013) or between CoQ_{10} levels in seminal plasma and sperm morphology (Eroglu et al. 2014). In varicocele patients, a significantly higher proportion of total CoQ_{10} was present in seminal plasma when compared to the other patients (Mancini et al. 1998, 2003). Also in varicocele patients, there is an altered pattern of distribution of CoQ_{10} between sperm cells and seminal plasma (Mancini and Balercia 2011; Mancini et al. 2003).

14.4.1.1 Supplementation of CoQ_{10} and Levels in Serum and Semen

In men with idiopathic infertility, there were no significant differences with regards to fertile men for values in serum total antioxidant capacity, and serum and seminal CoQ_{10}; also, no relationship was found between serum and seminal levels of CoQ_{10} (Eroglu et al. 2014). As a result of daily 300 mg CoQ_{10} supplementation for 26 weeks, blood and seminal plasma CoQ_{10} increased (seminal plasma CoQ_{10} plateauing by week 8). Two weeks after withdrawal, CoQ_{10} concentrations, both in blood and seminal plasma, had dropped to pretreatment values (Safarinejad 2009). This behavior was observed in different studies (Balercia et al. 2004, 2009a, b; Mancini et al. 2005a), and further supported in the meta-analysis by Lafuente and collaborators (Lafuente et al. 2013). Also, doses of 200 mg/day of CoQ_{10} caused an increase in CoQ_{10} levels as well as increased $CoQ_{10}H_2$ levels in idiopathic asthenozoospermic infertile patients (Balercia et al. 2009a, b). With similar protocol, Nadjarzadeh and colleagues have found in idiopathic oligoasthenoteratozoospermic infertile men, increases in CoQ_{10} levels after 3-months treatment with 200 mg CoQ_{10}. This increase was accompanied by higher antioxidant activity (catalase and superoxide dismutase) (Nadjarzadeh et al. 2014) and evidence of lower oxidative stress through lower concentrations of thiobarbituric-acid-reactive substances (TBARS) (Nadjarzadeh et al. 2011).

14.4.1.2 CoQ_{10} and Sperm Concentration

With regards to sperm concentration, in idiopathic oligoasthenoteratozoospermic infertile men, a dose of 300 mg CoQ_{10} given orally daily for a 26-week period resulted in significant improvement in sperm density (Safarinejad 2009). Later on, the same group found similar results using 300 mg for 12 months (Safarinejad 2012) and even only with 200 mg CoQ_{10} (Safarinejad et al. 2012).

Other studies revealed no significant differences in sperm concentration with doses of 200 and 100 mg of CoQ_{10} in asthenoteratozoospermic or oligoasthenoteratozoospermic men (Balercia et al. 2009a; Cakiroglu et al. 2014; Nadjarzadeh et al. 2011). A later systematic review and other reviews seem to support the evidence that CoQ_{10} supplementation may produce an increase in sperm concentration (Lafuente et al. 2013; Littarru and Tiano 2010).

14.4.1.3 CoQ_{10} and Sperm Motility

With regards to sperm motility, this parameter was improved in asthenozoospermic samples from patients undergoing ICSI incubated with 50 μM CoQ_{10} (Lewin and Lavon 1997). Only samples from asthenozoospermic patients exhibited increased motility while samples from normozoospermic patients did not (Lewin and Lavon 1997; Littarru and Tiano 2010). In idiopathic oligoasthenoteratozoospermic infertile men, 300 mg CoQ_{10} given orally daily for a 26-week period resulted in significant improvement in sperm motility (Safarinejad 2009). Later on, the same group found similar results using 300 mg for 12 months and even only with 200 mg CoQ_{10} (Safarinejad 2012; Safarinejad et al. 2012). In idiopathic asthenozoospermic infertile patients, a significant difference was found in forward (class a + b) motility of sperm cells after 6 months of 200 mg CoQ_{10} dietary supplementation (Balercia et al. 2004). In a later study, the same group reported with a greater sample size the same results giving further support to the idea that CoQ_{10} significantly improves sperm cell total motility and forward motility (Balercia et al. 2009a). The same results have been observed with lower dose (100 mg) in asthenoteratozoospermic men (Cakiroglu et al. 2014). It is noteworthy to point out that treatment may be more beneficial for patients with lower motility and lower CoQ_{10} level.

In patients with varicocele, sperm motility was inversely correlated to CoQ_{10} levels, contrary to what is observed in non-varicocele asthenozoospermic patients; this association was only partially reversed by surgical treatment and CoQ_{10} was able to improve sperm motility (Balercia et al. 2002, 2009a, b; Mancini et al. 2003, 2005b, 2012).

In addition, systematic review seems to support the evidence that CoQ_{10} supplementation produces an increase in sperm motility (Lafuente et al. 2013). Nevertheless, it must be noted that other authors like Nadjarzadeh et al. do not show significant changes in sperm motility in oligoasthenoteratozoospermic patients (Nadjarzadeh et al. 2011, 2014).

14.4.1.4 CoQ_{10} and Sperm Morphology

With regards to sperm morphology, 300 mg CoQ_{10} given orally daily to idiopathic oligoasthenoteratozoospermic infertile men for a 26-week period resulted in an increase in the percent of normal forms, albeit non-significant, after treatment a follow-up period of 30 weeks resulted in (Safarinejad 2009). The same group found similar results using 300 mg for 12 months and even only with 200 mg CoQ_{10} (Safarinejad 2012; Safarinejad et al. 2012). In asthenoteratozoospermic men, 100 mg increases the percentage of normal forms (Cakiroglu et al. 2014). Although, other studies did not reveal significant differences in sperm morphology (Balercia et al. 2004; Nadjarzadeh et al. 2011).

14.4.1.5 CoQ_{10} and Fertilization Potential

Fertilization potential has been significantly improved with administration of 60 mg of CoQ_{10}, even when no other parameters showed significant changes (Lewin and Lavon 1997; Sinclair 2000). In ICSI patients, the administration of 50 μM CoQ_{10} resulted in improved fertilization rates (Lewin and Lavon 1997). Acrosome reaction had significantly increased after CoQ_{10} supplementation by the end of the treatment phase whereas no changes were observed in semen volume or antisperm-antibody binding (Lafuente et al. 2013; Safarinejad 2009).

However, a systematic review by Lafuente and associates provides evidence that CoQ_{10} supplementation does not increase pregnancy rates when given to infertile men (Lafuente et al. 2013), and no available data regarding live births exists. In contrast, studies using other antioxidants have reported improved pregnancy rates (spontaneous and through assisted reproduction (Comhaire et al. 2005; Lafuente et al. 2013; Tremellen et al. 2007). In a study by Balercia and colleagues, 200 mg/day of CoQ_{10} resulted in improved spontaneous pregnancy rate, fact that suggests that this therapeutic approach is beneficial (Balercia et al. 2009a; Lafuente et al. 2013); it must be noted however that the sample number in this study was rather small not permitting to obtain significant results.

14.4.1.6 Treatment with CoQ_{10} and Other Agents in Sperm Capacity

Other studies have assessed the effect of CoQ_{10} in conjunction with other agents in sperm parameters. In two studies, CoQ_{10} was used along with zinc and D-Asp (Giacone et al. 2017; Talevi et al. 2013). Zinc has been used for improving male fertility due to its antioxidant and anti-apoptotic features, also involving DNA-related mechanisms (Ebisch et al. 2007) and seems to be efficient in asthenozoospermic men in improving sperm concentration, motility, integrity and thus, fertilization and pregnancy rates (Omu et al. 1999). On the other hand, D-Aspartic acid, an endogenous amino acid, has been shown to be present in different locations and cells of the reproductive tract, and its concentration is diminished in oligoasthenoteratozoospermic patients (D'Aniello et al. 2005).

A combination of 120 mg CoQ_{10} with vitamins E and C given daily for 3 and 6 months to oligoasthenozoospermic men improved sperm concentration and motility and resulted in pregnancies, some of which were spontaneous (Kobori et al. 2014). Further, a combination of CoQ_{10} with vitamins E and C, and L carnitine given daily for 3 and 6 months to infertile men improved sperm concentration and diminished sperm anomalies and oxidative stress, increasing the percentage of pregnancies (Gvozdjakova et al. 2015). A commercially available formulation (20 mg CoQ_{10}) given for 4 months to idiopathic asthenoteratozoospermic men resulted in increases in progressive motility but not in other sperm parameters; moreover, some of the patients achieved pregnancy while enrolled in the study (Busetto et al. 2012). In the same type of patients, another combination containing the same amount of CoQ_{10} resulted in the increase of all sperm parameters as well as improved DNA integrity (Abad et al. 2013).

In infertile patients with grade I varicocele showing greater number of DNA-damaged sperm than other infertile men, a mixture containing different vitamins and CoQ_{10} significantly reduced sperm DNA fragmentation and numbers of degraded sperm cells and also resulted in increased sperm concentration (Gual-Frau et al. 2015). In azoospermic patients with maturational arrest, a combination of CoQ_{10}, multivitamins and micronutrients, resulted in the appearance of sperm with progressive motility and normal morphology, even if slight, after 1 month of treatment and two pregnancies after 3 months (Singh et al. 2010).

14.5 Mechanism of Action of CoQ_{10} in Protection of Male Fertility

As it is easily understandable, the observed improvement CoQ_{10} administration produces in sperm motility is related to the dual role of CoQ_{10} as a main factor in mitochondrial bioenergetics and as membrane antioxidant (Balercia et al. 2009b; Mancini and Balercia 2011). Due to the kinetic properties of the energetic process, a relevant increase in respiratory activity can be achieved even with small increases in CoQ_{10} (Balercia et al. 2009b; Mancini and Balercia 2011). Furthermore, this effect could be also due to the antioxidant properties of CoQ_{10} fighting against oxidative stress (Balercia et al. 2009b; Giacone et al. 2017; Lafuente et al. 2013; Mancini and Balercia 2011; Mancini et al. 2012; Nadjarzadeh et al. 2011). This is supported by the fact most studies have reported improved motility but not increased sperm concentration as a result of CoQ_{10} administration.

CoQ_{10} is one of the most important antioxidants in seminal plasma. CoQ_{10} levels significantly relates to sperm motility and count as expected from the cellular compartmentalization observed. While not completely understood, CoQ_{10} distribution seems to be actively guided between intra- and extra-cellular compartments; such compartmentalization seems to be greatly altered in patients with varicocele (Balercia et al. 2009b; Lafuente et al. 2013; Mancini and Balercia 2011; Mancini et al. 2012).

Oxidative stress, as mentioned before, may be responsible for sperm dysfunction; this associated is more easily addressed in patients with varicocele who exhibit an enhanced ROS generation and high levels of nitric oxide (Mancini et al. 1998) and provide an excellent study model (Aksoy et al. 2000; Balercia et al. 2009b; Hendin et al. 1999). Other patients exhibiting greater levels of ROS are smokers (Saleh et al. 2002) and men with chronic prostatitis (Balercia et al. 2009b; Pasqualotto et al. 2000), and oligospermic patients that show longer time in the epididymis with longer exposure to ROS (Ford and Whittington 1998; Lafuente et al. 2013). In fact, some trials have shown that antioxidant supplementation improves sperm quality in male factor infertility (Lenzi et al. 1998), even increasing fertilization potential in some cases (Geva et al. 1996; Kessopoulou et al. 1995; Lafuente et al. 2013). It has been demonstrated that CoQ_{10} supplementation results in increased levels of $CoQ_{10}H_2$, and this in turns inhibits organic peroxide formation, protein and DNA oxidation and lipid peroxidation, thereby reducing the oxidative stress that attack to sperm cells (Alleva et al. 1997; Littarru and Tiano 2010).

Although the main mode of action of CoQ_{10} seems to be its role in mitochondrial function and its antioxidant properties, other possible mechanisms involve gene expression related to cell signaling, metabolism, and transport have been hypothesized to exist (Garrido-Maraver et al. 2014; Groneberg et al. 2005; Littarru and Tiano 2010).

14.6 CoQ_{10} and Female Fertility

14.6.1 Deterioration of Female Reproductive Capacity

Female fertility is one of the major human physiological functions affected by the natural process of aging (Ben-Meir et al. 2015). The optimal time in the reproductive age of women is around the age of 25, with a rapid decline from around 32–37 years of age, until the end of a woman's reproductive life, at about 50 years (Ben-Meir et al. 2015; Bentov et al. 2014). Female reproductive capacity deteriorates with age as manifested by a reduction in ovarian reserve (Ben-Meir et al. 2015; Vital-Reyes et al. 2006); that is, there is a decrease in ovarian capacity to perform maturation and follicular development, and steroid hormones synthesis (Vital-Reyes et al. 2006). This is due to the accelerated decrease in oocyte quality and quantity (Bentov and Casper 2013), with a progressive and non-linear loss and mutation of primordial follicles (Ben-Meir et al. 2015; Bentov et al. 2014; Vital-Reyes et al. 2006). In addition, this deterioration may be related to alterations in different functional structures of female reproductive tract and provokes a decrease in the reproductive time of the woman and therefore female infertility (Ben-Meir et al. 2015; Bentov and Casper 2013; Bentov et al. 2011; Meldrum et al. 2016; Zhang et al. 2006). The size and count of antral follicles, in basal conditions, is used as an indicator to know if there is an adequate ovarian reserve and a proper response to ovarian hyperstimulation in women submitted to assisted fertilization, number of follicles of size ranging from 2 to 10 mm of follicular diameter indicate ovarian reserve and likely outcome (Jayaprakasan et al. 2010; Vital-Reyes et al. 2006), with an important hallmark for triggering ovulation with exogenous hormones when the dominant follicle reaches about 18 mm of diameter (Escudero et al. 2005; Humaidan et al. 2005; Kuang et al. 2014; Olivennes et al. 2000).

Female fertility is influenced by oxidative phosphorylation capacity, since this has a critical importance in early stages of development of the oocyte and embryo (Bartmann et al. 2004; Bentov et al. 2014; Gendelman and Roth 2012; Meldrum et al. 2016; Schon et al. 2000), due to high energy oxygen consumption levels (Ben-Meir et al. 2015). Therefore, any alteration in this process would produce a decrease in energy availability and alterations in the reproductive system.

Additionally, polycystic ovary syndrome (PCOS) and endometriosis are two complicated female reproductive pathologies. They are associated with pathologies such as diabetes, obesity, hyperandrogenism and hypercholesterolemia (El Refaeey et al. 2014). Oxidative stress and reactive oxygen species (ROS) are potentially involved in PCOS (Melo et al. 2016; Zhang et al. 2016) and endometriosis (Donnez et al. 2016). In fact, follicular fluid of women with endometriosis undergoing assisted reproductive techniques has been shown to have higher levels of oxidative stress (de Lima et al. 2017).

Low CoQ_{10} levels have been observed in female with fertility problems showing different ovarian and embryonic alterations such as aneuploid oocyte and embryos (Ben-Meir et al. 2015). This diminution occurs around 30 years of age. Aneuploid oocyte and embryos can be found, even in young women, between 62% and 65% of the cases (Bentov and Casper 2013; Bentov et al. 2014). There is also evidence that aging is accompanied by oocyte mitochondrial dysfunction (Keefe et al. 1995; Ramalho-Santos et al. 2009), a higher incidence of oocyte and embryo aneuploidies (Bentov and Casper 2013; Bentov et al. 2014; Kuliev et al. 2005; Pellestor et al. 2005), decreased oocyte quality, lower ovarian stimulation response and lower implantation rate (Bartmann et al. 2004; Bentov et al. 2014; Meldrum et al. 2016; Schon et al. 2000).

14.6.2 Coenzyme Q_{10} and the Female Reproductive Tract

Ovarian reserve shows a fast decline as women age (Ben-Meir et al. 2015; Faddy 2000); among the different reasons, oocyte damage, as the DNA repair systems decrease with age, should be highlighted (Ben-Meir et al. 2015; Titus et al. 2013). Two possible theories have been proposed to explain the decrease of the oocyte quality in relation to aging. The first one indicates that during the first years of fertile life there is a selection of oocytes that present better quality. The second theory reflects how the aging process itself influences a woman's own ovarian reserve (Bentov and Casper 2013).

As in all tissues, in the ovary, CoQ_{10} production takes place in the inner mitochondrial membrane (Ben-Meir et al. 2015). The oocyte has an important capacity to produce high levels of CoQ_{10} (El Refaeey et al. 2014), due to high numbers of mitochondria in mature normal oocytes (Ben-Meir et al. 2015; El Refaeey et al. 2014), and displays adequate exogenous absorption of this coenzyme from external sources (Ben-Meir et al. 2019; Bentov et al. 2014). Therefore, development of a mitochondrial dysfunction may cause deficiency of CoQ_{10} (Ben-Meir et al. 2015, 2019).

During oocyte maturation, nuclear, cytoplasmic and epigenetic changes take place in order to form the meiotic spindle. These processes require energy through oxidative phosphorylation, since other energy processes such as glycolysis may not be as effective in the oocyte (Ben-Meir et al. 2015). The process of oocyte maturation may be detained by alterations in bioenergetics or mitochondria (Bartmann et al. 2004; Ben-Meir et al. 2015; Bentov et al. 2014). This alteration can also provoke a decrease in oocyte quality, sometimes even leading to oocyte death (El Refaeey et al. 2014), and risk of embryonic mal-development (Ben-Meir et al. 2015; Meldrum et al. 2016; Ritter et al. 2015; Schon et al. 2000; Takeuchi et al. 2005; Zhang et al. 2006). The relationship to embryo development is due to the fact that mitochondria are maternally inherited (Bentov and Casper 2013; Meldrum et al. 2016), as the paternal ones are destroyed by lysosomal degradation after fertilization (Sato and Sato 2012). As such, oocyte-derived mitochondria are responsible for providing the necessary energy for early embryonic development (Meldrum et al. 2016). According to some studies, an alteration in the functional state of the energy producing structures in oocyte mitochondria can influence the correct configuration of the meiotic spindle and the chromosomal misalignment (El Refaeey et al. 2014; Meldrum et al. 2016; Ritter et al. 2015; Schon et al. 2000). As a result, decreased preimplantation embryo developmental potential has been observed in mice (Meldrum et al. 2016; Zhang et al. 2016).

Therefore, it is of great importance, in the assisted reproduction setting, to observe the correlation between the levels of CoQ_{10} and oocyte and embryo grade maturation (Ben-Meir et al. 2015). In fact, a recent study demonstrates how, in patients undergoing in vitro fertilization, the combination of CoQ_{10} and dehydroepiandrosterone improves the ovarian response without observing any difference in the clinical outcome (Gat et al. 2016).

14.6.3 *Effect of Supplementation of CoQ_{10}*

Supplementation with CoQ_{10} in aged animals and humans has been shown to promote fertility as it may increase ovarian reserve and ovarian mitochondrial energy levels of ATP, while reducing levels of ROS (Talukdar et al. 2015) and oocyte chromosome aneuploidies (Bentov and Casper 2013; Bentov et al. 2014; El Refaeey et al. 2014). Furthermore, CoQ_{10} supplementation has been observed to result in improved morphology of the meiotic spindle and the chromosomal alignment, avoiding the chromosomal dispersion (Ben-Meir et al. 2015; Meldrum et al. 2016; Ritter et al. 2015; Schon et al. 2000).

CoQ_{10} supplementation in several animal species has been found to favor ovarian activity and its development. These studies seem to be in agreement with the evidences found in humans. Favoring ovarian activity and its development in turn, improves embryo quality and fertility in older women (Ben-Meir et al. 2015; El Shourbagy et al. 2006; Meldrum et al. 2016). Aged mice and rats supplemented with CoQ_{10} showed improved ovulation rates and ovarian response (Ben-Meir et al. 2015, 2019; Bentinger et al. 2008). Twelve weeks of supplementation resulted in increased number of primordial follicles and ovulatory capacity, mitochondrial energy activity and DNA restoration, and delayed ovarian reserve depletion (Ben-Meir et al. 2015). Similar results have been observed in women with preeclampsia treated with CoQ_{10} and clomiphene citrate (El Refaeey et al. 2014).

Taking the importance of mitochondrial function into consideration, maternal supplementation with CoQ_{10} may lead to improvement of mitochondrial function and improvement of all the processes along fertilization and gestation (Ben-Meir et al. 2019; Lopez-Lluch 2019; Meldrum et al. 2016).

14.7 CoQ_{10} and Embryo Development

14.7.1 *CoQ_{10} and Early Gestation Events*

It has been shown that embryo implantation potential correlates with ATP embryo content (Bentov et al. 2014; El Refaeey et al. 2014; Van Blerkom et al. 1995). Supporting this concept, Bentov and colleagues have observed that supplementation with CoQ_{10} increases ATP levels in embryos helping implantation. This supplementation could help old patients with infertility problems (Bentov et al. 2010).

In agreement with the importance of CoQ_{10} in early embryonic development, Stojkovic and colleagues, studying in vitro culture supplementation in animal embryos with 30–100 μM of CoQ_{10}, observed a higher rate of early cleavage, embryo development to the blastocyst stage and energy molecules levels (Stojkovic et al. 1999). Other authors have found the same evidence in in vitro culture of bovine embryos, with a greater early embryo cleavage, greater blastocyst formation, an increase in numbers of expansive blastocysts and the size of inner cell mass (Bentov et al. 2010; Bentov et al. 2014; El Refaeey et al. 2014). Further, CoQ_{10} supplementation resulted in higher survival rates and blastocyst formation derived from porcine oocytes; the improvement was accompanied by with higher levels of mRNA expression indicating the importance of bioenergetics in the expression of genes in early embryonic stages (Hwang et al. 2016). Furthermore, fertilization with sperm incubated with CoQ_{10} reduced the number of apoptotic cells in blastocysts from bull oocytes (Gualtieri et al. 2014).

14.7.2 CoQ_{10} and Pregnancy

Studies performed in humans demonstrate that CoQ_{10} administration in women may increase pregnancy rate (Bentov et al. 2014). In addition, treatment with CoQ_{10} may improve also male fertility, aiding to improve pregnancy success (Arcaniolo et al. 2014; Safarinejad et al. 2012).

Both neuroendocrine and uterine factors play a role in pregnancy decline during aging. However, it has also been observed to be caused by a decrease in oocyte quality and metabolic activity (Chiang et al. 2011). As a result of aging, accumulation of alterations in chromosomes may affect embryonic development (Ben-Meir et al. 2015) and lead to lower implantation rate (Bentov et al. 2014). Plasma CoQ_{10} levels seem to be related to pregnancy outcome, and decreased CoQ_{10} plasma levels are associated with pregnancy termination (Ben-Meir et al. 2015). Spontaneous miscarriage, or risk of late term miscarriage and/or pre-term deliveries may be related to decreased plasma levels of CoQ_{10} (Noia et al. 1996, 1998, 1999). Noia and associates reported that the normal curve of plasma CoQ_{10} levels rises during each trimester whereas decreased CoQ_{10} levels were related with spontaneous abortion (Noia et al. 1996). These studies indicate that maternal plasma levels of CoQ_{10} may indicate pathological contractile uterine activity and risk of spontaneous abortion (Noia et al. 1999).

Other authors have determined the relationship existing between lifestyle and antioxidant levels in each trimester during pregnancy and 1 month postpartum. These studies show that CoQ_{10} serum levels increase during gestation. High levels were observed at all times during pregnancy in non-smokers and a significant increase in the third trimester in women who frequently perform physical exercise (Matsuzaki et al. 2014). High CoQ_{10} levels in pregnancy can be due to an increase in cholesterol, and more specifically LDL cholesterol, since this is the main carrier of CoQ_{10} in human plasma (Abo-Elmatty et al. 2012; Haruna et al. 2010; Teran et al. 2009).

During pregnancy, several complications may take place, some of them associated with low levels of CoQ_{10} in plasma. One of these complications is preeclampsia. Preeclampsia takes place in about 8–10% of pregnant women at around week 20 and it is associated with hypertension and proteinuria (Abo-Elmatty et al. 2012; Roland et al. 2010). Women experiencing preeclampsia show low CoQ_{10} and α-tocopherol plasma levels associated with reduced antioxidant defenses and higher endothelial cell damage (Palan et al. 2004; Teran et al. 2009). Interestingly, in these cases, CoQ_{10} levels in umbilical cord and placenta are increased probably indicating a compensatory mechanism (Teran et al. 2005, 2008). Therefore, determination of CoQ_{10} in plasma as gestation progressive can be a feasible mechanism to early detect preeclampsia (Abo-Elmatty et al. 2012; Roland et al. 2010). Further, preeclampsia, as well as other gestational pathologies, may be diminished, and even prevented, through CoQ_{10} administration (Abo-Elmatty et al. 2012; Littarru and Tiano 2010). However, a study performed in rats contraindicate the administration of CoQ_{10} during pregnancy by a putative CoQ_{10}-dependent oxidative stress (Staicu et al. 2011) but this study contradicts many other studies performed in animals and humans indicating safety and lack of oxidative stress after CoQ_{10} supplementation (Hernandez-Camacho et al. 2018; Lopez-Lluch 2019; Lopez-Lluch et al. 2010).

14.7.3 Plasma and Amniotic Fluid CoQ_{10} Levels

A correlation between maternal plasma levels of CoQ_{10} and fetal amniotic concentration of CoQ_{10} and birth weight has been established (Giannubilo et al. 2014; Haruna et al. 2010). Among the studies analyzed, it was observed that plasma levels of CoQ_{10} in the mother were greater than in amniotic fluid (up to 10 times lower) (Franke et al. 2013; Teran et al. 2011). At the same time, there is a negative correlation between the amniotic concentration of CoQ_{10} and birth weight. Higher amniotic concentration of CoQ_{10} is associated with lower-weight fetuses as a possible compensatory mechanism against oxidative stress (Giannubilo et al. 2014). However, other authors have reported that lower CoQ_{10} levels were associated with premature births (Giannubilo et al. 2014; Teran et al. 2011). Furthermore, CoQ_{10} fetal umbilical cord plasma levels at 38 weeks of gestation is lower than in the mother (Compagnoni et al. 2004; Franke et al. 2013; Garcia-Rodriguez et al. 2012; Teran et al. 2005).

All these differences can be due to the fact that placental transfer acts as a "filter", retaining higher level of CoQ_{10} in the placenta (Franke et al. 2013; Teran et al. 2005, 2008).

14.7.4 *Prevention of Oxidative Stress during Pregnancy*

During pregnancy, oxidative stress has been associated with the onset and progression of several complications (Matsuzaki et al. 2014). Oxidative stress plays an important role in the onset of embryonic alterations secondary to different conditions such as gestational diabetes mellitus, decreased blood flow or substance abuse, premature labor, labor or preeclampsia (Giannubilo et al. 2014; Matsuzaki et al. 2014; Roland et al. 2010; Sakamaki et al. 1999). Evidence indicates that CoQ_{10} levels are low in pregnancy pathologies showing high levels of oxidative stress (Matsuzaki et al. 2014). In turn, oxidative stress and free radicals may lead to fetal stress, thus influencing the possibility of achieving a viable pregnancy (Giannubilo et al. 2014; Staicu et al. 2011). Increased placental mitochondrial activity and ROS production during pregnancy results in increased oxidative stress. Different studies reveal a correlation between antioxidant levels, such as CoQ_{10}, and oxidative stress levels. Some evidence indicates an increase in antioxidant levels during pregnancy can counteract the increase in oxidative stress (Matsuzaki et al. 2014; Myatt and Cui 2004). That is, oxidative stress may provoke a response of elevating CoQ_{10}, in order to minimize or prevent the detrimental effect this status may cause (Giannubilo et al. 2014; Staicu et al. 2011).

14.8 Conclusion

Supplementation with CoQ_{10} seems to improve both male and female fertility, affecting gamete capacity, conception, embryo development and pregnancy outcome. The main mechanism by which CoQ affects fertilization capacity seems to be by protecting against ROS. CoQ_{10} may, as well, attenuate the negative impact of age on fertility.

Due to heterogeneity in populations, treatment and results, caution should be advised. Further well-designed and robust studies are needed to get additional insight on the benefits of CoQ_{10} administration on human fertility. Ultimately, the mode of action should be clarified. However, the preventive use of CoQ_{10} may be considered safe for consumption. However indiscriminate administration of CoQ_{10} is not advised and should only be given to subjects that can clearly benefit from it.

References

Abad C, Amengual MJ, Gosalvez J, Coward K, Hannaoui N, Benet J, Garcia-Peiro A, Prats J (2013) Effects of oral antioxidant treatment upon the dynamics of human sperm DNA fragmentation and subpopulations of sperm with highly degraded DNA. Andrologia 45(3):211–216. https://doi.org/10.1111/and.12003

Abo-Elmatty DM, Badawy EA, Hussein JS, Elela SA, Megahed HA (2012) Role of heme oxygenase, leptin, coenzyme Q10 and trace elements in pre-eclamptic women. Indian J Clin Biochem 27(4):379–384. https://doi.org/10.1007/s12291-012-0226-7

Agarwal A, Saleh RA, Bedaiwy MA (2003) Role of reactive oxygen species in the pathophysiology of human reproduction. Fertil Steril 79(4):829–843. https://doi.org/10.1016/s0015-0282(02)04948-8

Aitken RJ (1997a) The cell biology of fertilization. Adv Exp Med Biol 424:291–299. https://doi.org/10.1007/978-1-4615-5913-9_51

Aitken RJ (1997b) Molecular mechanisms regulating human sperm function. Mol Hum Reprod 3(3):169–173. https://doi.org/10.1093/molehr/3.3.169

Aitken RJ, Krausz C (2001) Oxidative stress, DNA damage and the Y chromosome. Reproduction 122(4):497–506. https://doi.org/10.1530/rep.0.1220497

Aitken RJ, De Iuliis GN, Finnie JM, Hedges A, McLachlan RI (2010) Analysis of the relationships between oxidative stress, DNA damage and sperm vitality in a patient population: development of diagnostic criteria. Hum Reprod 25(10):2415–2426. https://doi.org/10.1093/humrep/deq214

Aksoy H, Aksoy Y, Ozbey I, Altuntas I, Akcay F (2000) The relationship between varicocele and semen nitric oxide concentrations. Urol Res 28(6):357–359. https://doi.org/10.1007/s002400000132

Alleva R, Scararmucci A, Mantero F, Bompadre S, Leoni L, Littarru GP (1997) The protective role of ubiquinol-10 against formation of lipid hydroperoxides in human seminal fluid. Mol Asp Med 18(Suppl):S221–S228. https://doi.org/10.1016/s0098-2997(97)00040-x

Amaral S, Moreno AJ, Santos MS, Seica R, Ramalho-Santos J (2006) Effects of hyperglycemia on sperm and testicular cells of Goto-Kakizaki and streptozotocin-treated rat models for diabetes. Theriogenology 66(9):2056–2067. https://doi.org/10.1016/j.theriogenology.2006.06.006

Amaral S, Oliveira PJ, Ramalho-Santos J (2008) Diabetes and the impairment of reproductive function: possible role of mitochondria and reactive oxygen species. Curr Diabetes Rev 4(1):46–54. https://doi.org/10.2174/157339908783502398

Angelitti AG, Colacicco L, Calla C, Arizzi M, Lippa S (1995) Coenzyme Q: potentially useful index of bioenergetic and oxidative status of spermatozoa. Clin Chem 41(2):217–219

Arcaniolo D, Favilla V, Tiscione D, Pisano F, Bozzini G, Creta M, Gentile G, Menchini Fabris F, Pavan N, Veneziano IA, Cai T (2014) Is there a place for nutritional supplements in the treatment of idiopathic male infertility? Arch Ital Urol Androl 86(3):164–170. https://doi.org/10.4081/aiua.2014.3.164

Auger J, Eustache F, Andersen AG, Irvine DS, Jorgensen N, Skakkebaek NE, Suominen J, Toppari J, Vierula M, Jouannet P (2001) Sperm morphological defects related to environment, lifestyle and medical history of 1001 male partners of pregnant women from four European cities. Hum Reprod 16(12):2710–2717. https://doi.org/10.1093/humrep/16.12.2710

Balercia G, Arnaldi G, Fazioli F, Serresi M, Alleva R, Mancini A, Mosca F, Lamonica GR, Mantero F, Littarru GP (2002) Coenzyme Q10 levels in idiopathic and varicocele-associated asthenozoospermia. Andrologia 34(2):107–111. https://doi.org/10.1046/j.0303-4569.2001.00485.x

Balercia G, Mosca F, Mantero F, Boscaro M, Mancini A, Ricciardo-Lamonica G, Littarru G (2004) Coenzyme Q(10) supplementation in infertile men with idiopathic asthenozoospermia: an open, uncontrolled pilot study. Fertil Steril 81(1):93–98. https://doi.org/10.1016/j.fertnstert.2003.05.009

Balercia G, Buldreghini E, Vignini A, Tiano L, Paggi F, Amoroso S, Ricciardo-Lamonica G, Boscaro M, Lenzi A, Littarru G (2009a) Coenzyme Q10 treatment in infertile men with idiopathic asthenozoospermia: a placebo-controlled, double-blind randomized trial. Fertil Steril 91(5):1785–1792. https://doi.org/10.1016/j.fertnstert.2008.02.119

Balercia G, Mancini A, Paggi F, Tiano L, Pontecorvi A, Boscaro M, Lenzi A, Littarru GP (2009b) Coenzyme Q10 and male infertility. J Endocrinol Investig 32 (7):626–632. https://doi.org/10.3275/6301, https://doi.org/10.1007/bf03346521

Bartmann AK, Romao GS, Ramos Eda S, Ferriani RA (2004) Why do older women have poor implantation rates? A possible role of the mitochondria. J Assist Reprod Genet 21(3):79–83. https://doi.org/10.1023/b:jarg.0000027018.02425.15

Ben-Meir A, Burstein E, Borrego-Alvarez A, Chong J, Wong E, Yavorska T, Naranian T, Chi M, Wang Y, Bentov Y, Alexis J, Meriano J, Sung HK, Gasser DL, Moley KH, Hekimi S, Casper RF, Jurisicova A (2015) Coenzyme Q10 restores oocyte mitochondrial function and fertility during reproductive aging. Aging Cell 14(5):887–895. https://doi.org/10.1111/acel.12368

Ben-Meir A, Kim K, McQuaid R, Esfandiari N, Bentov Y, Casper RF, Jurisicova A (2019) Co-enzyme Q10 supplementation rescues cumulus cells dysfunction in a maternal aging model. Antioxidants (Basel) 8(3). https://doi.org/10.3390/antiox8030058

Bentinger M, Tekle M, Brismar K, Chojnacki T, Swiezewska E, Dallner G (2008) Stimulation of coenzyme Q synthesis. Biofactors 32(1–4):99–111. https://doi.org/10.1002/biof.5520320112

Bentov Y, Casper RF (2013) The aging oocyte--can mitochondrial function be improved? Fertil Steril 99(1):18–22. https://doi.org/10.1016/j.fertnstert.2012.11.031

Bentov Y, Esfandiari N, Burstein E, Casper RF (2010) The use of mitochondrial nutrients to improve the outcome of infertility treatment in older patients. Fertil Steril 93(1):272–275. https://doi.org/10.1016/j.fertnstert.2009.07.988

Bentov Y, Yavorska T, Esfandiari N, Jurisicova A, Casper RF (2011) The contribution of mitochondrial function to reproductive aging. J Assist Reprod Genet 28(9):773–783. https://doi.org/10.1007/s10815-011-9588-7

Bentov Y, Hannam T, Jurisicova A, Esfandiari N, Casper RF (2014) Coenzyme Q10 supplementation and oocyte aneuploidy in women undergoing IVF-ICSI treatment. Clin Med Insights Reprod Health 8:31–36. https://doi.org/10.4137/CMRH.S14681

Boni R, Gallo A, Cecchini S (2017) Kinetic activity, membrane mitochondrial potential, lipid peroxidation, intracellular pH and calcium of frozen/thawed bovine spermatozoa treated with metabolic enhancers. Andrology 5(1):133–145. https://doi.org/10.1111/andr.12259

Busetto GM, Koverech A, Messano M, Antonini G, De Berardinis E, Gentile V (2012) Prospective open-label study on the efficacy and tolerability of a combination of nutritional supplements in primary infertile patients with idiopathic astenoteratozoospermia. Arch Ital Urol Androl 84(3):137–140

Cakiroglu B, Eyyupoglu SE, Gozukucuk R, Uyanik BS (2014) Ubiquinol effect on sperm parameters in subfertile men who have astheno-teratozoospermia with normal sperm concentration. Nephrourol Mon 6(3):e16870. https://doi.org/10.5812/numonthly.16870

Chiang T, Schultz RM, Lampson MA (2011) Age-dependent susceptibility of chromosome cohesion to premature separase activation in mouse oocytes. Biol Reprod 85(6):1279–1283. https://doi.org/10.1095/biolreprod.111.094094

Comhaire FH, Christophe AB, Zalata AA, Dhooge WS, Mahmoud AM, Depuydt CE (2000) The effects of combined conventional treatment, oral antioxidants and essential fatty acids on sperm biology in subfertile men. Prostaglandins Leukot Essent Fatty Acids 63(3):159–165. https://doi.org/10.1054/plef.2000.0174

Comhaire FH, El Garem Y, Mahmoud A, Eertmans F, Schoonjans F (2005) Combined conventional/antioxidant "Astaxanthin" treatment for male infertility: a double blind, randomized trial. Asian J Androl 7(3):257–262. https://doi.org/10.1111/j.1745-7262.2005.00047.x

Compagnoni G, Lista G, Giuffre B, Mosca F, Marini A (2004) Coenzyme Q10 levels in maternal plasma and cord blood:correlations with mode of delivery. Biol Neonate 86(2):104–107. https://doi.org/10.1159/000078382

D'Aniello G, Ronsini S, Guida F, Spinelli P, D'Aniello A (2005) Occurrence of D-aspartic acid in human seminal plasma and spermatozoa: possible role in reproduction. Fertil Steril 84(5):1444–1449. https://doi.org/10.1016/j.fertnstert.2005.05.019

Datta U, Chandra Sekar M, Hembram ML, Dasgupta R (2009) Development of a new method to preserve caprine cauda epididymal spermatozoa in-situ at −10 degrees C with electrolyte free medium. J Assist Reprod Genet 26(8):467–473. https://doi.org/10.1007/s10815-009-9344-4

De Kretser DM, Baker HW (1999) Infertility in men: recent advances and continuing controversies. J Clin Endocrinol Metab 84(10):3443–3450. https://doi.org/10.1210/jcem.84.10.6101
de Lima CB, Cordeiro FB, Camargo M, Zylbersztejn DS, Cedenho AP, Bertolla RP, Lo Turco EG (2017) Follicular fluid lipid peroxidation levels in women with endometriosis during controlled ovarian hyperstimulation. Hum Fertil (Camb) 20(1):48–54. https://doi.org/10.1080/14647273.2016.1246753
Donnelly ET, O'Connell M, McClure N, Lewis SE (2000) Differences in nuclear DNA fragmentation and mitochondrial integrity of semen and prepared human spermatozoa. Hum Reprod 15(7):1552–1561. https://doi.org/10.1093/humrep/15.7.1552
Donnez J, Donnez O, Orellana R, Binda MM, Dolmans MM (2016) Endometriosis and infertility. Panminerva Med 58(2):143–150
Ducci M, Gazzano A, Tedeschi D, Sighieri C, Martelli F (2002) Coenzyme Q10 levels in pigeon (Columba livia) spermatozoa. Asian J Androl 4(1):73–76
Ebisch IM, Thomas CM, Peters WH, Braat DD, Steegers-Theunissen RP (2007) The importance of folate, zinc and antioxidants in the pathogenesis and prevention of subfertility. Hum Reprod Update 13(2):163–174. https://doi.org/10.1093/humupd/dml054
El Refaeey A, Selem A, Badawy A (2014) Combined coenzyme Q10 and clomiphene citrate for ovulation induction in clomiphene-citrate-resistant polycystic ovary syndrome. Reprod Biomed Online 29(1):119–124. https://doi.org/10.1016/j.rbmo.2014.03.011
El Shourbagy SH, Spikings EC, Freitas M, St John JC (2006) Mitochondria directly influence fertilisation outcome in the pig. Reproduction 131(2):233–245. https://doi.org/10.1530/rep.1.00551
Eroglu M, Sahin S, Durukan B, Ozakpinar OB, Erdinc N, Turkgeldi L, Sofuoglu K, Karateke A (2014) Blood serum and seminal plasma selenium, total antioxidant capacity and coenzyme q10 levels in relation to semen parameters in men with idiopathic infertility. Biol Trace Elem Res 159(1–3):46–51. https://doi.org/10.1007/s12011-014-9978-7
Escudero EL, Boerrigter PJ, Bennink HJ, Epifanio R, Horcajadas JA, Olivennes F, Pellicer A, Simon C (2005) Mifepristone is an effective oral alternative for the prevention of premature luteinizing hormone surges and/or premature luteinization in women undergoing controlled ovarian hyperstimulation for in vitro fertilization. J Clin Endocrinol Metab 90(4):2081–2088. https://doi.org/10.1210/jc.2004-1160
Eskenazi B, Kidd SA, Marks AR, Sloter E, Block G, Wyrobek AJ (2005) Antioxidant intake is associated with semen quality in healthy men. Hum Reprod 20(4):1006–1012. https://doi.org/10.1093/humrep/deh725
Faddy MJ (2000) Follicle dynamics during ovarian ageing. Mol Cell Endocrinol 163(1–2):43–48. https://doi.org/10.1016/s0303-7207(99)00238-5
Ford WC, Whittington K (1998) Antioxidant treatment for male subfertility: a promise that remains unfulfilled. Hum Reprod 13(6):1416–1419. https://doi.org/10.1093/oxfordjournals.humrep.a019707
Franke AA, Lai JF, Morrison CM, Pagano I, Li X, Halm BM, Soon R, Custer LJ (2013) Coenzyme Q10, carotenoid, tocopherol, and retinol levels in cord plasma from multiethnic subjects in Hawaii. Free Radic Res 47(9):757–768. https://doi.org/10.3109/10715762.2013.822495
Garcia-Rodriguez CE, Mesa MD, Olza J, Vlachava M, Kremmyda LS, Diaper ND, Noakes PS, Miles EA, Ramirez-Tortosa MC, Liaset B, Froyland L, Rossary A, Farges MC, Vasson MP, Aguilera CM, Helmersson-Karlqvist J, Godfrey KM, Calder PC, Basu S, Gil A (2012) Does consumption of two portions of salmon per week enhance the antioxidant defense system in pregnant women? Antioxid Redox Signal 16(12):1401–1406. https://doi.org/10.1089/ars.2012.4508
Garrido-Maraver J, Cordero MD, Oropesa-Avila M, Fernandez Vega A, de la Mata M, Delgado Pavon A, de Miguel M, Perez Calero C, Villanueva Paz M, Cotan D, Sanchez-Alcazar JA (2014) Coenzyme q10 therapy. Mol Syndromol 5(3–4):187–197. https://doi.org/10.1159/000360101

Gat I, Blanco Mejia S, Balakier H, Librach CL, Claessens A, Ryan EA (2016) The use of coenzyme Q10 and DHEA during IUI and IVF cycles in patients with decreased ovarian reserve. Gynecol Endocrinol 32(7):534–537. https://doi.org/10.3109/09513590.2015.1137095

Gendelman M, Roth Z (2012) Incorporation of coenzyme Q10 into bovine oocytes improves mitochondrial features and alleviates the effects of summer thermal stress on developmental competence. Biol Reprod 87(5):118. https://doi.org/10.1095/biolreprod.112.101881

Geva E, Bartoov B, Zabludovsky N, Lessing JB, Lerner-Geva L, Amit A (1996) The effect of antioxidant treatment on human spermatozoa and fertilization rate in an in vitro fertilization program. Fertil Steril 66(3):430–434. https://doi.org/10.1016/s0015-0282(16)58514-8

Geva E, Lessing JB, Lerner-Geva L, Amit A (1998) Free radicals, antioxidants and human spermatozoa: clinical implications. Hum Reprod 13(6):1422–1424. https://doi.org/10.1093/oxfordjournals.humrep.a019709

Ghanbarzadeh S, Garjani A, Ziaee M, Khorrami A (2014a) CoQ_{10} and L-carnitine attenuate the effect of high LDL and oxidized LDL on spermatogenesis in male rats. Drug Res (Stuttg) 64(10):510–515. https://doi.org/10.1055/s-0033-1361176

Ghanbarzadeh S, Garjani A, Ziaee M, Khorrami A (2014b) Effects of L-carnitine and coenzyme q10 on impaired spermatogenesis caused by isoproterenol in male rats. Drug Res (Stuttg) 64(9):449–453. https://doi.org/10.1055/s-0033-1361103

Giacone F, Condorelli RA, Mongioi LM, Bullara V, La Vignera S, Calogero AE (2017) In vitro effects of zinc, D-aspartic acid, and coenzyme-Q10 on sperm function. Endocrine 56(2):408–415. https://doi.org/10.1007/s12020-016-1013-7

Giannubilo SR, Tiano L, Ciavattini A, Landi B, Carnevali P, Principi F, Littarru GP, Mazzanti L (2014) Amniotic coenzyme Q10: is it related to pregnancy outcomes? Antioxid Redox Signal 21(11):1582–1586. https://doi.org/10.1089/ars.2014.5936

Griveau JF, Dumont E, Renard P, Callegari JP, Le Lannou D (1995) Reactive oxygen species, lipid peroxidation and enzymatic defence systems in human spermatozoa. J Reprod Fertil 103(1):17–26. https://doi.org/10.1530/jrf.0.1030017

Groneberg DA, Kindermann B, Althammer M, Klapper M, Vormann J, Littarru GP, Doring F (2005) Coenzyme Q10 affects expression of genes involved in cell signalling, metabolism and transport in human CaCo-2 cells. Int J Biochem Cell Biol 37(6):1208–1218. https://doi.org/10.1016/j.biocel.2004.11.017

Gual-Frau J, Abad C, Amengual MJ, Hannaoui N, Checa MA, Ribas-Maynou J, Lozano I, Nikolaou A, Benet J, Garcia-Peiro A, Prats J (2015) Oral antioxidant treatment partly improves integrity of human sperm DNA in infertile grade I varicocele patients. Hum Fertil (Camb) 18(3):225–229. https://doi.org/10.3109/14647273.2015.1050462

Gualtieri R, Barbato V, Fiorentino I, Braun S, Rizos D, Longobardi S, Talevi R (2014) Treatment with zinc, d-aspartate, and coenzyme Q10 protects bull sperm against damage and improves their ability to support embryo development. Theriogenology 82(4):592–598. https://doi.org/10.1016/j.theriogenology.2014.05.028

Gvozdjakova A, Kucharska J, Lipkova J, Bartolcicova B, Dubravicky J, Vorakova M, Cernakova I, Singh RB (2013) Importance of the assessment of coenzyme Q10, alpha-tocopherol and oxidative stress for the diagnosis and therapy of infertility in men. Bratisl Lek Listy 114(11):607–609. https://doi.org/10.4149/bll_2013_129

Gvozdjakova A, Kucharska J, Dubravicky J, Mojto V, Singh RB (2015) Coenzyme Q(1)(0), alpha-tocopherol, and oxidative stress could be important metabolic biomarkers of male infertility. Dis Markers 2015:827941. https://doi.org/10.1155/2015/827941

Haruna M, Matsuzaki M, Ota E, Honda Y, Tanizaki T, Sekine K, Tabata N, Yeo S, Murashima S (2010) Positive correlation between maternal serum coenzyme Q10 levels and infant birth weight. Biofactors 36(4):312–318. https://doi.org/10.1002/biof.104

Hendin BN, Kolettis PN, Sharma RK, Thomas AJ Jr, Agarwal A (1999) Varicocele is associated with elevated spermatozoal reactive oxygen species production and diminished seminal plasma antioxidant capacity. J Urol 161(6):1831–1834

Hernandez-Camacho JD, Bernier M, Lopez-Lluch G, Navas P (2018) Coenzyme Q10 supplementation in aging and disease. Front Physiol 9:44. https://doi.org/10.3389/fphys.2018.00044
Humaidan P, Bungum L, Bungum M, Hald F, Agerholm I, Blaabjerg J, Yding Andersen C, Lindenberg S (2005) Reproductive outcome using a GnRH antagonist (cetrorelix) for luteolysis and follicular synchronization in poor responder IVF/ICSI patients treated with a flexible GnRH antagonist protocol. Reprod Biomed Online 11(6):679–684. https://doi.org/10.1016/s1472-6483(10)61685-9
Humm KC, Sakkas D (2013) Role of increased male age in IVF and egg donation: is sperm DNA fragmentation responsible? Fertil Steril 99(1):30–36. https://doi.org/10.1016/j.fertnstert.2012.11.024
Hwang IS, Kwon D, Kwak TU, Lee JW, Im GS, Hwang S (2016) Improved survival and developmental rates in vitrified-warmed pig oocytes after recovery culture with coenzyme Q10. Cryo Letters 37(1):59–67
Irvine DS, Twigg JP, Gordon EL, Fulton N, Milne PA, Aitken RJ (2000) DNA integrity in human spermatozoa: relationships with semen quality. J Androl 21(1):33–44
Iwanier K, Zachara BA (1995) Selenium supplementation enhances the element concentration in blood and seminal fluid but does not change the spermatozoal quality characteristics in subfertile men. J Androl 16(5):441–447
Jayaprakasan K, Deb S, Batcha M, Hopkisson J, Johnson I, Campbell B, Raine-Fenning N (2010) The cohort of antral follicles measuring 2-6 mm reflects the quantitative status of ovarian reserve as assessed by serum levels of anti-Mullerian hormone and response to controlled ovarian stimulation. Fertil Steril 94(5):1775–1781. https://doi.org/10.1016/j.fertnstert.2009.10.022
Keefe DL, Niven-Fairchild T, Powell S, Buradagunta S (1995) Mitochondrial deoxyribonucleic acid deletions in oocytes and reproductive aging in women. Fertil Steril 64(3):577–583
Kessopoulou E, Powers HJ, Sharma KK, Pearson MJ, Russell JM, Cooke ID, Barratt CL (1995) A double-blind randomized placebo cross-over controlled trial using the antioxidant vitamin E to treat reactive oxygen species associated male infertility. Fertil Steril 64(4):825–831. https://doi.org/10.1016/s0015-0282(16)57861-3
Kobori Y, Ota S, Sato R, Yagi H, Soh S, Arai G, Okada H (2014) Antioxidant cosupplementation therapy with vitamin C, vitamin E, and coenzyme Q10 in patients with oligoasthenozoospermia. Arch Ital Urol Androl 86(1):1–4. https://doi.org/10.4081/aiua.2014.1.1
Kuang Y, Hong Q, Chen Q, Lyu Q, Ai A, Fu Y, Shoham Z (2014) Luteal-phase ovarian stimulation is feasible for producing competent oocytes in women undergoing in vitro fertilization/intracytoplasmic sperm injection treatment, with optimal pregnancy outcomes in frozen-thawed embryo transfer cycles. Fertil Steril 101(1):105–111. https://doi.org/10.1016/j.fertnstert.2013.09.007
Kuliev A, Cieslak J, Verlinsky Y (2005) Frequency and distribution of chromosome abnormalities in human oocytes. Cytogenet Genome Res 111(3–4):193–198. https://doi.org/10.1159/000086889
Lafuente R, Gonzalez-Comadran M, Sola I, Lopez G, Brassesco M, Carreras R, Checa MA (2013) Coenzyme Q10 and male infertility: a meta-analysis. J Assist Reprod Genet 30(9):1147–1156. https://doi.org/10.1007/s10815-013-0047-5
Lenzi A, Picardo M, Gandini L, Dondero F (1996) Lipids of the sperm plasma membrane: from polyunsaturated fatty acids considered as markers of sperm function to possible scavenger therapy. Hum Reprod Update 2(3):246–256. https://doi.org/10.1093/humupd/2.3.246
Lenzi A, Gandini L, Bagolan P, Nahum A, Dondero F (1998) Sperm parameters after early left varicocele treatment. Fertil Steril 69(2):347–349. https://doi.org/10.1016/s0015-0282(97)00474-3
Lewin A, Lavon H (1997) The effect of coenzyme Q10 on sperm motility and function. Mol Asp Med 18(Suppl):S213–S219. https://doi.org/10.1016/s0098-2997(97)00036-8
Li K, Shi Y, Chen S, Li W, Shang X, Huang Y (2006) Determination of coenzyme Q10 in human seminal plasma by high-performance liquid chromatography and its clinical application. Biomed Chromatogr 20(10):1082–1086. https://doi.org/10.1002/bmc.645
Littarru GP, Tiano L (2010) Clinical aspects of coenzyme Q10: an update. Nutrition 26(3):250–254. https://doi.org/10.1016/j.nut.2009.08.008

Lobascio AM, De Felici M, Anibaldi M, Greco P, Minasi MG, Greco E (2015) Involvement of seminal leukocytes, reactive oxygen species, and sperm mitochondrial membrane potential in the DNA damage of the human spermatozoa. Andrology 3(2):265–270. https://doi.org/10.1111/andr.302

Lopez-Lluch G (2019) The important role of CoQ_{10} in aging. Antioxidants (Basel) 8(12). https://doi.org/10.3390/antiox8120570

Lopez-Lluch G, Rodriguez-Aguilera JC, Santos-Ocana C, Navas P (2010) Is coenzyme Q a key factor in aging? Mech Ageing Dev 131(4):225–235. https://doi.org/10.1016/j.mad.2010.02.003

Macchia G, Topo E, Mangano N, D'Aniello E, Boni R (2010) DL-Aspartic acid administration improves semen quality in rabbit bucks. Anim Reprod Sci 118(2–4):337–343. https://doi.org/10.1016/j.anireprosci.2009.07.009

Mahfouz R, Sharma R, Thiyagarajan A, Kale V, Gupta S, Sabanegh E, Agarwal A (2010) Semen characteristics and sperm DNA fragmentation in infertile men with low and high levels of seminal reactive oxygen species. Fertil Steril 94(6):2141–2146. https://doi.org/10.1016/j.fertnstert.2009.12.030

Mancini A, Balercia G (2011) Coenzyme Q(10) in male infertility: physiopathology and therapy. Biofactors 37(5):374–380. https://doi.org/10.1002/biof.164

Mancini A, Conte G, Milardi D, De Marinis L, Littarru GP (1998) Relationship between sperm cell ubiquinone and seminal parameters in subjects with and without varicocele. Andrologia 30(1):1–4. https://doi.org/10.1111/j.1439-0272.1998.tb01374.x

Mancini A, Milardi D, Conte G, Bianchi A, Balercia G, De Marinis L, Littarru GP (2003) Coenzyme Q10: another biochemical alteration linked to infertility in varicocele patients? Metabolism 52(4):402–406. https://doi.org/10.1053/meta.2003.50083

Mancini A, De Marinis L, Littarru GP, Balercia G (2005a) An update of coenzyme Q10 implications in male infertility: biochemical and therapeutic aspects. Biofactors 25(1–4):165–174. https://doi.org/10.1002/biof.5520250119

Mancini A, Milardi D, Conte G, Festa R, De Marinis L, Littarru GP (2005b) Seminal antioxidants in humans: preoperative and postoperative evaluation of coenzyme Q10 in varicocele patients. Horm Metab Res 37(7):428–432. https://doi.org/10.1055/s-2005-870232

Mancini A, Festa R, Raimondo S, Silvestrini A, Giacchi E, Littarru GP, Pontecorvi A, Meucci E (2012) Biochemical alterations in semen of varicocele patients: a review of the literature. Adv Urol 2012:903931. https://doi.org/10.1155/2012/903931

Matsuzaki M, Haruna M, Ota E, Murayama R, Yamaguchi T, Shioji I, Sasaki S, Yamaguchi T, Murashima S (2014) Effects of lifestyle factors on urinary oxidative stress and serum antioxidant markers in pregnant Japanese women: a cohort study. Biosci Trends 8(3):176–184. https://doi.org/10.5582/bst.2014.01014

Meldrum DR, Casper RF, Diez-Juan A, Simon C, Domar AD, Frydman R (2016) Aging and the environment affect gamete and embryo potential: can we intervene? Fertil Steril 105(3):548–559. https://doi.org/10.1016/j.fertnstert.2016.01.013

Melo AS, Kliemchen J, Junior AA, Ferriani RA, Navarro PA (2016) Oxidative stress and polycystic ovary syndrome: evaluation during ovarian stimulation for ICSI. Reproduction. https://doi.org/10.1530/REP-16-0084

Myatt L, Cui X (2004) Oxidative stress in the placenta. Histochem Cell Biol 122(4):369–382. https://doi.org/10.1007/s00418-004-0677-x

Nadjarzadeh A, Sadeghi MR, Amirjannati N, Vafa MR, Motevalian SA, Gohari MR, Akhondi MA, Yavari P, Shidfar F (2011) Coenzyme Q10 improves seminal oxidative defense but does not affect on semen parameters in idiopathic oligoasthenoteratozoospermia: a randomized double-blind, placebo controlled trial. J Endocrinol Investig 34(8):e224–e228. https://doi.org/10.3275/7572

Nadjarzadeh A, Shidfar F, Amirjannati N, Vafa MR, Motevalian SA, Gohari MR, Nazeri Kakhki SA, Akhondi MM, Sadeghi MR (2014) Effect of coenzyme Q10 supplementation on antioxidant enzymes activity and oxidative stress of seminal plasma: a double-blind randomised clinical trial. Andrologia 46(2):177–183. https://doi.org/10.1111/and.12062

Nogueira BG, Sampaio BF, Souza MI, Costa ESEV, Zuccari CE (2015) Coenzyme Q10 and alpha-tocopherol prevent the lipid peroxidation of cooled equine semen. Reprod Domest Anim 50(6):1003–1010. https://doi.org/10.1111/rda.12627

Noia G, Littarru GP, De Santis M, Oradei A, Mactromarino C, Trivellini C, Caruso A (1996) Coenzyme Q10 in pregnancy. Fetal Diagn Ther 11(4):264–270. https://doi.org/10.1159/000264313

Noia G, Romano D, De Santis M, Mariorenzi S, Caruso A, Mancuso S (1998) Coenzyme Q10 fetal plasma levels. Fetal Diagn Ther 13(2):127–130. https://doi.org/10.1159/000020820

Noia G, Romano D, De Santis M, Cavaliere AF, Straface G, Alcaino S, Di Domenico M, Petrone A, Caruso A, Mancuso S (1999) The antioxidants (coenzyme Q10) in materno-fetal physiopathology. Minerva Ginecol 51(10):385–391

Olivennes F, Belaisch-Allart J, Emperaire JC, Dechaud H, Alvarez S, Moreau L, Nicollet B, Zorn JR, Bouchard P, Frydman R (2000) Prospective, randomized, controlled study of in vitro fertilization-embryo transfer with a single dose of a luteinizing hormone-releasing hormone (LH-RH) antagonist (cetrorelix) or a depot formula of an LH-RH agonist (triptorelin). Fertil Steril 73(2):314–320. https://doi.org/10.1016/s0015-0282(99)00524-5

Omu AE, Fatinikun T, Mannazhath N, Abraham S (1999) Significance of simultaneous determination of serum and seminal plasma alpha-tocopherol and retinol in infertile men by high-performance liquid chromatography. Andrologia 31(6):347–354. https://doi.org/10.1046/j.1439-0272.1999.00296.x

Palan PR, Shaban DW, Martino T, Mikhail MS (2004) Lipid-soluble antioxidants and pregnancy: maternal serum levels of coenzyme Q10, alpha-tocopherol and gamma-tocopherol in preeclampsia and normal pregnancy. Gynecol Obstet Investig 58(1):8–13. https://doi.org/10.1159/000077011

Palmeira CM, Santos DL, Seica R, Moreno AJ, Santos MS (2001) Enhanced mitochondrial testicular antioxidant capacity in Goto-Kakizaki diabetic rats: role of coenzyme Q. Am J Physiol Cell Physiol 281(3):C1023–C1028. https://doi.org/10.1152/ajpcell.2001.281.3.C1023

Parinaud J, Le Lannou D, Vieitez G, Griveau JF, Milhet P, Richoilley G (1997) Enhancement of motility by treating spermatozoa with an antioxidant solution (Sperm-Fit) following ejaculation. Hum Reprod 12(11):2434–2436. https://doi.org/10.1093/humrep/12.11.2434

Pasqualotto FF, Sharma RK, Potts JM, Nelson DR, Thomas AJ, Agarwal A (2000) Seminal oxidative stress in patients with chronic prostatitis. Urology 55(6):881–885. https://doi.org/10.1016/s0090-4295(99)00613-5

Pellestor F, Anahory T, Hamamah S (2005) Effect of maternal age on the frequency of cytogenetic abnormalities in human oocytes. Cytogenet Genome Res 111(3–4):206–212. https://doi.org/10.1159/000086891

Ramalho-Santos J, Varum S, Amaral S, Mota PC, Sousa AP, Amaral A (2009) Mitochondrial functionality in reproduction: from gonads and gametes to embryos and embryonic stem cells. Hum Reprod Update 15(5):553–572. https://doi.org/10.1093/humupd/dmp016

Ritter LJ, Sugimura S, Gilchrist RB (2015) Oocyte induction of EGF responsiveness in somatic cells is associated with the acquisition of porcine oocyte developmental competence. Endocrinology 156(6):2299–2312. https://doi.org/10.1210/en.2014-1884

Roland L, Gagne A, Belanger MC, Boutet M, Berthiaume L, Fraser WD, Julien P, Bilodeau JF (2010) Existence of compensatory defense mechanisms against oxidative stress and hypertension in preeclampsia. Hypertens Pregnancy 29(1):21–37. https://doi.org/10.3109/10641950902777689

Ross C, Morriss A, Khairy M, Khalaf Y, Braude P, Coomarasamy A, El-Toukhy T (2010) A systematic review of the effect of oral antioxidants on male infertility. Reprod Biomed Online 20(6):711–723. https://doi.org/10.1016/j.rbmo.2010.03.008

Safarinejad MR (2008) Evaluation of endocrine profile, hypothalamic-pituitary-testis axis and semen quality in multiple sclerosis. J Neuroendocrinol 20(12):1368–1375. https://doi.org/10.1111/j.1365-2826.2008.01791.x

Safarinejad MR (2009) Efficacy of coenzyme Q10 on semen parameters, sperm function and reproductive hormones in infertile men. J Urol 182(1):237–248. https://doi.org/10.1016/j.juro.2009.02.121

Safarinejad MR (2012) The effect of coenzyme Q(1)(0) supplementation on partner pregnancy rate in infertile men with idiopathic oligoasthenoteratozoospermia: an open-label prospective study. Int Urol Nephrol 44(3):689–700. https://doi.org/10.1007/s11255-011-0081-0

Safarinejad MR, Safarinejad S, Shafiei N, Safarinejad S (2012) Effects of the reduced form of coenzyme Q10 (ubiquinol) on semen parameters in men with idiopathic infertility: a double-blind, placebo controlled, randomized study. J Urol 188(2):526–531. https://doi.org/10.1016/j.juro.2012.03.131

Sakamaki H, Akazawa S, Ishibashi M, Izumino K, Takino H, Yamasaki H, Yamaguchi Y, Goto S, Urata Y, Kondo T, Nagataki S (1999) Significance of glutathione-dependent antioxidant system in diabetes-induced embryonic malformations. Diabetes 48(5):1138–1144. https://doi.org/10.2337/diabetes.48.5.1138

Sakkas D, Mariethoz E, Manicardi G, Bizzaro D, Bianchi PG, Bianchi U (1999) Origin of DNA damage in ejaculated human spermatozoa. Rev Reprod 4(1):31–37. https://doi.org/10.1530/ror.0.0040031

Saleh RA, Agarwal A, Sharma RK, Nelson DR, Thomas AJ Jr (2002) Effect of cigarette smoking on levels of seminal oxidative stress in infertile men: a prospective study. Fertil Steril 78(3):491–499. https://doi.org/10.1016/s0015-0282(02)03294-6

Sato M, Sato K (2012) Maternal inheritance of mitochondrial DNA: degradation of paternal mitochondria by allogeneic organelle autophagy, allophagy. Autophagy 8(3):424–425. https://doi.org/10.4161/auto.19243

Schon EA, Kim SH, Ferreira JC, Magalhaes P, Grace M, Warburton D, Gross SJ (2000) Chromosomal non-disjunction in human oocytes: is there a mitochondrial connection? Hum Reprod 15(Suppl 2):160–172. https://doi.org/10.1093/humrep/15.suppl_2.160

Sharma RK, Agarwal A (1996) Role of reactive oxygen species in male infertility. Urology 48(6):835–850. https://doi.org/10.1016/s0090-4295(96)00313-5

Sheweita SA, Tilmisany AM, Al-Sawaf H (2005) Mechanisms of male infertility: role of antioxidants. Curr Drug Metab 6(5):495–501. https://doi.org/10.2174/138920005774330594

Showell MG, Brown J, Yazdani A, Stankiewicz MT, Hart RJ (2011) Antioxidants for male subfertility. Cochrane Database Syst Rev 1:CD007411. https://doi.org/10.1002/14651858.CD007411.pub2

Sinclair S (2000) Male infertility: nutritional and environmental considerations. Altern Med Rev 5(1):28–38

Singh AK, Tiwari AK, Singh PB, Dwivedi US, Trivedi S, Singh SK, Agrawal NK, Deshpande SB (2010) Multivitamin and micronutrient treatment improves semen parameters of azoospermic patients with maturation arrest. Indian J Physiol Pharmacol 54(2):157–163

Smith R, Vantman D, Ponce J, Escobar J, Lissi E (1996) Total antioxidant capacity of human seminal plasma. Hum Reprod 11(8):1655–1660. https://doi.org/10.1093/oxfordjournals.humrep.a019465

Staicu ML, Muresan A, Tache S, Moldovan R (2011) Effects of exogenous antioxidants on oxidative stress in pregnancy. J Med Life 4(2):163–167

Stojkovic M, Westesen K, Zakhartchenko V, Stojkovic P, Boxhammer K, Wolf E (1999) Coenzyme Q(10) in submicron-sized dispersion improves development, hatching, cell proliferation, and adenosine triphosphate content of in vitro-produced bovine embryos. Biol Reprod 61(2):541–547. https://doi.org/10.1095/biolreprod61.2.541

Takeuchi T, Neri QV, Katagiri Y, Rosenwaks Z, Palermo GD (2005) Effect of treating induced mitochondrial damage on embryonic development and epigenesis. Biol Reprod 72(3):584–592. https://doi.org/10.1095/biolreprod.104.032391

Talevi R, Barbato V, Fiorentino I, Braun S, Longobardi S, Gualtieri R (2013) Protective effects of in vitro treatment with zinc, d-aspartate and coenzyme q10 on human sperm motility, lipid peroxidation and DNA fragmentation. Reprod Biol Endocrinol 11:81. https://doi.org/10.1186/1477-7827-11-81

Talukdar A, Sharma KA, Rai R, Deka D, Rao DN (2015) Effect of coenzyme Q10 on Th1/Th2 paradigm in females with idiopathic recurrent pregnancy loss. Am J Reprod Immunol 74(2):169–180. https://doi.org/10.1111/aji.12376

Tapia JA, Macias-Garcia B, Miro-Moran A, Ortega-Ferrusola C, Salido GM, Pena FJ, Aparicio IM (2012) The membrane of the mammalian spermatozoa: much more than an inert envelope. Reprod Domest Anim 47(Suppl 3):65–75. https://doi.org/10.1111/j.1439-0531.2012.02046.x

Teran E, Vivero S, Racines-Orbe M, Castellanos A, Chuncha G, Enriquez G, Moya W (2005) Coenzyme Q10 is increased in placenta and cord blood during preeclampsia. Biofactors 25(1–4):153–158. https://doi.org/10.1002/biof.5520250117

Teran E, Chedraui P, Racines-Orbe M, Vivero S, Villena F, Duchicela F, Nacevilla L, Schwager G, Calle A (2008) Coenzyme Q10 levels in women with preeclampsia living at different altitudes. Biofactors 32(1–4):185–190. https://doi.org/10.1002/biof.5520320122

Teran E, Hernandez I, Nieto B, Tavara R, Ocampo JE, Calle A (2009) Coenzyme Q10 supplementation during pregnancy reduces the risk of pre-eclampsia. Int J Gynaecol Obstet 105(1):43–45. https://doi.org/10.1016/j.ijgo.2008.11.033

Teran E, Racines-Orbe M, Toapanta J, Valdivieso L, Vega Z, Vivero S, Moya W, Chedraui P, Perez-Lopez FR (2011) Maternal plasma and amniotic fluid coenzyme Q10 levels in preterm and term gestations: a pilot study. Arch Gynecol Obstet 283(Suppl 1):67–71. https://doi.org/10.1007/s00404-011-1894-x

Titus S, Li F, Stobezki R, Akula K, Unsal E, Jeong K, Dickler M, Robson M, Moy F, Goswami S, Oktay K (2013) Impairment of BRCA1-related DNA double-strand break repair leads to ovarian aging in mice and humans. Sci Transl Med 5(172):172ra121. https://doi.org/10.1126/scitranslmed.3004925

Tremellen K, Miari G, Froiland D, Thompson J (2007) A randomised control trial examining the effect of an antioxidant (Menevit) on pregnancy outcome during IVF-ICSI treatment. Aust N Z J Obstet Gynaecol 47(3):216–221. https://doi.org/10.1111/j.1479-828X.2007.00723.x

Tulsiani DR (2000) Carbohydrates mediate sperm-ovum adhesion and triggering of the acrosome reaction. Asian J Androl 2(2):87–97

Van Blerkom J, Davis PW, Lee J (1995) ATP content of human oocytes and developmental potential and outcome after in-vitro fertilization and embryo transfer. Hum Reprod 10(2):415–424. https://doi.org/10.1093/oxfordjournals.humrep.a135954

Vital-Reyes V, Rodriguez-Burford C, Chhieng DC, Alvarado-Cabrero I, Reyes-Fuentes A, Grizzle WE (2006) Ovarian expression of markers associated with proliferation or apoptosis in women with diminished ovarian reserve. Fertil Steril 86(1):176–185. https://doi.org/10.1016/j.fertnstert.2005.12.075

Walczak-Jedrzejowska R, Wolski JK, Slowikowska-Hilczer J (2013) The role of oxidative stress and antioxidants in male fertility. Cent Eur J Urol 66(1):60–67. https://doi.org/10.5173/ceju.2013.01.art19

Wathes DC, Abayasekara DR, Aitken RJ (2007) Polyunsaturated fatty acids in male and female reproduction. Biol Reprod 77(2):190–201. https://doi.org/10.1095/biolreprod.107.060558

Zalata A, Hafez T, Comhaire F (1995) Evaluation of the role of reactive oxygen species in male infertility. Hum Reprod 10(6):1444–1451. https://doi.org/10.1093/humrep/10.6.1444

Zhang X, Wu XQ, Lu S, Guo YL, Ma X (2006) Deficit of mitochondria-derived ATP during oxidative stress impairs mouse MII oocyte spindles. Cell Res 16(10):841–850. https://doi.org/10.1038/sj.cr.7310095

Zhang J, Fang L, Lu Z, Xiong J, Wu M, Shi L, Luo A, Wang S (2016) Are sirtuins markers of ovarian aging? Gene 575(2 Pt 3):680–686. https://doi.org/10.1016/j.gene.2015.09.043

Part IV
Prolongevity Effectors and Coenzyme Q

Chapter 15
Caloric Restriction, Longevity and Coenzyme Q

María Isabel Burón, Cristina Parrado, José Antonio González-Reyes, Lucía Fernández del Río, Elena Gutiérrez-Casado, Miguel Calvo-Rubio, Sandra Rodríguez-López, and José Manuel Villalba

Abstract The relationship between diet, longevity and health is complex, considering that nutritional components affect several physiologic processes and play a regulatory role in metabolic pathways crucial for the survival of cells. Caloric restriction is well known to extend healthy, average, and maximum life span in an evolutionary conserved way, acting through effector molecules which reprogram energy metabolism in response to reduced energy availability. As a central component in energy metabolism, alterations in coenzyme Q biosynthesis and its cellular balance may participate in adaptive responses to physiological, experimental, or pathological conditions. In fact, coenzyme Q levels, expression of *COQ* genes and activity of coenzyme Q-dependent antioxidant systems are targets of caloric restriction in a tissue-specific way, with skeletal muscle exhibiting an early response to this intervention. In mammal models, maximal longevity is not affected by life-long administration of coenzyme Q_{10} when animals are fed normal diets, but dietary coenzyme Q_{10} supplementation can ameliorate deleterious alterations associated with aging and even extends lifespan of rats fed a potentially prooxidant diet. The role of additional factors, as duration of caloric restriction intervention, caloric restriction protocol, or the predominant fat source present in the experimental diets, among others, is still to be fully determined to understand the actual role of coenzyme Q alterations and their impact on this dietary intervention in relation with the physiology of aging.

Keywords Caloric restriction · Longevity · Mitochondria · Coenzyme Q · Diet · Nutrition

---

M. I. Burón · C. Parrado · J. A. González-Reyes · L. F. del Río · E. Gutiérrez-Casado
M. Calvo-Rubio · S. Rodríguez-López · J. M. Villalba (✉)
Departamento de Biología Celular, Fisiología e Inmunología, Universidad de Córdoba, Córdoba, Spain
e-mail: jmvillalba@uco.es

G. López Lluch (ed.), *Coenzyme Q in Aging*,
https://doi.org/10.1007/978-3-030-45642-9_15

15.1 Introduction

Both lifespan and health span are influenced by nutrition, since nutrients and their metabolites control energy balance, enzymatic activities, and genome stability throughout the lifecycle (Dato et al. 2016). The relationship between diet, longevity and health is complex, considering that nutritional components affect several physiologic processes and play a regulatory role in metabolic pathways crucial for the survival of cells such as inflammation or immune function (Santoro et al. 2014). It is also commonly accepted that the complex interactions of multiple polymorphisms play a key role in determining how individuals may respond to dietary interventions (nutrigenetics) or how some nutrients may affect gene expression (nutrigenomic) (Darnton-Hill et al. 2004).

Caloric restriction (CR) is a dietary intervention well known to extend healthy, average, and maximum life span in an evolutionary conserved way (Weindruch et al. 1988; Weindruch and Sohal 1997). Currently available data obtained from many experimental models (Ruetenik and Barrientos 2015; Fontana and Partridge 2015) indicate that a lifestyle modification in the form of CR, even if started during adulthood, can help to minimize the risk of age-associated disorders and promote healthy aging in humans, probably through molecular mechanisms largely conserved along evolution.

The underlying biological mechanism responsible for the life extension effect of CR is still not known, although many hypotheses have been proposed. Moreover, a wide range of interventions have been used to impose CR, even within single species, and the mechanisms through which they extend lifespan can differ. The pioneering studies on body's response to CR already pointed to different pathways which were not mutually exclusive, such as decreased oxidative stress (Sohal and Weindruch 1996), decreased glycation and glycosidation (Kristal and Yu 1992), decreased body temperature associated with hypometabolic state (Walford and Spindler 1997), neuroendocrine changes (Nelson et al. 1995), and alterations in gene expression and protein degradation (Xia et al. 1995). Cellular effector processes can include enhanced genomic stability and chromatin remodeling, improved chaperone-mediated protein homeostasis and cellular turnover processes, including autophagy, and increases in various forms of stress resistance. Investigations aimed to understand at a molecular level those mechanisms by which CR extends longevity and improves the quality of life have led to the proposal that this non-genetic intervention acts through effector molecules which reprogram energy metabolism in response to reduced energy availability (Anderson and Weindruch 2010). Molecular effectors shown to mediate the effects of CR on health and longevity include forkhead box O (FoxO), target of rapamycin (TOR), AMP-activated protein kinase (AMPK), sirtuins (SIRT), heat shock factors (HSF), and nuclear factor E2-related factor-2 (NRF2) (see Fontana and Partridge 2005 for a review). In addition, studies carried out in several species have shown that the changes in energy metabolism induced by CR are dependent on the specific tissue (Heilbronn and Ravussin 2003; Park and Prolla 2005; Anderson and Weindruch 2010), which is consistent with a database of gene expression (AGEMAP. Atlas of Gene Expression in MouseAging Project) where a relevant tissue specificity was found in the transcriptional changes that are associated with age (Zahn et al. 2007).

Redox homeostasis of the cell is ensured by complex endogenous antioxidant defense systems, which include antioxidant enzymes and non-enzymatic compounds like glutathione, proteins (ferritin, transferrin, ceruloplasmin, and even albumin) and low molecular weight scavengers, like uric acid, coenzyme Q (CoQ), and lipoic acid (Dröge 2002). In addition, exogenous antioxidants present in fruits and vegetables complement the activity of the mentioned endogenous antioxidative defense. The theory of free radicals in aging (Harman 1956) continues to be among the most popular and accepted theories to explain the causes of aging (Barja 2013, 2014). Thus, it has been investigated in depth if CR alters the production of reactive oxygen species (ROS) or cellular oxidative damage. The accumulation of oxidatively damaged biomolecules in aged animals can be prevented by CR (reviewed by Cerqueira and Kowaltowski 2013). In studies carried out on rodents, it has been shown that CR decreases oxidative damage to proteins (Sohal et al. 1994a; Lass et al. 1998), DNA (Sohal et al. 1994b; Kaneko et al. 1997), and lipids (Kim et al. 1996; Lass et al. 1998). It has also been demonstrated that CR slows ROS production in mitochondria (Sohal et al. 1994a; Bevilacqua et al. 2004; Hagopian et al. 2005). In sum, these findings support the idea that decreased oxidative stress may be a mechanism contributing, at least partially, to the delay of aging with CR.

Since elevated metabolic activity might lead to a higher generation of potentially pro-oxidant substances, CR has been also proposed to act through a reduction in metabolic rate, resulting in a decrease of toxic molecules from metabolism (Sohal and Weindruch 1996; Anderson and Weindruch 2010). During aging, induction of stress response appears a result of damage to proteins and other macromolecules caused by cellular energy decline and by the deficit in systems for the renewal of these molecules. The effects of CR are offset by those caused by aging, minimizing the stress response genes (Sohal and Forster 2014). Accordingly, CR promotes the reduction of metabolic genes involved in detoxification, DNA repair and response to oxidative stress, probably because substrate availability for these systems may be reduced under this situation. Furthermore, CR also triggers genes related with an increase in the biosynthesis and renewal of macromolecules (Weindruch et al. 2001; Park and Prolla 2005). The work carried out by Lee et al. (2004) documented that CR and dietary supplementation with antioxidants as CoQ and lipoic acid result in changes of gene expression patterns consistent with a decrease of the basal level of cellular oxidants, although dietary interventions with the antioxidants were not as effective as CR in inhibiting the aging process.

CoQ is a member of the mitochondrial respiratory chain present in all cells and membranes and carries out functions of great importance for the cellular metabolism. Due to the importance of CoQ in several different aspects of cell physiology, it can be considered not only as a key factor in metabolism, but also for antioxidant protection, signaling regulation and organelle activity. The biosynthesis of CoQ has been studied to a great detail in bacteria and yeast, and many of these observations have been useful to understand the process in animal models (Turunen et al. 2004).

CoQ distribution is not uniform among the various tissues and organs, and its redistribution between organs is insignificant. In mice, rats, and humans, maximal CoQ concentrations are present in kidney and heart, whereas lower amounts can be detected in liver, brain, and skeletal muscle (Turunen et al. 2004; Bentinger et al. 2010). Thus, CoQ levels may be adapted to the particular physiology of each tissue, which is probably determined by a coordinated balance between its synthesis and degradation, both of which occur in all tissues (Turunen et al. 2004). In addition, in a study carried out by our Group, focused on the relative abundance of the two main CoQ isoforms present in mouse tissues (CoQ_9 and CoQ_{10}), we found that in young mice the highest CoQ_9/CoQ_{10} ratios were observed in liver and skeletal muscle whereas the lowest ratios were found in kidney and brain. These data support the idea that the homeostatic range for CoQ isoforms also varies among tissues and organs, and the maintenance of a given CoQ_9/CoQ_{10} ratio may be important for proper function of each tissue (Parrado-Fernández et al. 2011). Furthermore, besides being synthesized in all organisms, CoQ can be also provided by diet, although its assimilation by tissues of vertebrates is not homogeneous and is modulated with aging (Aberg et al. 1992; Lass et al. 1999), with central nervous system and muscle tissues exhibiting less capacity to incorporate CoQ from the diet (Bentinger et al. 2003).

Tissue concentrations of CoQ are modified during aging and pathophysiological conditions, which may influence cellular functions. In this case, the extent of disturbances is dependent on the localization and the modified distribution of this antioxidant at cellular and membrane levels. The maintenance of CoQ levels in tissues as brain, muscle or liver is extremely important due to their high dependence of these tissues on oxidative metabolism. Kalen and co-workers detected an age-associated loss of CoQ in homogenates from several human tissues (Kalen et al. 1989), while a constant level of CoQ with age was reported for homogenates of rat brain and lung (Beyer et al. 1985; Sohal and Forster 2007).

As the only lipid-soluble antioxidant that can be synthesized *in vivo* by all organisms and can also be taken from dietary source, CoQ emerges as a significant molecular target whose levels are regulated by aging and dietary interventions (Wang et al. 2015).

15.2 Coenzyme Q, Calorie Restriction and Aging. Lessons from Different Animal Models

Recent studies carried out in unicellular and invertebrate model organisms, and in different mammal systems including rodents, monkeys, and humans, have clearly indicated that diet has a prominent role in modulating mechanisms of aging and its associated diseases (Fontana and Partridge 2015).

15.2.1 Dietary CoQ and Genetic Interventions Related with CoQ Biosynthesis

It has been suggested that a diet rich in antioxidants can bring health benefits and thus, a great interest has been directed towards assessing the antioxidant capacity of natural products (Pisoschi and Pop 2015). However, the excessive use of antioxidants in unnecessary conditions could impair the aging process (Linnane et al. 2007). For this reason, a right balance between the needs of the organisms and supplementation with antioxidants, such as CoQ, must be maintained to avoid both deficient and excess conditions. Although CoQ has been linked to lifespan, the effects of dietary CoQ supplement in lifespan and the actual role played by its endogenous biosynthesis in aging have been contradictory when comparing different animal models and/or experimental conditions.

Whereas the yeast aging model has uncovered many interconnections between aging and other relevant cellular processes, including mitochondrial function, which may be key determinants in more complex eukaryotes, the nematode *Caenorhabditis elegans* shares numerous processes and pathways with more complex animals and, in spite of its simplicity, it is a versatile model organism and a powerful tool to unravel the effect of genes on longevity, due to its amenability to genetic manipulation and short lifespan. Longevity is clearly affected by CoQ levels in *C. elegans*, although several studies focused on elucidating the effect of dietary CoQ on aging have yielded contradictory observations. Larsen and Clarke (2002) reported beneficial effects of CoQ deprivation on longevity of wild type *C. elegans*worms that were fed a CoQ_8-less diet. Life span extension induced by moderated CoQ depletion in this invertebrate model was explained on the basis of lower oxygen radical production in mitochondria. However, Ishii et al. (2004) reported later that exogenously supplied CoQ_{10} can play a significant anti-aging function both in wild-type strains of *C. elegans* and in *mev-1(kn1)* mutants that are hypersensitive to oxidative stress and age precociously, probably because of elevated superoxide anion production in mitochondria. Of note, it was proposed that the mechanism of action could involve the participation of CoQ either as an antioxidant to dismutate the free radical superoxide anion or by reducing the uncoupling of reactions during electron transport that could otherwise result in superoxide anion production.

Studies carried out with *clk-1(coq7)* mutants (Ewbank et al. 1997), as well as those set up to elucidate the effect of silencing several *coq* genes (Asencio et al. 2003; Rodriguez-Aguilera et al. 2005), have also indicated that lower levels of the endogenous CoQ_9 isoform can lead to a more optimized operation of the mitochondrial transport chain and generate less ROS in *C. elegans*. However, the longevity-extending phenotype of the *clk-1* mutation was only evident when the worms were fed a normal diet with CoQ_8-containing bacteria, but lost when the worms were fed CoQ_8-less bacteria (Jonassen et al. 2001). In accordance, severe CoQ depletion leads to developmental and reproductive inefficiency with the observation that knockouts in *coq-1*, *coq-2*, *coq-3* or *coq-8* genes showed deleterious defects that led to *C. elegans* early developmental arrest that was partially prevented by dietary CoQ

(Asencio et al. 2009; Gavilan et al. 2005; Hihi et al. 2002). In *Drosophila* it has been demonstrated that mutation of *sbo* gene (a functional homolog of *COQ2*) leads to an extension of lifespan by decreasing endogenous CoQ_9 and CoQ_{10} biosynthesis (Liu et al. 2011).

Mammals have a variety of efficient systems to maintain CoQ in a reduced state, which sustains a high antioxidant status in the tissues. Although total levels of CoQ have been reported to decrease with aging, at least in some tissues (López-Lluch et al. 2010), life-long administration of CoQ_{10} did not affect maximal longevity of rats that had been fed normal diets (Lönnrot et al. 1995, 1998; Sohal and Forster 2007). However, supplementation of the diet with CoQ_{10} did ameliorate deleterious alterations associated with aging and even extended lifespan of rats that had been fed a potentially prooxidant diet containing n-6 PUFA from sunflower oil as dietary fat source (Quiles et al. 2004; Varela-Lopez et al. 2016). Using this latter model, we also demonstrated that CoQ_{10} supplementation improved liver antioxidant defense systems by potentiating thiol-dependent mechanisms (Bello et al. 2005) and promoted an anti-aterogenic and anti-inflammatory pattern of plasma proteins in aged rats (Santos-González et al. 2007). In a human model, it was found that supplementation with CoQ_{10} improved several parameters associated with a better antioxidant status in healthy elderly men and women fed a Mediterranean diet, such as postprandial oxidative stress and the action of antioxidant systems and antioxidant gene expression (Yubero-Serrano et al. 2011, 2013), the expression of genes related with inflammatory response and endoplasmic reticulum stress (Yubero-Serrano et al. 2012), postprandial metabolism of advanced glycation end products (Lopez-Moreno et al. 2016), and postprandial changes of p53 in response to oxidative DNA damage (Gutierrez-Mariscal et al. 2012).

15.2.2 Alteration of Endogenous CoQ Levels by Dietary Interventions. The Effect of Calorie Restriction

Dietary CoQ is taken up from the intestine into the circulation with a low rate, and only about 2–4% of dietary CoQ can be recovered (Zhang et al. 1995), which is not equally delivered to all organs (Kwong et al. 2002). Since bioavailability of dietary CoQ is very low, it is thus very important to develop mechanisms to increase functional CoQ in tissues and organs in processes such as aging or several diseases (Bentinger et al. 2003). In this sense, attention has been paid to interventions able to upregulate the synthesis of endogenous CoQ using naturally occurring substances, nutritional supplements or some drugs which increase the cellular synthetic rate at the transcriptional and/or the translational level (Bentinger et al. 2008a, b). Tocotrienol epoxides have been proven to be very efficient in augmenting endogenous CoQ levels, and those having one epoxide in the side chain doubled or trebled CoQ synthesis while those bearing two epoxides additionally also inhibited cholesterol synthesis (Bentinger et al. 2008a, b). More recently, it has been reported that oxidosqualenes were also useful for stimulating the synthesis and levels of CoQ both *in vitro* and *in vivo* (Bentinger et al. 2014).

CR intervention may also affect differentially endogenous CoQ balance depending on the tissue and the model of study. In our research Group we have demonstrated that distribution of CoQ homologues and CoQ_9/CoQ_{10} ratio displayed significant variability among tissues and organs obtained from 3 month young mice that were subjected to short term (1 month) CR. It is remarkable that a clear response to this intervention (as measured by a significant alteration of CoQ levels in tissue homogenates) was found in skeletal muscle, where levels of both CoQ_9 and CoQ_{10} were significantly higher in comparison with the control group, but not in other tissues including liver (Parrado-Fernández et al. 2011). Previous reports documented that long-term CR increased CoQ_9 content of mitochondria isolated from skeletal muscle (Lass et al. 1999), liver, heart and kidney (Kamzalov and Sohal 2004). The increase of CoQ_9 by CR in skeletal muscle mitochondria (Lass et al. 1999) agrees with our own observations (Parrado-Fernández et al. 2011), although we also observed an increase of the CoQ_{10} isoform. It is interesting to note that we detected these changes in skeletal muscle very early, after 1 month of CR, which agrees with Bevilacqua et al. (2004) who demonstrated that skeletal muscle (but not liver) mitochondria are rapidly adapted to short-term (2-week and 2-month) CR in the rat with a significant decrease of ROS generation (Ramsey et al. 2004). It is possible that CR effects on CoQ levels in liver are a late event since life-long CR indeed attenuated the increase of CoQ_9/CoQ_{10} ratio in rat liver plasma membrane from aged rats (de Cabo et al. 2004). On the other hand, it has been recently reported that a CR intervention based on every-other-day feeding procedure and physical exercise increased skeletal muscle CoQ levels in old but not in young animals (Rodríguez-Bies et al. 2015). These apparent discrepancies might highlight the importance of the protocol used to impose CR.

CR-elicited changes of CoQ content in skeletal muscle might be related with the beneficial effects reported for this intervention. In this sense, Leeuwenburgh et al. (1997) demonstrated previously that CR prevented the aging-associated increase of protein oxidation in the skeletal muscle and later, Bevilacqua et al. (2004) reported that skeletal muscle (but not liver) mitochondria are rapidly adapted to short-term CR in the rat with a significant decrease in reactive oxygen species (ROS) generation. Recent structural studies developed in mice subjected to CR for 6 month have also demonstrated that this intervention produces a better healthy state of skeletal muscle fibers, represented by increased cross-sectional area and decreased circularity in cross sections (López-Domínguez et al. 2013).

Recent investigations carried out with *C. elegans* have pointed out to the idea that CR could decrease CoQ_9 levels *via* gene expression in this model organism (Fischer et al. 2015). In this way, dietary restriction down-regulated the steady-state expression levels of several evolutionary conserved genes (i.e. coq-1) that encode key enzymes of the mevalonate and CoQ-synthesizing pathways, and also decreased the levels of total CoQ_9 and its reduced form (ubiquinol) in *C. elegans* (Fischer et al. 2015). However, it is noteworthy that supplementation with 4-hydroxybenzoate (4-HBA), the CoQ ring biosynthetic precursor, conferred increased longevity and stress resistance in *C. elegans* through a mechanism that involved not only its strong antioxidant capacity, but also SIRT1/SIRT2-mediated DAF-16/FoxO activation, independently of dietary restriction and insulin/IGF signaling pathway (Kim et al. 2014). The fact the CR may decrease CoQ_9 levels in *C. elegans*, resulting in a concomitant decrease of mitochondrial ROS, is in agreement with previous investigations based on mutants, genetic silencing or *coq* genes, and dietary CoQ deprivation in the same model organism, although this concept is in contrast with several investigations carried out in rat and mice where CR was found to consistently increase CoQ levels, at least in some tissues and at some ages (see above). It has to be noted, however, that one study carried out with rats showed that both CoQ_9 and CoQ_{10} isoforms were found to be decreased by CR in liver mitochondria (Armeni et al. 2003).

As a central component in energy metabolism, alterations in CoQ biosynthesis and its cellular balance may participate in adaptive responses to physiological, experimental, or pathological conditions (Turunen et al. 2004). In this sense, CoQ increases have been reported to occur under cold adaptation and with exercise, whereas CoQ levels may be decreased by aging, at least in some tissues (Bentinger et al. 2010). Due to the complex pathway leading to CoQ synthesis, simultaneous analysis of CoQ levels and several *COQ* mRNAs or proteins in various tissues can be a useful tool to understand the genetic factors that determine tissue-specific distribution of CoQ and its modulation during metabolic adaptation in mammal models. We have demonstrated that tissues obtained from young mice (3 month) also display specific patterns of *mCOQ* biosynthesis transcripts, and these patterns can be modulated by CR (Parrado-Fernández et al. 2011). Interestingly, CR increased *mPDSS2* mRNA in skeletal muscle, although *mCOQ7* was decreased. In contrast, as it happened for CoQ_9 levels, most *mCOQ* transcripts were significantly decreased by 1 month CR in heart. CR also modified CoQ_9/CoQ_{10} ratio, which was increased in kidney but decreased in heart without alterations of *mPDSS1* or *mPDSS2* transcripts. Finally, CR did not change CoQ or steady-state levels of any of the *mCOQ* transcripts in brain. The fact that CR can increase both steady-state levels of *mPDSS2* and total CoQ in skeletal muscle may be of great importance because skeletal muscle is one of the specific sites of CoQ deficiency and tissue damage caused by a mutation in the PDSS2 gene in humans (Lopez et al. 2006).

15.2.3 The Effect of Calorie Restriction on Plasma Membrane CoQ-Dependent Redox Systems

The mitochondrial (Boveris and Chance 1973) and plasma (O'Donnell and Azzi 1996) membranes are predominant sites of ROS production and oxidative damage targets, and CoQ content in plasma membrane can influence its function by regulating the redox balance of lipid bilayer. Thus, sphingomyelin-depending signaling in plasma membrane has been reported to be modulated by CoQ through the reduction of lipoperoxidation (Fernández-Ayala et al. 2000). Furthermore, besides affecting CoQ levels and CoQ biosynthesis (see above), CR could also affect the activities of plasma membrane CoQ-dependent oxidoreductases. Therefore, several studies have postulated that a central role of CR may involve membrane alterations to promote a decrease in the production of ROS and/or oxidative damage, which could be related not only with a change in CoQ levels but also with modification in CoQ-dependent enzymatic systems. In this sense, the activity of plasma membrane redox system (PMRS) changes in liver during aging and CR modulates these changes. By this mechanism, CR maintains a higher antioxidant capacity in liver plasma membrane of old animals by increasing the activity of CoQ-dependent reductases (Lopez-Lluch et al. 2005). In the same way, age-related increases in plasma membrane lipid peroxidation, protein carbonyls, and nitrotyrosine were attenuated by CR in an *in vitro* model in which brain cells were incubated in medium containing serum from animals fed *ad libitum* (AL) or CR diets (de Cabo et al. 2003). Levels of PMRS enzyme activities were higher, and markers of oxidative stress were lower in cultured neuronal cells treated with CR serum compared with those treated with AL serum (Hyun et al. 2006). CR is thus suggested to play a protective role by enhancing CoQ-dependent PMRS enzyme activities, which results in decreased levels of ROS-mediated damage to membrane lipids and proteins. CR-induced increases of CoQ levels, as well as activation of CoQ dependent antioxidant system, have been also reported to occur in liver and brain plasma membranes from old rats, which takes place in parallel to the attenuation of age-related oxidative damage, although CR did not alter plasma membrane CoQ levels in young rats (de Cabo et al. 2004; Hyun et al. 2006) or in young mice (Lopez-Lluch et al. 2005). This is consistent with the absence of changes for both CoQ isoforms and their ratio observed in sarcolemmal membranes from CR young-adult (6 month) mice (López-Domínguez et al. 2013).

15.3 The Crosstalk Between Calorie Restriction and Dietary Lipid Source

The Membrane Theory of Aging proposes that life span is inversely related to the level of unsaturation in membrane phospholipids (Sohal and Weindruch 1996; Pamplona et al. 1998, 2002; Portero-Otin et al. 2001; Hulbert 2005). Carbon atoms that form a double bond in polyunsaturated fatty acids (PUFA) are most susceptible to oxidative attack by free radicals and the products of this lipid peroxidation, as hydroxynonenal, are themselves reactive and can cause further damage to other cellular components (Halliwell and Gutteridge 1984). In accordance with the Membrane Theory of Aging, comparisons between birds and mammals of similar body sizes have indicated that birds have greater longevity and fewer PUFA in phospholipids skeletal muscle (Hulbert 2003) and heart (Pamplona et al. 1999) than mammals. This decrease in the number of double bonds of the fatty acids of the membranes can be an adaptation of the longest-lived species to prevent the development and accumulation of oxidative damage with time (Pamplona et al. 2002).

CR might act in a similar manner since this intervention is associated with a reduction in the mitochondrial content of long chain PUFA and an increase of the degree of membrane saturation in rat liver (Laganiere and Yu, 1989, 1993; Faulks et al. 2006). Additional studies carried out in rat spleen (Laganiere and Fernandes 1991; Venkatraman and Fernandes 1992), cerebral cortex (Tacconi et al. 1991), and heart (Lee et al. 1999, 2002) have also supported the idea that CR decreases unsaturation of membrane lipids. Thus, the decrease in the degree of unsaturation of the fatty acids and the concomitant increase of resistance against oxidative damage, which could contribute positively to the extension of longevity, has been proposed as a mechanism to underlay the action of CR in cell membranes (Yu et al. 2002; Yu 2005). More recently, a lipidomic study has allowed us to better understand the molecular basis of this diminished susceptibility of membranes to peroxidation, which relies on a redistribution in the type of unsaturation: CR increased monounsaturated fatty acids (MUFA) in liver, whereas the levels of PUFA were decreased without any observed changes in saturated fatty acids (SFA). These specific changes may be the result of a metabolic reprogramming leading to lower levels of oxidative damage which could contribute to the increased lifespan of CR mice (Jové et al. 2014).

These observations have supported the hypothesis that dietary fatty acids might be nutritional components that influence life span in CR animals and thus, manipulation of membrane fatty acids by feeding CR animals with diets containing different lipid compositions might be a valuable strategy to determine their specific role in determining the longevity extension effect of CR intervention (Villalba et al. 2015). Since it was previously reported that CR dampens dietary fat-induced changes in liver plasma membrane phospholipid composition (Cha and Jones 2000), subsequent research carried out in our Group was focused towards elucidating if alterations in dietary lipids could lead to changes in mitochondrial phospholipid fatty acid composition by feeding CR mice diets containing lard (high in saturated and monounsaturated fats), soybean oil (high in linoleic acid) or fish oil (high in n-3 PUFAs) as the predominant fat source. Overall, these investigations have demonstrated that skeletal muscle and liver mitochondrial phospholipid fatty acids were readily changed to reflect the dietary fat source by both short- and long-term CR (Chen et al. 2012, 2014; López-Domínguez et al. 2015). These alterations in phospholipid fatty acid composition were associated with changes in mitochondrial ROS production, proton leak, electron transport chain (ETC) enzyme activities, and other membrane-linked processes. In skeletal muscle, ROS production was decreased in long-term CR mice fed a diet containing lard compared to CR mice consuming either soybean- or fish oil-based diets (Chen et al. 2014). Also, liver and muscle mitochondrial proton leak was decreased in mice fed CR-lard compared to the other diet groups (Chen et al. 2013, 2014).

Interestingly, life span was increased in CR mice consuming lard *versus* fish oil, with CR mice consuming soybean oil showing life spans intermediate to the lard or fish oil-containing diets (López-Domínguez et al. 2015). These observations are consistent with the Membrane Theory of Aging and question the efficacy of feeding diets high in PUFA to CR animals (Villalba et al. 2015). The lack of differences in prevalence of neoplasms or other major measures of end-of-life pathology between CR dietary groups support that differences in lifespan were likely due to delay in onset of disease rather than preventing the occurrence of specific disease conditions (López-Domínguez et al. 2015). The fact that the CR-lard diet significantly increased MUFA levels in liver and muscle phospholipids and the recent demonstration that CR produces a redistribution in the type of unsaturation with a significant increase of MUFA in liver (Jové et al. 2014), make it very likely that MUFA increases may be a causal factor in the observed effects of CR-lard diet increasing life span (Villalba et al. 2015). The potential importance of macronutrient composition of the diets to the retardation of aging with CR is highlighted by the recent speculation that differences in life span between CR monkeys from different colonies used respectively in the Wisconsin and National Institute on Aging studies may be due to diet composition (Colman et al. 2009; Mattison et al. 2012).

Since CR increased CoQ levels in mouse skeletal muscle (see above), we were also interested in studying how dietary fat affected CoQ levels in this tissue, both in animals fed AL and under CR. As depicted in Fig. 15.1, dietary fat significantly modified the response of skeletal muscle to CR. The CR-dependent increase of CoQ_9 and CoQ_{10} levels that we had observed with diets containing soybean oil was not detected with diets containing lard or fish oil. Instead, both CoQ_9 and CoQ_{10} levels, and hence total CoQ, were unaffected by CR when fish oil was the predominant dietary fat source, and both CoQ isoforms were even decreased by CR when mice were fed a lard-based diet (see Fig. 15.1a–c). On the other hand, CoQ_9/CoQ_{10} ratio in skeletal muscle was not altered by CR with any of the diets (Fig. 15.1d). When analyzing the effect of dietary fat source within a given group of caloric intake (control or CR), in mice fed AL we found minimal levels of skeletal muscle CoQ when animals were fed a soybean-based diet in comparison with both lard- and fish oil-containing diets. However, in CR mice the lowest levels of CoQ in skeletal muscle were observed when mice were fed a lard-based diet. Interestingly, under CR the two PUFA-enriched diets (containing either soybean or fish oil) augmented CoQ_9 and total CoQ in comparison with the CR-lard diet, whereas CoQ_{10} levels were augmented with CR-fish but not with CR-soybean diet (see Fig. 15.1a–c). Although, as stated above, CR did not modify CoQ_9/CoQ_{10} ratio in skeletal muscle in any of the diets (Fig. 15.1d), dietary fat source indeed produced a significant effect on this parameter, with the highest CoQ_9/CoQ_{10} ratio being obtained with soybean oil, in comparison with lard- or soybean-based diets (Fig. 15.1d). Chen et al. (2012) reported that short term (1 month) consumption of CR-lard diet in very young mice (1 month of dietary intervention) induced a reduced mitochondrial proton leak whereas CR-fish diet had detrimental effect with an increased lipid peroxidation.

Interestingly, the protective effect of CR observed on skeletal muscle was enhanced by dietary fish oil in young-adult mice that were fed experimental diets for 6 month, as demonstrated by the significant improvement of fiber structural features. Additionally, when fish oil was the main lipid dietary source, a cellular

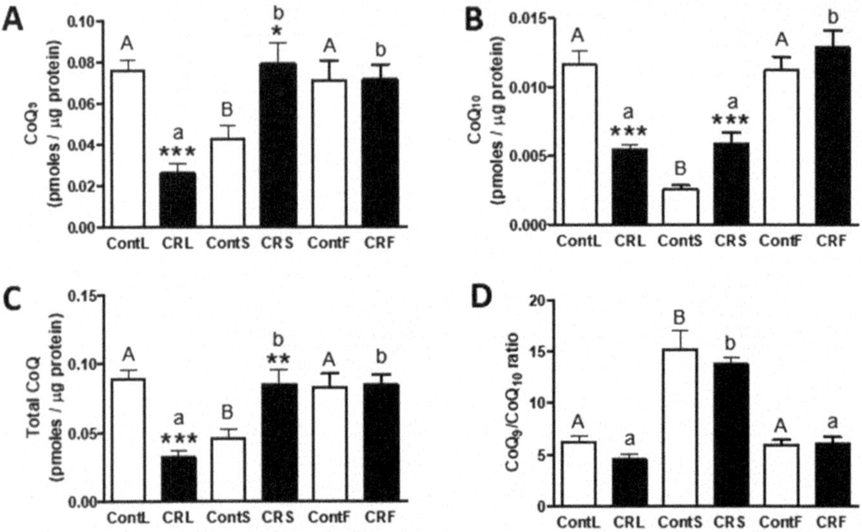

Fig. 15.1 The effect of CR and dietary fat on skeletal muscle CoQ_9 **(A)**, CoQ_{10} **(B)**, total CoQ **(C)** and CoQ_9/CoQ_{10} ratio **(D)**. Male mice (C57BL/6, 3 month-old) were fed experimental diets based on an AIN93G formulation for 1 month. The control groups were fed 95% of average *ad libitum* diary intake, whereas the CR groups consumed 60% of average diary intake. All diets were identical except for dietary lipid source, which was soybean oil (high in n-6 PUFA), fish oil (high in n-3 PUFA) or lard (high in saturated and monounsaturated fatty acids). Soybean oil (14% of total fat content) was added to the fish oil and lard diets to insure adequate intake of linoleic acid. Dietary groups are denoted as follows: ContL (control-lard), CRL (CR-lard), ContS (control-soybean oil), CRS (CR-soybean oil), ContF (control-fish oil) and CRF (CR-fish oil). Lipids were recovered by organic extraction from hind limb skeletal muscle samples with hexane, and separated and quantified by HPLC as described (Parrado-Fernández et al. 2011). Data are means ± SEM (n = 16). Data were analyzed by two-way ANOVA. Significant differences between control and CR for a given fat source are indicated by asterisks (***$p < 0.001$). Uppercase letters are used to denote statistical differences as a function of dietary fat within control groups, whereas lowercase letters are used to denote statistical differences as a function of dietary fat within CR groups. Groups that do not share letter exhibit statistically significant differences. **A**) $p < 0.05$ ContS *versus* ContL and ContF; $p < 0.001$ CRL *versus* CRS and CRF. **B**) $p < 0.001$ ContS *versus* ContL and ContF; $p < 0.001$ CRF *versus* CRL and CRS. **C**) $p < 0.01$ ContS *versus* ContL; $p < 0.05$ ContS *versus* ContF; $p < 0.001$ CRL *versus* CRS and CRF. **D**) $p < 0.001$ ContS *versus* ContL and ContF; $p < 0.001$ CRS *versus* CRL and CRF

anti-apoptotic environment was produced in skeletal muscle with a downregulation of components involved in the initial stages of apoptosis engagement, both at the plasma membrane and the mitochondria (López-Domínguez et al. 2013). It is tempting to speculate that these protective effects could be related with a CoQ increase. However, although fish oil attenuated skeletal muscle apoptotic signaling in young-adult CR mice, most of these changes were abolished or even reverted in aged mice, with a significant decrease of caspase-9 activity, a marker of mitochondrial apoptosis, in the CR-lard group (López-Domínguez et al. 2013).

15.4 Conclusions and Perspectives

CoQ levels significantly affect aging in model systems as *C. elegans*. Both genetic (mutations and gene silencing of *COQ* biosynthetic genes) and dietary (CR and feeding worms a CoQ_8-less diet) approaches have generally shown that decreased levels of CoQ in tissues may lead to extended lifespan in this model system, although exogenously supplied CoQ_{10} or CoQ-biosynthetic precursors can also play a significant anti-aging function. In mammal models (as rats), maximal longevity is not affected by life-long administration of CoQ_{10} when animals are fed normal diets, but dietary CoQ_{10} supplementation can ameliorate deleterious alterations associated with aging and even extend lifespan of rats that had been fed a potentially prooxidant diet. In humans, supplementation with CoQ_{10} improves several parameters associated with a better antioxidant status in healthy elderly men and women fed a Mediterranean diet, which supports that CoQ_{10} supplementation could be a valuable strategy to counteract several oxidative modifications in the elderly. However, given the low bioavailability of dietary CoQ, it is also very important to develop mechanisms to increase functional CoQ in tissues and organs along aging or disease states. In animal models, several oxidation products as tocotrienol epoxides and oxidosqualenes have proven very useful for stimulating CoQ synthesis. In addition, CoQ levels, *COQ* gene expression and CoQ-dependent antioxidant systems are CR targets in cellular systems and in mice in a tissue-specific way, with skeletal muscle exhibiting an early response to this intervention. However, whether CR increases, decreases or has no effect on CoQ levels, relies on additional factors such as duration of CR intervention, CR protocol or the predominant fat source present in CR diet, and probably additional factors still to be determined. In young mice fed under CR, levels of CoQ (and particularly those of CoQ_{10}) adapt to the prevailing oxidative status in the tissue, which can be modulated by dietary fat. Future investigations are warranted in order to elucidate how long-term CR intervention with different dietary fats affects CoQ levels in old animals.

Acknowledgements Research in J.M.V laboratory is supported by the Spanish Ministerio de Economía y Competitividad (BFU2011-23578, BFU2015-64630-R, and RTI2018-100695-B-100), Junta de Andalucía (CVI-276) and Universidad de Córdoba. LFR, EGC and MC-R were supported by FPU fellowships from the Spanish Ministerio de Educación, Cultura y Deporte and CVI-276. SRL was supported by a FPI fellowship from the Spanish Ministerio de Economía y Competitividad.

References

Aberg F, Appelkvist EL, Dallner G et al (1992) Distribution and redox state of ubiquinones in rat and human tissues. Arch Biochem Biophys 295:230–234

Anderson RM, Weindruch R (2010) Metabolic reprogramming, caloric restriction and aging. Trends Endocrinol Metab 21:134–141

Armeni T, Principato G, Quiles JL et al (2003) Mitochondrial dysfunctions during aging: vitamin E deficiency or calorie restriction--two different ways of modulating stress. J Bioenerg Biomembr 35:181–191
Asencio C, Rodriguez-Aguilera JC, Ruiz-Ferrer M et al (2003) Silencing of ubiquinone biosynthesis genes extends life span in Caenorhabditis elegans. FASEB J 17:1135–1137
Asencio C, Navas P, Cabello J et al (2009) Coenzyme Q supports distinct developmental processes in Caenorhabditis elegans. Mech Ageing Dev 130:145–153
Barja G (2013) Updating the mitochondrial free radical theory of aging: an integrated view, key aspects, and confounding concepts. Antioxid Redox Signal 20:1420–1445
Barja G (2014) The mitochondrial free radical theory of aging. Prog Mol Biol Transl Sci 127:1–27
Bello RI, Gómez-Díaz C, Burón MI et al (2005) Enhanced anti-oxidant protection of liver membranes in long-lived rats fed on a coenzyme Q10-supplemented diet. Exp Gerontol 40:694–706
Bentinger M, Dallner G, Chojnacki T et al (2003) Distribution and breakdown of labeled coenzyme Q10 in rat. Free Radic Biol Med 34:563–575
Bentinger M, Teklea M, Brismarb K et al (2008a) Stimulation of coenzyme Q synthesis. BioFactors 32:99–111
Bentinger M, Tekle M, Brismar K et al (2008b) Polyisoprenoid epoxides stimulate the biosynthesis of coenzyme Q and inhibit cholesterol synthesis. J Biol Chem 283:14645–14653
Bentinger M, Tekle M, Dallner G (2010) Coenzyme Q-biosynthesis and functions. Biochem Biophys Res Commun 396:74–79
Bentinger M, Kania M, Danikiewicz W et al (2014) Effects of various squalene epoxides on coenzyme Q and cholesterol synthesis. Biochem Biophys Acta 1841:977–986
Bevilacqua L, Ramsey JJ, Hagopian K et al (2004) Effects of short-and medium-term calorie restriction on muscle mitochondrial proton leak and reactive oxygen species production. Am J Phys 286:E852–E861
Beyer RE, Burnett BA, Cartwright KJ et al (1985) Tissue coenzyme Q (ubiquinone) and protein concentrations over the life span of the laboratory rat. Mech Ageing Dev 32:267–281
Boveris A, Chance B (1973) The mitochondrial generation of hydrogen peroxide. General properties and effect of hyperbaric oxygen. Biochem J 134:707–716
Cerqueira FM, Kowaltowski AJ (2013) Mitochondrial metabolism in aging: effect of dietary interventions. Ageing Res Rev 12:22–28
Cha MC, Jones PJ (2000) Energy restriction dilutes the changes related to dietary fat type in membrane phospholipid fatty acid composition in rats. Metabolism 49:977–983
Chen Y, Hagopian K, McDonald RB et al (2012) The influence of dietary lipid composition on skeletal muscle mitochondria from mice following 1 month of calorie restriction. J Gerontol A Biol Sci Med Sci 67:1121–1131
Chen Y, Hagopian K, Bibus D et al (2013) The influence of dietary lipid composition on liver mitochondria from mice following 1 month of calorie restriction. Biosci Rep 33:83–95
Chen Y, Hagopian K, Bibus D et al (2014) The influence of dietary lipid composition on skeletal muscle mitochondria from mice following eight months of calorie restriction. Physiol Res 63:57–71
Colman RJ, Anderson RM, Johnson SC et al (2009) Caloric restriction delays disease onset and mortality in rhesus monkeys. Science 325:201–204
Darnton-Hill I, Nishida C, James WPT (2004) A life course approach to diet, nutrition and the prevention of chronic diseases. Public Health Nutr 7:101–121
Dato S, Bellizzi D, Rose G et al (2016) The impact of nutrients on the aging rate: a complex interaction of demographic, environmental and genetic factors. Mech Ageing Dev 154:49–61
de Cabo R, Furer-Galban S, Anson RM et al (2003) An in vitro model of caloric restriction. Exp Gerontol 38:631–639
de Cabo R, Cabello R, Rios M et al (2004) Calorie restriction attenuates age-related alterations in the plasma membrane antioxidant system in rat liver. Exp Gerontol 39:297–304
Dröge W (2002) Free radicals in the physiological control of cell function. Physiol Rev 82:47–95

Ewbank JJ, Barnes TM, Lakowski B (1997) Structural and functional conservation of the Caenorhabditis elegans timing gene clk-1. Science 275:980–983

Faulks SC, Turner N, Else PL et al (2006) Calorie restriction in mice: effects on body composition, daily activity, metabolic rate, mitochondrial reactive oxygen species production, and membrane fatty acid composition. J Gerontol A Biol Sci Med Sci 61:781–794

Fernández-Ayala DJ, Martín SF, Barroso MP et al (2000) Coenzyme Q protects cells against serum withdrawal-induced apoptosis by inhibition of ceramide release and caspase-3 activation. Antioxid Redox Signal 2:263–275

Fischer A, Klapper M, Onur S et al (2015) Dietary restriction decreases coenzyme Q and ubiquinol potentially via changes in gene expressionelegans in the model organism C. elegans. Biofactors 41:166–174

Fontana L, Partridge L (2015) Promoting health and longevity through diet: from model organisms to humans. Cell 161:108–118

Gavilan A, Asencio C, Cabello J et al (2005) C. elegans knockouts in ubiquinone biosynthesis genes result in different phenotypes during larval development. Biofactors 25:21–29

Gutierrez-Mariscal FM, Perez-Martinez P, Delgado-Lista J et al (2012) Mediterranean diet supplemented with coenzyme Q10 induces postprandial changes in p53 in response to oxidative DNA damage in elderly subjects. Age 34:389–403

Hagopian K, Harper ME, Ramsey JJ et al (2005) Long term calorie restriction reduces proton leak and hydrogen peroxide production in liver mitochondria. Am J Phys 288:E674–E684

Halliwell B, Gutteridge JM (1984) Lipid peroxidation, oxygen radicals, cell damage and antioxidant therapy. Lancet 1:1396–1398

Harman DJ (1956) Aging: a theory based on free radical and radiation chemistry. J Gerontol 11:298–300

Heilbronn LK, Ravussin E (2003) Calorie restriction and aging: review of the literature and implications for studies in humans. Am J Clin Nutr 78:361–369

Hihi AK, Gao Y, Hekimi S (2002) Ubiquinone is necessary for Caenorhabditis elegans development at mitochondrial and non-mitochondrial sites. J Biol Chem 277:2202–2206

Hulbert AJ (2003) Life, death and membrane bilayers. J Exp Biol 206:2303–2311

Hulbert AJ (2005) On the importance of fatty acid composition of membranes for aging. J Theor Biol 234:277–288

Hyun DH, Emerson SS, Jo DG et al (2006) Calorie restriction up-regulates the plasma membrane redox system in brain cells and suppresses oxidative stress during aging. Proc Natl Acad Sci U S A 103:19908–19912

Ishii N, Senoo-Matsuda N, Miyake K et al (2004) Coenzyme Q10 can prolong C. elegans lifespan by lowering oxidative stress. Mech Ageing Dev 125:41–46

Jonassen T, Larsen PL, Clarke CF (2001) A dietary source of coenzyme Q is essential for growth of long-lived Caenorhabditis elegans clk-1 mutants. Proc Natl Acad Sci U S A 98:421–426

Jové M, Naudí A, Ramírez-Núñez O et al (2014) Caloric restriction reveals a metabolomic and lipidomic signature in liver of male mice. Aging Cell 13:828–837

Kalen A, Appelkvist EL, Dallner G (1989) Age-related changes in the lipid compositions of rat and human tissues. Lipids 24:579–584

Kamzalov S, Sohal R (2004) Effect of age and caloric restriction on coenzyme Q and α-tocopherol levels in the rat. Exp Gerontol 39:1199–1205

Kaneko T, Tahara S, Matsuo M (1997) Retarding effect of dietary restriction on the accumulation of 8-hydroxy-2′-deoxyguanosine in organs of Fischer 344 rats during aging. Free Radic Biol Med 23:76–81

Kim JD, Yu BP, McCarter RJ et al (1996) Exercise and diet modulate cardiac lipid peroxidation and antioxidant defenses. Free Radic Biol Med 20:83–88

Kim DK, Jeon H, Cha DS (2014) 4-Hydroxybenzoic acid-mediated lifespan extension in Caenorhabditis elegans. J Functional Foods 7:630–640

Kristal BS, Yu BP (1992) An emerging hypothesis: synergistic induction of aging by free radicals and Maillard reactions. J Gerontol 47:B107–B114

Kwong LK, Kamzalov S, Rebrin I et al (2002) Effects of coenzyme Q(10) administration on its tissue concentrations, mitochondrial oxidant generation, and oxidative stress in the rat. Free Radic Biol Med 33:627–638
Laganiere S, Fernandes G (1991) Study on the lipid composition of aging Fischer-344 rat lymphoid cells: effect of long-term calorie restriction. Lipids 26:472–478
Laganiere S, Yu BP (1989) Effect of chronic food restriction in aging rats. I. Liver subcellular membranes. Mech Ageing Dev 48:207–219
Laganiere S, Yu BP (1993) Modulation of membrane phospholipid fatty acid composition by age and food restriction. Gerontology 39:7–18
Larsen PL, Clarke CF (2002) Extension of life-span in Caenorhabditis elegans by a diet lacking coenzyme Q. Science 295:120–123
Lass A, Sohal BH, Weindruch R et al (1998) Caloric restriction prevents age-associated accrual of oxidative damage to mouse skeletal muscle mitochondria. Free Radic Biol Med 25:1089–1097
Lass A, Kwong L, Sohal RS (1999) Mitochondrial coenzyme Q content and aging. Biofactors 9:199–205
Lee CK, Klopp RG, Weindruch R et al (1999) Gene expression profile of aging and its retardation by caloric restriction. Science 285:1390–1393
Lee CK, Allison DB, Brand J et al (2002) Transcriptional profiles associated with aging and middle age-onset caloric restriction in mouse hearts. Proc Natl Acad Sci U S A 99:14988–14993
Lee CK, Pugh TD, Klopp RG et al (2004) The impact of alpha-lipoic acid, coenzyme Q10 and caloric restriction on life span and gene expression patterns in mice. Free Radic Biol Med 36:1043–1057
Leeuwenburgh C, Wagner P, Holloszy JO et al (1997) Caloric restriction attenuated dityrosine crosslinking of cardiac and skeletal muscle proteins in aging mice. Arch Biochem Biophys 346:74–80
Linnane AW, Kios M, Vitetta L (2007) The essential requirement for superoxide radical and nitric oxide formation for normal physiological function and healthy aging. Mitochondrion 7:1–5
Liu J, Wu Q, He D et al (2011) Drosophila sbo regulates lifespan through its function in the synthesis of coenzyme Q in vivo. J Genet Genomics 38:225–234
Lönnrot K, Metsa-Ketela T, Alho H et al (1995) The role of coenzyme Q10 in aging: a follow-up study on life-long oral supplementation Q10 in rats. Gerontology 41:109–120
Lönnrot K, Holm P, Lagerstedt A et al (1998) The effects of lifelong ubiquinone Q10 supplementation on the Q9 and Q10 tissue concentrations and life span of male rats and mice. Biochem Mol Biol Int 44:727–737
Lopez LC, Schuelke M, Quinzii CM et al (2006) Leigh syndrome with nephropathy and CoQ_{10} deficiency due to decaprenyl diphosphate synthase subunit 2 (PDSS2) mutations. Am J Hum Genet 79:1125–1129
López-Domínguez JA, Khraiwesh H, González-Reyes JA et al (2013) Dietary fat modifies mitochondrial and plasma membrane apoptotic signaling in skeletal muscle of calorie-restricted mice. Age 35:2027–2044
López-Domínguez JA, Ramsey JJ, Tran D et al (2015) The influence of dietary fat source on life span in calorie restricted mice. J Gerontol A Biol Sci Med Sci 70:1181–1188
Lopez-Lluch G, Rios M, Lane MA et al (2005) Mouse liver plasma membrane redox system activity is altered by aging and modulated by calorie restriction. Age 27:153–160
López-Lluch G, Rodríguez-Aguilera JC, Santos-Ocaña C, Navas P (2010) Is coenzyme Q a key factor in aging? Mech Ageing Dev 131:225–235
Lopez-Moreno J, Quintana-Navarro GM, Delgado-Lista J et al (2016) Mediterranean diet supplemented with coenzyme Q10 modulates the postprandial metabolism of advanced glycation end products in elderly men and women. J Gerontol A Biol Sci Med Sci. In press. https://doi.org/10.1093/gerona/glw214
Mattison JA, Roth GS, Beasley TM et al (2012) Impact of caloric restriction on health and survival in rhesus monkeys from the NIA study. Nature 489:318–321

Nelson JF, Karelus K, Bergman MD et al (1995) Neuroendocrine involvement in aging: evidence from studies of reproductive aging and caloric restriction. Neurobiol Aging 16:837–843
O'Donnell VB, Azzi A (1996) High rates of extracellular superoxide generation by cultured human fibroblasts: involvement of a lipid-metabolizing enzyme. Biochem J 318:805–812
Pamplona R, Portero-Otin M, Riba D et al (1998) Mitochondrial membrane peroxidizability index is inversely related to maximum life span in mammals. J Lipid Res 39:1989–1994
Pamplona R, Portero-Otin M, Requena JR et al (1999) A low degree of fatty acid unsaturation leads to lower lipid peroxidation and lipoxidation-derived protein modification in heart mitochondria of the longevous pigeon than in the short-lived rat. Mech Ageing Dev 106:283–296
Pamplona R, Barja G, Portero-Otin M (2002) Membrane fatty acid unsaturation, protection against oxidative stress, and maximum life span: a homeoviscous-longevity adaptation? Ann N Y Acad Sci 959:475–490
Park SK, Prolla TA (2005) Lessons learned from gene expression profile studies of aging and caloric restriction. Ageing Res Rev 4:55–65
Parrado-Fernández C, López-Lluch G, Rodríguez-Bies E et al (2011) Calorie restriction modifies ubiquinone and COQ transcript levels in mouse tissues. Free Radic Biol Med 50:1728–1736
Pisoschi AM, Pop A (2015) The role of antioxidants in the chemistry of oxidative stress: a review. Eur J Med Chem 97:55–74
Portero-Otin M, Bellmunt MJ, Ruiz MC et al (2001) Correlation of fatty acid unsaturation of the major liver mitochondrial phospholipid classes in mammals to their maximum life span potential. Lipids 36:491–498
Quiles JL, Ochoa JJ, Huertas JR et al (2004) Coenzyme Q supplementation protects from age-related DNA double-strand breaks and increased lifespan in rats fed on a PUFA-rich diet. Exp Gerontol 39:189–194
Ramsey JJ, Hagopian K, Kenny TM et al (2004) Proton leak and hydrogen peroxide production in liver mitochondria from energy-restricted rats. Am J Physiol Endocrinol Metab 286:E31–E40
Rodriguez-Aguilera JC, Gavilan A, Asencio C et al (2005) The role of ubiquinone in Caenorhabditis elegans longevity. Ageing Res Rev 4:41–53
Rodríguez-Bies E, Navas P, López-Lluch G (2015) Age-dependent effect of every-other-day feeding and aerobic exercise in ubiquinone levels and related antioxidant activities in mice muscle. J Gerontol A Biol Sci Med Sci 70:33–43
Ruetenik A, Barrientos A (2015) Dietary restriction, mitochondrial function and aging: from yeast to humans. Biochim Biophys Acta 1847:1434–1447
Santoro A, Pini E, Scurti M et al (2014) Combating inflammaging through a Mediterranean whole diet approach: the NU-AGE project's conceptual framework and design. Mech Ageing Dev 136–137:3–13
Santos-González M, Gómez Díaz C, Navas P et al (2007) Modifications of plasma proteome in long-lived rats fed on a coenzyme Q10-supplemented diet. Exp Gerontol 42:798–806
Sohal RS, Forster MJ (2007) Coenzyme Q, oxidative stress and aging. Mitochondrion 7(Suppl):S103–S111
Sohal RS, Forster MJ (2014) Caloric restriction and the aging process: a critique. Free Radic Biol Med 73:366–382
Sohal RS, Weindruch R (1996) Oxidative stress, caloric restriction, and aging. Science 273:59–63
Sohal RS, Ku HH, Agarwal S et al (1994a) Oxidative damage, mitochondrial oxidant generation and antioxidant defenses during aging and in response to food restriction in the mouse. Mech Ageing Dev 74:121–133
Sohal RS, Agarwal S, Candas M et al (1994b) Effect of age and caloric restriction on DNA oxidative damage in different tissues of C57BL/6 mice. Mech Ageing Dev 76:215–224
Tacconi MT, Lligona L, Salmona M, Pitsikas N, Algeri S et al (1991) Aging and food restriction: effect on lipids of cerebral cortex. Neurobiol Aging 12:55–59
Turunen M, Olssonc J, Dallner G (2004) Metabolism and function of coenzyme Q. Biochim Biophys Acta 1660:171–199

Varela-Lopez A, Bullon P, Battino M et al (2016) Coenzyme Q protects against age-related alveolar bone loss associated to n-6 polyunsaturated fatty acid rich-diets by modulating mitochondrial mechanisms. J Gerontol A Biol Sci Med Sci 71:593–600
Venkatraman J, Fernandes G (1992) Modulation of age-related alterations in membrane composition and receptor-associated immune functions by food restriction in Fischer 344 rats. Mech Ageing Dev 63:27–44
Villalba JM, López-Domínguez JA, Chen Y et al (2015) The influence of dietary fat source on liver and skeletal muscle mitochondrial modifications and lifespan changes in calorie-restricted mice. Biogerontology 16:655–670
Walford RL, Spindler SR (1997) The response to calorie restriction in mammals shows features also common to hibernation: a cross-adaptation hypothesis. J Gerontol 52:B179–B183
Wang Y, Oxer D, Hekimi S (2015) Mitochondrial function and lifespan of mice with controlled ubiquinone biosynthesis. Nat Commun 6:6393
Weindruch R, Sohal RS (1997) Seminars in medicine of the Beth Israel Deaconess Medical Center. Caloric intake and aging. N Engl J Med 337:986–994
Weindruch R, Naylor PH, Goldstein AL et al (1988) Influences of aging and dietary restriction on serum thymosin alpha 1 levels in mice. J Gerontol 43:B40–B42
Weindruch R, Kayo T, Lee CK et al (2001) Microarray profiling of gene expression in aging and its alteration by caloric restriction in mice. J Nutr 131:918S–923S
Xia E, Rao G, Van Remmen H et al (1995) Activities of antioxidant enzymes in various tissues of male Fischer 344 rats are altered by food restriction. J Nutr 125:195–201
Yu BP (2005) Membrane alteration as a basis of aging and the protective effects of calorie restriction. Mech Ageing Dev 126:1003–1010
Yu BP, Lim BO, Sugano M (2002) Dietary restriction downregulates free radical and lipid peroxide production: plausible mechanism for elongation of life span. J Nutr Sci Vitaminol 48:257–264
Yubero-Serrano EM, Delgado-Casado N, Delgado-Lista J et al (2011) Postprandial antioxidant effect of the Mediterranean diet supplemented with coenzyme Q10 in elderly men and women. Age 33:579–590
Yubero-Serrano EM, González-Guardia L, Rangel-Zúñiga O et al (2012) Mediterranean diet supplemented with coenzyme Q10 modifies the expression of proinflammatory and endoplasmic reticulum stress-related genes in elderly men and women. J Gerontol A Biol Sci Med Sci 67:3–10
Yubero-Serrano EM, Gonzalez-Guardia L, Rangel-Zuñiga O et al (2013) Postprandial antioxidant gene expression is modified by Mediterranean diet supplemented with coenzyme Q(10) in elderly men and women. Age 35:159–170
Zahn JM, Poosala S, Owen AB et al (2007) AGEMAP: a gene expression database for aging in mice. PLoS Genet 3:e201
Zhang Y, Alberg F, Appelkvist EL et al (1995) Uptake of dietary coenzyme Q supplement is limited in rats. J Nutr 125:446–453

Chapter 16
Age Dependent Changes of Coenzyme Q Levels and its Induction in Experimental Systems

Magnus Bentinger, Gustav Dallner, Kerstin Brismar, Ewa Swiezewska, and Michael Tekle

Abstract Coenzyme Q (CoQ) is required for normal metabolic functions in all tissues. In humans, the amount of this lipid increases in all organs during adolescence and reaches its highest peak during the first 2–3 decades of life. At 80 years of age, the amount is decreased to around half. This development is also observed in different regions of the brain. In rodents, a number of physical and chemical factors are known to increase CoQ in various organs. Depending on the type, duration and doses of the treatments applied, the amount of the lipid increased in different organs to variable extents. Vitamin E plays a regulatory role in CoQ synthesis. Low amount of vitamin E in the diet results in lower amount of CoQ in liver and blood, while high amount of this vitamin increase the amount. Several nuclear receptors are involved in the synthesis of the lipid. Mice knockout models of PPARα, RXRα, TRα, and LXRα receptors exhibit significantly reduced amounts of this lipid in the liver and spleen. The amount of CoQ increases in a number of rodent and human pathological conditions, while it decreases in human liver cancer and cardiomyopathy.

Keywords Coenzyme Q · Aging · Lipid induction · Nuclear receptors · Human diseases

M. Bentinger · G. Dallner · K. Brismar
Department of Molecular Medicine and Surgery, Karolinska Institutet, Rolf Luft Research Center for Diabetes and Endocrinology, Stockholm, Sweden

E. Swiezewska
Department of Biochemistry and Biophysics, Polish Academy of Sciences, Warsaw, Poland

M. Tekle (✉)
Department of Molecular Medicine and Surgery, Karolinska Institutet, Rolf Luft Research Center for Diabetes and Endocrinology, Stockholm, Sweden

Department of Clinical Pharmacology, Karolinska University Hospital, Stockholm, Sweden
e-mail: Michael.Tekle@sll.se

G. López Lluch (ed.), *Coenzyme Q in Aging*,
https://doi.org/10.1007/978-3-030-45642-9_16

16.1 Introduction

It is now well established that coenzyme Q (CoQ), a neutral lipid and one of the products of the mevalonate pathway, has many important functions in the cell and regulates a number of processes. It plays a central role in the mitochondrial function and it is one of the main factors in mitochondrial supercomplexes and necessary for the supramolecular organization of the respiratory chain (Genova and Lenaz 2011). It is one of the regulators of the mitochondrial permeability transition pool and it was also proposed that it participates in the function of uncoupling proteins (Echtay et al. 2000; Azzolin et al. 2010). It is quenching free radicals and is our only lipid soluble antioxidant that is synthesized endogenously (Ernster and Forsmark-Andree 1993). Regulation of cell growth and differentiation requires the function of the CoQ dependent NADH-oxidase in the plasma membrane (Crane et al. 1985; Buron et al. 1993). This lipid has also turned out to be an immunological factor as it is affecting the expression of β2-integrins, TNFα and NF-κB dependent genes regulating release of mediators and signal substances from monocytes and lymphocytes into the circulation and exhibit multiple anti-inflammatory effects (Turunen et al. 2002b; Groneberg et al. 2005; Schmelzer and Doring 2010).

A new development is the elucidation of the biosynthetic mechanism of CoQ which is now characterized in detail. At least 12 genes are required and the proteins are present in a biosynthetic complex (Tran and Clarke 2007; Acosta et al. 2016). Various factors which regulate the state and form of the complex are of great interest. During the recent years our knowledge of the occurrence of genetic mutations affecting CoQ expression has increased. A number of primary mutations in the CoQ synthetizing enzymes have now been identified (Rotig et al. 2007; Quinzii and Hirano 2010). These mutations lead to functional deficiencies early in life and involve mainly the brain, cerebellum, muscles and kidney. Emerging studies indicate the involvement of miRNA in the regulation of lipid metabolism and it remains to establish their role in CoQ metabolism (Allen and Vickers 2014).

Even a moderate decrease in CoQ amount reduces the antioxidant protection and has negative consequences. Several diseases such as liver cancer, cardiomyopathy and complex myopathies exhibit lower CoQ content in the respective organs and is considered to be part of the disease process (Turunen et al. 2004). During aging many changes occur in cellular constituents and functions which influence organ metabolism. A highly discussed issue is the increased production of free radicals which would require an increased antioxidant protection. Taking into consideration that CoQ is not only an antioxidant, but also has several other functions, changes in the cellular amount during aging could influence a number of cellular mechanisms.

16.2 Decrease of CoQ During Aging

CoQ is present in all tissues and cells of mammals and has an important role in respiration, antioxidant action and regulation of a number of other biological processes. Hence, its amount is of great importance for a normal cellular function. Development and aging affect most cellular components and it is not surprising that the amount of this lipid is greatly changed during the lifetime of the organism. In humans, the amount is substantial already at birth and in general it further increases in the first 20 years of life (Table 16.1) (Kalen et al. 1989). A gradual decrease takes place in the following years, a process that continues to older age. At the age of 80, CoQ in pancreas and spleen decreases to one-third compared to younger counterparts. In all other organs there is also a CoQ reduction, but not to the extent of the organs mentioned above.

In human brains, a great variation in the amount of CoQ between different regions is found and the levels are between 5 and 20 μg/g wet weight (Fig. 16.1). Remarkably, the amount of CoQ in the brain is stable in all regions until the age of 65–70, but at the age of 90 and older, a decrease is obvious in all parts of the brain, especially in the medulla oblongata and hippocampus, where the amount decreases by about 50% (Soderberg et al. 1990).

The situation in rodent brain is different during aging (Fig. 16.2). In the initial months of life there is a continuous increase of both CoQ_9 and CoQ_{10} contents (Zhang et al. 1996a). The levels are stable after 5 months of age and remain unchanged during the rest of life. In mice and rat, the main form of the lipid is composed of 9 isoprenes. However in the brain, 30% of the total CoQ content consists of CoQ_{10} (10 isoprenes), which is the main form in humans. CoQ_{10} is also found in other organs of rodents, but in considerably smaller amounts with the exception of the spleen and intestine that also possess about 30% CoQ_{10} (Aberg et al. 1992). Like the human brain, the various regions of the rat brain contain different levels of CoQ, which obviously is dependent on the amount and distribution of the cell types (Zhang et al. 1996a).

Table 16.1 Age dependent changes of coenzyme Q_{10} concentration in various human organs

	1 year	20 years	40 years	80 years (Age)
Adrenal	58	16	12	9
Heart	79	110	75	47
Kidney	53	98	71	64
Liver	45	61	58	51
Lung	6	6	7	3
Pancreas	38	21	19	7
Spleen	30	33	29	13

Units are in μg/g wet weight. Reference Kalen et al. (1989)

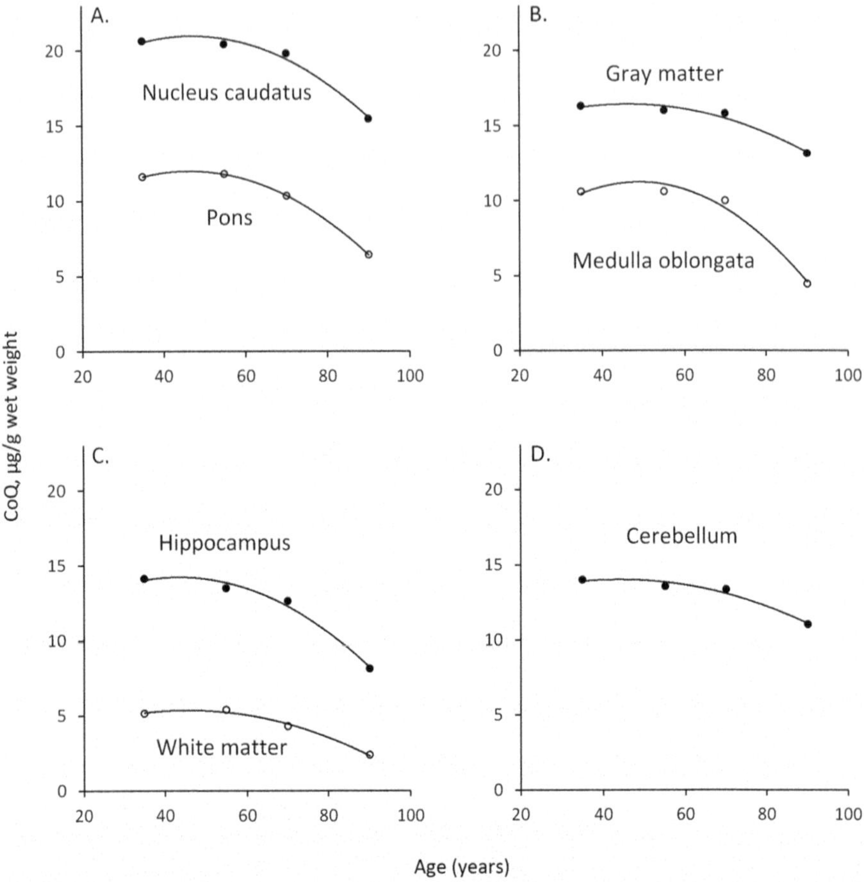

Fig. 16.1 Coenzyme Q content in different regions of the human brain during aging. (**a**) Nucleus Caudatus and pons. (**b**) Gray matter and medulla oblongata. (**c**) Hippocampus and white matter. (**d**) Cerebellum. From reference Soderberg et al. (1990)

16.3 Induction of CoQ Biosynthesis

At present, several conditions and treatments are known to induce CoQ biosynthesis in rodents. There are a number of publications in the literature that are discussing if there are similar effects in humans. For obvious reasons it is very difficult to evaluate these conditions in humans.

The conditions that elevate the CoQ amount in rodents are summarized in Table 16.2. Exercise is considered as an effective inducer of CoQ in rat muscle, white and brown adipose tissues and liver mitochondria. The increase of the lipid may be caused by a general elevation of the total concentration in all membranes or on an increase in the number of mitochondria, since this organelle contains the

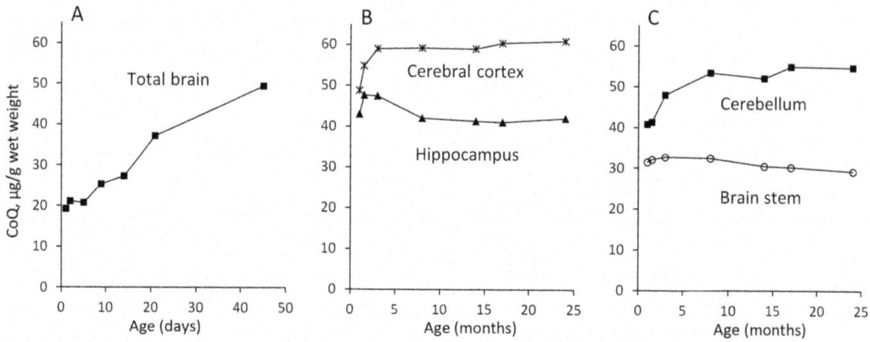

Fig. 16.2 Total coenzyme Q (CoQ_9 + CoQ_{10}) content in rat brain during development. (**a**) Total CoQ content in whole rat brain during the first 45 days. (**b**) CoQ level in the cerebral cortex and hippocampus during 1–24 months of life. (**c**) CoQ content in the cerebellum and brain stem during 1–24 months of life. From reference Zhang et al. (1996a)

Table 16.2 Treatments that stimulate CoQ biosynthesis in rodents

Conditions	Organ	CoQ % of control	References
Exercise	Rat muscle, brown and white adipocyte tissues	160–200	Gohil et al. (1985)
	Rat liver mitochondria	180–230	Quiles et al. (1994)
Cold exposure	Rat liver	250	Aithal et al. (1968)
Carnosine	Mouse liver	160	Bentinger et al. (2003)
	Mouse liver	131	Santos et al. (unpublished data)
Thyroxin	Rat liver, heart, muscle	115–170	Pedersen et al. (1963)
Dehydroepiandrosterone	Mouse liver	180	Tekle (unpublished)
	Rat liver, kidney, muscle	120–182	Aberg et al. (1996)
Vitamin A deficiency	Mouse liver	130	Sohlenius-Sternbeck et al. (2000)
Vitamin E treatment	Rat liver	125	Zhang et al. (1996b)
Acetylsalicylic acid	Rat liver, kidney, muscle	129–140	Aberg et al. (1996)
Clofibrate	Rat liver, heart, muscle,	120–250	Aberg et al. (1994)
Fluorine	Rat brain	125	Guan et al. (1998)
Mevinolin	Rat muscle	248	Low et al. (1992)
	Rat blood	159	Low et al. (1992)
Aminotriazole	Rat liver, kidney, muscle	105–155	Aberg et al. (1996)
DEHP	Rat liver, heart, muscle	170–440	Aberg et al. (1994)
2-Ethylhexanoic acid	Rat liver	196	Aberg et al. (1996)

major part of cellular CoQ. Cold exposure is also another non-drug inducer that stimulates the biosynthesis.

A number of substances are known as inducers of CoQ biosynthesis. Carnosine is a dipeptide that plays an important role in a number of biological functions, as antioxidant, anti-inflammatory and anti-senescence agent. Additionally, it also increases CoQ in mice liver upon dietary administration. Among the hormones thyroxin and dehydroepiandrosterone are described to increase CoQ amount in various tissues. Vitamin A deficiency and administration of vitamin E elevates the amount of CoQ. Acetylsalicylic acid, clofibrate and mevinolin are established drugs that induce CoQ synthesis. Fluorine in a low concentration increases CoQ in rat brain. Aminotriazole, di(2-ethylhexyl)phthalate (DEHP) and 2-ethylhexanoic acid are examples of chemical substances present in the environment which are known as inducers. It is demonstrated in tissue culture systems that camptothecin, an anticancer agent upregulates CoQ synthesis. This process is dependent on the activation of the transcription factor NF-κB (Brea-Calvo et al. 2009).

16.4 Nuclear Receptors and CoQ Biosynthesis

Nuclear receptors are regulators of a number of genetic signaling pathways involved in the biosynthetic processes. In rodents, peroxisome proliferator activating receptor alpha (PPARα), retinoid x receptor alpha (RXRα) and liver x receptor alpha (LXRα) are involved in the regulation of CoQ synthesis (Turunen et al. 2000; Bentinger et al. 2003; Bentinger et al. 2012). Since RXRα is the dimeric partner of a number of receptors, its involvement in the synthetic process is not clear. In PPARα knockout mice CoQ amount and synthesis is not disturbed but the induction with peroxisomal inducers is eliminated (Table 16.3). The CoQ synthesis increases upon cold exposure, however inducers such as DEHP has no longer an effect on the synthetic process in PPARα knockout mice. On the other hand, RXRα knockout mice did not respond to cold exposure but induction with DEHP is extensive. LXRα appears to have different roles in various organs. LXRα knockout mice have a decreased CoQ synthesis in the liver but increased in the spleen as well as in other organs. The LXRβ knockout mice have no influence concerning CoQ synthesis in any of the organs. Consequently, the LXR-αβ double knockout mice behave as the α knockout form. It is expected that in the near future the role of nuclear receptors in CoQ biosynthesis will be explored in detail.

The effectiveness of the inducers is demonstrated by the use of clofibrate and DEHP in rat (Aberg et al. 1996). Clofibrate is interesting since it is also used in human therapy in some type of hyperlipidemia. After 2 weeks of treatment the CoQ amount in organs is influenced and after 6 weeks a more than 100% increase is observed in the liver, 40% in muscle and heart and 20% in blood while the brain remains unaffected (Fig. 16.3).

DEHP is a chemical component that we encounter in our daily life. It is present everywhere as a consequence of release from polyvinylchloride (PVC) plastic where it is used as plasticizer. It is also inadvertently introduced to patients since the plastic bags and tubes used for blood storage and dialysis are composed of PVC plastics. When it is administered to rats, already after 3-weeks the CoQ amount is

Table 16.3 Nuclear receptors and CoQ amounts in mouse liver and spleen

Mice	Treatment	CoQ (% of wild-type)
Wild type, liver	None	100
	DEHP	226
	Cold exposure	157
PPARα-knock out, liver	None	98
	DEHP	98
	Cold exposure	149
RXRα – deficient, liver	None	49
	DEHP	555
Thyroxin, liver	Cold exposure	114
	None	100
	TRα−/−	166
	TRβ−/− (CoQ_9)	100
	TRβ−/− (CoQ_{10})	500
LXR, liver	None	100
α−/−	None	69
β−/−	None	98
αβ−/−	None	63
LXR, spleen	None	100
α−/−	None	132
β−/−	None	96
αβ−/−	None	148

From references Turunen et al. (2000), Bentinger et al. (2003), Bentinger et al. (2012) and Tekle (unpublished data)

increased substantially and after 6 weeks the elevation is 400% in the liver, 80% in the muscle and 20% in the heart and blood (Fig. 16.4).

16.5 Interference with Uptake of CoQ

Today the only way to increase CoQ in human blood and tissues is by dietary administration of the lipid (Littarru and Tiano 2005). The uptake is limited to a few percent but various formulations for increased uptake were proposed, such as supplementing the reduced form, a water soluble form and derivatization by hydroxylation or succinylation (Turunen et al. 1999; Miles et al. 2007; Bergamini et al. 2012). To find an optimal condition is of interest for future investigations. A Vitamin E deficient diet decreases the CoQ uptake to a limited extent, but enrichment of the diet with vitamin E increases uptake of CoQ considerably (Table 16.4).

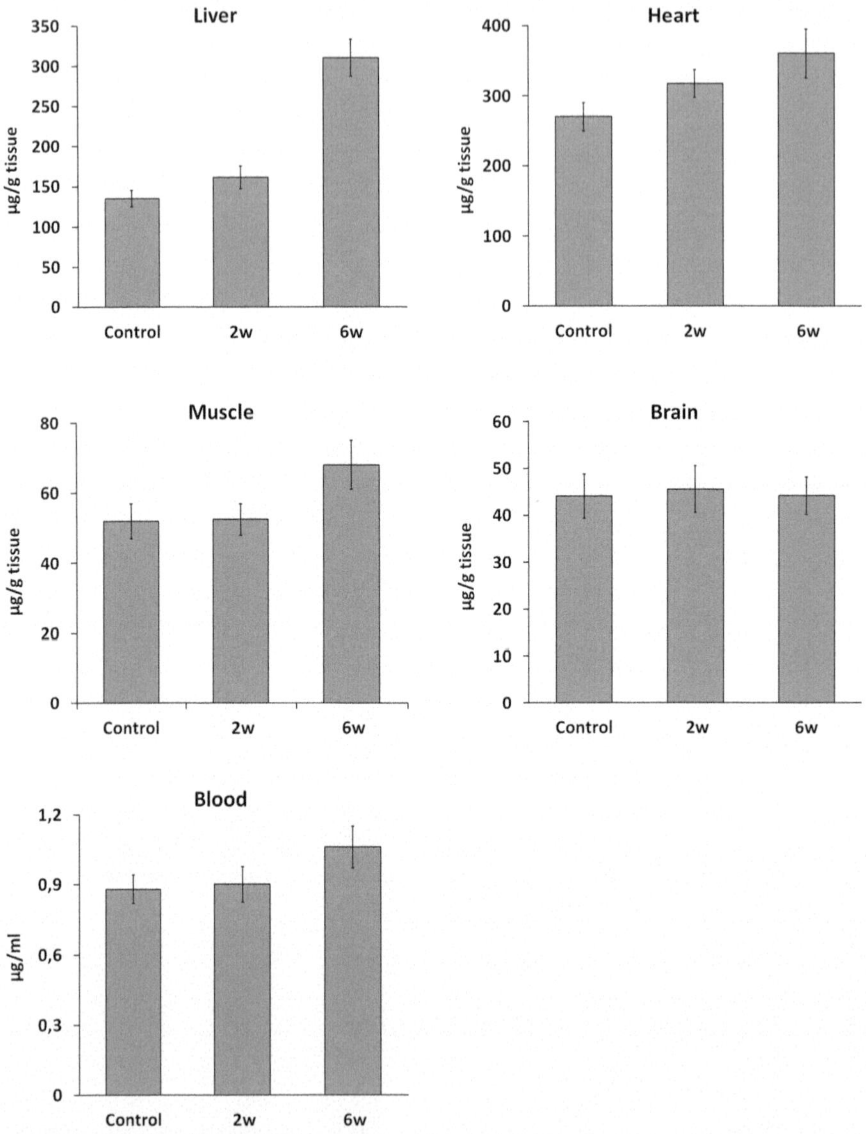

Fig. 16.3 Amounts of CoQ_9 in rat tissues and blood after 2 and 6 weeks of treatment with clofibrate. From reference Aberg et al. (1994)

16.6 Induction of Coenzyme Q Synthesis in Experimental Systems

In aging and in many diseases - including inborn errors - the optimal treatment procedure would be to use an inducer of the CoQ synthesis without toxic effects. In this

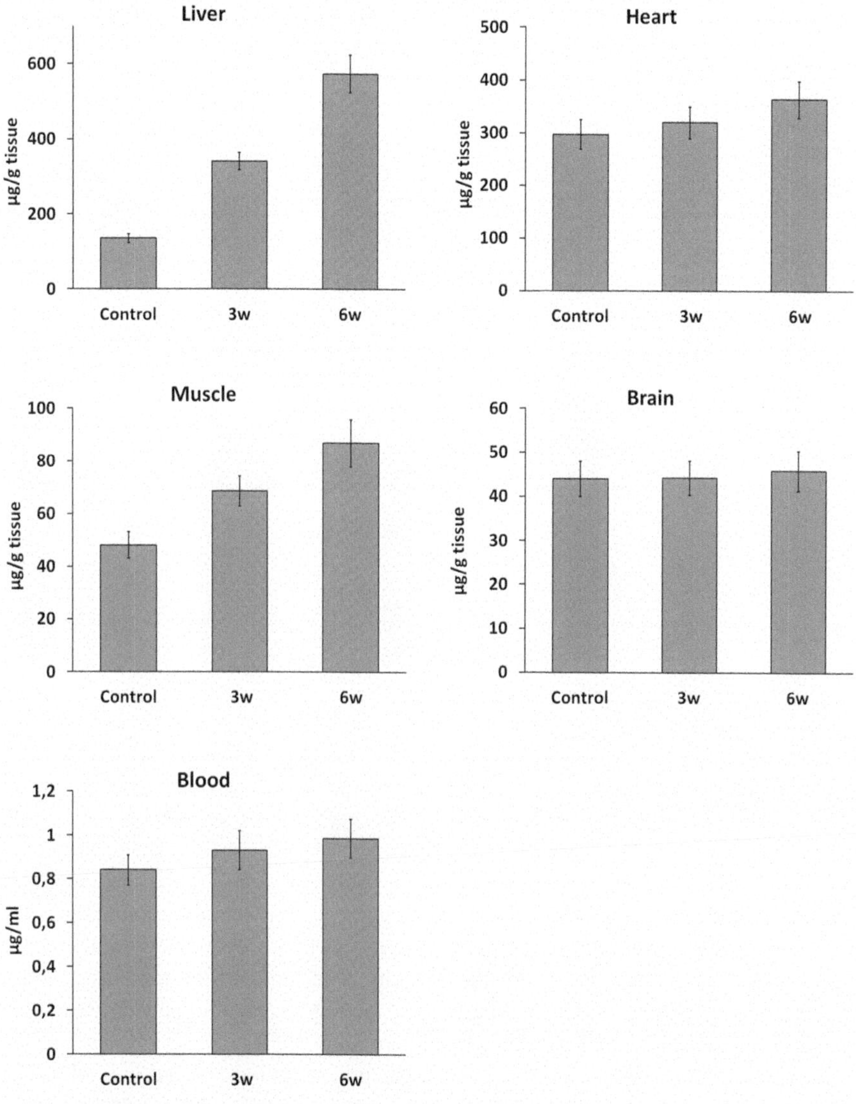

Fig. 16.4 Amounts of CoQ_9 in rat tissues and blood after 3 and 6 weeks treatment with di(2-ethylhexyl)phthalate. From reference Aberg et al. (1994)

respect modified forms of tocotrienols appear to be efficient (Bentinger et al. 2008). Both mono and diepoxidated forms of α, β, γ and δ tocotrienols greatly increased the incorporation of ^{3}H-mevalonate into CoQ when cultured with HepG2 cells (Fig. 16.5). Most of the lipids with *all-trans* isoprenoid side chains such as vitamin K2, squalene, solanesol, solanesyl-phosphate, solanesyl-acetate, solanesyl-phtalimide, epoxidated tocotrienols as well as epoxidated CoQ are efficient inducers in tissue cultures. It was also found that the COQ1 and COQ2 genes were

Table 16.4 Effects of vitamin E status on the dietary uptake of CoQ_{10} into liver and plasma of rats

	Vitamin E DD	Normal vitamin E diet CoQ_{10} pmol/mg protein	Vitamin E SD
Liver			
None	88	102	127
CoQ_{10} treated	481	520	674
Plasma			
None	68	74	84
CoQ_{10} treated	149	156	252

DD deficient diet, *SD* supplemented diet. From reference Zhang et al. (1996b)

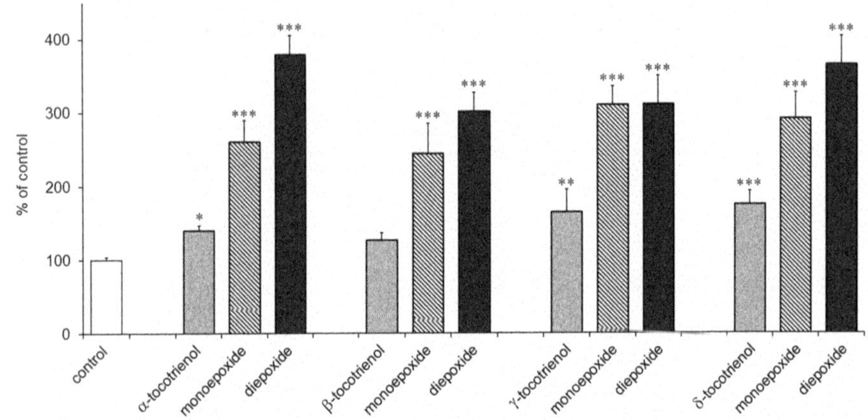

Fig. 16.5 Effects of tocotrienols and their epoxidated forms on CoQ biosynthesis. Incorporation of ^3H-mevalonate into CoQ in HepG2 cells was determined. From reference Bentinger et al. (2008)

induced in the cells. In prolonged incubation of the cells with these substances, a substantial elevation of not only the rate of synthesis, but also the total amount of the lipid was achieved. The various derivatives of tocotrienols are promising as future drugs since they have very low toxicity. The human diet contains a large amount of tocotrienols (Theriault et al. 1999), greatly exceeding the amount of the epoxidated form given to the rodents (Bentinger et al. 2008).

Squalene is one of the *all-trans* isoprenoids present in the cell and is a substrate for cholesterol synthesis. This intermediate can be modified by transferring the epoxide moiety to the second or third unsaturated bond. When 10, 11-oxidosqualene was injected to the mice during 6 days, the amount of cholesterol in the circulation was not influenced but the amount of circulating CoQ increased significantly (Fig. 16.6a, 16.6b). The validity of this finding was also proven by injecting the substance into ApoE deficient mice in which CoQ in blood was doubled already after 4 days (Fig. 16.6c). A similar finding was also observed in the type 2 diabetic model Goto-Kakizaki (GK) rats where the CoQ amount was almost doubled after 6 days of injection (Fig. 16.6d) (Bentinger et al. 2014).

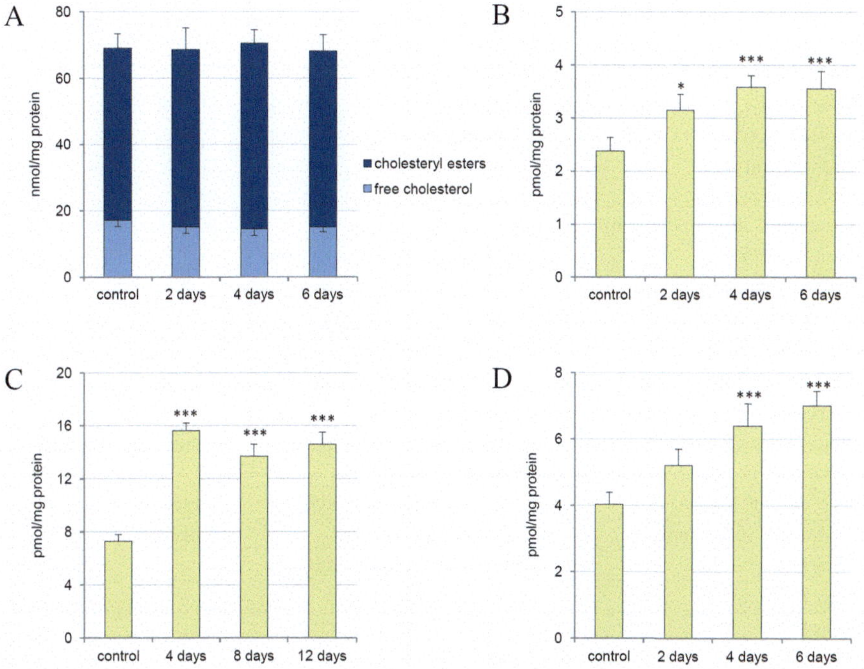

Fig. 16.6 The influence of treatment with 10,11-oxidosqualene on plasma levels of cholesterol and CoQ_9 in control mice (**a, b**), CoQ_9 in ApoE deficient mice (**c**) and CoQ_9 in GK-rats (**d**). From reference Bentinger et al. (2014)

16.7 Coenzyme Q in Diseases

A limited number of studies are available where CoQ is measured in various tissues in diseased conditions. This subject is of interest in order to find out the importance of this lipid for various normal functions. Prion disease develops as a consequence of protein conformational changes and is closely related to Creutzfeldt-Jakob disease. The mouse model of the disease has been investigated to a great extent and contributed to the identification of the changes causing the disease. Among the modifications is the greatly induced synthesis of CoQ resulting in a 2.5-fold increase of the lipid in the brain (Table 16.5) (Guan et al. 1996). Interestingly, the other mevalonate pathway lipids, dolichol and dolichylphosphate are also elevated by several folds in the brains of scrapie infected mice.

Chemical carcinogenesis can be performed by intermittent administration of 2-acetylaminofluorene to rats which in the initial period leads to the development of preneoplastic nodule in the liver. These nodules contain a greatly increased amount of CoQ before the development of cancer (Olsson et al. 1991). Animal models of diabetes type 2 are more or less similar to the human form of diabetes. One of the few available data concerning CoQ amount is that in the GK rat testis mitochondria

Table 16.5 Coenzyme Q in diseases

Increase	Control	Diseased	Unit	Reference
Scrapie disease (mouse)	40.3	99.9	μg/g ww	Guan et al. (1996)
Preneoplastic nodule (rat)	135	855	μg/mg p	Olsson et al. (1991)
Diabetes (testis mit) GK rat	1.6	2.1	nmol/mg p	Palmeira et al. (2001)
Liver, GK rat	158	226	nmol/g tissue	Grunler (unpublished data)
Alzheimer's disease (human)				Soderberg et al. (1992)
Frontal cortex	13.1	16.9	μg/g ww	
Frontal white matter	2.6	4.2		
Nucleaus caudatus	14.9	22.4		
Cerebellum	9.7	14.8		
Medulla oblongata	5.0	9.4		
Decrease				
Niemann-pick type C disease (mouse liver)	150	98	ng/mg p	Schedin et al. (1998)
Liver cancer (human)	40	18	μg/g ww	Eggens et al. (1989)
Cardiomyopathy (human)				Folkers et al. (1985)
Class I		0.40	μg/g dw	
Class II		0.34		
Class III		0.28		
Class IV		0.28		

ww wet weight, *dw* dry weight, *p* protein, *GK* Goto Kakizaki, *mit* mitochondria

where an increase of CoQ was observed (Palmeira et al. 2001). A similar finding was also observed in the liver of GK rat (Grunler, unpublished data).

The major form of human dementia is the Alzheimer's disease. This is a well characterized disease where a number of changes in the brain are found but no therapy is available today. In this disease, CoQ is greatly increased in all parts of human brain including the cerebellum and medulla oblongata (Soderberg et al. 1992).

Some diseases are associated with inhibition of the synthesis of this lipid. A decrease in CoQ amount was also observed in the liver of Niemann-Pick type C disease mice, a model commonly used to find out the factors behind this disease (Schedin et al. 1998).

In human liver cancer, the amount of CoQ per gram tissue is greatly decreased which is in contrast to what was found in the preneoplastic nodule of the rat (Eggens et al. 1989).

In some investigations, biopsies taken from human heart were analyzed in order to follow the effectiveness of dietary CoQ therapy (Folkers et al. 1985). When the disorders were followed from class I to class IV, which are classes introduced to classify the severity of the myocardial disorder where class I is milder and IV is an advanced form of the disease, the CoQ levels decreased accordingly. In fact it has been shown that a long term CoQ treatment during chronic heart failure improves heart function and adverse cardiovascular events (Alehagen et al. 2013; Mortensen et al. 2014).

16.8 Discussion

In this review, we describe the changes in CoQ amount that occur during aging, under various experimental conditions and in diseases. These data deals with events related to cells and tissues and do not include information concerning measurements in blood since the amount of CoQ found in the circulation does not necessarily reflect the metabolism and the levels in various organs. The major part of the data available in the literature originates from studies based on blood CoQ content. However the situation, in contrast to other blood lipids such as cholesterol, is different. Cholesterol is one of the major lipid constituents of the blood and reflects the metabolism both in blood and organs (Goldstein and Brown 1990; Johnson et al. 1997). It is synthesized in all organs but the daily requirement is very different from that of CoQ. For example, the adrenals synthesize large amounts of hormones using mainly external uptake of the substrate which is cholesterol. One third of our cholesterol originates from the diet as an important food constituent (Hussain 2014). The liver synthesizes large amounts of cholesterol that is secreted to the circulation and mixed with the dietary lipids bound to lipoproteins. This process involves equilibrium between the blood and metabolism in various organs. Consequently, the concentration of cholesterol in the blood reflects the levels characteristic for the general metabolism. The situation is different concerning CoQ. All organs and cells in the body synthesize this lipid in sufficient amounts to attain the cellular requirements and, in opposite to cholesterol, CoQ is not metabolized to be a substrate for downstream cellular functions (Elmberger et al. 1987). The uptake of CoQ from the diet is limited, not exceeding 3–4%, and the level is rapidly saturated (Zhang et al. 1995). It appears that the major part of CoQ in the blood is originating from synthesis in the liver where a low amount is excreted with the VLDL into the blood (Elmberger et al. 1989). Probably, the importance and main function of CoQ in the blood plasma is to maintain the required antioxidant defence of the blood lipoproteins and no tissue redistribution occurs under normal conditions (Thomas et al. 1997). Therefore, the blood level does not necessarily reflect the levels and metabolism of various organs. This situation is most clearly demonstrated in children with inborn errors of CoQ synthesis who require a continuous treatment with CoQ in order to survive (Rustin et al. 2004). Most of these children have the same lipid composition in the blood as healthy children. This fact does not exclude the usefulness of CoQ measurements in the blood for diagnosis of specific conditions, where changes are related to specific metabolic disorders. Another possibility to diagnose CoQ deficiency could be to isolate and measure lipid content in the blood monocytes since these cells behave in most aspects similar to the tissue cells in various organs (Turunen et al. 2002a). However, this analysis would be difficult to perform in a larger scale.

A comparison of CoQ amounts in the blood under various conditions and in diseases are often performed in extensive studies. This kind of comparison can be relevant if homogenous populations are compared. In one study, it was found that Polish and Swedish women selected from restricted regions had the same amount of

CoQ in the blood (Tekle et al. 2010). Women from the capital of Serbia had twice as much CoQ, while women from part of Serbia, which had been affected by chemical pollutant in a recent war, had only 20% of what was found in the Swedish and Polish cohorts. In another investigation, Kenyan rural (Samburu) and urban (Nairobi) populations were compared and those in the rural areas showed a three-fold higher CoQ amount (Theuri et al. 2013) probably due to enhanced physical activities. These facts should be kept in mind in future larger studies.

Another question of great interest is the redox status of CoQ which is often investigated. In spite of the fact that the body utilizes CoQ as an effective antioxidant and also for regenerating α-tocopherol, most of the lipid is found in reduced form. Now it is well established that a number of different reductive enzymes are capable of reducing the oxidized CoQ and these enzymes are in excess (Olsson et al. 1999; Montano et al. 2015). It is also established that reduced CoQ is rapidly auto-oxidized when exposed to air depending on a number of conditions (Aberg et al. 1992). Consequently, the type of the solvents used for extraction and the time consumed for isolation of the lipid is of great importance and influences the redox status extensively. As a result, it is difficult to obtain reproducible data and the real status of the lipid in the tissues can be different. This fact is also reflected when analyzing a number of studies where the amount of reduced per oxidized CoQ is very different in spite of the similar experimental conditions. In a recent investigation it was shown that reversed mitochondrial electron transport from reduced CoQ to complex I leads to the generation of superoxide and oxidation of complex I proteins (Guaras et al. 2016). Consequently, CoQ redox status acts as a metabolic sensor which influences the efficiency of the respiratory chain.

During aging, not only in rodents but also in humans, CoQ is decreasing in all organs (Kalen et al. 1989). Since the major part of the lipid is in the inner mitochondrial membrane, it is expected that mitochondrial respiration and ATP production is reduced (Kalen et al. 1990). The requirement for oxidative phosphorylation is decreased during aging and the importance of this change is not completely elucidated. Most or all of the procedures described to increase the amount of this lipid are working in experimental systems but none of them have so far proved to be efficient in humans. Today, the only way to increase CoQ action in human is by taking the dietary administration route (Littarru and Tiano 2005). In doses of 100–200 mg, part of this lipid is taken up into the circulation and affects human metabolism (Brauner et al. 2014). This occurs in spite of the established fact that the uptake into the organs from the blood is limited. However this limitation does not interfere with the findings that organ functions are influenced. Dietary CoQ is present in monocytes and influences the production of β2-integrins, complement receptors, TNFα secretion and a number of genes involved in cell signaling and cellular metabolism. For these reasons CoQ may up or down regulate a large number of metabolic processes by binding to cell surface receptors that induce intracellular metabolic pathways. These effects do not require the entrance of the lipid into the cell.

References

Aberg F, Appelkvist EL, Dallner G, Ernster L (1992) Distribution and redox state of ubiquinones in rat and human tissues. Arch Biochem Biophys 295(2):230–234

Aberg F, Zhang Y, Appelkvist EL, Dallner G (1994) Effects of clofibrate, phthalates and probucol on ubiquinone levels. Chem Biol Interact 91(1):1–14

Aberg F, Zhang Y, Teclebrhan H, Appelkvist EL, Dallner G (1996) Increases in tissue levels of ubiquinone in association with peroxisome proliferation. Chem Biol Interact 99(1–3):205–218

Acosta MJ, Vazquez Fonseca L, Desbats MA, Cerqua C, Zordan R, Trevisson E, Salviati L (2016) Coenzyme Q biosynthesis in health and disease. Biochim Biophys Acta 1857(8):1079–1085. https://doi.org/10.1016/j.bbabio.2016.03.036

Aithal HN, Joshi VC, Ramasarma T (1968) Effect of cold exposure on the metabolism of ubiquinone in the rat. Biochim Biophys Acta 162(1):66–72

Alehagen U, Johansson P, Bjornstedt M, Rosen A, Dahlstrom U (2013) Cardiovascular mortality and N-terminal-proBNP reduced after combined selenium and coenzyme Q10 supplementation: a 5-year prospective randomized double-blind placebo-controlled trial among elderly Swedish citizens. Int J Cardiol 167(5):1860–1866. https://doi.org/10.1016/j.ijcard.2012.04.156

Allen RM, Vickers KC (2014) Coenzyme Q10 increases cholesterol efflux and inhibits atherosclerosis through microRNAs. Arterioscler Thromb Vasc Biol 34(9):1795–1797. https://doi.org/10.1161/ATVBAHA.114.303741

Azzolin L, von Stockum S, Basso E, Petronilli V, Forte MA, Bernardi P (2010) The mitochondrial permeability transition from yeast to mammals. FEBS Lett 584(12):2504–2509. https://doi.org/10.1016/j.febslet.2010.04.023

Bentinger M, Turunen M, Zhang XX, Wan YJ, Dallner G (2003) Involvement of retinoid X receptor alpha in coenzyme Q metabolism. J Mol Biol 326(3):795–803

Bentinger M, Tekle M, Brismar K, Chojnacki T, Swiezewska E, Dallner G (2008) Polyisoprenoid epoxides stimulate the biosynthesis of coenzyme Q and inhibit cholesterol synthesis. J Biol Chem 283(21):14645–14653. https://doi.org/10.1074/jbc.M710202200

Bentinger M, Tekle M, Dallner G, Brismar K, Gustafsson JA, Steffensen KR, Catrina SB (2012) Influence of liver-X-receptor on tissue cholesterol, coenzyme Q and dolichol content. Mol Membr Biol 29(7):299–308. https://doi.org/10.3109/09687688.2012.694484

Bentinger M, Kania M, Danikiewicz W, Kaczorowska E, Wojcik J, Brismar K, Dallner G, Chojnacki T, Swiezewska E, Tekle M (2014) Effects of various squalene epoxides on coenzyme Q and cholesterol synthesis. Biochim Biophys Acta 1841(7):977–986. https://doi.org/10.1016/j.bbalip.2014.03.007

Bergamini C, Moruzzi N, Sblendido A, Lenaz G, Fato R (2012) A water soluble CoQ_{10} formulation improves intracellular distribution and promotes mitochondrial respiration in cultured cells. PLoS One 7(3):e33712. https://doi.org/10.1371/journal.pone.0033712

Brauner H, Luthje P, Grunler J, Ekberg NR, Dallner G, Brismar K, Brauner A (2014) Markers of innate immune activity in patients with type 1 and type 2 diabetes mellitus and the effect of the anti-oxidant coenzyme Q10 on inflammatory activity. Clin Exp Immunol 177(2):478–482. https://doi.org/10.1111/cei.12316

Brea-Calvo G, Siendones E, Sanchez-Alcazar JA, de Cabo R, Navas P (2009) Cell survival from chemotherapy depends on NF-kappaB transcriptional up-regulation of coenzyme Q biosynthesis. PLoS One 4(4):e5301. https://doi.org/10.1371/journal.pone.0005301

Buron MI, Rodriguez-Aguilera JC, Alcain FJ, Navas P (1993) Transplasma membrane redox system in HL-60 cells is modulated during TPA-induced differentiation. Biochem Biophys Res Commun 192(2):439–445

Crane FL, Sun IL, Clark MG, Grebing C, Low H (1985) Transplasma-membrane redox systems in growth and development. Biochim Biophys Acta 811(3):233–264

Echtay KS, Winkler E, Klingenberg M (2000) Coenzyme Q is an obligatory cofactor for uncoupling protein function. Nature 408(6812):609–613. https://doi.org/10.1038/35046114

Eggens I, Elmberger PG, Low P (1989) Polyisoprenoid, cholesterol and ubiquinone levels in human hepatocellular carcinomas. Br J Exp Pathol 70(1):83–92

Elmberger PG, Kalen A, Appelkvist EL, Dallner G (1987) In vitro and in vivo synthesis of dolichol and other main mevalonate products in various organs of the rat. Eur J Biochem 168(1):1–11

Elmberger PG, Kalen A, Brunk UT, Dallner G (1989) Discharge of newly-synthesized dolichol and ubiquinone with lipoproteins to rat liver perfusate and to the bile. Lipids 24(11):919–930

Ernster L, Forsmark-Andree P (1993) Ubiquinol: an endogenous antioxidant in aerobic organisms. Clin Investig 71(8 Suppl):S60–S65

Folkers K, Vadhanavikit S, Mortensen SA (1985) Biochemical rationale and myocardial tissue data on the effective therapy of cardiomyopathy with coenzyme Q10. Proc Natl Acad Sci U S A 82(3):901–904

Genova ML, Lenaz G (2011) New developments on the functions of coenzyme Q in mitochondria. Biofactors 37(5):330–354. https://doi.org/10.1002/biof.168

Gohil K, Rothfuss L, Lang J, Packer L (1985) Effect of exercise training on tissue vitamin E and ubiquinone content. J Appl Physiol 63(4):1638–1641

Goldstein JL, Brown MS (1990) Regulation of the mevalonate pathway. Nature 343(6257):425–430. https://doi.org/10.1038/343425a0

Groneberg DA, Kindermann B, Althammer M, Klapper M, Vormann J, Littarru GP, Doring F (2005) Coenzyme Q10 affects expression of genes involved in cell signalling, metabolism and transport in human CaCo-2 cells. Int J Biochem Cell Biol 37(6):1208–1218. https://doi.org/10.1016/j.biocel.2004.11.017

Guan Z, Soderberg M, Sindelar P, Prusiner SB, Kristensson K, Dallner G (1996) Lipid composition in scrapie-infected mouse brain: prion infection increases the levels of dolichyl phosphate and ubiquinone. J Neurochem 66(1):277–285

Guan ZZ, Wang YN, Xiao KQ, Dai DY, Chen YH, Liu JL, Sindelar P, Dallner G (1998) Influence of chronic fluorosis on membrane lipids in rat brain. Neurotoxicol Teratol 20(5):537–542

Guaras A, Perales-Clemente E, Calvo E, Acin-Perez R, Loureiro-Lopez M, Pujol C, Martinez-Carrascoso I, Nunez E, Garcia-Marques F, Rodriguez-Hernandez MA, Cortes A, Diaz F, Perez-Martos A, Moraes CT, Fernandez-Silva P, Trifunovic A, Navas P, Vazquez J, Enriquez JA (2016) The CoQH2/CoQ ratio serves as a sensor of respiratory chain efficiency. Cell Rep 15(1):197–209. https://doi.org/10.1016/j.celrep.2016.03.009

Hussain MM (2014) Intestinal lipid absorption and lipoprotein formation. Curr Opin Lipidol 25(3):200–206. https://doi.org/10.1097/MOL.0000000000000084

Johnson R, McNutt P, MacMahon S, Robson R (1997) Use of the Friedewald formula to estimate LDL-cholesterol in patients with chronic renal failure on dialysis. Clin Chem 43(11):2183–2184

Kalen A, Appelkvist EL, Dallner G (1989) Age-related changes in the lipid compositions of rat and human tissues. Lipids 24(7):579–584

Kalen A, Appelkvist EL, Dallner G (1990) The effects of inducers of the endoplasmic reticulum, peroxisomes and mitochondria on the amounts and synthesis of ubiquinone in rat liver subcellular membranes. Chem Biol Interact 73(2–3):221–234

Littarru GP, Tiano L (2005) Clinical aspects of coenzyme Q10: an update. Curr Opin Clin Nutr Metab Care 8(6):641–646

Low P, Andersson M, Edlund C, Dallner G (1992) Effects of mevinolin treatment on tissue dolichol and ubiquinone levels in the rat. Biochim Biophys Acta 1165(1):102–109

Miles MV, Patterson BJ, Chalfonte-Evans ML, Horn PS, Hickey FJ, Schapiro MB, Steele PE, Tang PH, Hotze SL (2007) Coenzyme Q10 (ubiquinol-10) supplementation improves oxidative imbalance in children with trisomy 21. Pediatr Neurol 37(6):398–403. https://doi.org/10.1016/j.pediatrneurol.2007.08.003

Montano SJ, Grunler J, Nair D, Tekle M, Fernandes AP, Hua X, Holmgren A, Brismar K, Ungerstedt JS (2015) Glutaredoxin mediated redox effects of coenzyme Q10 treatment in type 1 and type 2 diabetes patients. BBA Clin 4:14–20. https://doi.org/10.1016/j.bbacli.2015.06.001

Mortensen SA, Rosenfeldt F, Kumar A, Dolliner P, Filipiak KJ, Pella D, Alehagen U, Steurer G, Littarru GP (2014) The effect of coenzyme Q10 on morbidity and mortality in chronic

heart failure: results from Q-SYMBIO: a randomized double-blind trial. JACC Heart Fail 2(6):641–649. https://doi.org/10.1016/j.jchf.2014.06.008
Olsson JM, Eriksson LC, Dallner G (1991) Lipid compositions of intracellular membranes isolated from rat liver nodules in Wistar rats. Cancer Res 51(14):3774–3780
Olsson JM, Xia L, Eriksson LC, Bjornstedt M (1999) Ubiquinone is reduced by lipoamide dehydrogenase and this reaction is potently stimulated by zinc. FEBS Lett 448(1):190–192
Palmeira CM, Santos DL, Seica R, Moreno AJ, Santos MS (2001) Enhanced mitochondrial testicular antioxidant capacity in Goto-Kakizaki diabetic rats: role of coenzyme Q. Am J Physiol Cell Physiol 281(3):C1023–C1028
Pedersen S, Tata JR, Ernster L (1963) Ubiquinone (coenzyme Q) and the regulation of basal metabolic rate by thryoid hormones. Biochim Biophys Acta 69:407–409
Quiles JL, Huertas JR, Manas M, Battino M, Cassinello M, Littarru GP, Lenaz G, Mataix FJ (1994) Peroxidative extent and coenzyme Q levels in the rat: influence of physical training and dietary fats. Mol Asp Med 15(Suppl):s89–s95
Quinzii CM, Hirano M (2010) Coenzyme Q and mitochondrial disease. Dev Disabil Res Rev 16(2):183–188. https://doi.org/10.1002/ddrr.108
Rotig A, Mollet J, Rio M, Munnich A (2007) Infantile and pediatric quinone deficiency diseases. Mitochondrion 7(Suppl):S112–S121. https://doi.org/10.1016/j.mito.2007.02.008
Rustin P, Munnich A, Rotig A (2004) Mitochondrial respiratory chain dysfunction caused by coenzyme Q deficiency. Methods Enzymol 382:81–88. https://doi.org/10.1016/S0076-6879(04)82005-6
Schedin S, Nilsson M, Chojnacki T, Dallner G (1998) Alterations in the biosynthesis of cholesterol, dolichol and dolichyl-P in the genetic cholesterol homeostasis disorder, Niemann-pick type C disease. Biochim Biophys Acta 1394(2–3):177–186
Schmelzer C, Doring F (2010) Identification of LPS-inducible genes downregulated by ubiquinone in human THP-1 monocytes. Biofactors 36(3):222–228. https://doi.org/10.1002/biof.93
Soderberg M, Edlund C, Kristensson K, Dallner G (1990) Lipid compositions of different regions of the human brain during aging. J Neurochem 54(2):415–423
Soderberg M, Edlund C, Alafuzoff I, Kristensson K, Dallner G (1992) Lipid composition in different regions of the brain in Alzheimer's disease/senile dementia of Alzheimer's type. J Neurochem 59(5):1646–1653
Sohlenius-Sternbeck AK, Appelkvist EL, DePierre JW (2000) Effects of vitamin a deficiency on selected xenobiotic-metabolizing enzymes and defenses against oxidative stress in mouse liver. Biochem Pharmacol 59(4):377–383
Tekle M, Gromadzinska J, Joksic G, Antic R, Nilsson R, Dallner G, Unden AL, Brismar K (2010) Plasma levels of insulin-like growth factor-I, insulin-like growth factor binding protein-1, coenzyme Q10 and vitamin E in female populations from Poland, Serbia and Sweden. Environ Int 36(2):188–194. https://doi.org/10.1016/j.envint.2009.11.003
Theriault A, Chao JT, Wang Q, Gapor A, Adeli K (1999) Tocotrienol: a review of its therapeutic potential. Clin Biochem 32(5):309–319
Theuri G, Dallner G, Brismar K, Tekle M (2013) Effects of lifestyle on plasma levels of the IGF system and the antioxidants coenzyme Q10 and vitamin E in Kenyan rural and urban populations. Growth Hormon IGF Res 23(3):68–75. https://doi.org/10.1016/j.ghir.2013.01.003
Thomas SR, Neuzil J, Stocker R (1997) Inhibition of LDL oxidation by ubiquinol-10. A protective mechanism for coenzyme Q in atherogenesis? Mol Aspects Med 18(Suppl):S85–S103
Tran UC, Clarke CF (2007) Endogenous synthesis of coenzyme Q in eukaryotes. Mitochondrion 7(Suppl):S62–S71. https://doi.org/10.1016/j.mito.2007.03.007
Turunen M, Appelkvist EL, Sindelar P, Dallner G (1999) Blood concentration of coenzyme Q(10) increases in rats when esterified forms are administered. J Nutr 129(12):2113–2118
Turunen M, Peters JM, Gonzalez FJ, Schedin S, Dallner G (2000) Influence of peroxisome proliferator-activated receptor alpha on ubiquinone biosynthesis. J Mol Biol 297(3):607–614. https://doi.org/10.1006/jmbi.2000.3596

Turunen M, Swiezewska E, Chojnacki T, Sindelar P, Dallner G (2002a) Regulatory aspects of coenzyme Q metabolism. Free Radic Res 36(4):437–443

Turunen M, Wehlin L, Sjoberg M, Lundahl J, Dallner G, Brismar K, Sindelar PJ (2002b) beta2-integrin and lipid modifications indicate a non-antioxidant mechanism for the anti-atherogenic effect of dietary coenzyme Q10. Biochem Biophys Res Commun 296(2):255–260

Turunen M, Olsson J, Dallner G (2004) Metabolism and function of coenzyme Q. Biochim Biophys Acta 1660(1–2):171–199

Zhang Y, Aberg F, Appelkvist EL, Dallner G, Ernster L (1995) Uptake of dietary coenzyme Q supplement is limited in rats. J Nutr 125(3):446–453

Zhang Y, Appelkvist EL, Kristensson K, Dallner G (1996a) The lipid compositions of different regions of rat brain during development and aging. Neurobiol Aging 17(6):869–875

Zhang Y, Turunen M, Appelkvist EL (1996b) Restricted uptake of dietary coenzyme Q is in contrast to the unrestricted uptake of alpha-tocopherol into rat organs and cells. J Nutr 126(9):2089–2097

Chapter 17
Effects of Coenzyme Q_{10} Supplementation on Elderly People

Elena M. Yubero-Serrano, Francisco M. Gutierrez-Mariscal, and Jose Lopez-Miranda

Abstract Coenzyme Q_{10} (CoQ_{10}) is an essential component of the electron transport system and the only lipid-soluble compound synthesized endogenously present in all cell membranes with bioenergetics and antioxidant properties.

Aging, neurodegenerative disorders, cardiovascular disease and other aged-related diseases, as well as genetic mutations, have been associated with CoQ_{10} deficiency. Since both limited uptake and low bioavailability of dietary CoQ_{10} might influence in this deficiency, supplementation with CoQ_{10} must be considered in those cases as therapeutic solution. However, more research is needed in order to identify the appropriate dose, the effectiveness and the bioavailability of orally-administered CoQ_{10}. Furthermore research must be developed in order to design therapeutic agents to induce the endogenous synthesis CoQ_{10} specially in elderly people.

This review will focus in the most relevant biochemical characteristics of this important antioxidant, including its main functions, levels and distribution in human organism and the therapeutic potential of CoQ_{10}, especially, during aging and the associated diseases.

Keywords Coenzyme Q_{10} · Aging · Oxidative stress · Antioxidant · Aging-related diseases · Therapeutic approach

---

Elena M. Yubero-Serrano and Francisco M. Gutierrez-Mariscal contributed equally with all other contributors.

E. M. Yubero-Serrano (✉) · F. M. Gutierrez-Mariscal · J. Lopez-Miranda
Lipids and Atherosclerosis Unit. Instituto Maimónides de Investigación Biomédica de Córdoba (IMIBIC); Reina Sofia University Hospital, University of Cordoba, and CIBER Fisiopatologia Obesidad y Nutricion (CIBEROBN), Instituto de Salud Carlos III, Spain, Córdoba, Spain
e-mail: jlopezmir@uco.es

G. López Lluch (ed.), *Coenzyme Q in Aging*,
https://doi.org/10.1007/978-3-030-45642-9_17

17.1 Coenzyme Q_{10}

Coenzyme Q_{10}, also known as CoQ_{10}, vitamin Q_{10}, ubiquinone, and ubidecarenone, is a benzoquinone compound, identified as a component of the mitochondrial respiratory chain (Crane et al. 1989; Schultz and Clarke 1999). It has been isolated and characterized as an ubiquitous quinone substance that received the name of ubiquinone (Festenstein et al. 1955). After its isolation, CoQ_{10} was identified as an essential electron carrier in the inner membrane of mitochondria as member of the mitochondrial electron transport chain (Festenstein et al. 1955; Crane et al. 1957).

CoQ_{10} is the short name of 2,3-dimethoxy-5-methyl-6-decaprenyl-1,4-benzoquinone. It is a lipid-soluble quinone with a very high biological activity. CoQ_{10} has two different parts, a polar benzoquinone ring and a lipidic isoprenoid side chain whose length depends on the organism. Its structure is similar to vitamin E. The name CoQ_{10} indicates that the quinone ring is bound to ten isoprenyl subunits that are part of this compound's structure. This is the predominant human form of this molecule. Moreover, the term "coenzyme" denotes it as an organic (contains carbon atoms), non-protein molecule necessary for the proper functioning of its protein partner (an enzyme or an enzyme complex) (Jeya et al. 2010). The principal characteristic of CoQ_{10} is its presence in three redox states: the fully oxidized *ubiquinone* form, a *semiquinone* form (that acts as free radical) and the fully reduced *ubiquinol* (Fig. 17.1) (Alcazar-Fabra et al. 2016). CoQ_{10} is found in all cell membranes but the highest presence is in the inner membrane of mitochondria in every cell in the human organism. In mitochondria, CoQ_{10} is essential as a cofactor in the

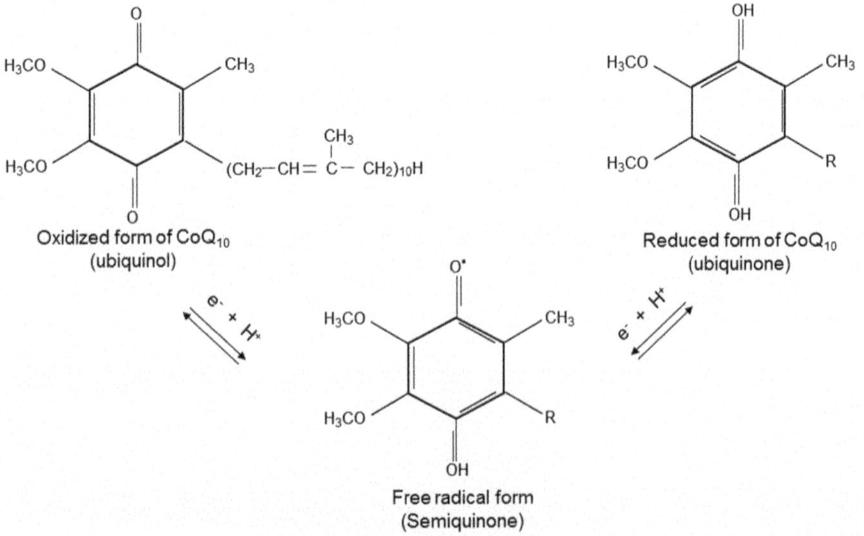

Fig. 17.1 Chemical structure of different forms of coenzyme Q_{10}
Chemical structure of different forms of coenzyme Q_{10} (2,3-dimethoxy-5-methyl-6-decaprenyl-1,4-benzoquinone). Ubiquinone is reduced to ubiquinol through a semiquinone intermediate

mitochondrial electron transport chain, and them, indispensable for aerobic cellular respiration and for the production of ATP and cell bioenergetics in aerobic organisms (Alcazar-Fabra et al. 2016; Acosta et al. 2016).

17.1.1 Coenzyme Q_{10} Functions

As member of the mitochondrial electron transport chain, CoQ_{10} accepts electrons from different donors named reductases, mainly NADH-coenzyme Q oxidoreductase (Complex I) and succinate-dehydrogenase (Complex II). The reduced form is later oxidized by transferring electrons to ubiquinol-cytochrome c reductase complex (Complex III) (Fig. 17.2). Its electron transport activity is accompanied by a pumping capacity that transfers protons from mitochondrial matrix to the intermembrane space contributing to create a proton gradient between mitochondrial matrix and cytosol (Crane 2001). This redox activity permits to mitochondria to participate in cell growth and maintenance (Overvad et al. 1999). Through this process, CoQ_{10} maintains a permanent redox equilibrium between the reduced form (ubiquinol) and

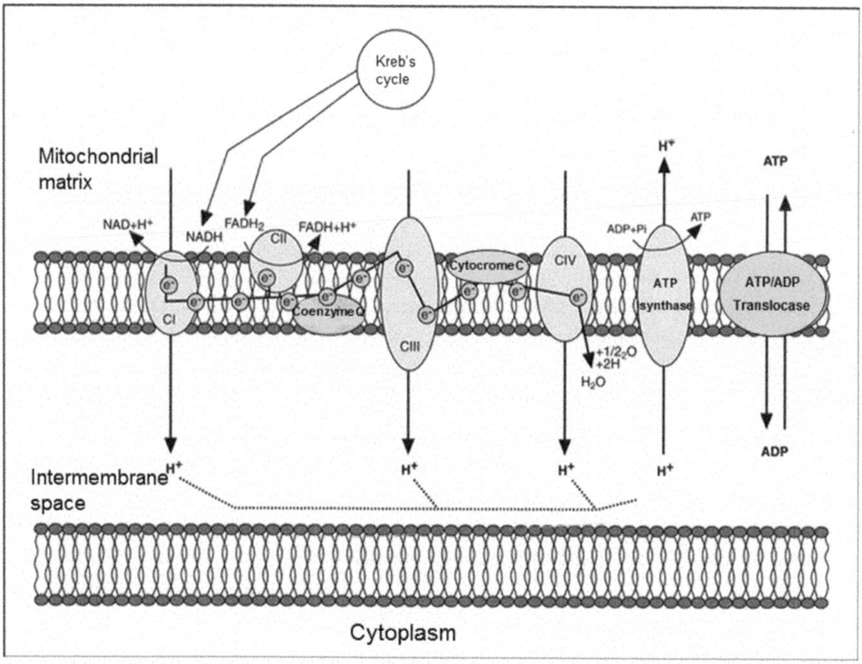

Fig. 17.2 Electron transport chain
The electron transport chain uses the electrons from electron carriers to generate an electrochemical gradient that will be used by ATP synthase to produce ATP. Coenzyme Q_{10} accepts the electrons from both complex I and complex II and delivers them to complex III

the oxidized form (ubiquinone). This equilibrium is maintained in mitochondria mainly by the activity of complexes I and II as electron donors and Complex III as acceptor. Other researches have revealed that CoQ_{10} is also a co-factor for the function of uncoupling proteins. For these reasons, CoQ_{10} is essential in the control of bioenergetics homeostasis in cells (Littarru and Tiano 2007; Potgieter et al. 2013).

In other cell membranes, at least three enzymes are known as CoQ_{10}-reductases: NADH/NADPH oxidoreductase (DT diaphorase), NADH cytochrome b_5 reductase and NADPH coenzyme Q reductase (Villalba and Navas 2000). In these membranes, ubiquinol acts as a potent antioxidant protecting cells from oxidative damage and contributing to the stability of the cell membranes, proteins, glycoproteins and DNA. Further reduced CoQ_{10} form has also been reported to protect LDL from oxidation (Lopez-Lluch et al. 2010). LDLs tend suffer more oxidation during aging probably by the reduction of the levels of CoQ_{10}. For this reason, CoQ_{10} supplementation could be a good therapy to decrease LDL oxidation reducing the high risk of cardiovascular disease during aging (Yubero-Serrano et al. 2011).

In addition to direct antioxidant radical scavenging, CoQ_{10}, and particularly the semiquinone intermediate (Fig. 17.1), recycles and regenerates other membrane antioxidants, such as α-tocopherol and also cytosolic and extracellular antioxidants such as ascorbic acid. CoQ_{10} is essential to maintain them in their reduced and activity state (Navas et al. 2007). All these activities make CoQ_{10} as the main lipidic antioxidant, more powerful than vitamin E, present in relative high concentrations and able to regenerate intracellular reducing mechanisms (Forsmark-Andree et al. 1995).

17.1.2 Levels and Distribution of Coenzyme Q_{10} in the Humans

All human cells studied so far can synthesize CoQ_{10}. Its amount in these cells depends on the organs and tissues. In humans, CoQ_{10} ranges from 8 μg/g in lung to 114 μg/g in heart. In some determinations, a shorter form, CoQ_9 has been found but only in small quantities (2–7%) (Jeya et al. 2010). In general, ubiquinol levels are higher than the levels of the oxidized form, ubiquinone, in most of the human tissues except in the case of lung and brain (Bhagavan and Chopra 2006). In the case of brain, the increase in the ratio ubiquinol/ubiquinone can be associated with neurological diseases due to mitochondrial dysfunction (Spinazzi et al. 2019).

Generally, tissues with high metabolic activity, such as the heart, kidney, liver and muscle, contain relatively high concentrations of CoQ_{10} (Ernster and Dallner 1995). At the cellular levels, most of the CoQ_{10} (40–50%) is localized at the mitochondrial inner membrane. It is present in the rest of cell membranes although at smaller amounts in the other organelles and in the cytosol.

Cell and tissue CoQ_{10} is coming from endogenous synthesis although it can be also obtained from food intake or oral supplementation. Interestingly, the range of CoQ_{10} concentration in humans show a high range and also depend on age, sex and race, and on the health of the individual (Sohal and Forster 2007). In healthy young individuals, total body content of CoQ_{10} is around 0.99 ± 0.3 mg/L (from 0.55 mg/L and 1.87 mg/L). However, these levels decrease during aging, and in age-related diseases such as in patients with cardiomyopathies, congestive heart failure and degenerative diseases (Fotino et al. 2012; Shetty et al. 2012).

17.1.3 Biosynthesis and Transport of CoQ_{10}

The main although not unique source of CoQ_{10} in humans is the endogenous synthesis. This synthesis depends on the mevalonate pathway (Fig. 17.3). CoQ_{10} synthesis Shares the mevalonate pathway with cholesterol, dolichol, dolychil-phoshate and isoprene chains that bind to aminoacid residues in proteins (Villalba et al. 2010). *De novo* CoQ_{10} synthesis in humans is initiated by the union of the benzoquinone ring precursor, 4-hydroxybenzoate, and the isoprenoid side chain produced from farnesyl pyrophosphate. These two molecules are condensed by the polyprenil-4-hydroxybenzoate transferase, COQ2. After that, a complex with many other components, at least eight enzymes (encoded by COQ3–10) modify the benzene ring with subsequent methylation, decarboxylation and hydroxylation reactions to (Quinzii et al. 2007; Turunen et al. 2004).

It seems clear that CoQ_{10} synthesis is located in mitochondria and from this, it is distributed in all the subcellular compartments. Then, a transport system from the mitochondria to the rest of cellular membranes must exist. Using *in vivo* labeling

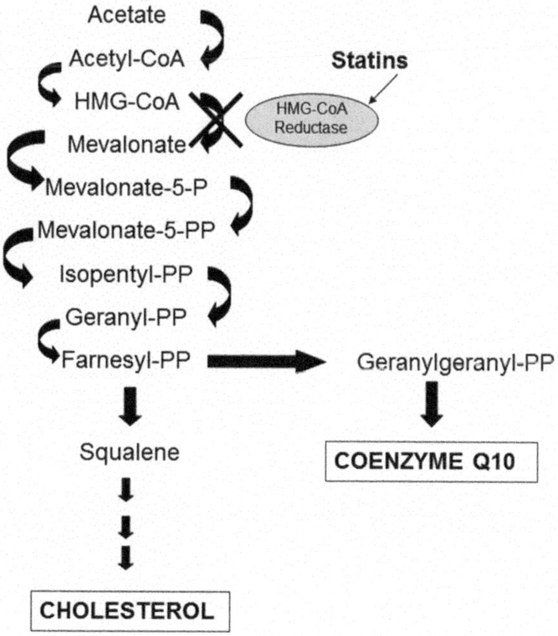

Fig. 17.3 Cholesterol and Coenzyme Q_{10} pathways share Mevalonate intermediate and Statin interruption **3-Hydroxy-3-methylglutaryl-Coenzyme A (HMG-CoA) is changed into Mevalonate by the action of the enzyme HMG-CoA reductase. From here, Mevalonate can be used to synthesize both cholesterol and coenzyme Q_{10}. Statins work by inhibiting the action of HMG-CoA reductase, thereby decreasing the amount of mevalonate available to make either cholesterol or coenzyme Q_{10}**

and cell fractionation in spinach leaves, it was demonstrated that CoQ_{10} is transported from the endoplasmic reticulum to other compartments through a vesicle-mediated process involving the Golgi system (Wanke et al. 2000). This cellular transport system was also found in human cells in culture (Fernández-Ayala et al. 2005a). Interestingly, exogenous CoQ_{10} can enter the cell through plasma membrane and incorporate to cell organelles including mitochondrial inner membrane (Fernández-Ayala et al. 2005b).

17.1.4 Uptake and Distribution of CoQ_{10}

CoQ_{10} is found in many dietary sources including animals and vegetables and can be also obtained from many dietary supplements. Large amounts are present in food from animal sources such as chicken legs, heart, liver and herrings. In comparison with meat and fish, lower levels are found in vegetables probably by the lower amount of mitochondria in comparison with animal cells (Table 17.1). In general, dietary intake of CoQ_{10} has been estimated as 3–5 mg/day. However, this intake is not necessary in situations without endogenous CoQ_{10} synthesis dysfunction in which the quinone reaches a saturation level in cells and tissues (Bhagavan and Chopra 2006).

Table 17.1 Content in coenzyme Q_{10} in food

Food Sources		CoQ_{10} content (μg/g)
Meat	Pork heart	203
	Pork liver	3.1
	Pork ham	20
	Beef heart	41
	Beef liver	19
	Lamb leg	2.9
	Chicken leg	17
Fish	Trout	11
	Sardines	64
	Red mackerel	43–67
	Tuna canned	0.3
Vegetables	Spinach	2.3
	Pea	0.1
	Cauliflower	0.6
Fruits	Orange	2.2
	Strawberry	0.1
	Apple	0.2
Cereals	Bread (rye)	4.7
	Bread (wheat)	2.1

To understand the distribution of CoQ_{10} in tissues after oral ingestion it is necessary to take into consideration its lipophilic nature. The absorption of CoQ_{10} is enhanced in the presence of lipids, then supplementation with CoQ_{10} must be performed with fat-rich meals. Due its biochemical characteristics, CoQ_{10} is absorbed slowly from the small intestine, possibly because it has a high molecular weight and is not very water soluble, passes into the lympha, and finally to the blood bound to chylomicrons and further to tissues, mainly liver. This mechanism is the same than vitamin E used to be incorporated from dietary sources (Zhang et al. 1995). In blood plasma, the reduced form, ubiquinol, is bound to lipoproteins, mostly to LDL (Bhagavan et al. 2007; Zhang et al. 1995). It has been considered that circulating concentrations of CoQ_{10} may be a putative biomarker to indicate their general status in the body and for monitoring the bioavailability of CoQ_{10} supplementation.

Accordingly, the number of different formulations developed to improve the incorporation of CoQ_{10} into human body, the importance of the vehicle and the solubility of CoQ_{10} in the preparation is clear in order to increase its bioavailability (López-Lluch et al. 2019). During the last decade, CoQ_{10} supplements have been developed as oil-based, softgel or powder-filled capsules and hard tablets. Comparisons between studies have indicated that CoQ_{10} bioavailability is influenced by the type of formulation, and that it is better to take CoQ_{10} with fatty foods (Villalba et al. 2010).

17.2 CoQ_{10} and Aging

We can define aging as the normal decline in survival suffering by all organisms along time. Understanding the molecular and cellular mechanisms underlying aging process would permit to develop strategies to resolve the problems associated to the increase of aged population that affects the whole world. Aging events have been studied from different points of view. One of the most accepted theories suggest that aging, is associated with the increase of oxidative damage in cells and tissues that drives aging and the age-related degenerative diseases (Su et al. 2010). In this theory, reactive oxygen species (ROS) are the factors that trigger the deleterious, irreversible changes and macromolecular damage associated with aging (Fig. 17.4) (Miquel 1998; Sohal et al. 2002).

Normal aerobic cell metabolism releases ow amounts of ROS as results of the partial reduction of molecular oxygen. These low ROS levels have beneficial effects maintaining antioxidant machinery. However, when there is a ROS overproduction, the accumulation of these reactive species produces oxidative modifications affecting many cell components including all the organic molecules present in cells (Valko et al. 2006). To avoid oxidative damage, organisms contains antioxidant mechanisms (Fig. 17.4) such as enzymes (superoxide dismutase, catalase, glutathione peroxidase) and small hydrophilic and hydrophobic molecules that act directly as non-enzymatic antioxidants (ascorbic acid, tocopherol, glutathione, CoQ_{10}, and others). Thus, we can define oxidative stress as the damage produced by the

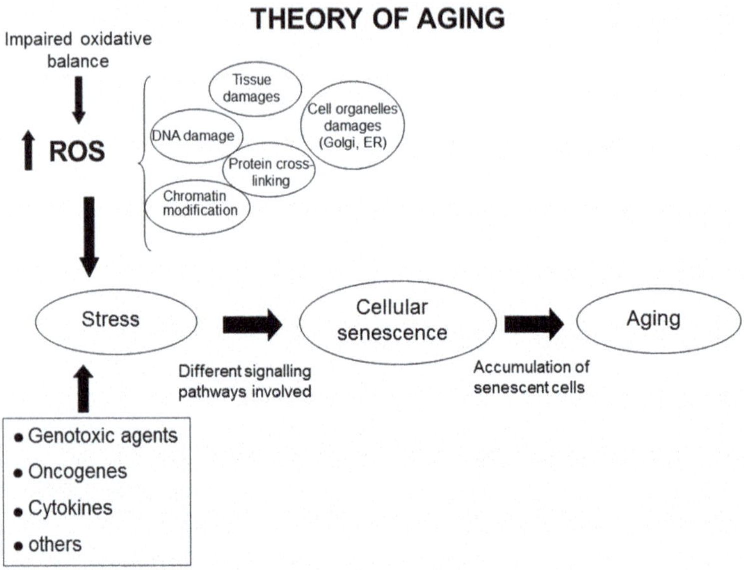

Fig. 17.4 Theory of aging
One of the most accepted theory of aging suggest that aging is produced by the deleterious, irreversible changes and macromolecular damage produced by ROS. Some modifications (mainly those related to DNA) are not completely repaired and thus accumulate, leading to cell death, organism malfunction, and the "aging phenotype"

imbalance between ROS production rate and the capacity to eliminate ROS by and antioxidant defences, in favour of ROS. This imbalance seems to be a hallmark of aging since oxidative stress has been associated with many age-associated diseases such as chronic-degenerative disease, such as cancer, metabolic and disease cardiovascular diseases.

Oxidative stress causes some of the modifications that cannot be completely repaired by antioxidant and cell turnover mechanisms and thus accumulate. The accumulation of oxidized structures in cells leads to cell senescence, cell death, organism malfunction, and the "aging phenotype." Nowadays, a version of the free-radical theory to explain aging, related with mitochondria as the main source and, at the same time target for ROS-dependent damage, is one of the most popular theories of aging (Barja 2007; Miquel et al. 1980). This theory postulates that mitochondrial DNA (mtDNA) suffer higher oxidative damage as the organism ages and this leads to the accumulation of mtDNA. This accumulation produces a vicious cycle in which an initial ROS-induced mtDNA damage increases oxidants production that, in turn, leads to more mitochondrial damage that produces more mtDNA damage (Gilmer et al. 2010).

According to this concept, mitochondria and mitochondrial ROS would play an important role in the development of strategies to delay and improve the aging process. These strategies must be focused on extending lifespan and/or retarding age-associated biological changes, including age-related diseases (Lee et al. 2004).

Among these strategies, nutritional and pharmacological interventions studied in several model organisms, including *yeast, flies,* mice and rats, as well as monkeys. Accordingly, with this strategy, some antioxidants have proved to be useful as dietary antiaging therapies (Duntas 2011; Lopez-Dominguez et al. 2012).

17.2.1 Coenzyme Q_{10} Deficiency in Aging

As it has been indicated before, all cells in the organism synthesize CoQ_{10} (Schultz and Clarke 1999). Along human life, CoQ_{10} increases until 20 years; however, it seems that the organism begin to lose its ability to synthesise CoQ_{10} during maturity and aging when and the coenzyme becomes deficient (Blatt and Littarru 2011; Gutierrez-Mariscal et al. 2011; Ochoa et al. 2007). Besides a decrease in biosynthetic capacity, other factors or situations may affect the levels of CoQ_{10}, including an increase in its degradation (Nakamura et al. 1999) or changes in membrane composition as occurs in different age-related diseases (Kagan and Quinn 1996). However, it is difficult to determine the importance of the changes in CoQ_{10} levels during aging since they are tissue- and organ-dependent. It has been shown that levels of CoQ_{10} in mitochondria of old rat brain increase (Battino et al. 1997) whereas they decrease in skeletal muscle (Lass et al. 1999). These differences make very complex the study of the importance of CoQ_{10} in aging process and further research is needed in order to clarify the importance of CoQ_{10} in aging progression.

Furthermore, dietary supplementation with CoQ_{10} does not affects all the organs. In young and healthy rodents, dietary CoQ_{10} is easily incorporated into liver and spleen; however, in older animals supplemental CoQ_{10} seems to restore normal levels (Beal 1999; Rosenfeldt et al. 1999).

This decrease in CoQ_{10} during aging has been related to a higher oxidative stress associated with aging and its related diseases. Thus, oral CoQ_{10} supplementation could be an effective antioxidant strategy to many age-associated diseases such as neurodegenerative disorders, diabetes, and cancer, muscular and cardiovascular diseases in which oxidative stress is an important factor.

17.3 Therapeutic Uses of CoQ_{10} in Age-Related Diseases

The fundamental role of CoQ_{10} in mitochondria, bioenergetics and antioxidant protection is the base of the therapeutic importance of CoQ_{10} supplementation. The studies performed in animals demonstrate that large doses of CoQ_{10} can reach all tissues and subcellular components including heart and brain mitochondria. This capacity has implications in therapies for many human diseases in which oxidative stress is a main factor. Several evidence have been recorded about the beneficial effects in cardiovascular, neurodegenerative and many other aged-related diseases

(Bhagavan and Chopra 2006; Villalba et al. 2010; Gonzalez-Guardia et al. 2015; Gutierrez-Mariscal et al. 2012; Gutierrez-Mariscal et al. 2014; Yubero-Serrano et al. 2013).

17.3.1 CoQ_{10} and Cardiovascular Disease

Cardiovascular disease (CVD) is one of the major causes of death and disability worldwide. We can suspect that the importance of disease will increase with the increase of elderly and the higher levels of obesity and sedentary lifestyles. Today, around 17 million deaths per year are associated to CVD (Flowers et al. 2014).

One of the main priorities in public health systems is to design a strategy to prevent CVD by modifying lifestyle. In this strategy, diet plays an important role. Several dietary factors have been related with a rise in the risk to suffer CVD, such as a low consumption of fruit and vegetables, a high intake of saturated fat and salt (Eilat-Adar et al. 2013). In the pathogenesis of CVD, oxidative stress plays a central role.

Oxidative stress has been also associated with congestive heart failure, hypertension and ischemic heart disease. The high-energy requirements and high mitochondria amount in heart muscle cells is the main cause of the high levels of CoQ_{10} found in this cell type. In samples from human heart it has been detected a significant decrease of the CoQ_{10} content in cardiomyopathies. This deficiency showed a direct correlation with the severity of disease (Folkers et al. 1985; Nobuyoshi et al. 1984). A recent meta-analysis demonstrated that CoQ_{10} supplementation in the clinical treatment of CVD shoes improvement of congestive heart failure, indicated by best left ventricular ejection fraction (LVEF). Further, the New York Heart Association classification (NYHA), showed that subjects treated with CoQ_{10} supplements improved in the ejection fraction in comparison with controls (placebo) (Fotino et al. 2013). Moreover, a prospective, randomized, double-blind, placebo-controlled, multicentre trial in which CoQ_{10} (Q-SYMBIO) is used as an adjunctive treatment of chronic heart failure has demonstrated that the treatment with this quinone is safe, well tolerated, and associated with a reduction in general symptoms (Mortensen et al. 2014). These clinical results could be based on the important molecular functions of CoQ_{10}, as integral component of mitochondrial respiratory chain (Littarru and Tiano 2010), and the only lipid-soluble antioxidant that slows lipid peroxidation in the circulation (Littarru and Tiano 2007).

CVD in elderly is accompanied with other complications such as diabetes and hypercholesterolemia. Certain drugs used in hypercholesterolemic disease can cause depletion of CoQ_{10} in particular statins that affect the first enzyme (hydroxylmethylglutaryl-coenzyme A reductase; HMG-CoA reductase) in cholesterol and CoQ_{10} synthesis pathway (Schaars and Stalenhoef 2008) (Fig. 17.3). Statins are widely prescribed to reduce cholesterol levels by inhibiting HMG-CoA reductase (Folkers et al. 1990). Chronic statin treatment can reduce endogenous-synthesized cholesterol levels but, at the same time, they also lower CoQ_{10} levels. This can be the reason why chronic statin treatment is associated with muscle-related symptoms, pain or myopathies that can be improved with CoQ_{10} supplementation (Caso et al. 2007).

In relationship with the treatment of heart diseases several studies have concluded that supplementation with CoQ_{10} (50–300 mg/day) can be the safe and optimal dose, although higher doses such as 1200 mg/day have been also safely used (Gao et al. 2011). The majority of these clinical studies indicate that the treatment with CoQ_{10} significantly improve the heart muscle function, increasing ATP synthesis and enhancing myocardial contractility (Folkers et al. 1985). Importantly, these treatments have demonstrated no adverse effects or drug interactions (Kaikkonen et al. 2002).

17.3.2 CoQ_{10} and Hypertension

Hypertension is also associated with aging. Hypertension is also a key risk factor for stroke, myocardial infarction, congestive heart failure, kidney failure, and peripheral vascular disease. Although many pharmacological treatment have shown efficacy lowering blood pressure and modestly decrease stroke, myocardial infarction, and mortality, hypertension remains with high prevalence especially in old population and additional treatments are needed (Musini et al. 2009). The health effects of CoQ_{10} as additional treatment have been investigated in several controlled intervention studies in human subjects in a range of CoQ_{10} doses from 100 mg to 200 mg/day (Young et al. 2011; Yubero-Serrano et al. 2011, 2013). CoQ_{10} affects vasodilatation by improving endothelium and vascular smooth muscle activity, counteracting vasoconstriction and lowering blood pressure. Among treated patients, it has been reported a decrease in systolic blood pressure ranged from 11–17 mmHg and 8 mmHg decrease in diastolic blood pressure after the treatment with CoQ_{10}. These results indicate a putative role of CoQ_{10} as a hypotensive agent and probably a safe adjuvant in the treatment with conventional anti-hypertensive pharmacological products.

17.3.3 CoQ_{10} and Endothelial Function

The progression and clinical manifestations of atherosclerosis and cardiovascular diseases depends on the dysfunction of endothelium. Several studies have determined the effect of oral CoQ_{10} supplementation on the physiology of endothelium in patients suffering coronary artery disease or diabetes mellitus or in elderly people (Gao et al. 2011; Tiano et al. 2007; Watts et al. 2002). In most of the individuals treated with CoQ_{10}, endothelial function, determined by flow-mediated dilation (FMD) or by nitro-glycerine-mediated dilation (NMD) and the activity of extracellular superoxide dismutase, improved. This effect has been associated with the antioxidant and anti-inflammatory activity of CoQ_{10} (Yubero-Serrano et al. 2012). CoQ_{10} treatment decreases the rate of production of peroxynitrite from nitric oxide (NO) that reacts with superoxide radicals. Likely, under conditions of oxidative stress, CoQ_{10} can reduce the levels of superoxide radicals (Yubero-Serrano et al. 2011, 2013; Tiano et al. 2007). Furthermore, *in vitro* studies have demonstrated that CoQ_{10} can efficiently prevent apoptosis of endothelial cells produced by high glucose and the adhesion to monocytes. This effect is very important in the prevention of the development of atherosclerosis (Tsuneki et al. 2007). Further, the inhibition of the LDL-oxidation by ubiquinol adds an important factor in the prevention of atherogenesis (Thomas et al. 1997).

17.3.4 CoQ_{10} and Renal Failure

Chronic kidney (CKD) and end-stage renal (ESRD) diseases are also associated with oxidative stress (Himmelfarb and Hakim 2003). Five hundred thousand patients in the United States receive maintenance haemodialysis for ESRD, with life expectancies less than 17–34% than those of the general population (US Renal Data System 2013). In these patients, the balance between ROS and antioxidants is disturbed and oxidative stress is produced. The high ratio of mortality has been attributed to an increased risk of cardiovascular disease produced by this high oxidative stress (Kuchta et al. 2011). The putative role of CoQ_{10} in CKD patients has been studied only in a few studies. Lippa et al. (1994) determined the levels of CoQ_{10} in 48 patients under chronic haemodialysis, in comparison with 15 uremic patients and a control group of healthy subjects (Lippa et al. 2000). In this study, CoQ_{10} levels were significantly lower in CKD patients. In a recent study, the levels of CoQ_{10} and oxidative stress biomarkers were determined in CKD, haemodialysis and peritoneal dialysis (PeD) patients. Contrary to the study of Lippa et al. (1994), in this study, researchers did not find differences in CoQ_{10} levels between those of CKD and haemodialysis groups. However, they did observe higher levels of members of the antioxidant system in patients undergoing PeD in comparison with CKD patients (Gokbel et al. 2011).

In other study, supplementation with CoQ_{10} (120 mg/day) to patients with CKD, reduced the number of patients on dialysis in comparison with placebo after 28 days of treatment (Singh et al. 2000). As in other cases, the tolerability and safety of oral CoQ_{10} administration was determined until doses as higher as 1800 mg/day in CKD patients (Yeung et al. 2015). Authors conclude that in these patients, CoQ_{10} supplementation is important since it reduces systemic oxidative stress and improves mitochondrial function in patients receiving haemodialysis (Yeung et al. 2015).

17.3.5 *CoQ_{10} and Neurodegenerative Diseases*

Mitochondrial dysfunction is a common characteristic of the neurodegenerative diseases. This dysfunction is accompanied by abnormal energy metabolism and higher oxidative stress. As we have previously described, CoQ_{10} levels in the brain and other tissues in humans and animals have been shown to decline with age. As an antioxidant molecule, CoQ_{10} has been involved in the development of neurodegenerative diseases such as Parkinson's disease (PD), Huntington's disease (HD) and other neurodegenerative disorders (Koroshetz et al. 1997; Shults et al. 1997). There is, therefore, a robust scientific rationale for testing this agent in neuroprotective therapies.

CoQ_{10} has shown protective effects in the nigrostriatal dopaminergic system in many preclinical studies of PD (Liu et al. 2011). Mitochondrial dysfunction has been strongly associated with PD and, probably for this reason, CoQ_{10} levels are significantly lower in mitochondria from PD patients. A randomised, placebo controlled and double-blind study performed in these patients demonstrated that a CoQ_{10} dose as high as 1200 mg/day, was safe and reduced the worsening of PD (Shults et al. 2002). Treatment with CoQ_{10} was accompanied by significant increases in plasma levels of CoQ_{10} and a higher NADH-cytochrome C reductase activity in white blood cells (Shults et al. 2002). Everyday activities of the patients, such as dressing, bathing and feeding, were significantly improved with the treatment with CoQ_{10}. Accordingly with this study, another placebo-controlled, double-blind trial showed that CoQ_{10} supplementation provided a mild but significant benefit on PD patients in comparison with (Muller et al. 2003).

In the case of HD, strong evidence indicate the existence of early oxidative stress. As in the other cases, this oxidative stress is coupled with mitochondrial dysfunction and the impairment of energy metabolism in which the deficiency in CoQ_{10} can be an important factor (Stack et al. 2008). In HD patients, CoQ_{10} doses, ranging from 600 to 1200 mg/day were tested during six-months in an open-label trial. Although no significant effect on clinical scores were found, the treatment with CoQ_{10} significantly decreased the levels of cortical lactate concentrations. This decrease was reversed following withdrawal of therapy indicating a protective effect of CoQ_{10} supplementation (Delanty and Dichter 1998). Again, the effect of CoQ_{10} in the behaviour of metabolic markers indicate the bioenergetics effect of oral CoQ_{10} in the mitochondrial metabolism of brain (Koroshetz et al. 1997).

17.4 Conclusions

The importance of CoQ_{10} is because this quinone is not just an agent in the transit of electrons in energy transduction in mitochondria. CoQ_{10} is a strong antioxidant able to regenerate the redox capacity in many tissues and organs. In normal conditions, its biosynthesis in mitochondria and endoplasmic reticulum provides sufficient CoQ_{10}, but in some conditions such as genetic failure or aging and age-related diseases, CoQ_{10} deficiency can be an important factor in the progression of incapacity or disease. A number of studies have demonstrated that CoQ_{10} can be easily and safely used as nutritional supplement to delay and mitigate the effects caused by its depletion.

The clear beneficial effects of CoQ_{10} are reinforced because its excellent safety record. CoQ_{10} is very well tolerated even at high doses and for prolonged periods with null or very limited side effects. However, it is necessary to increase the research to determine the appropriate dose, effectiveness, and to increase the bioavailability of orally administered CoQ_{10}, specially, in the elderly. Further, a promising strategy can be centred in the design of therapeutic agents that increment the endogenous synthesis of CoQ_{10}.

References

Acosta MJ, Vazquez Fonseca L, Desbats MA, Cerqua C, Zordan R, Trevisson E, Salviati L (2016) Coenzyme Q biosynthesis in health and disease. Biochim Biophys Acta 1857(8):1079–1085. https://doi.org/10.1016/j.bbabio.2016.03.036

Alcazar-Fabra M, Navas P, Brea-Calvo G (2016) Coenzyme Q biosynthesis and its role in the respiratory chain structure. Biochim Biophys Acta 1857(8):1073–1078. https://doi.org/10.1016/j.bbabio.2016.03.010

Barja G (2007) Mitochondrial oxygen consumption and reactive oxygen species production are independently modulated: implications for aging studies. Rejuvenation Res 10(2):215–224

Battino M, Svegliati Baroni S, Littarru GP, Bompadre S, Leone L, Gorini A, Villa RF (1997) Coenzyme Q homologs and vitamin E in synaptic and non-synaptic occipital cerebral cortex mitochondria in the ageing rat. Mol Asp Med 18(Suppl):S279–S282

Beal MF (1999) Coenzyme Q10 administration and its potential for treatment of neurodegenerative diseases. BioFactors (Oxford, England) 9(2–4):261–266

Bhagavan HN, Chopra RK (2006) Coenzyme Q10: absorption, tissue uptake, metabolism and pharmacokinetics. Free Radic Res 40(5):445–453

Bhagavan HN, Chopra RK, Craft NE, Chitchumroonchokchai C, Failla ML (2007) Assessment of coenzyme Q10 absorption using an in vitro digestion-Caco-2 cell model. Int J Pharm 333(1–2):112–117

Blatt T, Littarru GP (2011) Biochemical rationale and experimental data on the antiaging properties of CoQ(10) at skin level. BioFactors (Oxford, England) 37(5):381–385

Caso G, Kelly P, McNurlan MA, Lawson WE (2007) Effect of coenzyme q10 on myopathic symptoms in patients treated with statins. Am J Cardiol 99(10):1409–1412. https://doi.org/10.1016/j.amjcard.2006.12.063

Crane FL (2001) Biochemical functions of coenzyme Q10. J Am Coll Nutr 20(6):591–598

Crane FL, Hatefi Y, Lester RL, Widmer C (1957) Isolation of a quinone from beef heart mitochondria. Biochim Biophys Acta 25(1):220–221

Crane FL, Hatefi Y, Lester RL, Widmer C (1989) Isolation of a quinone from beef heart mitochondria. 1957. Biochim Biophys Acta 1000:362–363

Delanty N, Dichter MA (1998) Oxidative injury in the nervous system. Acta Neurol Scand 98(3):145–153

Duntas LH (2011) Resveratrol and its impact on aging and thyroid function. J Endocrinol Investig 34(10):788–792

Eilat-Adar S, Sinai T, Yosefy C, Henkin Y (2013) Nutritional recommendations for cardiovascular disease prevention. Nutrients 5(9):3646–3683. https://doi.org/10.3390/nu5093646

Ernster L, Dallner G (1995) Biochemical, physiological and medical aspects of ubiquinone function. Biochim Biophys Acta 1271(1):195–204

Fernández-Ayala DJ, López-Lluch G, García-Valdés M, Arroyo A, Navas P (2005a) Specificity of coenzyme Q10 for a balanced function of respiratory chain and endogenous ubiquinone biosynthesis in human cells. Biochim Biophys Acta 1706(1-2):174–83. https://doi.org/10.1016/j.bbabio.2004.10.009

Fernández-Ayala DJ, Brea-Calvo G, López-Lluch G, Navas P (2005b) Coenzyme Q distribution in HL-60 human cells depends on the endomembrane system. Biochim Biophys Acta 1713(2):129–37. https://doi.org/10.1016/j.bbamem.2005.05.010

Festenstein GN, Heaton FW, Lowe JS, Morton RA (1955) A constituent of the unsaponifiable portion of animal tissue lipids (lambda max. 272 m mu). Biochem J 59(4):558–566

Flowers N, Hartley L, Todkill D, Stranges S, Rees K (2014) Co-enzyme Q10 supplementation for the primary prevention of cardiovascular disease. Cochrane Database Syst Rev 12:CD010405. https://doi.org/10.1002/14651858.CD010405.pub2

Folkers K, Vadhanavikit S, Mortensen SA (1985) Biochemical rationale and myocardial tissue data on the effective therapy of cardiomyopathy with coenzyme Q10. Proc Natl Acad Sci U S A 82(3):901–904

Folkers K, Langsjoen P, Willis R, Richardson P, Xia LJ, Ye CQ, Tamagawa H (1990) Lovastatin decreases coenzyme Q levels in humans. Proc Natl Acad Sci U S A 87(22):8931–8934

Forsmark-Andree P, Dallner G, Ernster L (1995) Endogenous ubiquinol prevents protein modification accompanying lipid peroxidation in beef heart submitochondrial particles. Free Radic Biol Med 19(6):749–757

Fotino AD, Thompson-Paul AM, Bazzano LA (2012) Effect of coenzyme Q10 supplementation on heart failure: a meta-analysis. Am J Clin Nutr

Fotino AD, Thompson-Paul AM, Bazzano LA (2013) Effect of coenzyme Q(1)(0) supplementation on heart failure: a meta-analysis. Am J Clin Nutr 97(2):268–275. https://doi.org/10.3945/ajcn.112.040741

Gao L, Mao Q, Cao J, Wang Y, Zhou X, Fan L (2011) Effects of coenzyme Q10 on vascular endothelial function in humans: a meta-analysis of randomized controlled trials. Atherosclerosis 221(2):311–316

Gilmer LK, Ansari MA, Roberts KN, Scheff SW (2010) Age-related changes in mitochondrial respiration and oxidative damage in the cerebral cortex of the Fischer 344 rat. Mech Ageing Dev 131(2):133–143

Gokbel H, Atalay H, Okudan N, Solak Y, Belviranli M, Turk S (2011) Coenzyme Q10 and its relation with oxidant and antioxidant system markers in patients with end-stage renal disease. Ren Fail 33(7):677–681

Gonzalez-Guardia L, Yubero-Serrano EM, Delgado-Lista J, Perez-Martinez P, Garcia-Rios A, Marin C, Camargo A, Delgado-Casado N, Roche HM, Perez-Jimenez F, Brennan L, Lopez-Miranda J (2015) Effects of the Mediterranean diet supplemented with coenzyme q10 on metabolomic profiles in elderly men and women. J Gerontol 70(1):78–84. https://doi.org/10.1093/gerona/glu098

Gutierrez-Mariscal FM, Perez-Martinez P, Delgado-Lista J, Yubero-Serrano EM, Camargo A, Delgado-Casado N, Cruz-Teno C, Santos-Gonzalez M, Rodriguez-Cantalejo F, Castano JP, Villalba-Montoro JM, Fuentes F, Perez-Jimenez F, Lopez-Miranda J (2011) Mediterranean diet supplemented with coenzyme Q10 induces postprandial changes in p53 in response to oxidative DNA damage in elderly subjects. Age (Dordr) 34(2):389–403

Gutierrez-Mariscal FM, Perez-Martinez P, Delgado-Lista J, Yubero-Serrano EM, Camargo A, Delgado-Casado N, Cruz-Teno C, Santos-Gonzalez M, Rodriguez-Cantalejo F, Castano JP, Villalba-Montoro JM, Fuentes F, Perez-Jimenez F, Lopez-Miranda J (2012) Mediterranean diet supplemented with coenzyme Q10 induces postprandial changes in p53 in response to oxidative DNA damage in elderly subjects. Age (Dordr) 34(2):389–403. https://doi.org/10.1007/s11357-011-9229-1

Gutierrez-Mariscal FM, Yubero-Serrano EM, Rangel-Zuniga OA, Marin C, Garcia-Rios A, Perez-Martinez P, Delgado-Lista J, Malagon MM, Tinahones FJ, Perez-Jimenez F, Lopez-Miranda J (2014) Postprandial activation of p53-dependent DNA repair is modified by Mediterranean diet supplemented with coenzyme Q10 in elderly subjects. J Gerontol 69(7):886–893. https://doi.org/10.1093/gerona/glt174

Himmelfarb J, Hakim RM (2003) Oxidative stress in uremia. Curr Opin Nephrol Hypertens 12(6):593–598

Jeya M, Moon HJ, Lee JL, Kim IW, Lee JK (2010) Current state of coenzyme Q(10) production and its applications. Appl Microbiol Biotechnol 85(6):1653–1663. https://doi.org/10.1007/s00253-009-2380-2

Kagan VENH, Quinn JP (1996) Coenzyme Q: its role in scavenging and generation of radicals in membranes. In: Cardenas E, Packer LM (eds) Handbook of antioxidants. M Dekker, New York, pp 157–201

Kaikkonen J, Tuomainen TP, Nyyssonen K, Salonen JT (2002) Coenzyme Q10: absorption, antioxidative properties, determinants, and plasma levels. Free Radic Res 36(4):389–397

Koroshetz WJ, Jenkins BG, Rosen BR, Beal MF (1997) Energy metabolism defects in Huntington's disease and effects of coenzyme Q10. Ann Neurol 41(2):160–165

Kuchta A, Pacanis A, Kortas-Stempak B, Cwiklinska A, Zietkiewicz M, Renke M, Rutkowski B (2011) Estimation of oxidative stress markers in chronic kidney disease. Kidney Blood Press Res 34(1):12–19. https://doi.org/10.1159/000321508

Lass A, Kwong L, Sohal RS (1999) Mitochondrial coenzyme Q content and aging. BioFactors (Oxford, England) 9(2–4):199–205

Lee CK, Pugh TD, Klopp RG, Edwards J, Allison DB, Weindruch R, Prolla TA (2004) The impact of α-lipoic acid, coenzyme Q10, and caloric restriction on life span and gene expression patterns in mice. Free Radic Biol Med 36:1043–1057

Lippa S, Colacicco L, Bondanini F, Calla C, Gozzo ML, Ciccariello M, Angelitti AG (2000) Plasma levels of coenzyme Q(10), vitamin E and lipids in uremic patients on conservative therapy and hemodialysis treatment: some possible biochemical and clinical implications. Clin Chim Acta Int J Clin Chem 292(1–2):81–91

Lippa S, Colacicco L, Callà C, Sagliaschi G, Angelitti AG (1994) Coenzyme Q10 levels, plasma lipids and peroxidation extent in renal failure and in hemodialytic patients. Mol Aspects Med 15(Suppl):s213–9. https://doi.org/10.1016/0098-2997(94)90031-0

Littarru GP, Tiano L (2007) Bioenergetic and antioxidant properties of coenzyme Q10: recent developments. Mol Biotechnol 37(1):31–37

Littarru GP, Tiano L (2010) Clinical aspects of coenzyme Q10: an update. Nutrition 26(3):250–254. https://doi.org/10.1016/j.nut.2009.08.008

Liu J, Wang L, Zhan SY, Xia Y (2011) Coenzyme Q10 for Parkinson's disease. Cochrane Database Syst Rev 12:CD008150. https://doi.org/10.1002/14651858.CD008150.pub2

Lopez-Dominguez JA, Khraiwesh H, Gonzalez-Reyes JA, Lopez-Lluch G, Navas P, Ramsey JJ, de Cabo R, Buron MI, Villalba JM (2012) Dietary fat modifies mitochondrial and plasma membrane apoptotic signaling in skeletal muscle of calorie-restricted mice. Age (Dordr)

Lopez-Lluch G, Rodriguez-Aguilera JC, Santos-Ocana C, Navas P (2010) Is coenzyme Q a key factor in aging? Mech Ageing Dev 131(4):225–235. https://doi.org/10.1016/j.mad.2010.02.003

López-Lluch G, Del Pozo-Cruz J, Sánchez-Cuesta A, Cortés-Rodríguez AB, Navas P (2019) Bioavailability of coenzyme Q10 supplements depends on carrier lipids and solubilization. 57:133–140. https://doi.org/10.1016/j.nut.2018.05.020. Epub 2018 Jun 27

Miquel J (1998) An update on the oxygen stress-mitochondrial mutation theory of aging: genetic and evolutionary implications. Exp Gerontol 33(1–2):113–126

Miquel J, Economos AC, Fleming J, Johnson JE Jr (1980) Mitochondrial role in cell aging. Exp Gerontol 15(6):575–591

Mortensen SA, Rosenfeldt F, Kumar A, Dolliner P, Filipiak KJ, Pella D, Alehagen U, Steurer G, Littarru GP, Investigators QSS (2014) The effect of coenzyme Q10 on morbidity and mortality in chronic heart failure: results from Q-SYMBIO: a randomized double-blind trial. JACC Heart Fail 2(6):641–649. https://doi.org/10.1016/j.jchf.2014.06.008

Muller T, Buttner T, Gholipour AF, Kuhn W (2003) Coenzyme Q10 supplementation provides mild symptomatic benefit in patients with Parkinson's disease. Neurosci Lett 341(3):201–204

Musini VM, Tejani AM, Bassett K, Wright JM (2009) Pharmacotherapy for hypertension in the elderly. Cochrane Database Syst Rev 4:CD000028. https://doi.org/10.1002/14651858.CD000028.pub2

Nakamura T, Ohno T, Hamamura K, Sato T (1999) Metabolism of coenzyme Q10: biliary and urinary excretion study in Guinea pigs. BioFactors (Oxford, England) 9(2–4):111–119

Navas P, Villalba JM, de Cabo R (2007) The importance of plasma membrane coenzyme Q in aging and stress responses. Mitochondrion 7(Suppl):S34–S40. https://doi.org/10.1016/j.mito.2007.02.010

Nobuyoshi M, Saito T, Takahira H, Yamano Y, Kanazawa T (1984) Levels of coenzyme Q10 in biopsies of left ventricular muscle and influence of administration of coenzyme Q10. In: Folkers K, Yamamura Y (eds) Biomedical and clinical aspects of coenzyme Q, vol 4. Elsevier, Amsterdam, pp 221–229

Ochoa JJ, Quiles JL, Lopez-Frias M, Huertas JR, Mataix J (2007) Effect of lifelong coenzyme Q10 supplementation on age-related oxidative stress and mitochondrial function in liver and skeletal muscle of rats fed on a polyunsaturated fatty acid (PUFA)-rich diet. J Gerontol 62(11):1211–1218

Overvad K, Diamant B, Holm L, Holmer G, Mortensen SA, Stender S (1999) Coenzyme Q10 in health and disease. Eur J Clin Nutr 53(10):764–770

Potgieter M, Pretorius E, Pepper MS (2013) Primary and secondary coenzyme Q10 deficiency: the role of therapeutic supplementation. Nutr Rev 71(3):180–188. https://doi.org/10.1111/nure.12011

Quinzii CM, DiMauro S, Hirano M (2007) Human coenzyme Q10 deficiency. Neurochem Res 32(4–5):723–727

Rosenfeldt FL, Pepe S, Ou R, Mariani JA, Rowland MA, Nagley P, Linnane AW (1999) Coenzyme Q10 improves the tolerance of the senescent myocardium to aerobic and ischemic stress: studies in rats and in human atrial tissue. BioFactors (Oxford, England) 9(2–4):291–299

Schaars CF, Stalenhoef AF (2008) Effects of ubiquinone (coenzyme Q10) on myopathy in statin users. Curr Opin Lipidol 19(6):553–557. https://doi.org/10.1097/MOL.0b013e3283168ecd

Schultz JR, Clarke CF (1999) In: Cardenas E, Packer L (eds) Mitochondria, oxidants and ageing. M Dekker, New York, pp 95–118

Shetty RA, Forster MJ, Sumien N (2012) Coenzyme Q(10) supplementation reverses age-related impairments in spatial learning and lowers protein oxidation. Age (Dordr)

Shults CW, Haas RH, Passov D, Beal MF (1997) Coenzyme Q10 levels correlate with the activities of complexes I and II/III in mitochondria from parkinsonian and nonparkinsonian subjects. Ann Neurol 42(2):261–264

Shults CW, Oakes D, Kieburtz K, Beal MF, Haas R, Plumb S, Juncos JL, Nutt J, Shoulson I, Carter J, Kompoliti K, Perlmutter JS, Reich S, Stern M, Watts RL, Kurlan R, Molho E, Harrison M, Lew M (2002) Effects of coenzyme Q10 in early Parkinson disease: evidence of slowing of the functional decline. Arch Neurol 59(10):1541–1550

Singh RB, Khanna HK, Niaz MA (2000) Randomized, double-blind, placebo-controlled trial of coenzyme Q10 in chronic renal failure: discovery of a new role. J Nutr Environ Med 10:281–288

Sohal RS, Forster MJ (2007) Coenzyme Q, oxidative stress and aging. Mitochondrion 7(Suppl):S103–S111. https://doi.org/10.1016/j.mito.2007.03.006

Sohal RS, Mockett RJ, Orr WC (2002) Mechanisms of aging: an appraisal of the oxidative stress hypothesis. Free Radic Biol Med 33(5):575–586

Spinazzi M, Radaelli E, Horré K, Arranz AM, Gounko NV, Agostinis P, Maia TM, Impens F, Morais VA, Lopez-Lluch G, Serneels L, Navas P, De Strooper B (2019) PARL deficiency in

mouse causes Complex III defects, coenzyme Q depletion, and Leigh-like syndrome. Proc Natl Acad Sci U S A 116(1):277–286. https://doi.org/10.1073/pnas.1811938116
Stack EC, Matson WR, Ferrante RJ (2008) Evidence of oxidant damage in Huntington's disease: translational strategies using antioxidants. Ann N Y Acad Sci 1147:79–92. https://doi.org/10.1196/annals.1427.008
Su Y, Sun H, Fang J, Hu G, Xiao M (2010) Brain mitochondrial dysfunction in ovariectomized mice injected with D-galactose. Neurochem Res 35(3):399–404
Tiano L, Belardinelli R, Carnevali P, Principi F, Seddaiu G, Littarru GP (2007) Effect of coenzyme Q10 administration on endothelial function and extracellular superoxide dismutase in patients with ischaemic heart disease: a double-blind, randomized controlled study. Eur Heart J 28(18):2249–2255
Thomas SR, Neuzil J, Stocker R (1997) Inhibition of LDL oxidation by ubiquinol-10. A protective mechanism for coenzyme Q in atherogenesis? Mol Aspects Med 18(Suppl):S85–103. https://doi.org/10.1016/S0098-2997(97)00031-9
Tsuneki H, Sekizaki N, Suzuki T, Kobayashi S, Wada T, Okamoto T, Kimura I, Sasaoka T (2007) Coenzyme Q10 prevents high glucose-induced oxidative stress in human umbilical vein endothelial cells. Eur J Pharmacol 566(1–3):1–10. https://doi.org/10.1016/j.ejphar.2007.03.006
Turunen M, Olsson J, Dallner G (2004) Metabolism and function of coenzyme Q. Biochim Biophys Acta 1660(1–2):171–199
US Renal Data System UADRAoE-sRDitUSB, MD (2013) National Institutes of Health, National Institute of Diabetes and Digestive and Kidney Diseases; www.usrds.org
Valko M, Rhodes CJ, Moncol J, Izakovic M, Mazur M (2006) Free radicals, metals and antioxidants in oxidative stress-induced cancer. Chem Biol Interact 160(1):1–40
Villalba JM, Navas P (2000) Plasma membrane redox system in the control of stress-induced apoptosis. Antioxid Redox Signal 2(2):213–230
Villalba JM, Parrado C, Santos-Gonzalez M, Alcain FJ (2010) Therapeutic use of coenzyme Q10 and coenzyme Q10-related compounds and formulations. Expert Opin Investig Drugs 19(4):535–554
Wanke M, Dallner G, Swiezewska E (2000) Subcellular localization of plastoquinone and ubiquinone synthesis in spinach cells. Biochim Biophys Acta 1463(1):188–194
Watts GF, Playford DA, Croft KD, Ward NC, Mori TA, Burke V (2002) Coenzyme Q(10) improves endothelial dysfunction of the brachial artery in type II diabetes mellitus. Diabetologia 45(3):420–426
Yeung CK, Billings FT, Claessens AJ, Roshanravan B, Linke L, Sundell MB, Ahmad S, Shao B, Shen DD, Ikizler TA, Himmelfarb J (2015) Coenzyme Q10 dose-escalation study in hemodialysis patients: safety, tolerability, and effect on oxidative stress. BMC Nephrol 16:183. https://doi.org/10.1186/s12882-015-0178-2
Young JM, Florkowski CM, Molyneux SL, McEwan RG, Frampton CM, Nicholls MG, Scott RS, George PM (2011) A randomized, double-blind, placebo-controlled crossover study of coenzyme Q10 therapy in hypertensive patients with the metabolic syndrome. Am J Hyper 25(2):261–270
Yubero-Serrano EM, Delgado-Casado N, Delgado-Lista J, Perez-Martinez P, Tasset-Cuevas I, Santos-Gonzalez M, Caballero J, Garcia-Rios A, Marin C, Gutierrez-Mariscal FM, Fuentes F, Villalba JM, Tunez I, Perez-Jimenez F, Lopez-Miranda J (2011) Postprandial antioxidant effect of the Mediterranean diet supplemented with coenzyme Q10 in elderly men and women. Age (Dordr) 33(4):579–590. https://doi.org/10.1007/s11357-010-9199-8
Yubero-Serrano EM, Gonzalez-Guardia L, Rangel-Zuniga O, Delgado-Lista J, Gutierrez-Mariscal FM, Perez-Martinez P, Delgado-Casado N, Cruz-Teno C, Tinahones FJ, Villalba JM, Perez-Jimenez F, Lopez-Miranda J (2012) Mediterranean diet supplemented with coenzyme Q10 modifies the expression of proinflammatory and endoplasmic reticulum stress-related genes in elderly men and women. J Gerontol 67(1):3–10. https://doi.org/10.1093/gerona/glr167
Yubero-Serrano EM, Gonzalez-Guardia L, Rangel-Zuniga O, Delgado-Casado N, Delgado-Lista J, Perez-Martinez P, Garcia-Rios A, Caballero J, Marin C, Gutierrez-Mariscal FM, Tinahones FJ,

Villalba JM, Tunez I, Perez-Jimenez F, Lopez-Miranda J (2013) Postprandial antioxidant gene expression is modified by Mediterranean diet supplemented with coenzyme Q(10) in elderly men and women. Age (Dordr) 35(1):159–170. https://doi.org/10.1007/s11357-011-9331-4
Zhang Y, Aberg F, Appelkvist EL, Dallner G, Ernster L (1995) Uptake of dietary coenzyme Q supplement is limited in rats. J Nutr 125(3):446–453

Index

A

Age-related diseases, ix, x, 81, 94, 97, 101, 102, 116, 117, 142, 149, 172, 182, 272, 273, 351, 354–360
Aging, ix, 3–5, 30, 92–103, 114–117, 131–138, 141–151, 157–166, 170–172, 174, 175, 177, 180, 181, 192–212, 228, 231, 234, 252, 254, 255, 270–277, 285, 288, 294, 295, 297, 312–316, 318, 319, 321, 323, 330–332, 336, 350, 351, 353–355, 357, 360
Aging-related diseases, 102, 117
Alzheimer's disease (AD), 242–257
Anti-aging, 18, 172, 176–181, 184, 185, 276, 315, 323
Antioxidants, ix, 4–6, 12, 14, 15, 17, 18, 24, 25, 33, 34, 79–81, 92–94, 96, 97, 99, 100, 118, 135, 137, 138, 145, 148, 151, 164, 176–178, 181–183, 206, 208–210, 212, 229–231, 233–235, 242, 245–248, 250, 255–257, 273, 275–277, 284–290, 292, 293, 297–299, 313–316, 318, 319, 323, 330, 331, 334, 350, 353–356, 358, 359
Apoptosis, 4, 25, 34, 80, 99, 100, 102, 103, 148, 161, 162, 203, 207, 243, 245, 246, 251–255, 271, 273, 322, 358

B

Bioenergetics, ix, 6, 12, 51, 60, 74, 92, 93, 101, 102, 143, 170, 205, 229, 234, 235, 243, 244, 246, 271, 273, 276, 289, 293, 295, 297, 349, 350, 355, 359

C

Caenorhabditis elegans, ix, 4, 18, 20, 131–138, 157–159, 162–164, 315, 317, 318, 323
Caloric restriction (CR), 81, 270, 275, 276, 312, 313, 317–323
Channelling, 55, 59–64, 67, 69–71, 75
Clk-1, 4, 132–138, 157–166, 315
Coenzyme Q (CoQ), ix, 4, 5, 13, 24–28, 51–77, 92, 95, 117, 118, 120, 131–138, 141–151, 157–166, 170, 192–212, 249, 313, 318, 323, 330
Coenzyme Q_{10} (CoQ_{10}), ix, x, 92–103, 113–121, 170–185, 227–235, 270–278
$CoQ_{10}H_2$, 170–181, 183, 184, 231, 233
Cytochrome b5 reductase, 95, 350

D

Diets, 31, 32, 100, 133, 135, 143, 148, 151, 171, 175, 180, 182, 183, 200, 208–212, 273, 275, 285, 312, 314–316, 319–323, 335, 338, 356
Drosophila melanogaster, ix, 141–151, 274

E

Electron transport, ix, 12, 15, 16, 18, 33, 53, 65, 92, 94, 96, 100, 101, 142, 143, 145, 149, 151, 183, 193, 195, 234, 244, 246, 271, 315, 320, 348, 349
Exercise, 5, 116, 144, 183, 229, 270, 272, 276, 297, 317, 318, 332, 333

G. López Lluch (ed.), *Coenzyme Q in Aging*,
https://doi.org/10.1007/978-3-030-45642-9

F

Fertility, ix, 5, 133–135, 149, 204, 284–299

H

Human diseases, 142, 181, 194, 355
HuR, 25, 114–116, 118

I

Immunosenescence, ix, 116, 117, 270–278
Insulin resistance, 147, 148, 211, 229, 234

L

Lifespans, 29, 32, 34, 101, 142–151, 157–159, 162–166, 177, 198, 201, 205, 209, 270, 312, 315, 316, 320, 321, 323, 354
Lipid induction, 337, 339
Lipids, 4, 13–15, 18, 28, 30, 32, 33, 35, 52–55, 59, 60, 62, 69, 74, 75, 79, 80, 92–94, 97–99, 117, 118, 121, 143, 165, 171, 173, 176, 177, 181–183, 208, 211, 212, 229, 234, 244, 247, 252, 271, 275, 285–288, 293, 313, 318–322, 330–332, 335, 337–340, 353, 356
Longevity, ix, 4, 18, 97, 118, 131–136, 138, 143, 145–147, 149, 150, 157, 159, 199, 208, 211, 312–323

M

Metabolic syndrome, ix, 212, 227–235
Metabolism, 5, 12, 23, 24, 29, 34, 75, 92, 94, 98, 101, 114, 121, 132, 137, 138, 141–144, 146, 148, 149, 192, 193, 202, 211, 235, 252, 272, 284, 286–288, 293, 312–314, 316, 318, 330, 353, 359
Mice, ix, 4, 5, 29, 31–33, 52, 74, 75, 102, 117, 120, 121, 147, 157–166, 170–185, 199–201, 203–205, 231, 235, 244–248, 252–255, 257, 273, 275, 276, 296, 313, 314, 317–323, 331, 333–335, 338–340, 355
Mitochondria, ix, 4, 12, 15, 16, 18, 19, 26, 27, 29, 30, 32–35, 52, 55, 57, 59, 60, 62, 63, 65, 66, 69–74, 79–81, 92–98, 101, 117, 119–121, 134–138, 142, 143, 147, 158, 161, 164, 170, 177–181, 183, 192, 193, 195–199, 201–206, 208–212, 234, 235, 242–244, 246, 248, 252–255, 271–273, 275, 276, 288, 295, 313, 315, 317, 318, 322, 332, 333, 339, 348–352, 354–356, 359
Mitochondrial diseases, 5, 30, 66, 142, 205, 206, 273
Mitochondrial functions, 5, 32, 101, 102, 118, 132, 143, 147, 161, 163–166, 170–185, 192, 194, 201, 203, 206–209, 235, 246, 249, 271–273, 293, 296, 315, 330, 358
mtDNA mutations, 192, 194–196, 198–206, 211, 271

N

NAD(P)H:quinone oxidoreductase 1 (NQO1), 4, 79, 92, 93, 95–97, 99, 102, 103, 170, 233
Neurodegeneration, 5, 98, 148, 243–244, 247, 257
Nicotinamide adenine dinucleotide (NAD), 4, 34, 68, 70–72, 79, 92, 95, 97, 98, 100, 101, 103, 144, 170, 181
Nuclear receptors, 334, 335
Nutrition, 207, 208, 312

O

Oocytes, 284, 285, 288, 294–297
Oxidative stress, 17, 18, 25, 29, 34, 74, 95, 98–100, 103, 118, 135, 138, 145, 148, 171, 176, 177, 184, 192, 199, 203, 207–212, 229–231, 234, 235, 242–248, 250–255, 257, 272, 277, 285–290, 292–294, 298, 299, 312, 313, 315, 316, 319, 353–356, 358, 359

P

Parkinson's disease (PD), 242–257, 359
Peroxisome proliferator-activated receptor c coactivator 1α (PGC-1α), 81, 177, 178, 182–184, 212, 245
Plasma-membrane redox system, 92, 319
Protein regulation, 116, 254, 255
Protein stability, 120

R

Reactive oxygen species (ROS), 18, 25, 81, 96, 98, 101, 117, 136, 137, 141–145, 147, 148, 164, 165, 178, 179, 181–183, 192, 198, 199, 203, 205, 207–212, 235, 242–246, 254, 255, 271, 272, 285–288, 293, 294, 296, 299, 313, 315, 317–320, 353, 354, 358
Redox, 3, 6, 12, 14, 17, 31, 34, 52, 53, 55, 59–61, 63, 67, 74, 78, 79, 92, 93, 95, 96, 98–101, 103, 137, 138, 144, 164, 242, 250, 313, 318, 348, 349, 359

Redox regulation, 96, 101
Reproduction, ix, 136, 284–299
Respiratory chains, 3, 5, 16, 18, 23, 52, 54–57, 59, 62–64, 66, 69, 71–73, 75, 77, 79, 80, 145, 163, 179, 182, 193, 198, 199, 202–205, 207, 208, 210, 256, 271, 284, 313, 330, 348, 356
RNA binding proteins (RBPs), 25, 114–120

S
Senescence-Accelerated Mouse Prone 1 (SAMP1) mice, 81, 172–180, 182–184
Sirtuins, 81, 93, 101, 103, 142, 176, 177, 312
Sperm, 193, 284–293, 297
Supercomplexes, 3, 16, 56, 59, 70–72, 136, 330

T
Therapeutic approach, 292

U
Ubiquinol, 14, 15, 52–54, 60, 62, 65, 68, 70, 72, 76–80, 92, 95, 97–99, 145, 170, 246, 247, 250, 288, 289, 317, 348–350, 353, 358
Ubiquinones, 3, 12–14, 16, 18, 30, 52–55, 60–62, 65, 66, 69, 72, 74, 76, 78–80, 92, 95–99, 118, 132, 133, 170, 247, 250, 274, 277, 289, 348, 350

W
Worms, 132–138, 147, 150, 315, 323

The manufacturer's authorised representative in the EU is Springer Nature Customer Service Centre GmbH, Europaplatz 3, 69115 Heidelberg, Germany. If you have any concerns regarding our products, please contact ProductSafety@springernature.com

Printed and bound by CPI Group (UK) Ltd, Croydon, CR0 4YY
30/09/2026
02225248-0012